Global Hydrology:
processes, resources and environmental management

J A A Jones

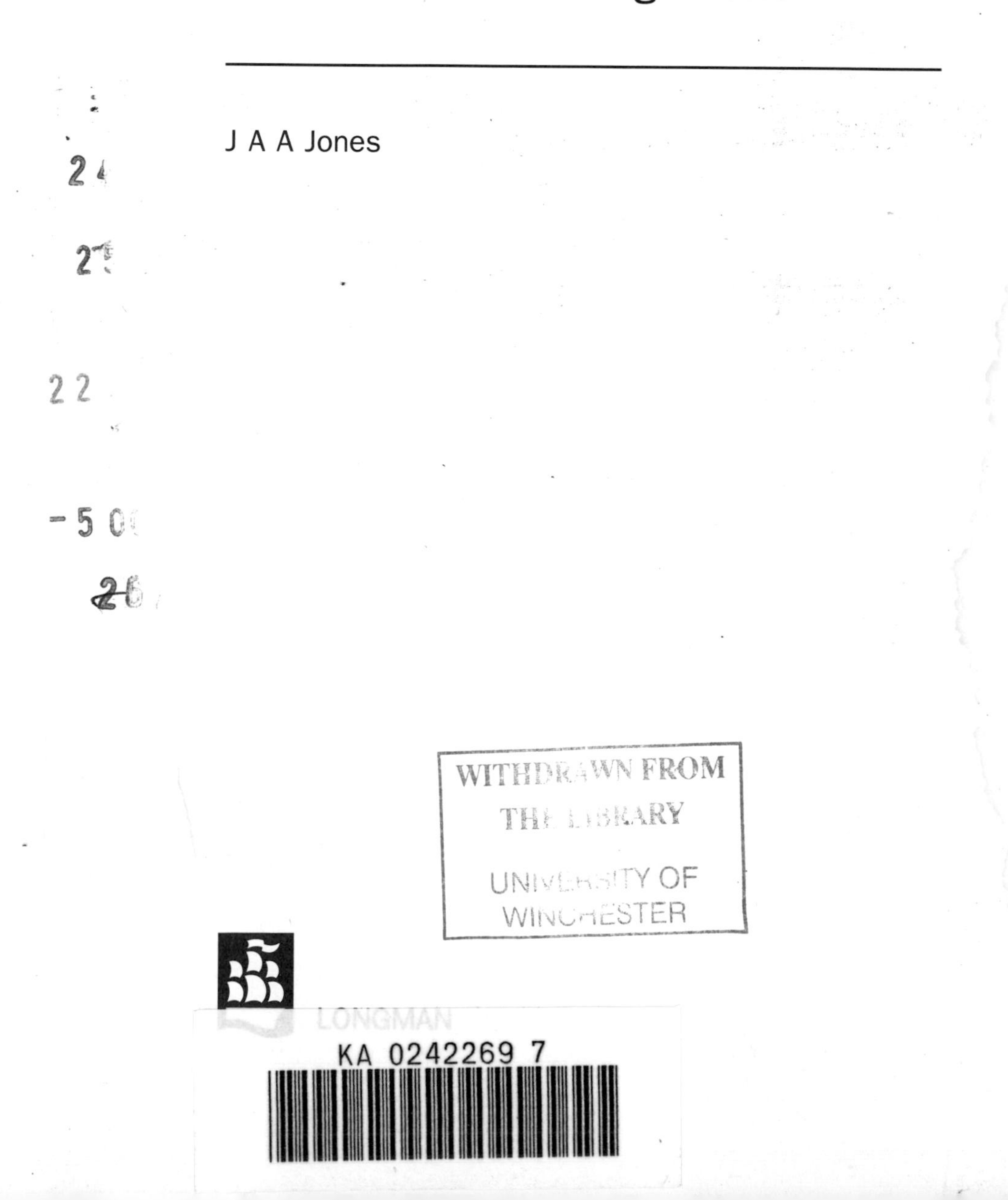

LONGMAN

Addison Wesley Longman Limited
Edinburgh Gate, Harlow
Essex CM20 2JE
England

and Associated Companies throughout the World

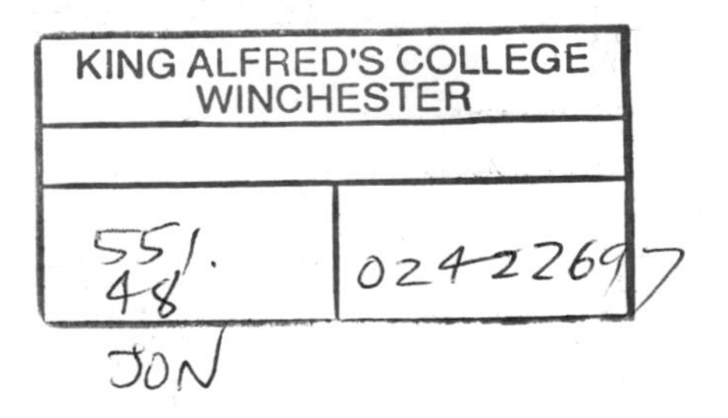

First published 1997

ISBN 0 582 09861 0

British Library Cataloguing-in-Publication Data
A catalogue record for this book is
available from the British Library.

Library of Congress Cataloging-in-Publication Data
A catalog entry for this title is
available from the Library of Congress.

Set by 7 in 9/11 times
Produced by Longman Singapore Publishers (Pte) Ltd.
Printed in Singapore

To Jennifer, Claire and Mark

Contents

Foreword

As the end of the twentieth century approaches, freshwater is an issue which is moving up the international agenda. There is growing concern that the rapidly rising demand is likely to double or even treble the global consumption of water before AD 2050. There is also concern for the declining knowledge of the world's water resources, knowledge which is vital to meeting this demand. At the same time floods are taking an increasing toll of life, droughts are becoming widespread, while pollution of surface water and groundwater is diminishing the resource available for use.

Global Hydrology addresses the science that is the key to combating these and similar problems concerned with water. It does this through a discussion of the global hydrological cycle and its component processes and by considering the monitoring and modelling needed to understand them. Among other topics the text also describes a number of contemporary controversies, such as the Aral Sea and the Aswan Dam, where water or lack of it is the overriding concern. In concluding, *Global Hydrology* questions a sustainable future and makes comment on how future climate change may move the hydrological goal posts.

Of course a text entitled *Global Hydrology* would have been impossible to publish twenty or even ten years ago. Then the science was strongly dedicated to the river basin, with few hydrologists able to acknowledge that hydrology extends beyond the individual watershed towards a global dimension. But for a number of reasons this has all changed during the last decade and the science has 'gone global'. Hence it is timely and appropriate that *Global Hydrology* should be published. Dr Jones is to be praised for his initiative and for making his expertise available through the medium of the printed page. *Global Hydrology* is a very welcome addition to the literature.

Professor John C. Rodda
President,
International Association of Hydrological Sciences,
Institute of Hydrology,
Wallingford,
Oxfordshire

Preface

My overriding aim in writing this textbook has been to present the physical science of hydrology in the context of its application to real-world problems. Water issues in all their many guises have become major elements in present-day public health and environmental protection, and even in modern politics and economics. Decision-making on these public issues needs to be firmly based upon a full appreciation of current scientific knowledge, including its weak areas as well as its strong points. In Britain, recognition of the need to integrate the water sciences within the broader environmental realm has been expressed during 1996 with the merging of the National Rivers Authority into the new Environment Agency and of the Institute of Hydrology into the new Centre for Ecology and Hydrology.

In this book, I have sought to provide an introduction to the science of hydrology which does not eschew the technical aspects, like monitoring and modelling, but which presents them in a digestible or 'user-friendly' manner. It assumes a scientific background that is no more advanced than the average university entrant in geography or environmental sciences, sometimes perhaps less. And mathematical aspects have been limited to what I regard as just sufficient for the reader to grasp the fundamentals of the arguments. No prior mathematical skills are required beyond the basics for university matriculation, and I have endeavoured to explain the essential additions in a simple way.

As an introduction, inevitably much remains unsaid. I have tried to provide the reader with a guide to the essential and up-to-date specialist texts on each aspect covered at the end of every chapter. I have also included a set of questions arising from each chapter, which I hope will stimulate thought and further exploration.

Finally, a word about the title. Of course, all science is 'global'. But in a world where 'technology transfer' is accelerating, we should be aware of the differences as well as the similarities in regional environments. Many past and present problems have been created by ignoring these differences. One of the most challenging aspects of contemporary hydrology is to obtain data from the many under-monitored regions of the world and to adapt our models to encompass these environments.

J.A.A. Jones
University of Wales
Aberystwyth

Acknowledgements

We are grateful to the following for permission to reproduce copyright material:

Ambio for Figs 1.6 and 1.7 from Lvovich (1977); American Journal of Science for Fig. 3.22 (Bunting, 1961). Reprinted by permission of the American Journal of Science; Benjamin Cummings for Fig. 7.10 (Chires, 1994), reprinted by kind permission; Elsevier Science for Fig. 3.20 (Conacher and Dalrymple, 1977), Fig 3.21 (Anderson and Burt, 1977a), Fig 5.13 (Hendrick and Comer, 1970), Figs 7.1 and 7.2 (Bosch and Hewlett, 1982), Fig. 7.16 (Hey, 1986). Reprinted with kind permission from Elsevier Science – NL Sara Burgerhartstraat 25, 1055 KV Amsterdam, The Netherlands; Bruce Findley for Figs 5.6, 5.9a and 5.18; W.H. Freeman and Company for Figs 3.2, 4.8, 4.9, 4.10, 4.11 and 6.19 (Dunne and Leopold, 1978) © 1978 by W.H. Freeman and Company. Reproduced with permission; Geo Books for Fig 3.18 (Anderson, Bosworth and Kneale, 1984) and Fig. 3.19b (McCaig, 1984); IAHS Press for Fig. 6.20 (Marsalek, 1977), Fig. 7.7 (Huff, 1977), Fig. 7.8 (Yperlaan, 1977), Fig. 8.7 (Roberts and Marsh, 1987), Fig. 8.9 (Betton, Webb and Walling, 1991) and Fig. 8.16 (Rang, Kleijn and Schouten, 1986); International Thomson Publishing Services Ltd for Fig. 4.3 (Shaw, 1994); John Lewin for Fig. 7.15; Macmillan Press Ltd for Fig. 4.5 (Wilson, 1990); Oxford University Press and UNESCO for Fig. 11.11 (Hufschmidt, 1993); Rijkswaterstaat Luid-Holland for Fig. 8.20 (Bielsma and Kuijpers, 1989); US Army Corps of Engineers for Fig. 3.27 (Wankiewicz, 1979); US Geological Survey for Fig. 6.21 (modified after Anderson 1970).

Whilst every effort has been made to trace the owners of copyright material, in a few cases this has proved impossible, and we take this opportunity to offer our apologies to any copyright holders whose rights we may have unwittingly infringed.

The author would like to thank David Griffiths and Geraint Hughes for photographic work and Mike Jones, Ian Gulley and Anthony Smith for cartography. Many thanks also to colleagues who have supplied me with recent literature, including John Rodda, Des Walling, Bill Edwards, Jim Hudson and Gareth Roberts.

Figure 10.2 is developed from an original idea of J. Allen.

CHAPTER 1

Water for the world

Topics covered

This book aims to analyse the major problems of matching supply and demand in the modern world from the basis of current scientific knowledge of natural processes and of man's effect upon them. It is also very much concerned with the environment and the effects of man's alteration of hydrological regimes upon that environment. Its principal thesis is that harmonious, and ultimately successful and sustainable, development requires a sound and thorough understanding of hydrological processes both on land and in the atmosphere.

1.1 The basic issues

Water is the most fundamental ingredient of all life on Earth; it is more universally necessary even than free oxygen. It is the most abundant component of any organism and the medium for most biochemical reactions. It plays an essential role in the exchange of material between an organism and its environment. The Earth–atmosphere system contains approaching 1500 million gigatonnes (Gt)[1] of free water compared with little more than 12 million gigatonnes of free oxygen, including oxygen dissolved in water. Despite this, in most environments outside the oceans and the great lakes water is more likely to be in short supply than oxygen. For most terrestrial life forms the supply of water in the local environment determines where they may live, where they may move and where they may die.

Man is the exception. The ability of the human race to manipulate water in the environment, to redistribute it in time and space, is one of the major characteristics that sets civilized man apart from the rest of the living world. Mankind took a giant step about 6000 years ago in Mesopotamia by diverting riverwater for use in irrigated agriculture. These first hydraulic engineers freed man from subsistence agriculture and enabled tradable agricultural surpluses to be created. In turn this permitted the job specialization and extensive socio-economic reorganization that led to the development of the first cities, the Urban Revolution, and the development of writing to manage these new city states and their trade. Civilization was born. Even when urban civilization spread to better-watered regions, it still required dams, aqueducts, pipes and drains to service man in his built environment.

Throughout history, water engineering has lain at the very foundation of civilization itself. States have fallen for lack of adequate provision. In the modern world, we can see the retreat of settlements from the desert margins as the mismanagement of soil and water resources combines with natural climatic stress. Indeed, the dual impact of agriculture and urban society has permanently altered global hydrology. Wherever they have spread, they have so modified the surface of the Earth that the flows and exchanges of water bear their imprint, in some cases millennia after the impact occurred. Water has been an important element in the many-sided conflict between civilization and nature, and the roots of the conflict were present at the dawn of history.

1 One gigatonne is a billion tonnes, i.e. 1×10^9 tonnes.

The twentieth century has seen the most extensive manipulation of water: large dams to provide water supply, generate hydropower or regulate riverflow, extensive groundwater schemes often linked to complex multi-source centralized public water supply systems, sewage disposal and in latter decades a rapid and vast expansion in the area under irrigated agriculture. About 13 per cent of global riverflow is now controlled by mankind, but our influence extends far wider. Though some of the grandest schemes, like the former Soviet proposal to reverse Arctic rivers, may never come to fruition, in some areas of the world impacts have gone beyond the stage of mere alteration and are reaching the point of crisis – for the environment as well as for humans.

1.1.1 Global concerns

As the world prepares to enter the twenty-first century, three issues dominate: (1) the provision of sufficient water for a growing and increasingly demanding population; (2) the impact of water development, and of other developments involving hydrological effects, on the environment; and (3) anticipated problems with climatic change and the unreliability of water resources.

The overwhelming bulk of water on Earth can be used neither by man nor by the majority of terrestrial plants and animals, because it is either too salty or frozen. Even so, the remainder is no mean quantity and it does have the advantage that it is naturally recycled, so that although it is *finite* in quantity it is ultimately *unlimited* in supply. Practical limits to both the quantity and quality are nevertheless imposed by *recycling times*.

The need for conservation and the careful control of exploitation is now almost universally acknowledged. This recognition owes much to fears that resources in general may be running out, to the rise of the environmental movement and to practical experience over recent decades, particularly with catastrophes and failures, both man-made and natural.

Limited and shifting resources The realization that mankind's use of many resources is approaching global limits hit the world community very forcibly in the early 1970s as a result of an unusual conjunction of events: the Club of Rome's computer simulations published in *The Limits to Growth* (Meadows *et al.* 1972), the 1973 oil crisis and a series of droughts and agricultural disasters, especially the series of Sahelian famines between 1968 and 1973, and the USSR's disastrous harvest in 1972.

The droughts had immediate repercussions on water management, including rainmaking activities in West Africa, a plan to divert the Congo to enlarge Lake Chad, and a Soviet scheme to divert major rivers of the Arctic basin south to irrigate the grain lands of the Ukraine and adjoining republics (Box 9.2). The 1970 drought in the eastern United States and the 1976 drought in the UK also had major impacts upon management procedures. The 1988 drought caused a 30 per cent shortfall in the North American grain harvest. Above all, the stress put on world agriculture by the combination of population growth and climatic events has led to a vast expansion in irrigation.

The national water surveys in Canada and the USA in 1978 and 1981 both foresaw critical shortages before 2000. The US report expressed particular concern for southern and western states. This view was reiterated in the global survey by the Worldwatch Institute of Washington, DC, in 1984 entitled *Water – the next resource crisis?*, which also focused on North Africa and the Middle East as regions potentially in great danger without very careful planning. Even in England and Wales, the National Rivers Authority's water resources strategy indicates concern over regional water shortages (NRA 1994).

The expansion of water pollution, also predicted in the *The Limits* report, could be as crucial for world water resources. Less well publicized calculations by the Soviet hydrologist M I Lvovich (1977, 1979) suggested that if water pollution remained unchecked the world would essentially run out of usable water resources by the year 2000 (section 1.2.4). He proposed strict pollution controls and widespread recycling of water as solutions. Many have criticized the details of these calculations, as with *The Limits* predictions. Nevertheless, the fact remains that, despite rigorous pollution controls in Western Europe and North America, the problem has barely been contained. In Eastern Europe, the effects of non-containment were plain for all to see after the collapse of Communism in 1989–90 (Peterson 1993).

The problems of quantity and quality in water supply are now seen as a single entity and Lvovich himself must take part of the credit for the emphasis that has ensued, despite some criticism of his assumptions (section 1.2.4). The US Congress set up the Environmental Protection Agency in 1970 to control water pollution. The European Union has issued a series of Directives to control water pollution since the 1974 Bathing Waters Directive, and the National Rivers Authority (NRA) was established in England in 1989 to protect both the quantity and the quality of water in supplies and in the environment. Globally, the UN World Health Organization (WHO) designated the years 1981–90 the International Drinking Water Supply and Sanitation Decade. The WHO aimed to provide *potable* (literally 'drinkable') water supply for 1.8 billion additional people in the Developing Countries during the decade along with access to proper sanitation for 2.4 billion.

Reorganization of water industries to combine responsibilities for all stages of water use and recycling was seen as an important means of tackling these problems, at least since the publication of the UN revised guidelines (UN 1970). These proposed that the hydrological concept of an *integrated circulating system* from source to reclamation should take precedence over existing ad hoc administrative and commercial development. The report also established the principle that the *natural river basin should be the basic unit* for administration and development. Reorganizations like those in England and Wales in 1973 and in Spain in 1985 followed these guiding principles.

By recognizing the essential unity of the cycle of water use within the administrative structure, it may be possible to select alternative solutions more easily, or to reduce reprocessing costs. Kaminskii (1977) strongly urged the Soviet government to consider the advantages of water recycling before implementing the vast plan for diverting the Arctic rivers. This unity of interest is underlined by the late-twentieth-century Western vogue for using rivers as open pipelines for public water supply, by regulating their flow in order to guarantee a minimum discharge at certain abstraction points downstream. Regulated water supply rivers cannot logically be used as a dumping ground for wastes in the old way (section 9.3).

Unfortunately, international boundaries will continue to disregard natural drainage basin units. Globally, one-third of rivers pass through more than one country. Good international collaboration *is* possible in these circumstances, as between Canada and the USA on the Columbia River or between the states along the Rhine. But all too frequently there are conflicts of interest (section 11.4). Indeed, halting cross-boundary riverflow might even be used as a military weapon, as the USA reportedly discussed with Turkey during the Gulf War, and as Saddam Hussein appears to have used against the Marsh Arabs (section 11.4). Similar problems can arise with shared groundwater reserves.

The most recent fear has been the potential impact of an enhanced greenhouse effect on global and regional resources (sections 2.6 and 11.5). The Intergovernmental Panel on Climate Change, set up in 1988 to analyse available evidence, clearly established the need for concern. The UK Climate Change Impacts Review Group (1991) demonstrated that, even in a country where overall annual rainfall is expected to increase, severe stresses could be put on resources in certain regions, at certain times of the year and on specific users.

From a political viewpoint, Sir Crispin Tickell (1977, 1991), former British ambassador to the UN, has predicted that future climatic change will so strain agriculture and water resources in many of the Least Developed Countries that civil disobedience, political coups and massive international migration will occur.

Environmentalism In contrast, the philosophy of the environmental movement, developed largely from the opposing, geocentric rather than anthropocentric viewpoint. Rachel Carson's book *Silent Spring* (1962), which highlighted the effects of pesticides on wildlife, is often cited as a founding text for the modern movement. Numerous conflicts since have focused on the environmental impact of large dams, including destruction of ancient ecosystems and the displacement of native populations (section 9.2.1).

What began as a concern for preserving vanishing nature, however, has grown into a belief that mankind's continued existence depends upon maintaining important natural balances. The 1971 Stockholm Conference on Man and his Environment focused world attention especially on the effects of acid rain on the biosphere. As a result of that conference, the UN Environment Programme (UNEP), the World Meteorological Organization (WMO), the World Health Organization (WHO) and UNESCO agreed to collaborate in a Global Environmental Monitoring System (GEMS), of which GEMS/WATER is a major arm.

More recent WMO conferences, such as the 1985 Villach Conference on Carbon Dioxide and Climate Change and the Second Global Climate Conference in 1990, have expressed deep concern over the effects of burning fossil fuels and the felling of native forests on the greenhouse effect and resulting shifts in water resources. The UNFPA (1992) report on *Population, Resources and the Environment* continues to underline the dangers of current trends in exploitation.

The concerns of environmentalists have therefore expanded from biological preservation to the 'global commons', the atmosphere and the hydrosphere. The links are highlighted in James Lovelock's (1979) **Gaia Hypothesis**, which proposed that life both creates and maintains a suitable environment for itself. The hypothesis remains unproven, but much important feedback has been noted over the past decade, including the possibly critical role of oceanic phytoplankton in reducing global warming either by absorbing excess atmospheric carbon or releasing dimethyl sulphide (DMS), which stimulates cloud formation (sections 2.3.1 and 2.6).

One of the more apparent results of pressure from environmentalists on water resources engineering has been the trend towards so-called 'non-structural solutions'. These consist of replacing traditional hardware solutions, such as impounding dams or flood control structures, with more environmentally friendly 'software' solutions, or at least solutions which require less hardware. Software solutions often use computer-based monitoring and modelling systems to provide early warning of floods, or to control pumping and flow regulation in relation to 'real-time' inputs and demands. A less sophisticated software solution might simply involve planning controls on floodplain development, which either replaces or supplements physical constraints on riverflow.

As engineering costs have risen dramatically over recent decades, so cost–benefit analyses have increasingly favoured the non-structural solutions. The result has been a marked dovetailing of once opposing interests.

At the same time, hydrologists themselves are becoming increasingly concerned with environmental planning and protection. Dunne and Leopold's (1978) book *Water in Environmental Planning* was an early voice in this field. The trend is exemplified in Britain by the merging of the NRA and Her Majesty's Inspectorate of Pollution and Waste Regulation Authorities to form a new Environmental Agency, and of the Institute of Hydrology into the Centre for Ecology and Hydrology, in 1996.

Catastrophes and failures Engineering and environmental costs also loom large in responses to many catastrophes and failures. The logical trend would seem to be to follow the non-structural route as far as possible and to minimize interference with nature wherever this is feasible.

Third World countries are often least able to withstand natural or man-made disasters, because of their financial, technological or social limitations. They also tend to occupy regions prone to natural disasters and with the most vulnerable environments. As the Brandt Report of 1980 from the Independent Commission on International Issues said of these countries:

> Without irrigation and water management, they are afflicted by droughts, floods, soil erosion and creeping deserts, which reduce the long-term fertility of the land. Disasters, such as droughts, intensify the malnutrition and ill-health of their people and they are all affected by endemic diseases which undermine their vitality.

The 1981 UN Special Conference on the Least Developed Countries (LDCs) in Paris set out to double living standards in these countries by 1990. The provision of an adequate water supply and of surveys to establish the possibilities of cultivating marginal land, perhaps by irrigation, were seen as crucial to this task. At the same time, the UK, like many other Western nations before and since, announced a new aid package for the 31 LDCs with special provision for water supply, sanitation, energy and agriculture.

Unfortunately, aid has not been the panacea for all. Indeed, it has aggravated some problems. Western aid has often been used for expensive water storage schemes which have failed to deliver the intended benefits. Some dams in the Middle East have failed to reach capacity or have taken very much longer to fill than expected, because the original estimates of available water resources were overestimates based on inadequate hydrological data. Other schemes have failed because they ignored the time-honoured institutional use of water in that environment, e.g. the Aswan High Dam (Box 9.1) or because they have ignored the fact that the *socio-political infrastructure* for the effective distribution of the water does not exist.

A common cause of *technical failure* has been silting up long before the planned 'design life' of the reservoirs. Typically, engineers working in these environments have no data on sediment transport in the rivers and, equally typically, the rivers tend to carry large sediment loads. Other more disastrous technical failures have involved bursting of the dams. This problem particularly afflicts earth dams, which are often all that local budgets can afford.

In addition, many developing countries took loans from Western banks for water resource development, which contributed to the severe *financial stress* these countries found themselves in under the high interest rates of the 1980s and early 1990s.

Solutions to many of these problems lie outside the competence of hydrologists alone. Nevertheless, hydrologists have particular responsibility for establishing available water resources, the frequency and magnitude of extreme storms and droughts, water quality and the possible impact of environmental change. They should also share the responsibility for ensuring that engineering works operate in sympathy with the environment.

1.1.2 Why study process?

The pure scientist's response to such a question is unequivocal: 'We study processes in order to understand how natural systems work.' The engineer, however, may well say that when pragmatic solutions are needed to pressing, practical problems, the rule of thumb, the experienced judgement or statistical models may well be quicker and easier to use and possibly more reliable than techniques which attempt to consider all the processes involved, especially if data are limited. In part, this must be a reflection of limitations in the current level of theoretical understanding in hydrological science, but lack of observational measurements will probably remain an even greater constraint in the immediate future.

There is no doubt that advances in theory, in observational techniques and in computer systems for communication and analysis will continue to shift the frontiers forward in favour of process-based methods. Nevertheless, the point at which hydrological systems reach 'physical indeterminacy' or require observational data of a type which is beyond our ability to obtain may well be with us for ever: even modern physics has come across this barrier. We will look at such problems more closely in Chapter 6, but at this juncture it is important to emphasize the benefits of observing processes and of building predictive models based upon those observations.

At the very least, the discipline of *'getting into the system'* to see how it works heightens our appreciation not only of the mechanics of the system as it currently operates but also of how it will respond to different inputs or changes in individual components within that system in the future. A predictive model based on a knowledge of these processes should be more flexible than one based on data describing simply the inputs and outputs of the system, such as rainfall and riverflow.

This gives models based on physical processes two overriding advantages. First, they are more *transferable*. A transferable model is one that can be applied in many different drainage basins with minimum adjustment, by simply adjusting the values of a certain number of environmental parameters within the model to fit the key characteristics of the basin, rather than altering the structure of the model itself. Transportability is highly desirable. Secondly, they allow more realistic simulation *experimentation*. Experimentation is done by setting up 'what if' scenarios and investigating the hydrological consequences; for example, what if annual rainfall increased by 10 per cent or the hillslopes were planted with trees?

Process-based models are proving invaluable for solving many environmental problems, such as the movement of pollutants or assessing the environmental impact of changes in land use or climate. Models combining physical, chemical and biological processes are particularly valuable in monitoring and managing river pollution (sections 8.1.3 and 8.3.2). Physical process models provide the best basis for operating modern river regulation schemes (sections 6.3 and 9.1.1). The Mackenzie River Management Scheme in Canada is an excellent example in which a combination of physical process models is used to monitor and predict changes in discharges and water quality, and forms the cornerstone for basin-wide planning control (Box 10.4).

Process studies are therefore crucial to effective impact assessment. They also allow the development of efficient hydrological models that can be used around the world, especially where insufficient records exist.

1.1.3 What is a water resource?

Water resources are certainly no easier to define than most other types of resource. Indeed, they can be more difficult because they fluctuate so much in both amount and quality. What is more, they possess a rather special relationship with *hazards*. One of the most common ways of increasing water resources is to harness more *flood* water. The two time-honoured roles of the water engineer, of providing protection against hazards and of developing water resources, are therefore closely related. Providing protection against *drought* hazard and providing a reliable water supply are one and the same. In each case, the aim of the engineer is to even out the uneven temporal distribution of natural flows as much as is deemed necessary or affordable according to human demand. Although a large proportion of the world population still relies on direct access to natural waters for its supplies, most of the water that is supplied through centralized systems relies on engineered resources, i.e. the enhancement of natural flows by storage and redistribution. Here, water resources are largely created by engineering.

As water is a renewable resource, the average rate of renewal *normally* provides the ultimate limit to exploitation. But as technology improves and previously unusable sources like seawater may be used, even this does not provide an absolute limit. The only absolute is the total amount of the water substance on Earth.

It seems reasonable to distinguish three levels of water resource: (1) actual available resources, as currently used or 'developed'; (2) present potential resources, defined within the framework of available technology, i.e. what could be utilized if finance and value judgements were not constraints; and (3) the future potential, which is to all practical purposes undefinable. In every case, the amount of water that can actually be exploited is likely to be markedly less than the natural renewal rates, principally because of the demands of the natural environment, particularly the biosphere. Even so, as recycling technology is adopted and improved, so mankind will be increasingly released from the constraints of natural renewal rates and from the water demands of the living environment.

For the foreseeable future, however, it is important to remember that the act of developing water resources is bound to affect the natural environment. The loss to that environment *should* always be weighed carefully against the human advantages.

In practical terms, at any particular location, we can define the present level of potentially **available water resources (AWR)** as the dependably renewed supply minus the requirements of downstream users and the natural ecology. For river-based supplies, this may be represented as:

$$\text{AWR} = \text{Dependable} - \text{MAF}$$

where MAF is the **minimum acceptable flow**. Unfortunately, neither the 'dependable' input nor the MAF is absolute. The 'dependable' renewals are really based on probabilities. They are average or 'characteristic' values based upon records or models. They can also change over the years as climate changes. Similarly, the MAF is largely defined by the size of demand and the type of use downstream. Agriculture, for example, tends to need less security of supply than domestic or industrial users. It may also depend on the time of year. Irrigated crops need higher security during the growing season. MAFs set to safeguard fish stocks during the summer may be overgenerous prior to the breeding season, and it has been suggested that they might be cut down in order to maximize springtime levels in the reservoirs (Jack and Lambert 1992). Even where MAFs are set as a legal requirement to protect the environment and downstream users, it seems inevitable that pressure such as this to utilize more of the dependable supplies will require greater vigilance and in some cases greater scientific understanding from freshwater biologists and environmental scientists in general.

1.1.4 Environmental management

Water management has been with us since the dawn of civilization, but environmental management is a development of the late twentieth century.

For the hydrologist, environmental management is concerned primarily with two interrelated issues. First, whether utilizing water resources, disposing of waste, or controlling flood hazard, *water* should be managed in such a way as *to minimize the interference with nature and to maximize the benefits for nature*. This means managing water use and manipulation in such a way as to preserve and even enhance the water needed by wildlife and the environment. Second, *the environment* should be managed in such a way as *to minimize adverse impacts on and maximize benefits for water resources or flood hazard*. The water resources engineer must commonly balance both of these against the general need to optimize the supply of water for human requirements. Fortunately, both these principles can also lead to improvements in the water resources themselves.

This approach is largely a product of the environmental movement, but for the water engineer it is acquiring a new logic, that of selecting more stable or resilient and cost-effective solutions. In some cases, admittedly, additional costs are incurred to meet environmental requirements with no direct benefit to the developer. But in many others the extra discipline imposed by having to consider environmental effects can lead to the discovery of novel and better solutions and be a stimulus to the development and application of more sophisticated systems.

Environmental management need not be simply protective. Water management may be undertaken in order to create positive improvements for nature. Similarly, watershed management may be undertaken to improve water quality or yields. An outstanding case of environmental management is the control of water levels in the Florida Everglades for the benefit of wildlife, although unfortunately this positive action has been made necessary by overexploitation of the water for human needs (section 11.3.1). It is conceivable that similar management techniques could be used in the future to protect wetlands threatened by climatic change.

Similarly, it is possible to manage an environment so as to increase water yield or to reduce flooding. Indeed, managing environments for the benefit of local hydrology is now well established and has been at the basis of many large-scale research programmes in hydrology (section 7.1). The US Forest Service has suggested that vegetation management could be used to make up much of the shortfall in the Colorado River (section 10.3). The UK Institute of Hydrology owes its existence to questions on the impact of afforestation on water yields arising from the work of Law (1956) (section 3.1.2). Ironically, this arose from applying the American philosophy of planting trees in reservoir catchment areas in order to control soil erosion and prevent the silting up of reservoirs, in a country where sediment yield is hardly on the same scale.

Both the idea of planting trees to control sediment production and the idea of felling trees to increase water yields were developed for purposes of exploitation rather than conservation. They are examples of environmental management that predate the rise of environmentalism. Yet by stimulating process-based research, the forest management approach has probably been the single most important source of knowledge on the mechanics of environmental impact.

Improvements in our knowledge of the processes of water loss from plants and the processes of inducing rainfall have now enabled environmental management to be extended to the atmospheric environment. Making rain or limiting evapotranspirational losses are now considered the bases of the 'atmospheric option' among many American water managers (section 10.3).

Indeed, the management possibilities are proving too much for some environmentalists. In his book *The End of Nature*, Bill McKibben (1990) takes pollution of the atmosphere with *greenhouse gases* as the ultimate event that in his view has led to the end of independent nature. Whereas this ending was largely accidental, he says the most recent one is planned through *planetary management* and *genetic engineering*. He is particularly scathing about planetary managers, to whom he attributes the idea that man might become 'a sort of collective mind for the earth', the head of a Gaian superorganism. For McKibben the only stable solution is for man to abandon his expansive numbers and expansive habits and embrace 'deep ecology'. Though he exaggerates and is overcritical of managers, clearly we must solve the problems of overpopulation and overconsumption to survive.

It is essential that the new generation of scientists understand their responsibility to both society and the environment. For the hydrologist and the water resources engineer the main issues seem to be:

1. Is it wrong in the long run to continue applying science and technology to water management in order to improve supplies, if this allows man to multiply and to become more affluent and to put

greater stress on the environment or fellow human beings?

2. Are we to remain merely servants of our present society and its habits, or should we encourage change?

We may not like to hear it, but wild nature has been retreating since civilization began. A more optimistic view is that modern management techniques offer nature real protection for the first time in history. Properly done, this protected nature does not have to resemble a theme park or a zoo.

In his book *Earth in the Balance*, US Vice-President Al Gore (1992) is in no doubt as to the political necessity of changing the wasteful and destructive habits of modern society. Water scientists and technologists *can* alter societies' view of water and its use, either by informing and spreading understanding or if necessary by compulsion, by restricting consumption, even though the deeper issues lie outside our control.

The reader may care to consider these broader questions as they read on.

1.2 The demand curve

Rising demand is the main source of current and anticipated problems. Worldwide demand has increased 35-fold over the last 300 years and has now reached over 3500 km^3 per year (Figure 1.1). During this century, demand in the USA and the CIS or Commonwealth of Independent States (the former Soviet Union) has increased by over 400 per cent, while it has doubled in Europe. Fortunately, in most of the Developed Countries there are signs of lower increases during the 1980s and 1990s than were predicted around 1970. Here increases of 2–3 per cent a year are now generally expected through the 1990s, compared with 4–8 per cent a year in the 1950s. In Britain, the Kielder Reservoir in Northumbria was planned during the 1970s at a time when some estimates were projecting a 50 per cent increase in national demand by 2000. Yet by its completion in 1982 Britain was in a recession, which coincided with the beginnings of major restructuring in many industries. British Steel, a large German chemical company and others either never took up their options on the water, because they decided not to move to the area, or else they curtailed expansion and drastically reduced their requirements. Demand in Northumbria has not recovered and is currently increasing at only 0.75 per cent p.a. Kielder's potential yield of 900 Ml day^{-1} provides a huge overcapacity which gives the region almost twice the developed resources that it needs to meet average demand (NRA 1992). Even in southern and eastern England, demand is projected to rise at only 1.0 to 1.5 per cent p.a. between 1990 and 2021.

Even so, demand is still rising. Moreover, current rates of increase in the Developing Countries are more comparable to those of the 1950s in the West. And each percentage increase is on a higher absolute amount.

This ever rising human demand for water is driven by four fundamental pressures: growing population and urbanization, the demands for greater production from agriculture and industry, and wasteful practices.

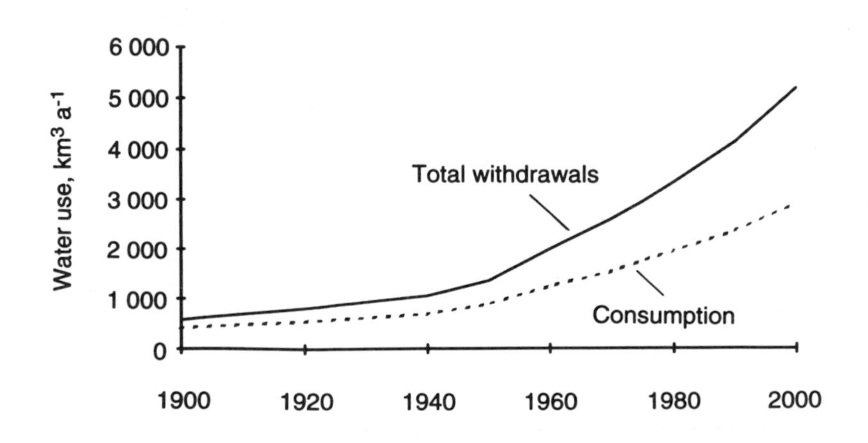

Figure 1.1 Global demand for water

1.2.1 The population factor and urbanization

Much of the increase in demand comes as what economists call 'multipliers' of the explosion in world population: improvements in the quality of life and affluence, the expansion of cities and centralized sewerage, and the demand for more industrial, commercial and agricultural products.

In 1900 world population stood at 1527 million. By 1990 it had passed 5 billion. By the turn of the millennium, it will exceed 6 billion and UN projections suggest that it will be approaching 8.5 billion by 2025 (Figure 1.2).

This cannot be translated directly into water demand. There are vast differences in human demands around the world, reflecting inequalities in cultural and economic development. Our basic metabolism requires a mere 1 litre a day, some of which could be derived from food. Fortunately, relatively few people have to survive at this level. Nevertheless, two-thirds of the world population live on less than 50 litres a day. In rural areas in Africa, Asia and Latin America, overall demand is around 20 to 30 litres per capita per day. Globally, rural consumption is typically no more than 100 to 200 litres per capita per day.

Most of the current expansion in world population is occurring in the Least Developed Countries. In 1950 the LDCs contained 67 per cent of world population, in 1990 76 per cent, but by 2025 this could rise to 84 per cent. If most of this expansion were in rural communities, then the strain on water resources would not be so great.

In reality, it is likely to bolster the drift to the cities. In 1950 just under 30 per cent of the total world population lived in cities and the LDCs had little more than half this percentage in cities. The figures now exceed 40 per cent globally and 34 per cent in the LDCs, but by 2025 this could exceed 60 per cent globally and 57 per cent in the LDCs (UN 1989). However, these figures disguise the fact that 79 per cent of the world urban community could be in the LDCs by 2025. The focus of urbanism is shifting rapidly during the 1990s in favour of the LDCs after passing the 50 per cent mark around 1975.

Urbanization leads to very different patterns of water consumption. Consumption in cities is typically 300 to 600 litres per person per day (cap^{-1} day^{-1}). Major cities now consume water at rates equivalent to the flow in some of the larger rivers in the world. London's 8 million people consume 400 l cap^{-1} day^{-1}, which is

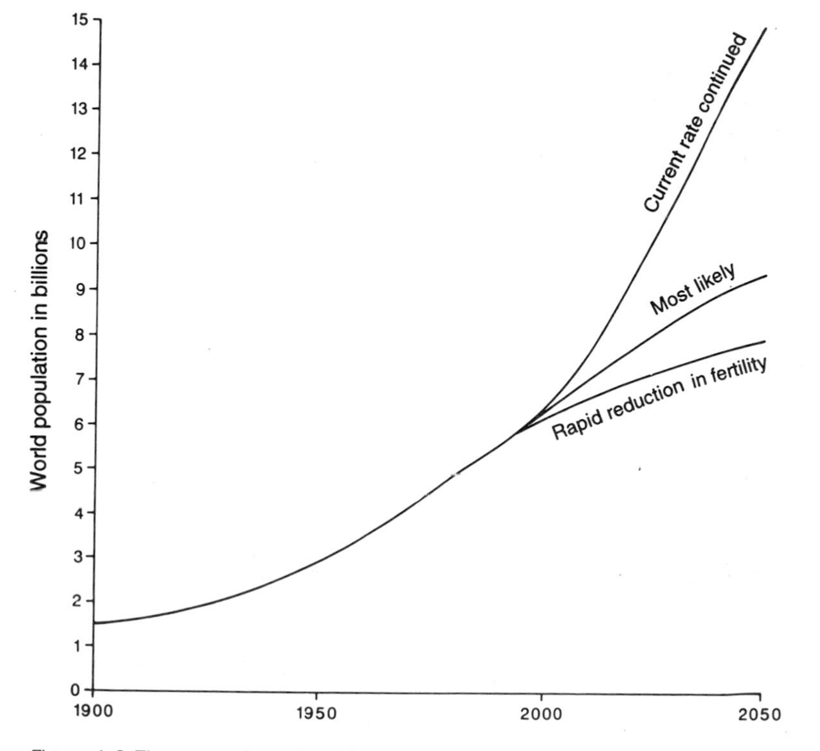

Figure 1.2 The expansion of world population

equivalent to a reliable discharge of 37 $m^3 s^{-1}$. This is comparable to the estimated natural mean discharge of the River Thames, and about 50 per cent more than now flows in the river. The Thames actually ran dry at Teddington weir during the 1976 drought. Cities demand reliable supplies.

Thus, modern cities normally require sizeable volumes of water to be imported into the area. As such *interbasin transfers* of water are increasingly needed, so they add to the tendency for drainage basins to become effectively man-made. Not surprisingly, many cities have been experiencing crises of supply. This is not new. Indeed, it is a persistent feature of city growth. Limited water supply has often restricted urban growth until a new source is laid on. It can be seen in the first cities of Mesopotamia, and Egypt and India have had many historical examples of boom and lean years, of cities expanding to the limits of the new water supply followed by stagnation or even atrophy. Even twentieth-century New York seems to have displayed some of this, with periods of population expansion after new reservoir supplies have been brought on stream.

Tokyo, the world's largest city, has been wrestling with a water shortage for at least 20 years. Part of Tokyo's problem stems from the fact that almost all of the low flow discharge in the Tone River was claimed for irrigation before 1900. In the 1960s, dams had to be built on the river to augment supply. Since then, the Tone River has provided three-quarters of the city's water in normal years, but the drought of 1973 combined with a doubling in peak demand since 1960 revealed its inadequacy. A permanent solution, even with the relatively modest aim of maintaining water supply through a drought of the severity that occurs once in ten years, would require reservoir storage of over 12 billion m^3. Such a large dam presents a desperate engineering prospect, especially in an earthquake zone. The solution has been to attack the problem on a number of fronts. First, interbasin transfers have been developed which may be used as and when needed, beginning with the Tama River to the south. Secondly, metering has been introduced and frugal use of water has been encouraged in a variety of ways. Even so, Tokyo has been abnormally slow to introduce centralized sewerage systems outside the main business centres and the lack of water has clearly hindered their expansion. Indeed, Japan as a whole has the poorest provision of sewage treatment plants among the OECD nations: barely 40 per cent of the population is linked to a treatment plant compared with 73 per cent in North America and between 80 and 100 per cent in Britain, The Netherlands, West Germany, Austria, Switzerland, Sweden and Denmark (World Resources Institute 1992).

The increasing concentration of mankind in urban areas is therefore exacerbating the problems caused by overall population growth. In the CIS, a 30 per cent growth in population during the final quarter of the twentieth century will be marked by a dramatic rise in urbanization, from 50 per cent to 70–75 per cent of the population. Even in North America, where the concentration is not changing much in percentage terms, the actual numbers in urban areas are due to double over the same period. An American study undertaken in the 1970s suggests the consequences for the demand for water. Of a 30 per cent increase in demand during the decade, 16 per cent was estimated to arise directly from population growth, 11 per cent from increasing demands from the existing population as affluence and expectations increase, and the remainder was due to increasing demands from the new additions to the population.

In global terms, only 4 per cent of the population consumed more than 300 litres per capita per day in the 1980s for domestic and municipal purposes. By the year 2000, this is likely to reach around 17 per cent. At the same time, the proportion of the world's population surviving on less than 50 litres a head is due to halve, to around 30 per cent.

These trends pose two serious problems. First, domestic and municipal consumption requires the highest quality water and it is becoming increasingly difficult to find and maintain extra supplies of potable water. Secondly, on present trends they will result in even greater quantities of **wastewater** that is polluted by organic and inorganic substances. Typically, 70–80 per cent of domestic and municipal supplies are returned as wastewater. The greatest increase in urban population is occurring in Asia, where wastewater production is likely to triple during the 1990s. By 2000, Asian wastewater production could equal the total amount of wastewater produced in the whole of Europe and North America combined in 1990. The greatest stress on water supplies is likely to be in the regions that are least well endowed and least able to cope. The prospect of a significant proportion of the new urban dwellers being in informal settlements with no proper water supply or sanitation may limit water demand, but this will be at the expense of health and pollution (section 1.2.4).

1.2.2 The industrial factor

About a quarter of all water withdrawals are made for industrial use. The global industrial consumption of 760 km^3 is second only to agriculture, which uses three times as much as industry worldwide. Only in Europe does industrial use actually exceed agricultural use. In Europe, industry withdraws almost twice that of agriculture and is equal to all other uses put together.

The trends in industrial use are complex. The overwhelming trend has been a rapid rise this century, with North America showing a 20-fold increase in industrial demand. North America and the CIS dominate the world pattern and account for half of all industrial withdrawals.

Within this overall pattern, there are major regional and local differences, which are determined by the stages of economic and industrial development, by the particular mix of industries, by different technological levels and by climate. Individual industries have very different water requirements: even within the clothing industry, it takes 250 m^3 to produce 1 tonne of linen against 5000 m^3 for the same amount of synthetic fibre. The chemical and metal industries use large quantities of water: 4000 m^3 are needed to smelt 1 tonne of nickel. Amounts are generally higher with older technologies, with heavy manufacturing industry, and in warmer climates. This means that the current industrialization of the Third World is causing higher rates of consumption than in the Developed World, where signs of a decline in industrial demand became visible in the 1970s. Industrial use in England and Wales fell by 50 per cent during the 1970s, accompanied by a 31 per cent fall in the water used in power generation. In contrast, domestic and municipal demand increased 12 per cent over the decade. The decline continues. Whereas industry and power took about two-thirds as much as domestic and municipal sources in the 1960s, in the 1990s it takes under one-third.

The decline in industrial consumption in Britain has had many causes: the demise of many traditionally water-intensive heavy industrial plants, economic recession, official encouragement for greater economy in use, especially via water pricing policy, new 'dry' technologies, and the 1976 drought. The drought, estimated at a once in 200 to 400 years event, impressed many industries with their vulnerability in terms of the cost of lost production and their reliance upon an outside supplier that had previously been considered totally dependable. Many reduced consumption and/or sought alternative supplies. Car manufacturers in the Midlands introduced 'dry' technology into the paint shops. Other manufacturers turned away from total dependence on centrally supplied sources or at least supplemented them from disused groundwater wells; many of these sources predated licensing and were therefore outside the official statistics, thus exaggerating the apparent fall in consumption. Nevertheless, problems for the Liverpool underground railway caused by rising groundwater levels during the 1980s were attributed to a real reduction in consumption by traditional local industries. The 1984, 1988–93 and 1995 droughts in England did not have the same impact largely because the water industry has improved its supply systems, but they have confirmed the wisdom of frugality.

While industrial use in Europe and North America as a whole may increase by only 20–30 per cent during the last two decades of this century, it is likely to rise 300–500 per cent in the Developing Countries. As with population and urbanization, the Brandt Report's 'South' is the region of growing stress.

Power generation Water use by the power industry follows the patterns of the rest of industry only to a certain degree. In recent decades it has tended to increase along with a shift away from traditional fossil fuels towards nuclear power and hydroelectricity.

At this point, the distinction must be made between **gross** and **consumptive use**. Consumptive use is the amount of water that is not returned to the rivers after use. In this respect, nuclear and hydropower are very different. Like fossil fuel power stations, nuclear power stations use water for cooling, although they require 50–100 per cent more than conventional stations and hence need to be sited near to a good supply like the sea, a large river or a lake. The cooling process is achieved mainly by evaporation. Hence consumptive losses are high. In contrast, hydropower is the largest *gross* user in the USA, yet it accounts for only 0.5 per cent of total consumptive use or **irrecoverable losses.** The main losses from hydropower operations are due to evaporation from reservoir surfaces.

The relative attraction of hydropower has increased over the last two decades, largely through the decreasing attractiveness of fossil and nuclear fuels. The increased cost of fossil fuels, fears of acid rain and global warming, revelations of the true cost of decommissioning nuclear plants and of treating or storing nuclear waste products, together with fears for the effects of radiation leaks on the environment and human health, have all helped to shift the balance.

In many parts of the Developing World, it has been seen as the only feasible local source of power for industrialization, especially when supported by foreign aid and the World Bank. Asia contains some 30 per cent of all the world's potential hydropower capacity, largely on the great rivers that issue from the Himalayas and the Tibetan plateau. Africa holds a further 20 per cent. Yet in the 1980s only about 15 per cent of world potential had been developed (Albertson 1983). Despite mounting opposition from environmentalists, more and more of this potential is bound to be converted into practical capacity in the coming years, like the Pergau Dam in Malaysia due to be completed early in the 2000s with about £250 billion in British aid.

Two interesting features of the current trend are: (1) a shift to 'low-head' sites; and (2) a reappraisal of the value of small-scale schemes, even in Developed Countries. Canada's James Bay development is an example of a lowland scheme that would not have been considered as a potential source in the era of high-head, alpine-style hydropower (Box 10.3). Equally, both the USA and Britain have been re-evaluating small rural dam sites. The US Federal Power Commission found enough such sites to provide power for 40 million people. A survey in Wales in the 1980s found a total hydropower potential equivalent to 80 kt of oil in small sites capable of generating 25 kW and over. Six new hydropower schemes were announced in Scotland during the 1980s, when only years earlier the North of Scotland Hydro Electricity Board nearly dropped 'Hydro' from its title. The environmental effects of hydropower dams are discussed in section 9.2.1.

1.2.3 The agricultural factor

Globally speaking, agriculture is the main user of water supplies. The overwhelming bulk of this consumption is for **irrigation**. Irrigation can double crop yields; indeed, it may be needed to guarantee any crop at all. If the world is to stand any hope of feeding the extra millions of people in coming decades, irrigated agriculture will have to continue to expand until a suitable alternative is found. In contrast, livestock use only 2 per cent of all withdrawals, although they can be important in major beef producing countries. Livestock account for over 60 per cent of total water usage in Lesotho and Brazil.

In addition to the large consumption of public water supplies for irrigation, ordinary **'rainfed'** agriculture has numerous direct and indirect effects upon the evaporation and runoff of natural rainfall. The effects may be due simply to the modification of key properties of the land, such as vegetation and soil drainage. These alone can cause problems for water managers, for example by increasing the losses of valuable water or by aggravating flood problems by speeding up runoff (section 7.1). But agriculture is also one of the main sources of 'non-point' water pollution and its increasing contribution to this is causing mounting concern (section 1.2.4 and Chapter 8).

Irrigation Irrigation is water intensive. One tonne of grain, enough to feed six people for one year, can require 3200 m^3 of irrigation water. Yet about 50 per cent of world food production by value is grown on irrigated land, which covers just 15 per cent of the cultivated land area. In all, 2206 km^3 a year were

Figure 1.3 Traditional irrigated agriculture in the Himalayan mountains diverts small streams across cultivation terraces

withdrawn for irrigation in the 1980s. This is due to increase to 2585 km^3 by the end of the century, although curiously, because of a 150–200 per cent rise in global domestic and industrial use, the proportion consumed by irrigation will fall from 63 per cent to 55 per cent during the 1990s (World Resources Institute 1992).

Irrigation is the most spatially concentrated of all water consumers. The spatial distribution is very much constrained by physical geography, by adequate water supplies and suitable soils, if not so much by naturally flat land. Land can be terraced and the water recycled from terrace to terrace, as in Figure 1.3, but extensive flat floodplains offer the easiest and most efficient locations for producing tradable surpluses. Nearly 60 per cent of worldwide withdrawals for irrigation are in Asia. Asia contains three-quarters of the world's irrigated area, 140 million ha, and one-third of all Asia's cultivated land is irrigated. China, India, Pakistan and Bangladesh alone contain more than half the world's irrigated lands. Small wonder that over 80 per cent of public water in Asia goes to irrigated agriculture. Withdrawals are particularly concentrated in the great river valleys, like the Indus, Ganges, Brahmaputra, Yangtze and Hwang-ho. Although high evaporation losses and a lack of local rainfall may militate against rainfed agriculture on their lowland floodplains, the rivers are fed by huge discharges from the snowmelt, icemelt and monsoon rains in their headwaters. Indeed, more than one-third of the world's irrigation water is snowmelt, and meltwater is a major source for 28 countries.

Human population is similarly concentrated on these floodplains, and is both the driving force behind the irrigated agriculture and the consequence of it. Some of the highest regional population densities in the world, in excess of 100 people km^{-2}, are to be found here.

North America has the second largest concentration of irrigated land, 29 million ha, with snowmelt from the Rocky Mountains being a major contributor. The CIS accounts for a further 20 million ha. This leaves less than 40 million ha, which absorbs a little over 300 km^3 of water, distributed throughout the rest of the world. Even relatively small areas of irrigation in global terms can, nevertheless, be significant in regional terms: Australasia has by far the smallest irrigated area of the UNEP regions (2 million ha), yet it has one of the highest areas per capita.

The second major feature of irrigated agriculture is its rapid expansion. Global demand for irrigation has expanded ten-fold this century, compared with a five-fold increase in the total demand for water. The irrigated area has doubled during the last 30 years (Figure 1.4).

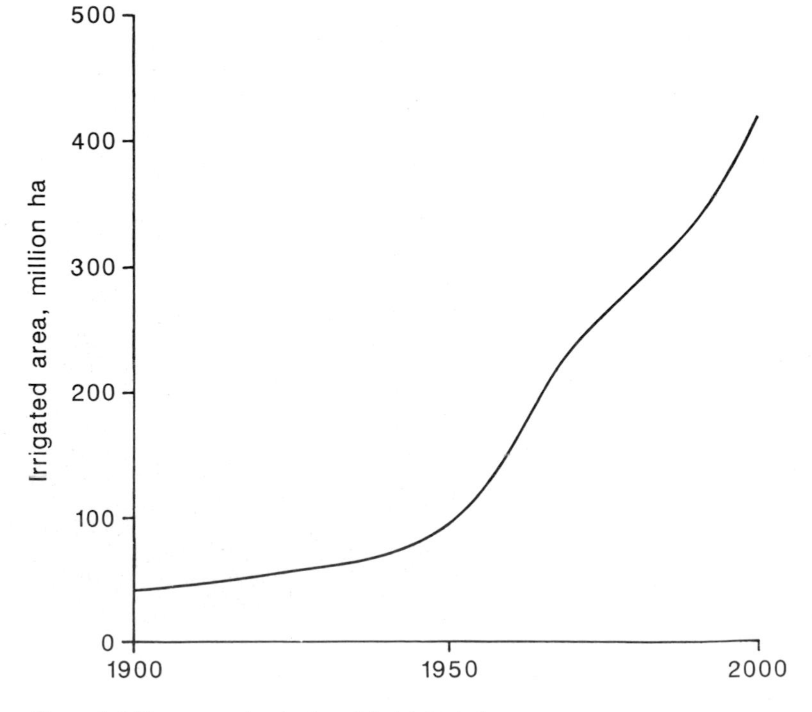

Figure 1.4 The expansion in the global irrigated area

The total irrigated area will continue to rise for some decades, but the peak period of global expansion seems to have passed, along with the main wave of grandiose Western aid schemes and the demise of pan-Soviet planning. In the 1970s, almost half a million hectares were being added each year. By the late 1990s this will have fallen to just 100 000 ha a year. Economic recession in the West, the demise of post-colonial symbioses and re-evaluation of political alignments, together with not a few critical analyses of aided projects, have all contributed to this reduction. Similarly, the break-up of the centrally planned economies of Eastern Europe after 1989 seems to have halted plans for major expansion in Europe (Box 9.2).

Yet quite aside from the politico-economic factors, it is estimated that by the early part of the twenty-first century the area of irrigated land will be approaching the environmental potential of 470 million ha. As the limits imposed by soil fertility, topography and altitude are reached, so we can expect a gradual net reduction to begin. But this will not be a planned reduction or an environmentally sound reduction. It is likely to follow the inexorable logic of past schemes that irrigation tends to destroy soils. An estimated 25 million hectares have already been destroyed by salinization and waterlogging since irrigation began (Box 7.1).

Fortunately, this *will* be accompanied by an environmentally sound trend towards reducing water consumption. Traditional methods of irrigation have tended to be very wasteful, particularly when based on flooding, as in rice paddy fields, or using unlined, open canals, especially in warm climates. In 1970, M I Lvovich calculated that the amount of water lost from irrigated land by evaporation would be three times as great in 2000 as in 1965. He was not, however, allowing for the rapid expansion of new technology in the form of sprinkler or drip irrigation (Figure 1.5), or lined and covered distribution channels. Concrete lining and covering can reduce seepage and evaporation losses by up to 60 per cent, drip irrigation by up to 75 per cent.

Table 1.1 illustrates the wide range in the water efficiency of irrigation systems around the world caused by differences in technology, culture and climate. The key figures are the **specific water use** and the **return rate**. Specific water use is the intake per hectare per year, and the return rate indicates the amount that is returned directly to the rivers. The highest rates of return are found in North America and the CIS, which are twice as high as the world average. However, for the CIS this efficiency of return is outweighed by the high level of initial intake. This gives the CIS a net consumptive use of 75 000 $m^3\ ha^{-1}\ a^{-1}$, which is more than ten times that of India. The extreme inefficiency of the old Soviet irrigated agriculture contributed to the approval of plans in 1982 to undertake a 50-year programme to divert the Arctic rivers southwards to bolster supplies. Even within Asia there are strong contrasts between the most highly efficient use of water

Figure 1.5 The high water consumption of traditional irrigation can be reduced by efficient modern methods, like this drip irrigation in a Rumanian vineyard – the perforated hosepipe delivers water direct to the roots of the vines

Table 1.1 Gross and consumptive use of water for irrigation in representative regional economies

	India	Mexico	CIS	USA	East Europe	France
Proportion used in irrigation	94%	90%	57%	37%	30%	25%
Gross intake ($km^3\ a^{-1}$)	310	30	136	165	13	14
Specific water use ($m^3\ ha^{-1}\ a^{-1}$)	8000	8500	125 000	7500	4000–6000	
Return rate	20%	40%	40%	40%	25–30%	
Net consumptive use ($m^3\ ha^{-1}\ a^{-1}$)	6400	5100	75 000	4500	2800–4500	

Source: Based on data from UNESCO (1978).

in Israel and more than double the amount per hectare used in Iraq.

1.2.4 Wasted and polluted water

There is now a very real fear that we are gradually reducing the available resources of freshwater by high levels of wastage, through inefficient distribution and use, and particularly by returning to the environment polluted waters that can critically reduce the quality of rivers, lakes and groundwater.

Old and leaking water distribution systems can cause remarkable amounts of drinkable-quality water to be lost. In Britain such losses are commonly of the order of 20 per cent, although locally they can be much higher. Wasteful domestic, industrial and agricultural practices either use more water than necessary or else pollute more water than necessary. Experiments with domestic metering suggest up to a further 20 per cent might be used unnecessarily in the average Western household (section 11.2).

Reducing 'irrecoverable' losses through evaporation and leakage, and encouraging a more frugal use of water in general are both key issues that the world is beginning to address. Both promise benefits for the environment, in so far as they will reduce the need to expand provisions for impounding and abstracting water. And both clearly offer parallel economic benefits.

However, the greatest threat at present comes from pollution. The first global assessment of freshwater quality produced by GEMS in 1988 found that contamination continues to rise throughout a large part of the world and that monitoring and control systems were generally inadequate.

Traditional methods of treating domestic and industrial wastewater effluents have relied upon initial dilution by the fresh receiving waters and subsequent self-cleansing processes (section 8.2.1). Unfortunately, these effluents can easily spoil up to 10 times their volume of clean water and the natural processes of purification are becoming increasingly stressed to the point of threatening life in some rivers. Globally, there is still enough clean natural water to dilute polluted wastewater by a factor of 25:1. Australasia even has 10 times this amount, but Europe has a ratio of only 8:1 and this will be slightly worse by the end of the century. Europe as a whole is therefore pushing the limit.

Lvovich's (1977) prognoses of a steep rise in river pollution from spoiling 15 per cent of global runoff in 1970 to 100 per cent in 2000 were fortunately exaggerated (Figure 1.6). But the general drift of his argument was correct. His solution of complete isolation of wastewater from natural water systems (Figure 1.7) may also be rather extreme, but his suggested means of achieving this by industrial recycling, dry technology, or spreading sewage on the land are already being undertaken with some success and explain some local reductions in the upward trend.

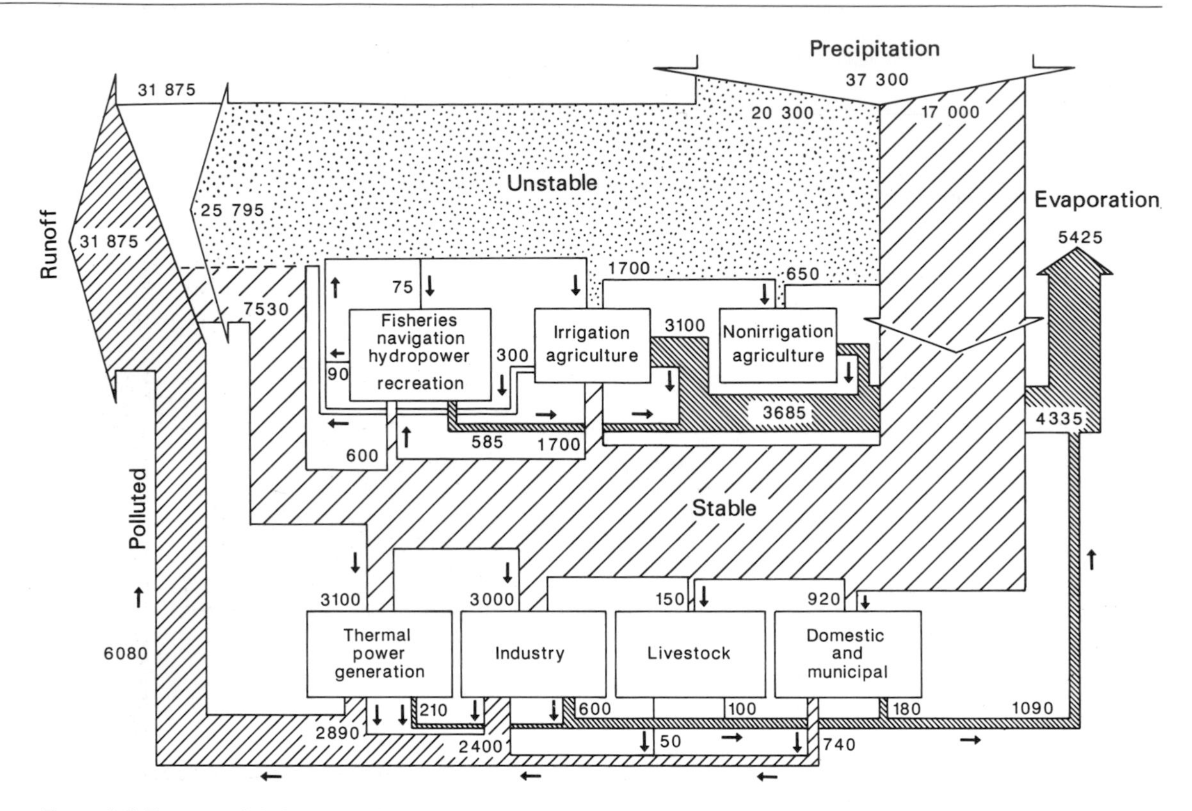

Figure 1.6 Forecast global water balance in the year 2000, assuming continuation of the consumption trends of the 1970s, according to Lvovich (1977)

Nevertheless, the preferred approach adopted in most Western nations is: (1) to restrict industrial effluents by licensing and using the 'polluter pays' principle to discourage infringements of the law; and (2) to increase the provision of 'tertiary' treatment for sewage.

Details will be discussed in Chapter 8, but this introduction would not be complete without a look at world trends. Figure 1.8 shows that pollution can be considered to have reached 'global scale' by the end of the nineteenth century. The main contributors throughout history have been organic wastes and the salinization of irrigation schemes. The last 150 years have seen the rise of new sources. Industry has introduced heavy metals, like lead and zinc, petrochemicals, and many new compounds. Urbanization has caused rivers to be overloaded with sewage and with petroleum products, lead and suspended solids from city streets. Urbanization and industrialization have also caused acid rain. And agriculture has introduced pesticides, herbicides and fertilizers, together with soil and slurry washed from the land.

Just 10 per cent of the world's rivers currently rate as 'organically polluted', but the problem is growing, particularly as people drift to the cities and overload or remain outside existing sewerage systems. Though predominantly a Third World problem, this also applies in parts of Europe. The same applies to rising salinity levels. Many tributaries of the Rio Ebro in northern Spain have experienced increases of 10–100 mg l^{-1} a^{-1} in **total dissolved solids** (a measure of salinity) as the irrigated area has doubled in the Ebro basin since 1945 (Figure 1.9). An extra factor here is solution of the underlying gypsum deposits. The polluted irrigation water drains back into the rivers and is used as intake by farms downstream. Locally, reduced fertility is already noticeable and some irrigated land on the lower Rio Gallego near Zaragoza may become unviable within 10 years. Pollution from cities like Zaragoza presents a greater problem (section 8.2.1), but the two problems are not entirely independent because increasing salinity may necessitate greater use of detergents.

One of the most widespread and troublesome problems at present is pollution by nitrates released from agricultural fertilizers (section 8.2.2). This has spread around the world with the 'Green Revolution' in

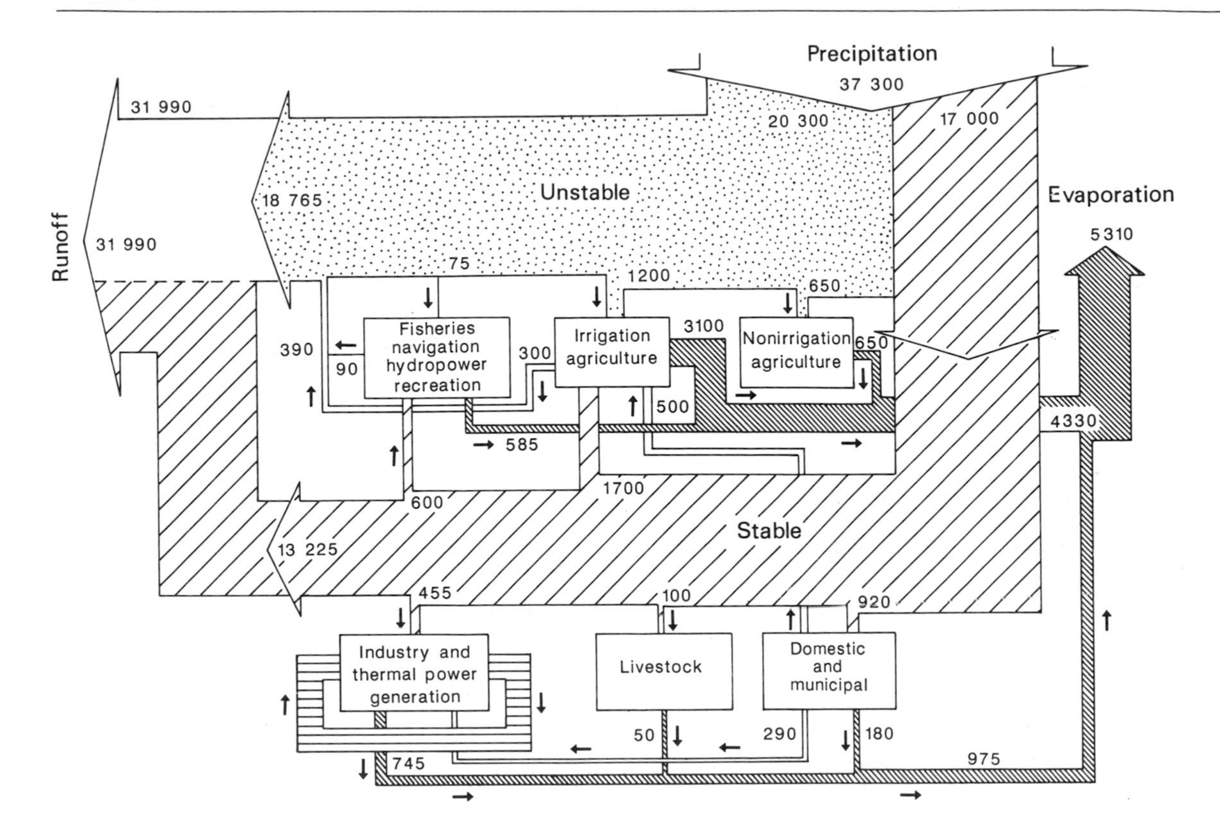

Figure 1.7 Lvovich's solution to world water supply problems: pollution and waste prevention measures are fully implemented, including industrial recycling

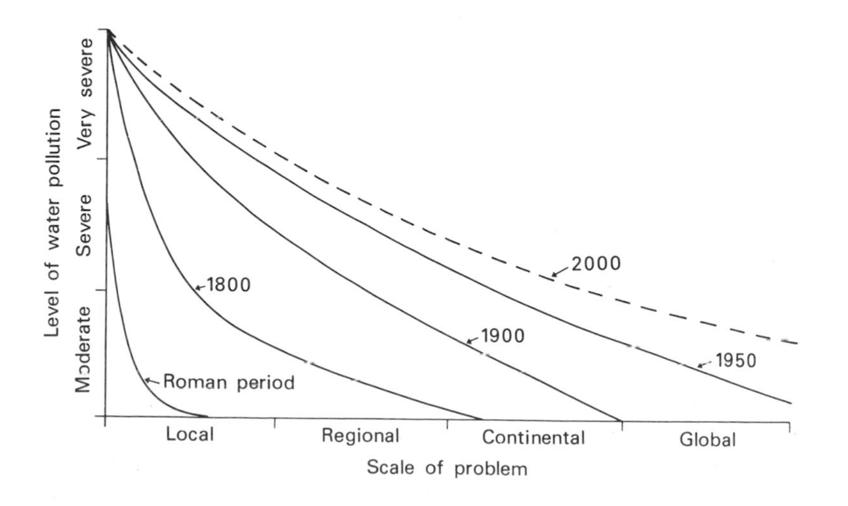

Figure 1.8 The global trend in water pollution. After Meybeck *et al.* (1990)

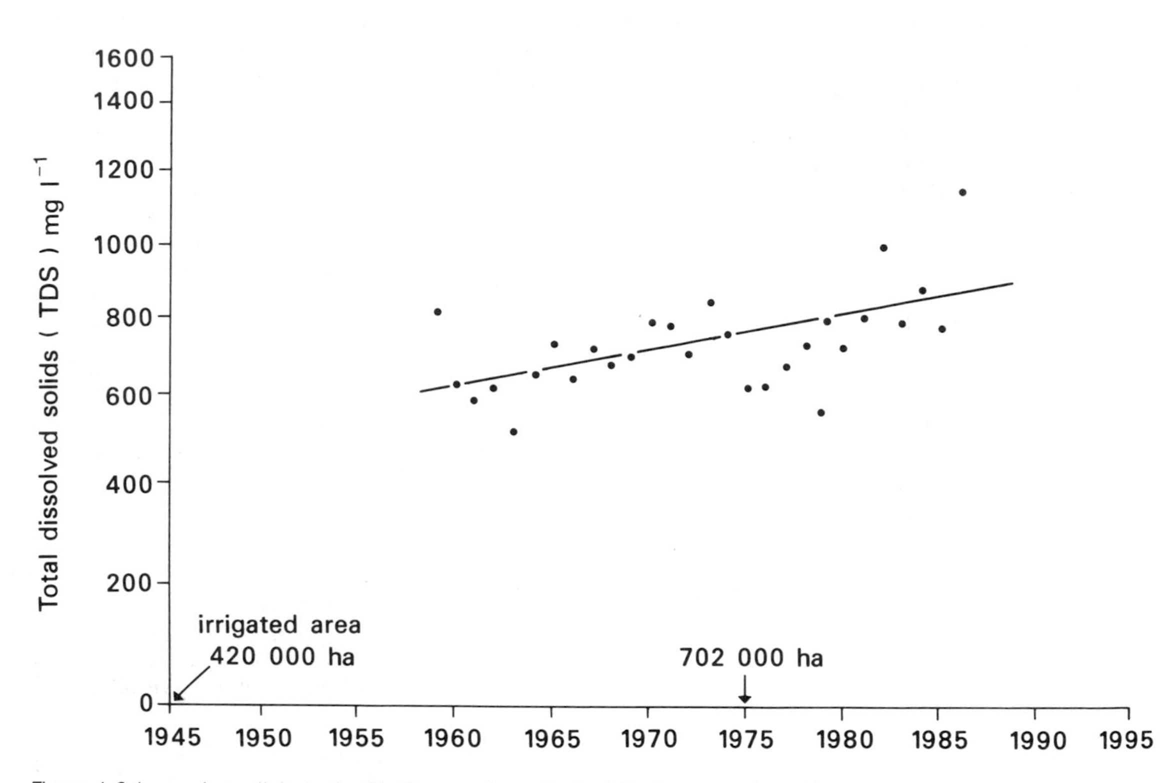

Figure 1.9 Increasing salinity in the Rio Ebro, northern Spain, following expansion of irrigation. After Alberto (unpublished)

agriculture since the 1960s, but it is particularly problematic in Europe, which uses five times the amount of nitrogen fertilizers per hectare as the Developing World and even 2.5 times that of the USA. Over 90 per cent of GEMS stations in Europe show some nitrate pollution. Worldwide, 10 per cent of stations fail WHO standards as unfit for human consumption.

Overall, the problems seem to be contained if not retreating in the richer OECD countries, despite lingering concerns over nitrates. In contrast, the former Communist countries of Europe must face a massive clean-up after 50 to 80 years without any policy of environmental protection. Even so, it is the Developing Countries that face the greatest challenge from lack of sewage treatment and from expanding industries, often using cheaper and dirtier technology. Despite the efforts of the International Drinking Water Supply and Sanitation Decade, large and increasing numbers of people still have no access to safe drinking water; over 1.2 billion people still live more than 15 minutes walk from a safe supply (Rodda 1994). The battle is on, but the problem is far from contained.

Box 1.1 International action

International collaboration plays a vital part in the advancement of scientific understanding of hydrological processes and the application of that knowledge, along with a wide range of economic and managerial skills, to solving water-related problems. Two international bodies are responsible for overseeing most of this cooperation and exchange of knowledge.

The International Association of Hydrological Sciences (IAHS) is primarily concerned with scientific aspects. It is part of the International Union of Geodesy and Geophysics (IUGG), itself part of the International Council of Scientific Unions (ICSU). Like fellow scientific unions and associations, the IAHS runs commissions and working groups that are established to promote research into key scientific areas. It holds major international conferences every four years, at which these groups report on progress in their selected area and where new groups may be formed to advance research frontiers in newly identified priority areas. The IAHS's International Hydrological Decade (1965–74) was the first internationally coordinated programme of hydrological

Box 1.1 (cont.)

experiments, which continues as UNESCO's International Hydrological Programme (IHP). Some areas lie beyond the expertise of the IAHS itself and involve inter-association or inter-union collaboration. Examples include the International Geosphere–Biosphere Programme (IGBP), an inter-union collaboration concerned with exchanges between the physical and biological environments, or the International Satellite Surface Climatology Project (ISLSCP), concerned to improve the use of satellite data for hydrology and climatology.

The World Meteorological Organization (WMO), an agency of the UN, is entrusted with the practical remit of overseeing water resources problems. Through its secretariat in Geneva, the WMO coordinates the collection and exchange of meteorological and hydrometric data, advises on instrumentation and network design, promotes the exchange of appropriate hydrological models and encourages the interchange of technical knowledge, especially towards the Developing Countries. In many practical ventures, it collaborates with fellow applied organizations, like the World Health Organization (WHO). In other cases, the WMO itself undertakes or commissions work of a fundamental, scientific nature, or collaborates with UNESCO's IHP, in order to support its primary role as a provider of information and expertise for the member countries of the United Nations Organization (UNO). The Global Energy and Water Cycle Experiment (GEWEX) is a good example, in which the WMO and the ICSU have collaborated within the World Climate Research Programme (WCRP) since 1993 to improve understanding of evaporation and precipitation processes at a global scale.

In 1977, realization that a lack of available freshwater resources was a major hindrance to economic and social development in many parts of the world led the UN Water Conference to be convened in Argentina. Out of this conference came the Mar del Plata Action Plan, which forms the basis for current UN strategy on water resources. The plan included eight key policies.

1. To promote **water resources assessment (WRA)**. In 1978 the WMO and UNESCO initiated a project to improve WRA globally and to assist countries in this work. This has involved cataloguing available information on WRA, providing data banks and mechanisms for the exchange of information, defining national requirements with particular regard for increasing demand, climate change and variability, reviewing their institutional arrangements for WRA and encouraging the trend towards integrated water management in each country. It has led to selective funding for improvements and the training of specialists and technicians.
2. To promote efficiency in water use.
3. To promote pollution control and improved protection of the environment and human health. A major result was the collaboration between the WMO and the WHO in the International Drinking Water and Sanitation Decade.
4. To improve water resources management and policy planning.
5. To improve protection against natural hazards. Major improvements in early warning of hurricanes via satellite, for example, in the Indian Ocean, have resulted from this strategy.
6. To encourage a programme of public education, disseminating information and promoting training and research.
7. To encourage regional cooperation.
8. To encourage international cooperation.

The strategy has been developed with the help of the Food and Agriculture Organization (FAO) for agriculture, the WHO for water quality, and the UN and WMO/UNESCO for economic aspects of the programme. By 1988, the WMO was able to publish a handbook to help countries determine the adequacy of their own WRA programme and to improve it.

However, the WMO/UNESCO (1991) review of the programme's achievements was less than encouraging. It noted cause for concern that data collection and analysis are not keeping up with present water development and management needs or with the pressing need to promote sustainable development. The worst areas are in Africa and the Middle East, where all aspects of the water balance are inadequately monitored and provisions are deteriorating (section 5.1). Even in North America, there are large areas with inadequate data.

Conclusions

The second UN Water Conference, held in Dublin in January 1992 to prepare reports for the Earth Conference in Rio, publicized the parlous state of water resources assessment throughout most of the world. It also pointed to the increasing pressures on resources, abstractions from rivers at rates approaching the renewal rate, an increasing number of aquifers that are being exploited at rates exceeding their recharge rate, population growth that is even further out of control than in 1977, fears of global warming, and perhaps most worrying of all the prospect of 'Water Wars' in areas of overexploitation and water shortage. If this looks like international failure, it is also the most powerful argument for strengthening the hand of international scientific and technical cooperation.

Further reading

Readable introductions to many of the general issues can be found in:

Gleick P H (ed.) 1993 *Water in crisis – a guide to the world's fresh water resources.* Oxford, Oxford University Press: 473pp. Topical reviews, plus extensive tabulated data.

National Geographic 1993 *Water – the power, promise, and turmoil of North America's fresh water.* National Geographic Special Edition: 120pp. Colourful and anecdotal.

Postel S 1992 *The last oasis: facing water scarcity.* London, Earthscan: 239pp. A thought provoking introduction by a vice-president of the Worldwatch Institute.

World Resources Institute 1994 *World Resources 1994–95.* Oxford, Oxford University Press: 400pp. Topical reviews, plus data. A series now updated biennially.

Young, G J, J C Dooge and J C Rodda 1994 *Global water resources issues.* Cambridge, Cambridge University Press: 194pp.

Water is discussed in a more general political context in:

Gore A 1992 *Earth in the balance.* Earthscan. Worthwhile, if not completely up to date in its hydrological information.

Discussion topics

1. Should water managers aim to satisfy society's demands for water or should they attempt to constrain demands?
2. Is international collaboration working? What more might be done to improve its effectiveness?
3. Produce an analysis of the water problems in your area.

CHAPTER 2 The global hydrological cycle

Topics covered

The circulation of water from the ocean and land surfaces into the atmosphere and back again creates a resource which is unlimited in global terms (Figure 2.1). It is the largest circulation of matter within the Earth–atmosphere system. As the ultimate recycling process, it washes away waste products in rainfall and runoff and it purifies supplies by evaporation. The *supply of freshwater* on Earth is entirely due to this hydrological cycle. In addition, the solar energy and the gradients in water concentration between surface and atmosphere that drive evaporation also drive transpiration, the process that causes the sap to rise in most terrestrial plants. In this sense, the hydrological cycle shares with solar radiation the role of *driving force behind primary biological production*, basic food production, on land.

The water cycle, however, has another vital role. By maintaining vapour in the atmosphere, it creates the climate for life. Water vapour creates a '*greenhouse effect*' by absorbing heat lost from the surface of the Earth. Without the present level of water in the atmosphere, surface air temperatures would be 16°C lower. Since the global average mean annual temperature of this air is only 14°C, it is clear that life, based upon the circulation of liquid water, could not exist as now over much of the Earth without the heat-conserving effect of atmospheric water, or an effective substitute gas. Water vapour nearly doubles the effect of the other greenhouse gases.

The cycle has a fourth vital role. It is a major *transporter of heat*. Nearly one-third of the energy that drives atmospheric circulation is transported and released as latent heat of condensation as clouds are formed, and recent estimates suggest that oceanic circulation transports slightly more energy polewards than does the atmosphere (Chahine 1992).

Activity rates within the hydrological cycle are determined by receipts of solar energy and by the composition of the atmosphere. Ultimately, it is astronomy that determines them. Hence the broad operation of the global hydrological cycle is controlled by the Sun's output and by the orbit, mass and composition of the Earth.

Our neighbouring planets are either too hot (Venus) or too cold (Mars) to maintain a cycle. On both Venus and Mars the low mass of the planets, respectively 80 per cent and 10 per cent of Earth's mass, creates a lower gravity force that has allowed lighter gases to escape into space, leaving behind heavy carbon dioxide (CO_2) as the principal constituent of the atmosphere. Venus, being 40 million kilometres nearer to the Sun, has higher radiation receipts, which combine with the dense CO_2 atmosphere to create an arid planet with surface temperatures of 475°C; any clouds are of sulphuric acid from volcanic eruptions, not water. Mars, 78 million kilometres further than Earth from the Sun, is also too harsh for life. There is water on Mars, but it is mostly frozen. Despite the CO_2 atmosphere, its low concentration and the planet's distance from the Sun mean that even on a summer's day at the equator air temperatures do not rise above –29°C. However, the dried-up river channels identified from the Mariner 9 and Viking space probes may be evidence of a warmer, wetter climate in the past, perhaps before the low gravity force allowed too much greenhouse gas to escape. Some of these channels may have been sculpted by water melted or released by volcanic activity or meteor impacts, but recent indications that the polar ice caps are composed of water rather than frozen CO_2, as previously thought, seem to suggest the former existence of a hydrological cycle on the planet. This is reinforced by observations of slight traces of water vapour in the present Martian atmosphere.

There is much stronger evidence for changes in the hydrological cycle on Earth. It is far from a fixed, isolated system. It is an open system that responds to changes in the astronomical, geological and biotic environments, as well as to human interference. Regular shifts in Earth's orbit alter annual and seasonal receipts of solar radiation and contribute to glacial cycles. Plate tectonics and continental drift alter the distribution of land and sea, and create mountain ranges, which affect rates of evaporation and precipitation. Gaseous exchange with plants and animals, and dust and gases from volcanoes alter the thermal properties of the atmosphere. Even humanity has probably been altering the cycle at the global scale for millennia, principally prior to this century by altering the vegetation.

As scientists attempt to predict future trends in water resources and the effects of human interference, two key facts have emerged over the last decade: (1) the centrality of the hydrological cycle within the global climatic system; and (2) the urgent need to understand the processes more fully.

2.1 The global system

The global hydrological system consists of four major reservoirs or stores, the ocean, terrestrial waters, terrestrial ice and the atmosphere, and the flows or 'fluxes' between them (Figure 2.1). The total volume of water within this **hydrosphere** is generally estimated at between 1.38 and 1.5 billion km^3. For practical purposes, this *volume* is finite, although the hydrological cycle makes total *supply* unlimited. The amount of water lost by chemical breakdown (hydrolysis) or created as 'juvenile' water in volcanic eruptions is negligible. Climate change merely alters the proportions held in different stores and the fluxes, not the total volume of water. However, over the billions of years of its existence the volume of free water has gradually increased by about 1 $km^3\ a^{-1}$, due to the gradual degassing of the Earth's mantle as water bound chemically and physically within rocks and minerals has been released by weathering. This source still contains a volume of water about 15 times that presently free in the hydrosphere.

The overwhelming proportion of all the world's free water lies in the oceans, estimated at 94–97.5 per cent, and it is saline. The next largest reservoirs are terrestrial ice and waters, although authorities disagree as to their relative size (Figure 2.2). All, however, agree that the atmosphere is by far the smallest of the 'big four' reservoirs.

Exchange between the stores takes place in all three material phases – gaseous, liquid and solid. Although Figure 2.3 shows that gaseous and liquid fluxes dominate exchanges, it is worth noting that the formation of ice within the atmosphere is an important precursor to rainfall and currently produces some 74 per cent of world rainfall. Solid exchange is more important in ice age climates.

The most critical flux for terrestrial water resources is the *net* transfer of 40 000 km^3 of water a year from the oceans to the land as precipitation. This is approximately equivalent to the total riverflow of the world and it controls the equilibrium recharge rate for groundwater supplies. For most practical purposes, it is

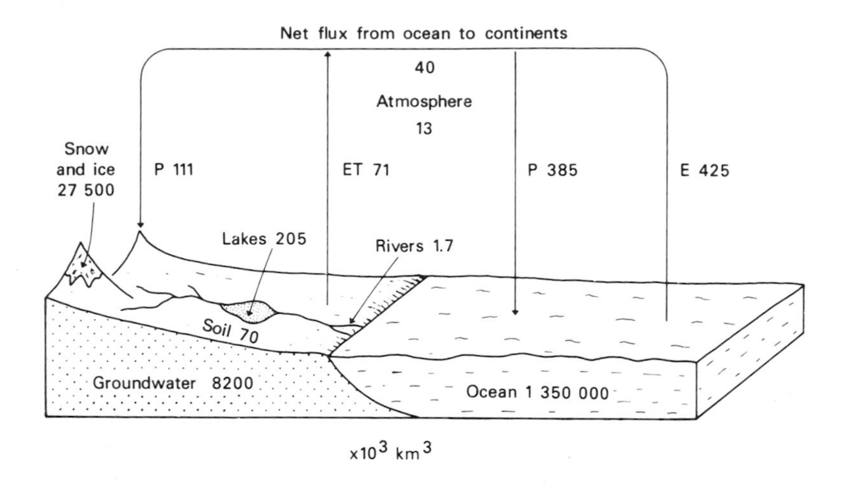

Figure 2.1 The global hydrological cycle

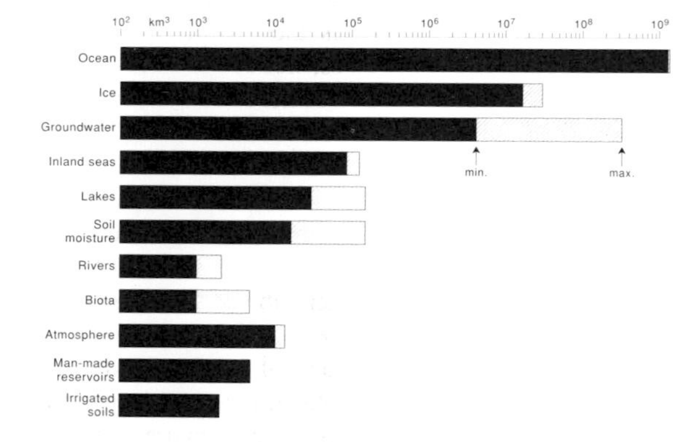

Figure 2.2 Major stores in the hydrological cycle. Maximum and minimum estimates shown

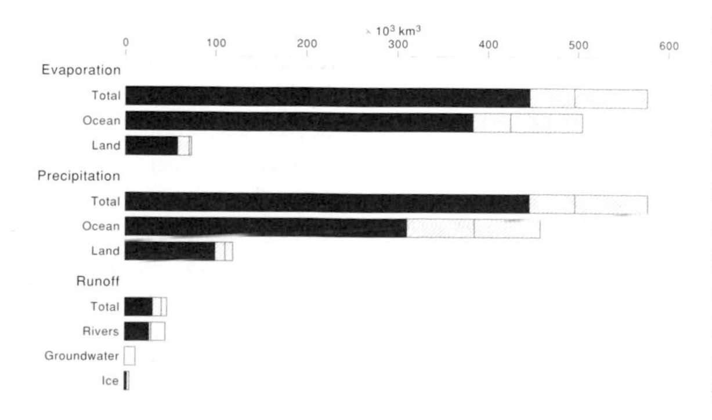

Figure 2.3 Annual flux rates within the global hydrological system. Maximum, minimum and mean estimates shown

still the limit of resource available to support terrestrial life.

The rate of turnover within stores varies considerably, from 10 000 to 15 000 years (15ka) for terrestrial ice to about 7 to 10 days for atmospheric moisture (Table 2.1). Turnover is 3–5 times faster in the freshwater stores than in the saline ones, with the exception of ice. This is a reflection of the need for rapid turnover to maintain freshness. The relative rates shown in Table 2.1 show the importance of exposure at the Earth's surface, which is the most active surface for interchange. The 'surface area to volume ratio' in Table 2.1 is a rough measure of the relative exposure of the water substance held in each store.

The ratio largely explains why average turnover is so slow in the oceans, despite the ocean surface being the source of 86 per cent of atmospheric moisture and one of the most active exchange interfaces in the whole system. It also demonstrates how exposed terrestrial surface waters are. Deviations from the trend show the importance of: (1) the amount of energy available at the interface; (2) the efficiency of energy use; and (3) the ease of exchange. Hence, turnover is relatively slow in ice sheets and glaciers because of the special properties of ice, which inhibit losses, and because of their location in low-energy (cold) climates. Turnover is slow for deep groundwater because of its remoteness from the Earth's surface and because of its depth within a relatively slowly permeable medium.

The *amount of energy available* controls the rate of exchange per square metre of exchange surface. This

Table 2.1 Storage and turnover in the major hydrological stores

Store	Best estimate volume (10^3 km^3)	Range in estimates (10^3 km^3)	Per cent range	Area (10^6 km^2)	Per cent total	Per cent freshwater	Turnover rate years	Surface area to volume ratio
Oceans	1 350 000	50 000	4	360	97.40	0	3000	0.26
Ice caps/ glaciers	27 500	12 700	81	16	1.98	85.9	8000– 15 000	0.67
Terrestrial waters	8477.8			134	0.61			
–groundwater	8200	325 990	7949		0.59	13.5	>5000-deep <330-active	
–inland seas	105	39.6	46		0.008			
–lakes	100	120.0	400		0.007	0.313	10	329.0 (lakes, soil moisture, rivers, biota)
–soil moisture	70	133.5	809		0.005	0.219	0.038–0.96	
–rivers	1.7	1.1	110		0.0001	0.005	0.038	
–biota	1.1	49.0	490		0.0001	0.003	0.077	
Atmosphere	13	3.5	33	510	0.001	0.04	0.027	>39 000
Man-made stores								
–reservoirs	5				0.0004	0.016		
–irrigated soil	2				0.0001	0.006		
TOTAL								
global	1 385 990.8							
freshwater	32 000				2.4			

Source: Compiled from Speidel and Agnew (1988) and Lvovich (1977) with additions.

includes thermal energy for evaporation and potential energy for gravitational flows. The availability of **thermal energy** depends not only upon the incoming radiative and thermal energy but also upon the radiative and thermal properties of the surface materials. For example, because snow and ice reflect 80–90 per cent of incoming solar radiation compared with 10 per cent for water, they create a cooler (lower energy) exchange surface. The amount of **potential energy** available for gravitational flows is largely determined by topography, but it may also be a function of rates of accumulation, especially in glaciers. Globally, however, gravitational flows at the surface are more than an order of magnitude smaller than evaporational exchanges.

The least *efficient use of thermal energy* occurs in evaporation from ice (strictly sublimation or vaporization), which requires 2800 J g^{-1} against 2500 J g^{-1} from liquid water and only 335 J g^{-1} for melting. Hence, meltwater occurs in far larger quantities than evaporation from snow and ice surfaces. The use of radiative and thermal energy at the surface is, however, complicated by the thermal and optical conductivity of the store, its thermal content and water circulation within it. In this regard, ice is again more resistant to loss than liquid water, with a lower thermal conductivity and thermal content and the lack of internal circulation all limiting the penetration of heat. In terms of the use of potential energy, the high viscosity of ice also reduces flow rates and discharges into the oceans.

Exchange is easiest where there is least resistance to flow. In general, this occurs at the Earth–atmosphere interface, which also happens to be the locus of maximum heat availability in the whole hydrosphere.

Of paramount importance for human water resources is the fact that humanity relies predominantly on the smallest stores, the surface or near-surface terrestrial waters and the atmosphere that serves them, which have the fastest rates of turnover and the greatest short-term fluctuations in supply. In trying to offset these vicissitudes, humanity has become no mean contributor to the system. As Figure 2.2 demonstrates, the amount of water stored in irrigated soils is comparable with, if not greater than, the amount naturally stored in all the world's rivers. The amount held in artificial reservoirs is even greater.

2.1.1 The world oceans

The oceans control the global system because they are: (1) the largest source of water; (2) the main reclamation

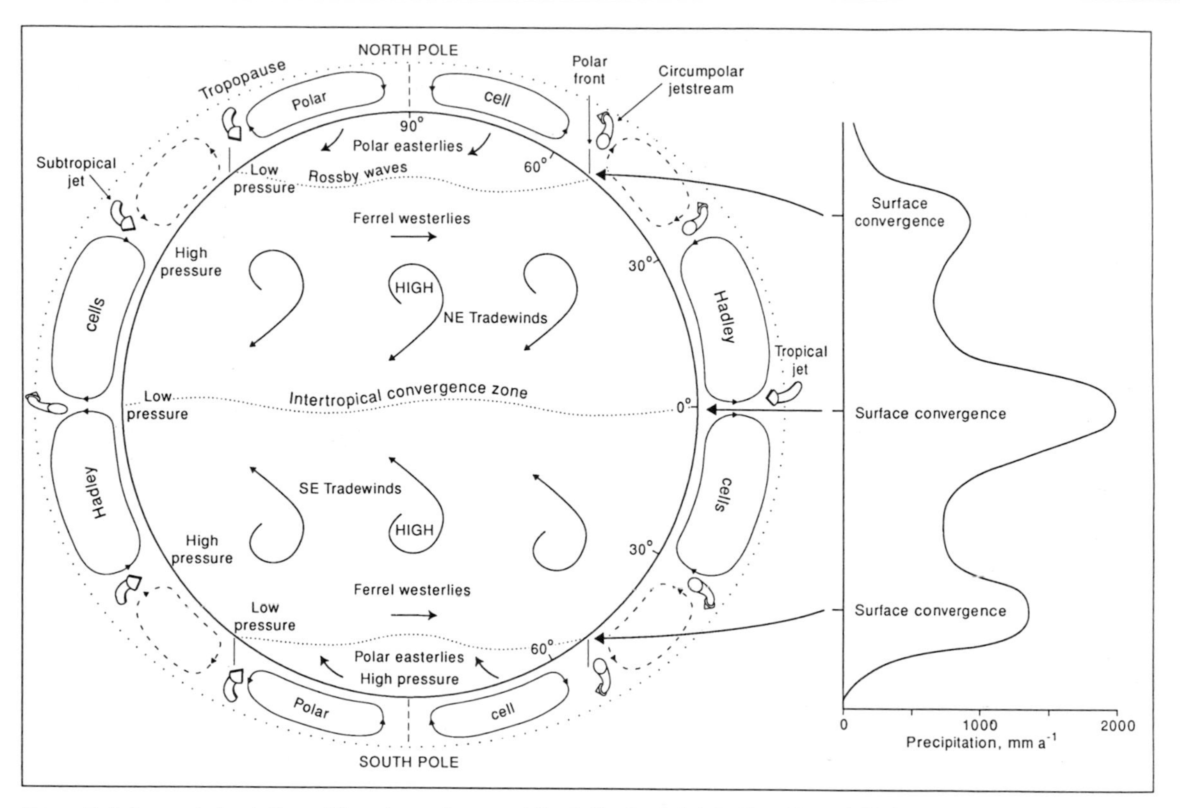

Figure 2.4 General circulation of the atmosphere and the latitudinal distribution of precipitation

and recycling plant; (3) a major heat store, which makes them the most important individual buffer to climate change; and (4) a major conveyor belt for heat. Their turnover rate is faster than the ice caps or deep groundwater primarily because they present such a large surface of liquid water in contact with the atmosphere, much of it within the tropics.

The world's oceans cover 71 per cent of the surface of the globe, over which evaporation normally takes place at full 'potential' rates, i.e. unhindered by any lack of available water as often occurs on land. The exception occurs in polar regions, where sea ice or floating ice shelves cover the surface and vaporization is very low. On average 7 per cent of the world's oceans are ice covered. In the northern hemisphere, sea ice can cover over 16 million km^2 in winter, retreating in summer to a core of about 11 million km^2, wholly within the 14 million km^2 of the Arctic Ocean.

Even so, nearly 90 per cent of all evaporation occurs from the ocean surface, nearly a quarter more than would be the case if evaporation rates were the same on land and sea. Since the ocean store is the lowest in altitudinal terms, water losses are almost exclusively via evaporation rather than gravitational losses, with the minor exception of small amounts of salt water seepage into terrestrial groundwater along permeable coastlines.

The oceans also cover about 73 per cent of the surface of the globe within the 'climatic tropics', the zone in which receipts of solar radiation exceed losses of radiation back to space. This is the powerhouse of world climate, the source of the great convective cells of atmospheric winds, the '**Hadley cells**', which directly control tropical climate and indirectly affect the rest of the world (Figure 2.4). The moisture evaporated from this area of the oceans plays a significant role in the poleward transfer of heat within the atmosphere and contributes to rainfall levels in mid-latitudes. It also creates important tropical rainfall systems, such as the monsoons, rainforest climates, the **intertropical convergence zone (ITCZ)** and hurricanes.

The oceans' own circulation system is an extremely important conveyor of heat. Until recently it was thought to account for up to one-third of poleward heat transfer, but recent oceanographic evidence suggests the proportion may be slightly more than half (Bryden *et al.* 1991; Chahine 1992). This is because: (1) oceans cover the bulk of the tropical Earth; (2) water has a higher heat capacity than air; and (3) warm ocean currents are

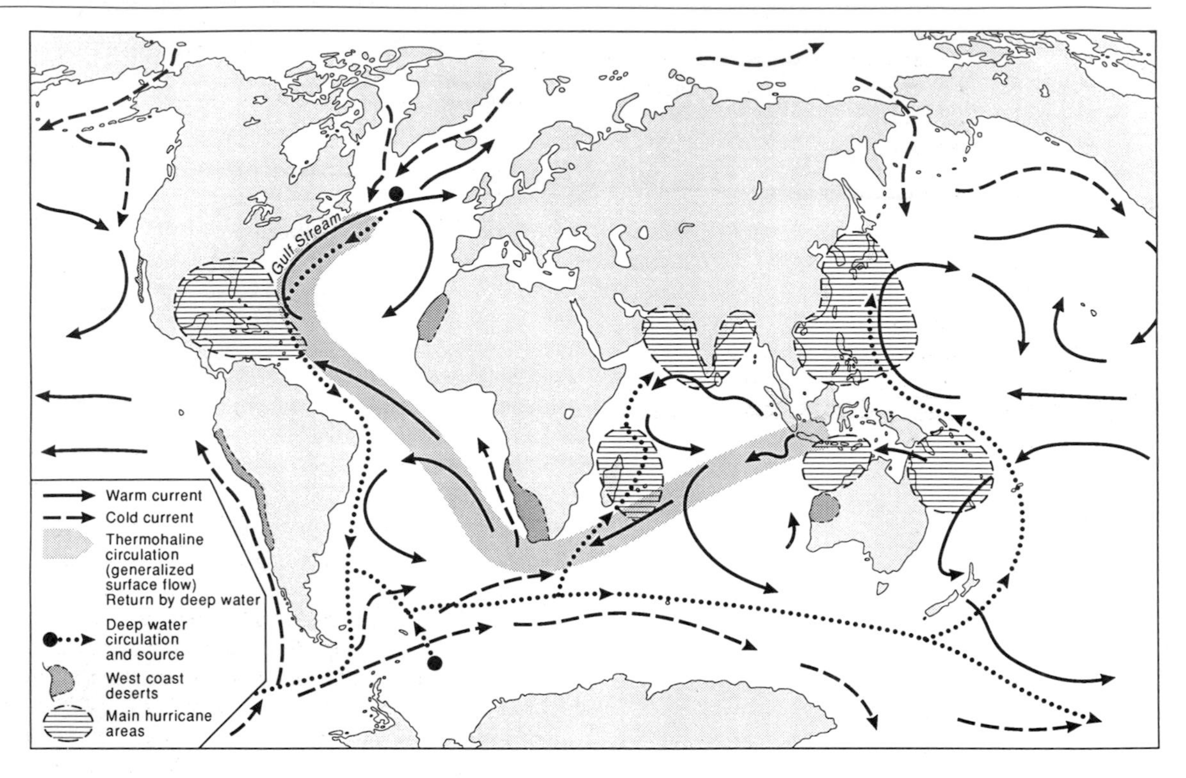

Figure 2.5 Circulation of the oceans, the main areas of hurricane activity and west coast deserts

free to transport heat out of the tropical realm everywhere except the north Indian Ocean. Where these ocean currents flow has important consequences for regional patterns of evaporation and rainfall in today's world (Figure 2.5). In the past, shifts in ocean currents may have initiated climate change, especially changes between glacial and non-glacial phases, altering the balance between the major hydrological stores.

Surface currents tend to mirror atmospheric circulation over the main zone of poleward transfer of heat between the equator and mid-latitudes. This is partly because they are driven by similar temperature gradients and partly because winds exert a frictional drag on the ocean surface (Perry and Walker 1979). Circulation is predominantly in a horizontal plane, bounded by the continents and set spinning by the Earth's rotation, the so-called **Coriolis force**, so that poleward transport tends to occur on the western side of the oceans. However, areas of **oceanic upwelling** or convective downwelling can also be related to surface wind-drag effects or density differences caused by salinity. These are increasingly recognized as important for precipitation over adjacent land areas, through their effect on sea surface temperatures (section 2.2.2). Since upwelling also raises nutrients and improves plankton productivity, it also has a significant effect on oceanic absorption of CO_2. Deeper circulation currents are driven more by differences in salt concentration (salinity) than by temperature gradients. The **thermohaline circulation** (Figure 2.5) transfers heat from the Indian Ocean into the North Atlantic. Broecker (1989) describes this as a conveyor belt for heat with a flow equal to 10 times total world riverflow. It is driven by high evaporation rates in the North Atlantic, which cause surface water to cool, increase in salinity and density, and so to sink and return south. Polar sea ice formation may add to this, as salt is expelled upon freezing. In the South Atlantic this deep current is drawn towards the Pacific in partial compensation for the outflow of lower density water at the surface flowing from the rainfall surplus in the Pacific (section 2.5). The 'ultimate' driving force is therefore the water balance of the oceans. But the thermohaline circulation also *supports* the pattern of rainfall and evaporation, and it links with surface currents driven by winds and temperature gradients, e.g. it adds heat to the Gulf Stream (Figure 2.5). It appears to have played a significant role in initiating glacial cycles. Periods of greater freshwater input, from precipitation, meltwater or more active outflow from the Arctic Ocean may create a surface capping of low salinity and slow down deep water formation.

Counterbalancing its role in triggering climate

change, the world's oceans have long been regarded as an important buffer to change, responsible for delaying response to external forcing. The oceans' resistance to change is due to the following:

1. Their great depth, averaging 3.75 km, and volume.
2. The high specific heat of water, i.e. the heat needed to raise its temperature (4.18 J g^{-1}°C^{-1}). Combined with the volume, this means that the oceans have a high **thermal capacity**, i.e. they can store a relatively large amount of heat energy with only a small change in temperature. Since rates of radiative and conductive cooling are proportional to surface temperature, they therefore lose heat to the atmosphere more slowly.
3. Optical penetration spreads solar heating over a greater depth than on land, adding to more efficient storage of heat.
4. Their fluid nature disperses heat and means that as water cools to the point of maximum density at 4°C it sinks and is replaced by warmer water from below, thus slowing the process of surface cooling.

The world's oceans are replenished from all of the other stores, receiving approximately 385 000 km^3 from precipitation, 38 000 to 41 500 km^3 from riverflow and groundwater and 2000 km^3 directly from calved icebergs each year (Table 2.2). The imbalance between evaporative losses and precipitation receipts is approximately equivalent to the annual input from terrestrial sources.

In the longer term, however, balances are continually shifting, particularly between the oceans and terrestrial ice. At maximum glaciation, 3.5 times as much water was locked up in ice (*c.* 1.2 × 10^{20} kg) as now. In the past 18 000 years since the peak of the last ice age, the ocean volume has increased by 10 million km^3 and sea level has risen 100 m. More than 100 000 km^3 has been added to the oceans this century alone, contributing to a rise in mean sea level of approximately 10 mm a decade. Rising sea level itself, however, does not necessarily imply increased water in the oceans. In the present case, it is partly due to thermal expansion caused by rising temperatures over the last two centuries since the end of the Little Ice Age. Oddly enough, the volume of polar ice also seems to have increased this century by about 40 000 km^3, a volume equivalent to total world runoff, so the source of the extra water in the oceans is in dispute. Is it perhaps due to human exploitation of deep groundwater, which is being returned to the active branch of the hydrological system?

Table 2.2 Rates of exchange in the global hydrological system (in thousands of cubic kilometres a year)

	Total	Oceans	Land
Evaporation	496	425	71
Precipitation	496	385	111

	Total	Rivers	Groundwater	Glacial meltwater
Runoff	41.5	27.0	12.0	2.5

Source: Based on Speidel and Agnew (1988).

2.1.2 Terrestrial ice

With approximately 27 500 km^3 of liquid water equivalent stored in them, polar ice sheets and glaciers constitute one of the largest reservoirs of water. Although not all authorities rank it next to the ocean (Figure 2.2), as far as freshwater is concerned it dominates global distribution to almost the same extent as the oceans dominate total water, storing 85 per cent of freshwater reserves. Terrestrial ice contains over 500 times the total volume of world riverflow in any one year.

The bulk of this water is locked up in the Antarctic (89.7 per cent) and Greenland (9.8 per cent) ice caps, which have the slowest turnover rate in the whole hydrological cycle – of the order of 15 000 years. The slow turnover reflects the special properties of ice already discussed, plus low surface-to-volume ratios and extremely cold environments. The extreme cold not only minimizes losses by melting and evaporation, it also reduces flow rates and limits recharge from snowfall. With little internal and basal meltwater to lubricate it, the ice behaves like a more viscous fluid. The annual yield from calved icebergs is equivalent to just 5 per cent of world riverflow. This amounts to only 8 per cent of the total stored in the polar ice sheets, whereas annual riverflow amounts to 2000 per cent of the water held in rivers. Low snowfall receipts derive from the cold sea surfaces, often covered in sea ice, the cold air and the lack of convective and frontal activity (sections 2.1.4 and 2.3).

Valley glaciers tend to have faster rates of turnover than ice caps, especially the **alpine glaciers**, a generic term applied to glaciers in all mountain areas outside polar regions. Whereas **high arctic** or **polar glaciers**, like those on Axel Heiberg Island, freeze to their beds for most of the year, alpine glaciers tend to maintain a lubricating layer of meltwater beneath them throughout

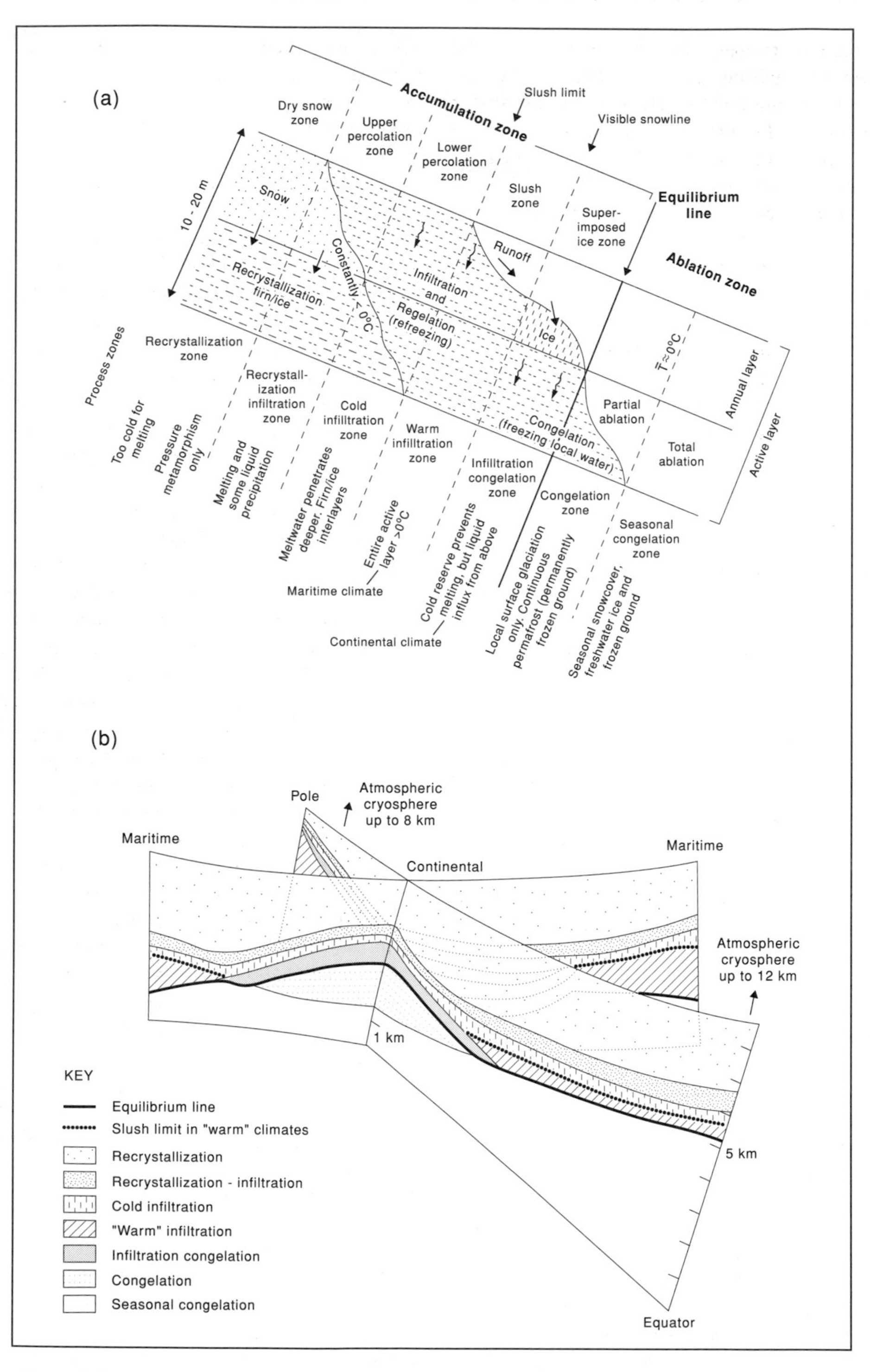

Figure 2.6 Geography of the cryosphere: (a) processes and zones in the layer of annual snow accumulation and the active layer of surface and subsurface ice; (b) global distribution of zones. Based on Shumskii (1964) and F Müller (pers. comm.).

the year, because temperatures are warm enough to allow **pressure melting** to occur (caused by the depression of freezing point temperature by the weight of the ice above). In summer, flow rates tend to be enhanced by meltwater that has entered the glacier via crevasses and sinkholes or 'moulins'. The role of such **englacial meltwaters** in glacier flow has been increasingly recognized in recent years (Hodge 1979) as an important supplement to the flow processes of ice deformation, sliding and pressure–melt–regelation encompassed in the 'classical' theories of glacier flow (Paterson 1994). Many **surging glaciers**, glaciers that suddenly increase velocity for short periods sometimes only occasionally and sometimes quite regularly, may be triggered by the build-up of englacial waters (Hodge 1979).

Because of their geographical location at lower latitudes and often nearer to population centres, alpine or warm-based glaciers tend to be more significant as suppliers of fresh meltwater to humanity and the living environment. Icemelt from alpine glaciers comprises about 1/200th of world river runoff – 15 per cent of the discharge of the Columbia River as it crosses from Canada into the USA is derived from the Columbia Icefield. This icefield is the largest ice mass equatorwards of the 59th parallel and also a major source of runoff for the Mackenzie and Sakatchewan Rivers. Spring meltwater from this and numerous other alpine ice bodies in the Rocky Mountains is a vital source of water for agriculture in the Canadian prairies. Large volumes of icemelt serve agriculture along the major rivers of the Himalayas. Indeed, Pakistan's national glacier inventory was undertaken as part of the planning exercise for the Tarbela Dam in the 1970s, because of the strong dependence of flow in the Indus on glacier meltwater. Glacial meltwaters also provide indispensable sources for hydropower in Norway and the Alps.

The most critical boundary line in the **terrestrial cryosphere**, the land-based section of the realm of ice, is the line which divides the **zone of net accumulation** of snow and ice from the zone of net loss by melting and evaporation, the **ablation zone** (Figure 2.6). Commonly referred to as the **annual snowline**, it is better termed the **equilibrium line**, since in the high Arctic snow can melt and refreeze on the surface of the glacier creating **superimposed ice**. This leaves the visible lower limit of new winter snow, the visible snowline, some way above the actual limit of net accumulation of mass.

Above the equilibrium line snow gradually metamorphoses into ice by a combination of pressure effects due to the weight of newly accumulated snow and, especially nearer the equilibrium line, of thermal melting, principally due to solar radiation and warm air bringing in advected heat. Partly metamorphosed snow from the previous winter, known as **firn** or **nevé**, may be exposed below the annual snowline. This is evidence that annual snowlines fluctuate from year to year as snowfall and temperature patterns vary.

One of the most critical parameters for assessing the net balance between accumulation and ablation in an ice body is the **accumulation area ratio**, the ratio of the area of net accumulation to the total area of the ice body, which has to be determined over a number of years because of fluctuations. This parameter is a surer index of the current 'state of health' of an ice body than observations of advance or retreat at the snout. This is because it is assessing present net balance over the glacier as a whole, whereas movements of the glacier snout may reflect local factors or historical accumulation patterns that have finally reached the snout. Unfortunately, because of the need for fairly detailed observations over a number of years, this is the most frequent piece of information missing in glacier inventories. With around 100 000 ice bodies containing more freshwater than the Great Lakes, Canada led the way in glacier inventory (Ommanney 1980). The World Glacier Inventory, based on the Canadian experience, has been one of the more arduous long-term projects of the IAHS International Commission on Snow and Ice over the past three decades.

The most dynamic part of the surface cryosphere is snowcover. Most of it disappears well within the year and so does not appear in annual statistics (Figure 2.2 and Table 2.1). Yet 60–70 per cent of the land mass of Asia, Europe and North America receives a stable seasonal snowcover in an average year, as do significant areas in the Australian and New Zealand Alps, the Atlas Mountains and the Andes (Figure 2.7). In addition, brief snowcovers regularly penetrate the milder climates of Western Europe, the north Mediterranean coast as far as the Lebanon, the tropical highlands of East Africa, New Guinea and Central Mexico, and the Drakensberg and Karroo of South Africa.

Most annual production of ice in the Earth–atmosphere system, 90 per cent of it, occurs in the atmosphere at altitudes of between 1 and 10 km. At present only 5 per cent of world precipitation reaches the surface in solid form. Nevertheless, this amounts to 2.5 million kg a^{-1}, and it is concentrated spatially so that, for example, 45 per cent of annual precipitation receipts in the Labrador–Ungavan Peninsula, northeast

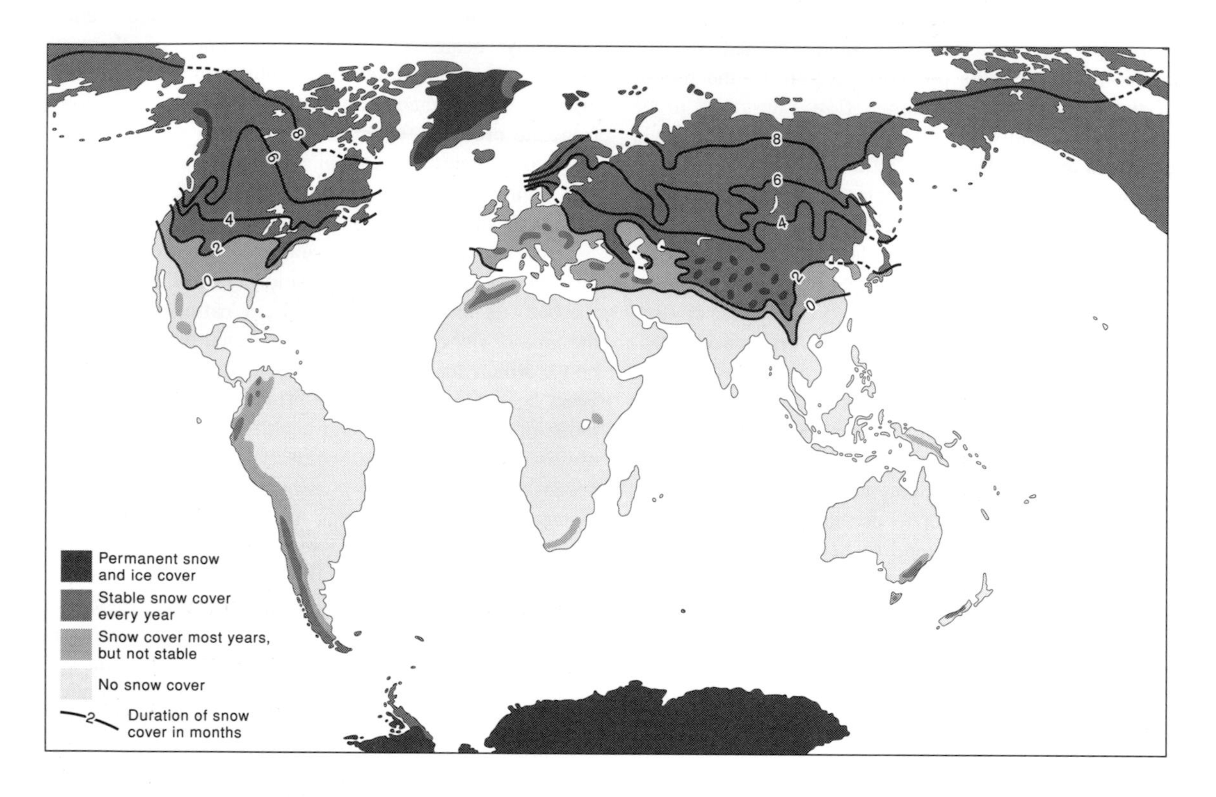

Figure 2.7 Global snowcover. Based on a map produced by the Canadian Inland Waters Directorate

Canada, are in the form of snow, a fact which has been extremely important in the design of the vast hydropower schemes on the Hamilton River and James Bay (Box 10.3).

About one-quarter of total annual snowfall is converted into semipermanent storage in glaciers and ice caps (Figure 2.8). The remainder melts. Half of this melts on land and contributes to surface runoff. This is a small fraction of total world runoff, 2 per cent, but its effects are disproportionate because of its spatial concentration and because it tends to contribute the bulk of meltwater in a short period during spring and early summer. This creates spring **freshets** or meltwater floods. Many twentieth-century impoundment schemes have been designed to alleviate this flood hazard and to utilize the meltwater. Two important advantages of this timetable of snow accumulation and spring melt are that: (1) it can convert relatively small monthly totals of precipitation into an exploitable volume; and (2) it releases water at the beginning of the growing season when agriculture and natural ecosystems need it most.

Snowmelt tends to be more efficient in producing runoff than rainfall because: (1) melt occurs relatively rapidly, quickly saturating the ground without significant evaporation losses; and (2) low vaporization losses make snowcover an efficient seasonal accumulator. Runoff production is often aided by frozen ground and other surfaces with low infiltration.

Most of the streamflow in the western USA is snowmelt. On the South Fork of the Ogden River, Utah, 75 per cent of annual discharge is derived from snowmelt in the six months after April. Snowmelt forms 90 per cent of the discharge of the Colorado River and 50 per cent in the plains rivers of the Mississippi–Missouri drainage system.

By no means all of the significant snowmelt contributions come from mountain regions. The James Bay Hydroscheme (Box 10.3) demonstrates the effectiveness of 8–9 months of accumulation in a lowland. Here the Great Lakes and Hudson Bay provide large sources of moisture and the frequent passage of weather fronts provides the mechanism for heavy snowfall.

Even quite shallow snowcover may be hydrologically significant. In the northern Great Plains an average annual accumulation of 300 mm of snow with a density

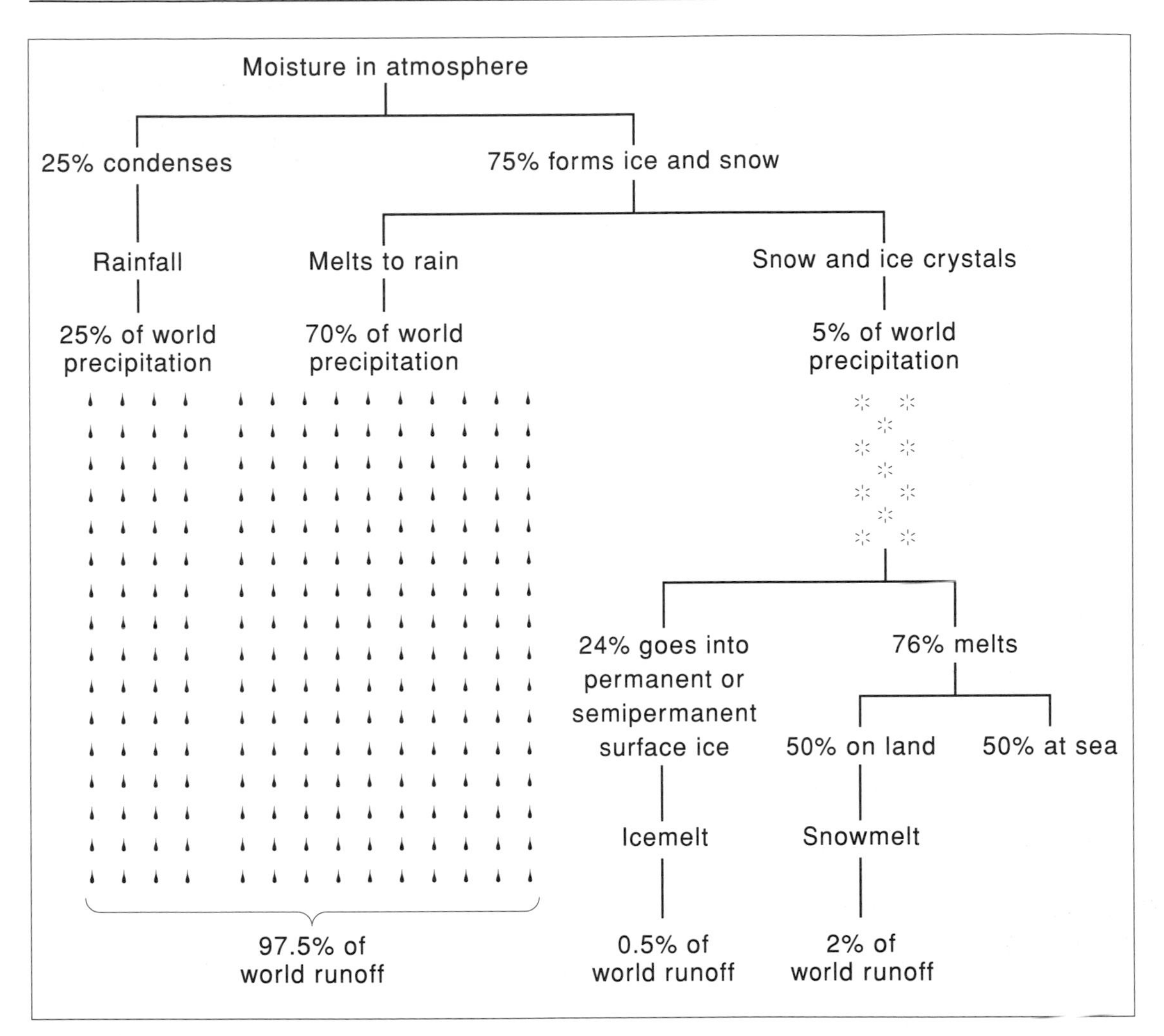

Figure 2.8 The snowfall cascade

of 375 kg m^{-3} covering the whole 1 865 000 km^2 area produces 213.5 billion m^3 of meltwater, equivalent to the total annual discharge of the Great Lakes into the St Lawrence River. With gentler slopes, deeper soils and greater infiltration than in the mountains, more of this meltwater is retained on the land to recharge soil moisture, and agricultural practices have recently been developed to enhance this natural tendency and improve soil moisture status in the early growing season (section 10.4.1).

Finally, like the oceans, snow and ice cover can play an active role in climate change. Its special thermal and radiative properties create a microclimate conducive to its own perpetuation. Not only does snow absorb less solar shortwave radiation, because of its high reflectivity or **albedo**, but it is also a very efficient radiator of longwave radiation, i.e. it has a high **emissivity**. Together these properties produce lower net radiation budgets, colder surfaces and lower surface air temperatures, making snowcover resistant to melting and possibly allowing more precipitation to reach the ground as snow. There is a pronounced positive feedback encouraging further accumulation, which can only be broken by a failure in snow supply or a very marked increase in radiation or heat supply.

The first critical threshold in the development of glaciation is for part of the winter accumulation to survive the summer. Personal observation of the tentative process of snow patch melting in interior Labrador led Jack Ives to propose that glaciation could be initiated in the lowlands by *in situ* accumulation from year to year, rather than advancing from the mountains (Ives 1957). The large volume of present-day rainfall that originates as snow which has been melted during its fall lends credence to the potential for a rapid onset of glaciation (sections 2.1.4 and 2.3).

2.1.3 Terrestrial waters

The reservoir of liquid water on land actually consists of a great range of sub-stores, from deep groundwater with a turnover rate of 5000 years or more to river networks with an average turnover of 11–14 days. There is relatively close agreement on the volumes stored in surface water bodies, but authorities differ widely on the assessment of the quantity of water held in the ground (Table 2.1). The most critical disagreement is on the amount of water held as deep groundwater, where estimates show a range of 8000 per cent. This is perhaps not so surprising, given the fact that finding much of this groundwater is a problem for geological survey akin to prospecting for oil.

Assessment of deep groundwater resources is the key issue in determining the balance between terrestrial waters and terrestrial ice (Figure 2.2). A number of recent Western estimates have rated deep groundwater as accounting for only about 0.5 per cent of global water (Pruppacher 1982; Speidel and Agnew 1988), whereas many East European estimates have given it over 4 per cent (Lvovich 1979). While this is clearly of scientific interest, however, it may not be so important for practical purposes. The recharge rates of both 'permanent' ice and deep groundwater are so small that making practical use of these stores for water supply is essentially mining a finite resource, with potentially disastrous environmental impact. Much of this groundwater may have been held in deep aquifers for up to 10 000 years; water in some north Saharan aquifers has survived since the wetter 'pluvial' climate at the end of the last glacial period. Such groundwater is termed **connate** and should not be confused with **fossil** or **juvenile** groundwater, which are respectively buried with the rocks or produced by chemical reactions during formation of the rocks. Although its age may be comparable to the ice in many alpine glaciers, it is less likely to be a usable resource from the quality standpoint after millennia of 'stale' deoxygenated and anaerobic conditions, chemical saturation and accumulation of pollutant solutes. Long-term mineralization processes will tend to convert the groundwater into brine. Lvovich (1977) has described it as a 'deadlock arc' in the cycle, comparable to the ocean and closed lakes or inland seas.

In contrast to deep groundwater, shallow groundwater is more likely to be fresh and to be actively replenished. It forms a major source of water supply in many parts of the world and turnover rates are 15 times more rapid than in the deeper waters. According to Lvovich's (1979) estimate, active groundwater accounts for 14 per cent of global freshwater resources, or 93 per cent of all freshwater in existence outside the ice caps and glaciers. Nevertheless, the cycling of water through shallow groundwater systems may take hundreds of years and caution is still required in exploitation.

A second area of doubt concerns the amount of water held within the soil. While '**groundwater**' is essentially defined as water forming a more or less continuous zone of saturation within the rocks, **soil moisture** consists of water held within the soil, often in an unsaturated state (section 3.2.2). It is not directly exploitable for water supply, although it may drain into rivers or groundwater. It is, however, extremely important for agriculture and for natural ecosystems. Again, it is a rather difficult component to assess and hence the range in published estimates is twice as much as the 'best estimate' value (Table 2.1).

Soil moisture is also an important component in soil–atmosphere interactions. The wetness of the soil affects evapotranspiration rates and the temperature and radiational properties of the soil surface. It is recognized as an important parameter for medium- to long-range weather forecasting, even affecting the course of monsoons in India. And it is a crucial element in the global climate models currently being perfected to study global warming. The urgent need for data to feed these models is leading to a considerable improvement in estimates of global soil moisture, based largely on satellite data (section 5.1).

Of all the terrestrial stores, **river channels** hold the least, and yet they are the prime source of human water supply. This apparent enigma becomes easier to understand when it is realized that:

1. It is not the quantity in store but the quantity that passes through the river system that is important. Viewed in this way, rivers can supply in one year a volume perhaps equivalent to half that available from all the soil moisture in the world, if not more, and 4–5 times that available from lakes (which have 50–60 times the storage). Even more remarkably, annual river discharge is more than 4 times the renewable annual yield from active groundwater.

2. Moreover, rivers tend to be more accessible and widely available than lakes or even large groundwater bodies.

3. Their rapid turnover and free draining tend to make them less vulnerable to pollutant accumulation.
4. They are easily dammed to increase the yield.

2.1.4 The atmosphere

The atmosphere is as vital to the hydrological cycle as the oceans, despite being the smallest of the major reservoirs. As with terrestrial surface waters, it is the yield, as precipitation, rather than the storage capacity that is most important in strict hydrological terms, excluding the important effects of water vapour and clouds on the heat budget of the earth.

The atmosphere's annual yield is about 500 000 km^3, which must be equal to the total evaporative yield from all the other, larger reservoirs combined. The fastest rate of turnover of all compensates for its low storage. The average water molecule remains airborne for no more than 10 days, equivalent to 40 or more complete changeovers a year.

Even so, this hyperactivity involves a very small portion of total global water, a mere 0.036 per cent at most (Table 2.1), and in reality much less because a large proportion is recycled within the year, the same water remaining at or near the surface and being evaporated and precipitated repeatedly. The hyperactivity is also laterally confined, predominantly within the tropics and subtropics: two-thirds of all precipitation occurs within 30° N or S. There is therefore a lot of 'dead' or inactive water even within ostensibly more active storage components.

Water vapour in the atmosphere on average accounts for about 4 per cent of the total volume of gases, but it is highly variable in both time and space, depending on evaporative sources, surface and air temperatures and wind characteristics. The amount of vapour that air can hold increases almost logarithmically with air temperature. Figure 2.9 shows this relationship expressed in terms of vapour pressure. **Vapour pressure** is the force exerted by the weight of water vapour in the air on each square centimetre of Earth surface. **Saturation vapour pressure (svp)** is the pressure due to the vapour when the air reaches saturation point, i.e. it can hold no more. The units are hectopascals (hPa), the Système International (SI) equivalent of the old millibars. The vapour content of air can also be viewed in terms of **absolute humidity**, which is the mass of vapour per unit volume of air, in g m^{-3}, or the **water vapour mixing ratio**, which is the mass of vapour as a percentage of the mass of dry air in a specified volume, in g kg^{-1}. Thus, at 30°C (a hot summer's day in Britain) the air has an svp of 42.43 hPa and can contain 30 g m^{-3} (absolute humidity), whereas at 0°C it can hold only 4.5 g m^{-3}, equivalent to an svp of 6.11 hPa.

About 90 per cent of the water vapour in the atmosphere is confined to the lower 6 km. This is because of the thermal structure of the atmosphere, the patterns of circulation that result from this, and the fact that air cools as it rises, acquiring a lower svp in the process.

Vapour is dispersed from an evaporating surface by air set rising by turbulent eddies caused by friction at the surface, by convection caused by local heating and expansion, or by frontal systems that force warm air to rise over colder air (sections 2.2 and 2.3). Because gravity concentrates the atmospheric gases near the surface, there is a rapid fall in pressure in the first few kilometres: at 5.5 km atmospheric pressure is about half that at the surface. Air rising into this lower pressure environment expands and cools '**adiabatically**', i.e. it cools purely as a result of expansion. When cooling causes vapour pressure to reach saturation vapour pressure, condensation begins.

An upper limit to rising air is imposed as a result of the ozone layer. The creation and presence of ozone (O_3) at altitudes of 25–50 km in the stratosphere converts ultravoilet radiation from the Sun into heat and creates a large-scale 'temperature inversion', the inverse of the temperature lapse in the troposphere below, which results from pressure-release cooling between the ground and 11 km. This effectively caps convection, because warm rising air loses its buoyancy, and it virtually confines the circulation of air from the surface to below the 'tropopause', at 11–20 km (Figure 2.4). As a consequence, water vapour is virtually non-existent in the stratosphere and the zone of weather production is limited to the troposphere.

Only the thinnest of clouds can normally form at heights of 6–12 km, because of the lack of vapour. Because of the low temperatures, they are composed of ice crystals. Only exceptionally strong convection or weather fronts can force larger quantities of vapour to reach these heights. On a broader scale, intense heating of the surface air in equatorial regions forces the tropopause to rise to about 16 km at the equator compared with 8 km at the poles (Figure 2.4). This allows greater rainfall intensities at lower latitudes and it permits the development of hurricanes.

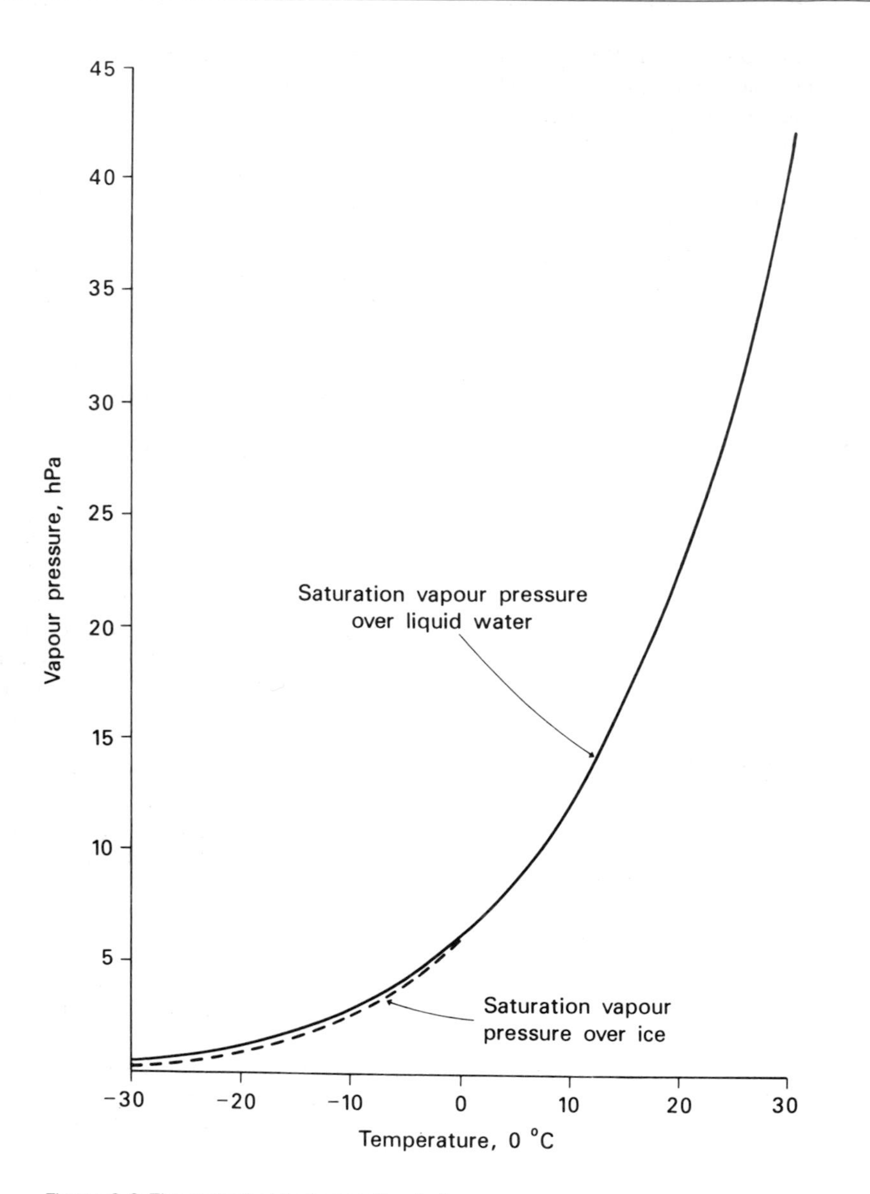

Figure 2.9 The water-holding capacity of air

2.2 Evaporation

2.2.1 Processes

Evaporation from water surfaces is a reasonably simple process to describe, and it is responsible for over 90 per cent of all atmospheric vapour. Describing evaporation from soil surfaces is more difficult, and transpiration from plants is even more complex because it involves interactive responses from vegetation (section 3.1). Fortunately, the complex processes of perspiration and expiration by mammals are not quantitatively important.

The British physicist John Dalton captured the essence of the process of evaporation early in the nineteenth century, when he propounded what has become known as **Dalton's Law**. According to this law, the evaporation rate, E, is controlled by two factors, the windspeed and the unfilled vapour-holding capacity of the air, known as the **saturation deficit**. The surface temperature of the water body clearly controls the overall number of molecules with sufficient energy to break free from the water surface, but it is the properties of the receiving air that determine how many of these molecules remain airborne and get swept away in the airstream. Dalton assumed that over a large body of water an equilibrium develops between the temperature of the surface and of the air immediately in contact with it, so that they can be regarded as the same. Hence, the vapour saturation deficit (determined by the temperature of the air and its pre-existing vapour content) can be regarded as controlling the uptake of water vapour into still air. Without any exchange of air, however, this would essentially halt once the air was saturated, as a kind of 'wet blanket' of air forms over the water. This is where the wind becomes important, in renewing the air in contact with the surface and in lifting the vapour upwards in turbulent eddies (Figure 2.10).

The equilibrium case can therefore be described by the formula:

$$E = u(e_s - e_a)$$

where u is a function of windspeed, e_a the current vapour pressure and e_s the saturation pressure at that temperature. Operational forms of this equation have been widely used to estimate oceanic evaporation, for example, by Budyko *et al.* (1962) and UNESCO (1978).

Evaporation and transpiration on land vary on a much more local scale, affected by highly variable land surface properties. It is therefore expedient to discuss these processes in detail in Chapter 3. The reader is referred to Shuttleworth (1983) for an analysis of evaporation models for estimating the global water budget.

Human manipulation of global vegetation over recent millennia must have had marked effects on evaporation rates and this topic is taken up in Chapter 7.

2.2.2 Global patterns

Evaporation is naturally higher from the oceans than from the land at all latitudes, except in areas covered by sea ice (Figure 2.11). It is generally greatest at low latitudes and over warmer oceans. The North Atlantic Drift raises ocean and air temperatures and with them evaporation rates well into the Barents Sea. Evaporation off the east coasts of the USA and Japan is also enhanced by the 'oasis effect', in which the westerly winds blowing over the warm Gulf Stream and Kuro Shio currents take up moisture more readily because they have been dried during their long passage over the continents. In contrast, cold currents tend to reduce evaporation on the eastern margin of many oceans, notably on the west coast of the Americas and southern Africa, contributing in some regions to coastal deserts. Low rates of actual evapotranspiration in mid-latitude continental interiors, despite high summer temperatures, are principally caused by the lack of rainfall.

Actual evaporation rates at sea are typically 2–3 times greater than on the adjacent land, although greater differences are associated with the deserts of the subtropical high-pressure belt. The least difference occurs on the margins of equatorial rainforests, especially around Gabon, where a cold current and oceanic upwelling reduce ocean surface temperatures. Increased cloud cover in the equatorial region also tends to cause a slight general lowering of evaporation rates (Figure 2.11).

Counterbalancing this are unexpectedly high evaporation rates in some parts of the equatorial oceans. A notable example is a 'plume' of evaporated moisture off the coast of Sumatra, which forms a persistent focus for intense convective activity.

Evaporation patterns in the equatorial Pacific fluctuate on a regular basis, in response to a cyclical

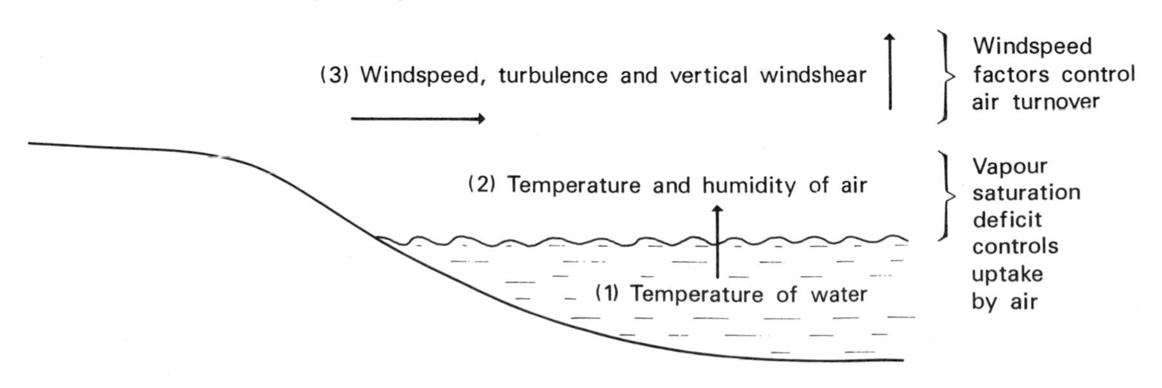

Figure 2.10 Evaporation from a water surface

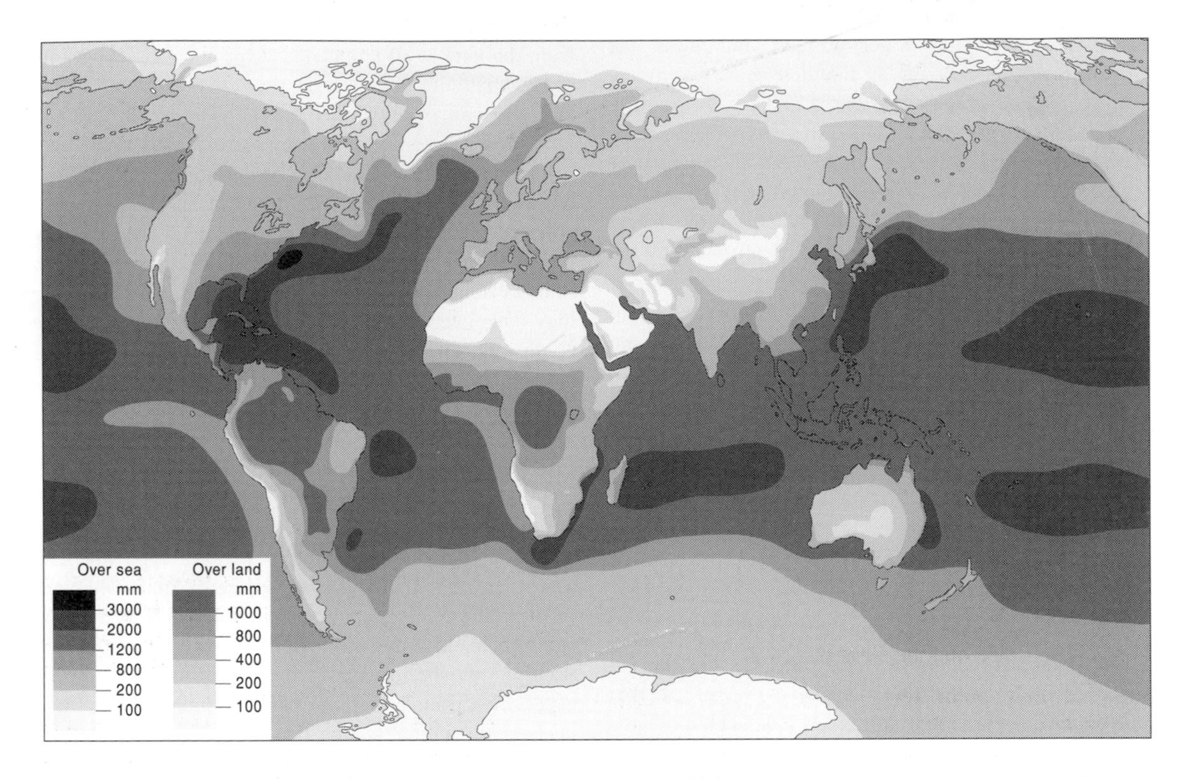

Figure 2.11 The global pattern of evaporation. Based on UNESCO (1978)

reversal of the normal east–west current. Every 5 years or so it flows from west to east as the **El Niño current**, bringing warmer waters (+7°C) to the coast of Ecuador accompanied by high evaporation rates and torrential rains and floods on the mainland (Figure 2.12). This happened dramatically in 1982. The significance of this reversal seems to extend well beyond the tropics. Its cause may be linked to fundamental shifts in the General Circulation of the atmosphere and it is being incorporated into recent climate change models (section 2.6).

2.3 Precipitation

2.3.1 Processes

Precipitation is formed in the atmosphere by two distinct groups of processes. At temperatures above 0°C, the warm-cloud process begins with condensation and the gradual growth of water droplets by condensation and coalescence. At lower temperatures, the cold-cloud process involves the formation and growth of ice crystals.

Neither condensation nor crystal formation is sufficient alone to generate precipitation. Hence many clouds do not produce precipitation. Many other clouds produce precipitation that is evaporated before it reaches the ground.

Two extra factors are needed: (1) a sufficient moisture supply; and (2) sufficient vertical motion to ensure continued cooling and collisions between droplets or crystals. The final threshold is reached when crystals or droplets are heavy enough to overcome Brownian motion in the air (i.e. random bombardment by air molecules) and updraught currents, and begin to fall from the cloud.

The warm-cloud process As moisture-laden air rises due to natural or forced convection, it cools at the **dry adiabatic lapse rate (DALR)**, 0.98°C per 100 m, until

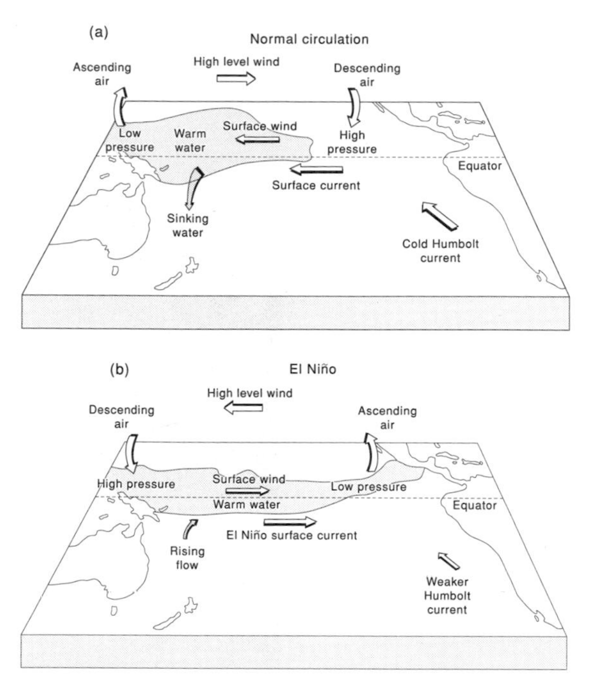

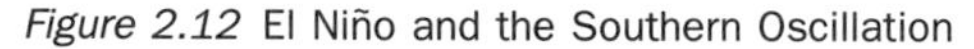

Figure 2.12 El Niño and the Southern Oscillation

it reaches **dew point temperature**, at which point condensation occurs. Thereafter, any further rise causes cooling at the **saturated adiabatic lapse rate (SALR)**, as condensation releases the latent heat of vaporization at a rate of approximately 2.5 million J kg^{-1} of condensed water. The SALR varies from 0.5°C per 100 m at higher temperatures to 0.9°C per 100 m, almost the DALR, at low temperatures. The latent heat released can often add a useful degree of extra buoyancy to the rising air, enhancing vertical motion. The more moisture that is condensed the greater is the likelihood of generating strong updraughts, and vice versa, so that a positive feedback loop can develop.

As a result, heavier rainfall is associated with cloudforms that have stronger vertical currents and hence greater vertical development. Because of this, Luke Howard's morphological classification of clouds, produced in the early nineteenth century before any of these processes were understood, provides a very useful guide to rain-producing potential. It has therefore been adopted as the international system of classification (Table 2.3).

In practice, two further factors are also important in the condensation process. These factors are microphysical, and very important for artificial rainmaking (section 10.3). Laboratory experiments in the 1890s by John Aitken and Charles Wilson showed that condensation does not occur at the theoretical dew point temperature in pure air. Either a degree of **supersaturation** is needed or particle impurities (aerosols), or both. Supersaturation occurs when the **relative humidity (RH)**, the vapour content expressed as a percentage of the maximum that can be held at a given temperature, exceeds 100 per cent. Wilson measured RHs of 700 per cent in pure air without condensation in his laboratory cloud chambers.

Such high levels of supersaturation are not necessary in nature because natural air contains myriads of tiny dust particles that act as foci or nuclei for condensation. Even so, some supersaturation is needed to prevent rapid re-evaporation of the initial droplets; commonly less than 1 per cent supersaturation is all that is needed, except among the tiniest droplets, less than 0.001 μm in radius, where RHs of 320 per cent are required.

Three types of aerosol have been identified according to their condensation properties: (1) non-wettable and insoluble nuclei, which are useless; (2) wettable but insoluble and non-hygroscopic, which are of some value because they effectively impersonate a water droplet once the surface is wetted; and (3) hygroscopic nuclei, the most efficient. **Hygroscopic nuclei** have such a strong affinity for water that they absorb water vapour at RHs below 100 per cent. Common salt, NaCl, can absorb water at an RH of 75 per cent and even as low as 50 per cent and the original nucleus eventually dissolves in the water.

The size of nuclei also plays an important role. Dust particles with radii of more than 10 μm drop out of the air quite quickly. Among the smaller particles that remain airborne: (1) giant nuclei, 1–10 μm in radius, are no use unless hygroscopic; (2) large nuclei, 0.1–1 μm, are more effective; but (3) Aitken nuclei, 0.01–0.1 μm, are the most effective, since many are electrically charged particles or ions, which attract negatively charged water to bond with a positively charged cation. Again, particles smaller than 0.001 μm are generally ineffective because they require high levels of supersaturation to initiate and retain condensation.

Suitable nuclei are derived from a variety of sources, either primary as (1) **dust** from the land, especially clay minerals, and (2) salt from **sea spray**, or secondary as (3) ionic attraction and **coagulation** of smaller particles, (4) **condensation** and sublimation of other vapours, (5) **reactions** between trace gases like methane, either with other gases or with solid aerosols, especially in photochemical reactions, or (6) by **dispersion** of larger particles or coagulations. Nuclei of sulphuric acid, H_2SO_4, can be formed in fog or cloud by the oxidizing of sulphur dioxide gas, SO_2, under sunlight, and so

Table 2.3 Clouds and precipitation

Altitude	Cloud type	Example	Formation process (typical)*	Temperature (°C)	Vertical motion (m s^{-1})	Form	Dominant precipitation process	Precipitation type	Duration pattern
Layer clouds:									
HIGH	*Cirrus* (Ci)		Convergence	<–25	0.05–	Regular	Nil	Nil	Nil
	Cirrocumulus (Cc)				0.01	Widespread			
	Cirrostratus (Cs)								
–7000 m–									
MIDDLE	*Altocumulus* (Ac)		Convergence	0 to –25	0.05–	Regular	Bergeron	Rain, snow	Intermittent
	Altostratus (As)		or orographic		0.1	Widespread	or warm	Rain, snow	Continuous
–2000 m–									
LOW	*Nimbostratus* (Ns)		Frontal uplift	>–5	0.05–	Regular	Bergeron	Rain, snow	Continuous
					0.2	Widespread			
	Stratocumulus (Sc)		Mixing air		<0.1	Irregular	or warm	Drizzle,	Continuous
			masses			Widespread		Snow grains,	
								Rain, snow	Intermittent
	Stratus (St)		Convergence		<0.1	Irregular			
			or orographic			Widespread		Drizzle,	Continuous
								Snow grains,	Intermittent
								rain, snow	
Vertically developed clouds:									
	Cumulonimbus (Cb)		Convection	summits	3–30	Large cell	Bergeron	Heavy rain, Showers	
				possibly				hail,	
				–50				snow pellets	
	Cumulus (Cu)		Convection	usually	1.5	Bubbles	Warm	Rain, snow	Showers
				>–5					
				summits			or Bergeron	Rain, snow	Showers
				possibly					
				–20					

*All cloud types can be generated by frontal uplift.

form the basis for acid rain. Dimethyl sulphide (DMS) released from marine plankton may stimulate cloud formation and could be a feedback factor in global warming (section 2.6).

Aitken nuclei are not only the most efficient condensation nuclei, they are also the most abundant, three to four times more so than giant nuclei. They are more abundant over land, whence most of the particles and source gases are derived: 5 million l^{-1} on land compared with 1 million l^{-1} at sea. This distribution is one of the factors creating more rainfall on land and therefore responsible for maintaining riverflow. However, despite assistance from other factors (section 2.3.2), the imbalance between oceanic and terrestrial precipitation is only a small fraction of the difference in Aitken particle concentrations, primarily because sea salt is such a superefficient condensation material. Even giant nuclei can initiate condensation if they contain sufficient salt.

Although condensation forms clouds, much more is needed to create rain. Condensation alone is unlikely to produce droplets more than 100 μm in radius, the normal maximum size of droplet in fog and most layer clouds like stratus. This is quite insufficient to create raindrops of 1 mm or hailstones of 10 mm. Many factors combine to limit growth by condensation, including: (1) the release of latent heat (heat released from the condensation of 1 g of water is sufficient to raise the temperature of 1 m^3 of air by 2.5°C) slowing down condensation; (2) the reduced efficiency of hygroscopic materials as they become dissolved and diluted; and especially (3) the need to maintain vapour supply, to keep the ambient vapour pressure in the surrounding air above the critical vapour pressure needed for condensation.

Updraughts, on the other hand, cause rapid adiabatic cooling, increasing RH and the likelihood of supersaturation. They also cause droplets to collide and coalesce. Mason (1962) quotes formulae which demonstrate the complementary effects of condensation and collision and give very natural drop size distributions. Droplets merge due to direct impact or to entrainment in the wake of falling drops. Droplets smaller than 19 μm in diameter are swept aside and avoid impact, but may be sucked into the wake.

The efficacy of collision and coalescence depends on the size of the convective cell and the strength of updraughts. In small convective clouds the updraught draws in drier, cooler air from around, which lowers vapour pressure, stifles condensation and reduces convective buoyancy. In large convective clouds, however, the surrounding air is likely to be relatively warm because it is part of the compensating, sinking air displaced by the convective bubble and is heated adiabatically, so it can add to buoyancy if it is entrained in the convection. In fact, a larger convective cell implies a stronger and more robust system in general. Hence the distinction between the small, rainless or 'fair-weather' cumulus clouds (*cumulus humilis*) and the large *cumulus* clouds that produce heavy showers (Table 2.3).

The intense rainfall and thunderstorms of the *cumulonimbus* tend to involve yet another process, confusingly termed **glaciation**. These clouds combine the characteristics of warm- and cold-cloud processes, but it is the formation of ice particles at the top of the cloud that creates the heavy downpours.

The cold-cloud process The key to the success of this group of processes is the fact that saturation vapour pressure (svp) over ice is lower than over liquid water (Figure 2.9). The difference is only about 0.25 hPa, but it is sufficient for ice crystals once formed to insidiously begin to capture moisture from the air before water droplets get a chance. Once begun, the process of sublimation from vapour to ice crystal lowers vapour pressure and further disadvantages the water droplets. Lower vapour pressure eventually causes the droplets to evaporate and the flow becomes one way from droplet to crystal until the droplets disappear (Figure 2.13).

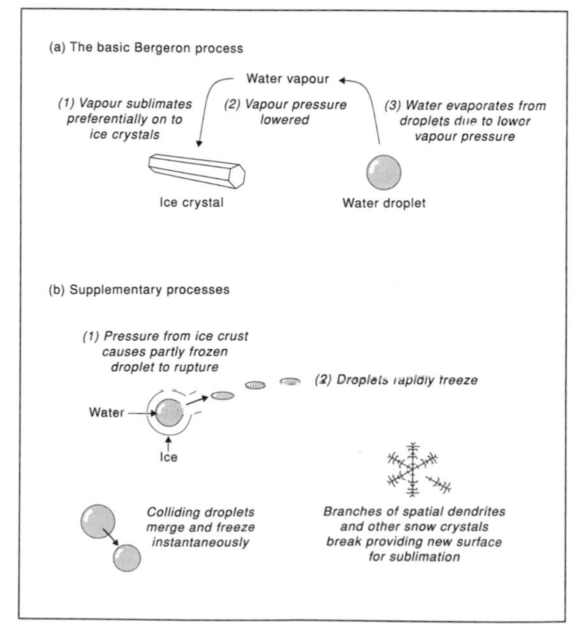

Figure 2.13 The cold-cloud or Bergeron–Findeisen process of precipitation growth

The process was originally discovered and researched by Tor Bergeron and Walter Findeisen in the 1930s and has become known as the Bergeron or **Bergeron–Findeisen process**. Because it only occurs at subzero temperatures, any water droplets must be supercooled, i.e. they exist as liquid below the normal freezing point because of impurities such as salt, or because it is more difficult for very small droplets with highly curved surfaces to freeze. If temperatures fall sufficiently, however, droplets can begin to freeze from the outside. Since ice has a density only 90 per cent that of water, expansion of the frozen shell exerts considerable pressure on the water trapped inside. The water, being incompressible, returns this pressure to the shell, shattering it and spewing out droplets that are instantaneously frozen to become extra freezing nuclei. In other cases, supercooled droplets may simply collide and the splattered droplets freeze. Alternatively, brisk currents or collisions may splinter the brittle ice crystals, creating more scavenging particles. This is particularly important in the –12 to –16°C range, where the fragile 'spatial dendrite' crystals tend to form.

Like condensation, sublimation and freezing both require nuclei. In pure air, spontaneous sublimation has been found not to occur until –41°C, in some experiments not till –70°C, and then only under high supersaturation. In natural air, spontaneous nucleation occurs at 0 to –10°C. Ice crystals themselves are very efficient nuclei, attracting water to the ends of the crystal lattices by free electrical charges. But other materials, especially clay minerals, also act as **freezing nuclei** (the term includes sublimation), and they are generally needed to start off the process. Water is particularly attracted to clay minerals with the same hexagonal crystal symmetry.

Freezing nuclei differ in two important respects from condensation nuclei: (1) they are far less numerous, 0 to 1 l^{-1} at 0 to –10°C; and (2) their numbers increase as temperature decreases. The lack of freezing nuclei around 0°C is one reason for the continued existence of water droplets in a supercooled state. However, any initial disadvantages are soon offset by the Bergeron process and possibly by the 'training' of nuclei.

Once some nuclei have acted as freezing nuclei, they can be reactivated at a higher temperature. B J Mason proposed that some clay minerals could be thus 'trained' in *cirrus* clouds at altitudes around 7 km. The low levels of convective turbulence at this altitude allow crystals to fall more readily. They usually evaporate before reaching the ground: *cirrus* clouds are not known for precipitation (Table 2.3). But in the process, the nuclei have moved to lower levels in the atmosphere where there is more moisture and more convection and they themselves have been modified so they are activated at higher temperatures the second time around. This reduces the shortage caused by lack of 'first time' nuclei in this critical temperature range.

Some cold-cloud processes can operate without supercooled water and even without a cloud being formed. In polar regions, even as far south as central Labrador–Ungava in winter, the air can be so still that crystals begin to fall before a visible cloud is formed. Although the intensity of this snowfall is very low, it may be the only source of precipitation in some areas, and recent observations in Antarctica have suggested that this is the main process of replenishment in the interior.

Precipitation amounts and intensities The intensity and amount of rain or snow that reaches the ground depend partly on the generating process and partly on what happens to it en route to the surface. The intensity and duration of precipitation leaving the cloud is determined by the type and intensity of precipitation-forming process, particularly whether part or all of the cloud is at subzero temperatures, the available moisture and the degree of atmospheric instability. However, the amount reaching the ground may be modified by melting, evaporation or sublimation (volatilization) in the air below.

As a water-collecting process the Bergeron process is more efficient, but this advantage is partly offset by the fact that there tends to be lower atmospheric instability and less vapour available in colder regions. Maximum yields from the Bergeron process are therefore found in mid-latitudes, where the upper air is sufficiently cold and the lower air is warm enough to hold sufficient moisture.

Precipitation intensity is the product of the number, size and fall velocity of falling drops or crystals. The **fall velocity** is the inherent rate of fall, the 'terminal velocity', less the effects of any updraughts. **Terminal velocity** is achieved when gravitational acceleration is counterbalanced by the friction of the air (aerodynamic drag) on the particle. Larger drops have higher terminal velocities and therefore create more intense rainfall. Hence, the typical raindrop of 0.5–5.0 mm diameter falls at 2 to 4 times the velocity of the largest droplet of drizzle and yields proportionately larger receipts (Table 2.4). Raindrops have higher terminal velocities for a given mass than snowflakes because of their compact form and automatic aerodynamic shaping as they fall. Snowflakes have particularly low terminal velocities because the crystals have a large surface area per gram weight and such awkward and inflexible shapes. Even

Table 2.4 Drop size and terminal velocity

Type of drop	Drop diameter (mm)		Terminal velocity (mm s^{-1})		
Cloud droplets	0.002		0.13		
	0.01	typical size	3.3	↓	24 h to reach ground from 300 m (e.g. *stratus* cloud)
	0.02		13.3		
Cloud droplets / Drizzle	0.1		279.0		
Drizzle				↓	precipitation likely
Drizzle / Raindrops	0.2		720		
Raindrops	0.5		2060		
	1.0		4030		
	2.0		6490		
	5.0	maximum size	9090	↓	1 minute to reach ground from 300 m

rounded snow pellets have relatively low terminal velocities compared with raindrops of the same diameter because of the lower density of ice and the internal air spaces that give the white appearance. However, frozen precipitation can compensate for lack of terminal velocity with larger size. Whereas raindrops break up at about 5 mm in diameter, snow can reach diameters of 40 mm and hailstones over 50 mm.

Drop-size distributions are broadly determined by the macrophysical properties of the clouds that initiate precipitation, but only a small fraction of the precipitation thus created actually reaches the surface in its original form – 70 per cent of present-day global precipitation is snow that has melted in warmer air beneath or within the cloud. This in-air melting process is extremely important for runoff production.

At the same time, large amounts of solid and liquid precipitation never reach the surface because it vaporizes. **Virga** or streaks beneath the cloud are visible evidence of this unsuccessful precipitation. The higher the cloud base and the lower the terminal velocity, the greater the opportunity for evaporative losses. Droplets of less than 100 μm radius fall so slowly they are unlikely to reach the ground without evaporating.

2.3.2 Global patterns

World precipitation shows two main peaks, at low and mid-latitudes, primarily associated with **zones of atmospheric convergence**. Horizontal air currents converging at the surface force moisture-laden air to be uplifted (Figure 2.4). At low latitudes this is the ITCZ, where northern and southern hemisphere wind systems converge. In mid-latitudes it is dominated by the frontal rainfall systems of moving depressions, in which moist southerly 'air masses' are forcibly uplifted and cooled above colder denser air from the north.

Conversely, zones dominated by atmospheric divergence, i.e. regions of high pressure in which air is descending and spreading out across the surface, are antipathetic to precipitation-forming processes. Here, descending air opposes updraughts and adiabatic heating reduces the buoyancy of any air rising convectively from the surface and increases saturation vapour pressure. There are four such zones, corresponding to the polar and subtropical high-pressure belts. The subtropical belts are responsible for the main natural deserts, of which the most intense and extensive are found on the larger land masses in the northern hemisphere.

Superimposed upon the broad latitudinal pattern created by the General Circulation of the atmosphere, the precipitation map (Figure 2.14) shows a secondary pattern largely caused by **land–sea distribution** and by **oceanic circulation**. The warm ocean currents that cause high evaporation rates in the North Atlantic and North Pacific also lead to increased rainfall. This occurs first in the subtropical hurricanes of Florida and the typhoons of Japan, and subsequently fuels the mid-latitude depressions that supply the western regions of Europe and North America. In the southern hemisphere, tropical storms are guided by the Coriolis force and warm ocean surfaces on to the coasts of northeast Australia and Brazil. Because of the different configuration of land masses in mid-latitudes in the southern hemisphere, the southern Chile coast is the only true parallel to the west coast rainfall regimes of the northern mid-latitudes, although New Zealand shows many similar characteristics.

A third-order pattern is added by major mountain systems, creating **orographic precipitation,** in which air is forced to rise and cool, and surface friction causes

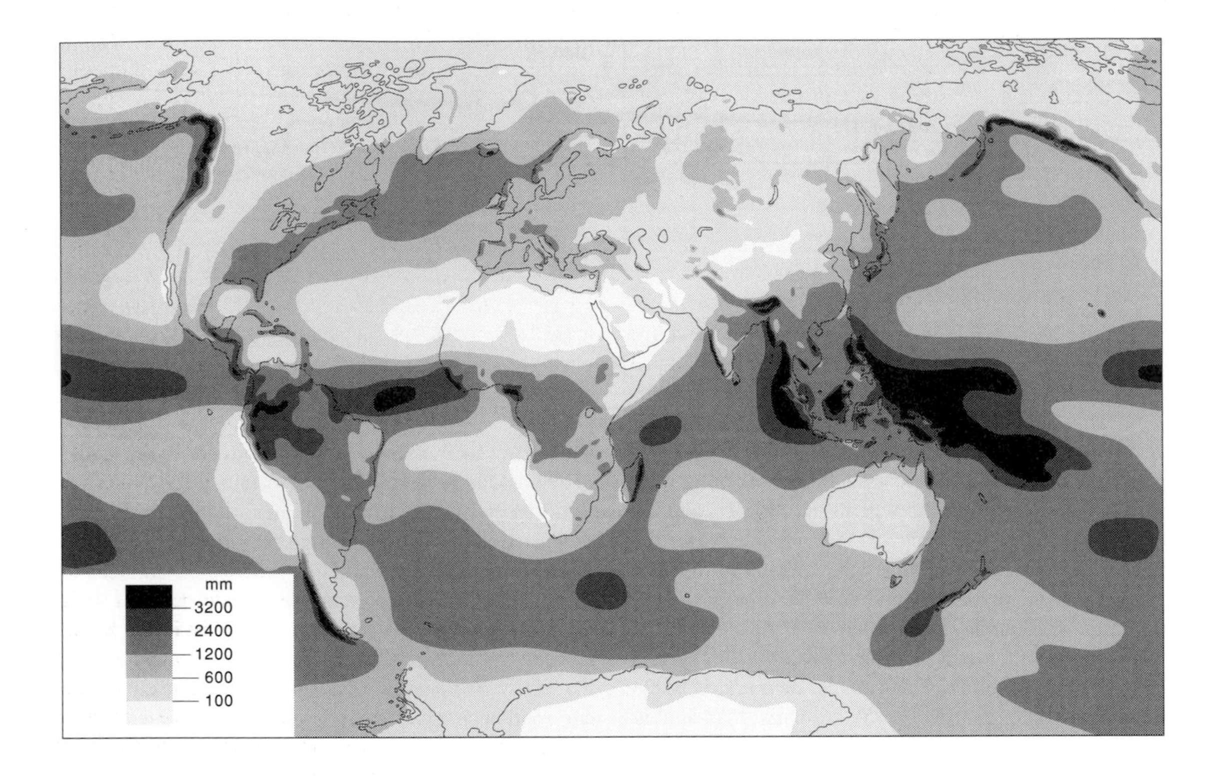

Figure 2.14 The global pattern of precipitation. Based on UNESCO (1978)

convergence, which adds to the potential for atmospheric instability. The mountains of the western cordillera of the Americas are the prime example at a global scale, but the effects of the Himalayas, Alps and Pyrenees are clearly visible.

Tropical storm systems Tropical rainfall systems are of two broad types, convective or wave-generated. Monsoonal rains tend to be a combination of both.

Monsoons are seasonal reversals in wind patterns. They result either from seasonal migrations in major atmospheric features, such as the ITCZ in West Africa or the subtropical jetstream in India, or from the development of thermal low pressure over the land in summer (Nieuwolt 1977). When the monsoon winds blow off a warm sea they bring the rainy season. Some tropical rains can actually result from incursions of mid-latitude depressions. In winter the westerly jet splits around the Himalayan mountain mass and the southern arm of this jet brings a rainy season to the Persian Gulf. These depressions can reach within 5° of the equator.

Both the Indian monsoon rains and the ITCZ contain many groups of convective cells and possibly some wave-generated cells. In the Bay of Bengal they are often catastrophically enhanced by hurricanes.

Hurricanes or typhoons are the most notorious storm system of the tropics. These are individual convective cells spawned over warm sea surfaces (26°C or more). Hurricanes are typified by very severe wind damage and torrential rains. A stationary hurricane can deliver 500 mm day^{-1}, a moving one 200 mm day^{-1}.

Convective cells have to be in the right place to develop into hurricanes. Many smaller convective cells develop and die without growing into hurricanes, partly because the spinning effect of the Coriolis force is negligible within 5° of the equator and partly because the tropical or trade wind inversion limits convective cloud growth, especially on the eastern side of oceans. This inversion is caused by air descending and spreading equatorwards out of the subtropical high-pressure cells. This air is heated adiabatically as it does so. Because air circulates clockwise out of these high-pressure cells and because these cells sit over the subtropical oceans, the air tends to return polewards on the western side of oceans and the inversion level is higher to the west.

The most intense and frequent hurricane development occurs on the western sides of oceans outside the equatorial zone with its low Coriolis effect (Figure 2.5). Only here can: (1) free convection release sufficient latent heat by condensation to fuel the instability and generate the winds; (2) sufficient Coriolis force cause the spiralling accumulation of clouds and moisture and create the centrifugal or 'cyclostrophic' flow that maintains the towering wall of cloud around the eye of the hurricane; and (3) warm ocean currents carry the crucial warm water surfaces into latitudes with higher Coriolis spin.

In contrast, most **wave-generated storms** are relatively weak affairs. They are caused principally by migrating gravity waves, in which the air is alternately expanded and compressed. The waves occur either in the surface easterly trade winds or the upper air westerly winds of the Hadley cell. Convective storm cells can get entrained in these waves and the erratic movements of many of these rain cells may be explained by alternating entrainment in waves within the easterly or westerly flows. More intense storms can result when waves in the two opposing flows fall directly over one another, especially when this allows convection to penetrate the trade wind inversion.

The **trade wind inversion** is a key feature in tropical rainfall patterns because of its control on the growth of storm cells. This affects not only the incidence of extreme storms, but also small-scale convection and long-term rainfall averages.

Perhaps the most dramatic effect on mean annual rainfall can be seen where mountains penetrate the inversion, as on Tenerife. In these instances, the general rule that annual rainfall increases with altitude because of orographic effects breaks down (Figure 2.15).

Mid-latitude storm systems Spatio-temporal patterns of precipitation in mid-latitudes reflect varying mixtures of (1) frontal storms and (2) free convective storms.

Thermal **convection** cells begin where overheating has raised the temperature of the local surface air about 10 per cent higher than the adjacent air. This occurs more frequently in summer than in winter, in lowlands more than uplands, and in continental areas more than maritime. At more equatorward latitudes, convective rainfall may be virtually the only type in summertime as the tracks of the frontal depressions shift polewards.

The localized nature of convective rainfall tends to produce greater spatial variability than the broad-scale frontal systems. Because upland surfaces tend to be cooler, convective activity is less intense in the uplands and the normal pattern of greater rainfall receipts in the uplands can sometimes disappear entirely in the British summer months. The most intense convective storms, especially thunderstorms, require considerable volumes of moisture and are typically found downwind of large, warm water bodies, as in summertime on the shores of the Bay of Biscay, the Great Lakes or the Gulf of Mexico. The contrast between dry continental air masses and the warm, moist air from the water body can be critical in triggering **conditional instability**, i.e. instability caused by the release of latent heat, as around the Great Lakes. However, lower rainfall

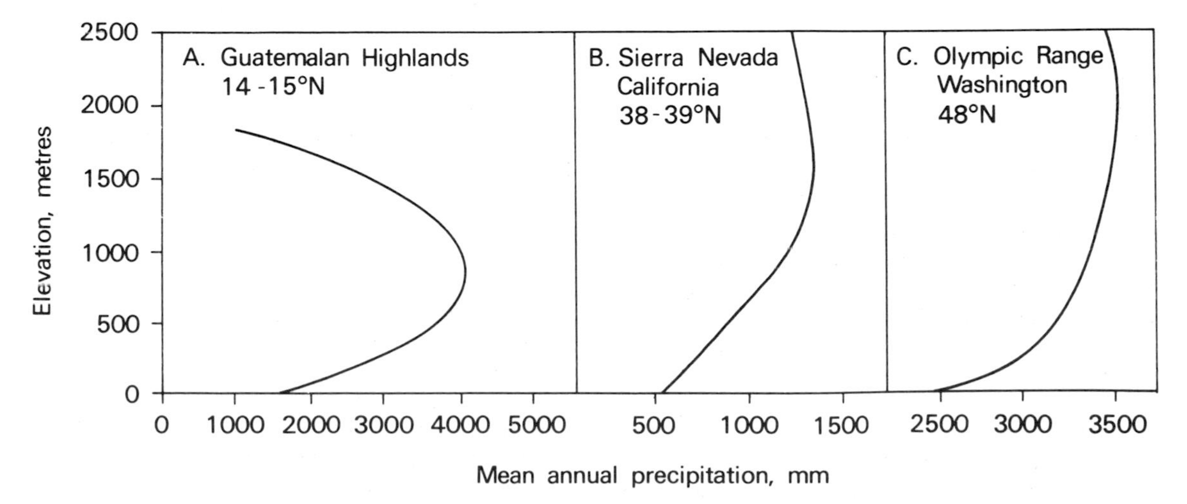

Figure 2.15 Varying rainfall–altitude gradients. Based on Barry and Chorley (1992)

receipts in summer tend to be the norm, even on the shores of the Mediterranean, and this reflects the overall dominance of frontal systems in winter as the main source of precipitation in mid-latitudes. The fact that annual precipitation patterns tend to reflect the surface relief also attests to the dominance of frontal systems.

Frontal storms result from the forced uplift of warm, moist air in depressions, or low-pressure systems. The depressions are generated as turbulent eddies in air sucked up by the jet streams of the Ferrel westerlies (Figure 2.16). The fronts are the zones of contact between cold polar air and warm, moist tropical air drawn into the depression. The depressions move in a generally easterly direction but they migrate latitudinally following the meridional waves in the jetstream. The larger of these so-called **Rossby waves** are stationary, allowing moist tropical air for example to be drawn further north over the North Atlantic and assuring Western Europe of higher rainfall receipts, or drawing cold, dry polar air further south in the centre of the North American land mass and causing frequent summer droughts and late spring frosts in the Canadian prairies. Other waves, with wavelengths of 2500 km or less, migrate eastwards along the jet causing short-term shifts in depression tracks.

The average depression takes about two days to pass over, with rainfall concentrated in the leading warm front and the trailing cold front. The primary mechanism controlling the intensity and distribution of precipitation is a low-level jet at an altitude of about 1 km, dubbed the 'conveyor belt'. This is comprised of warm moist air from the warm sector and runs polewards ahead of the cold front then veers equatorwards at the warm front. Convective instability plays an important part in producing the rainfall, but it is generally forced rather than free. It is generated when cold dry air in the mid-tropospheric westerly winds is being carried over warm moist air that is being forced to rise: condensation in the moist air can then cause instability. This creates localized cells and belts of more intense rainfall within the broad bands of rainfall extending up to 1000 km or more in length and 500 km in width.

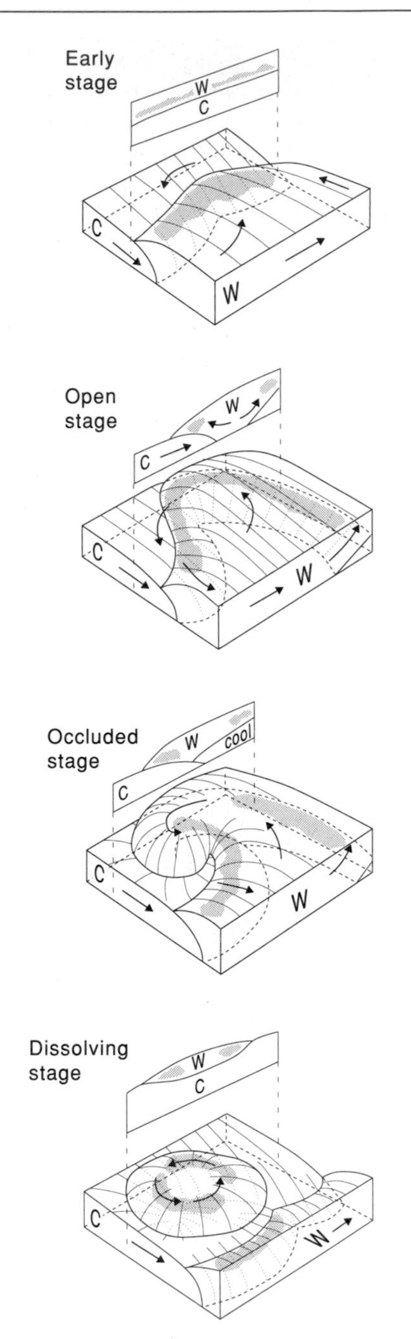

Figure 2.16 The mid-latitude depression: C = cold air, W = warm air. Based on Barry and Chorley (1992)

Rainfall tends to be heaviest and cover a wider area in the warm front, especially in the normal ana-front, where the tropically derived air in the conveyor belt is riding over the leading cold air. Occasionally, a kata-front occurs in which the underlying, surface air ahead of the front is moving faster, and the resultant subsiding air inhibits rainfall production. More commonly, only parts of a front are ana- or kata-, which adds to the spatial diversity in precipitation. The production of precipitation begins to cease once the warm sector has all been raised above the condensation level and thermal gradients cease to fuel instability: the occluded stage in Figure 2.16.

2.4 Runoff and runout

Figure 2.17 shows the global pattern of runoff based on the UNESCO (1978) atlas. The greatest obstacle to accurate estimation of global runoff is the lack of well-established gauging stations. At least one-third of total discharge into the sea occurs in regions where no measurements are available. The most poorly measured rivers tend to be in areas of high precipitation and high runoff. In addition, the more reliable methods of estimating runoff in ungauged basins tend to be more complex and difficult to apply than interpolation of precipitation or evaporation, because they require more knowledge of land surface characteristics (Chapter 6).

For these reasons, the so-called *climatological method* of estimating global runoff has generally been the only method available until quite recently and still remains at least as valid as that based on direct estimation of runoff, if not more so. This calculates runoff (*RO*) as the difference between precipitation (*P*) and evaporation (*E*):

$$RO = P - E$$

This was the method used to plot Figure 2.17. Unfortunately, even this method has the disadvantage that errors in the individual estimates of *P* and *E* can be additive. The range in estimates in proportion to the 'best estimates' in Table 2.2 is therefore higher, at 67 per cent for world riverflow and 100 per cent for groundwater flow, compared with only 26 per cent for world precipitation and evaporation rates.

The late Raymond Nace of the US Geological Survey was the first to attempt a detailed comparison of the two methods (Nace 1970). His climatological estimate produced a global discharge of 924 000 $m^3 s^{-1}$ (29 100 $km^3 a^{-1}$) against 700 000 $m^3 s^{-1}$ (22 080 $km^3 a^{-1}$) based on actual discharge measurements, a difference of 32 per cent. The climatological estimate produced by UNESCO (1978) was even higher, 1 490 360 $m^3 s^{-1}$ (47 000 $km^3 a^{-1}$).

It is certainly possible that climatological estimates overestimate runoff. In Western Europe, where direct measurements of riverflow, precipitation and evaporation are among the best in the world, climatological estimates have been found to overestimate by as much as 22 per cent.

Conversely, many of the major world rivers, like the Amazon, Brahmaputra, Ganges, Congo and Orinoco, are poorly gauged and, indeed, are virtually impossible to gauge during floods. Monitoring of the discharge of

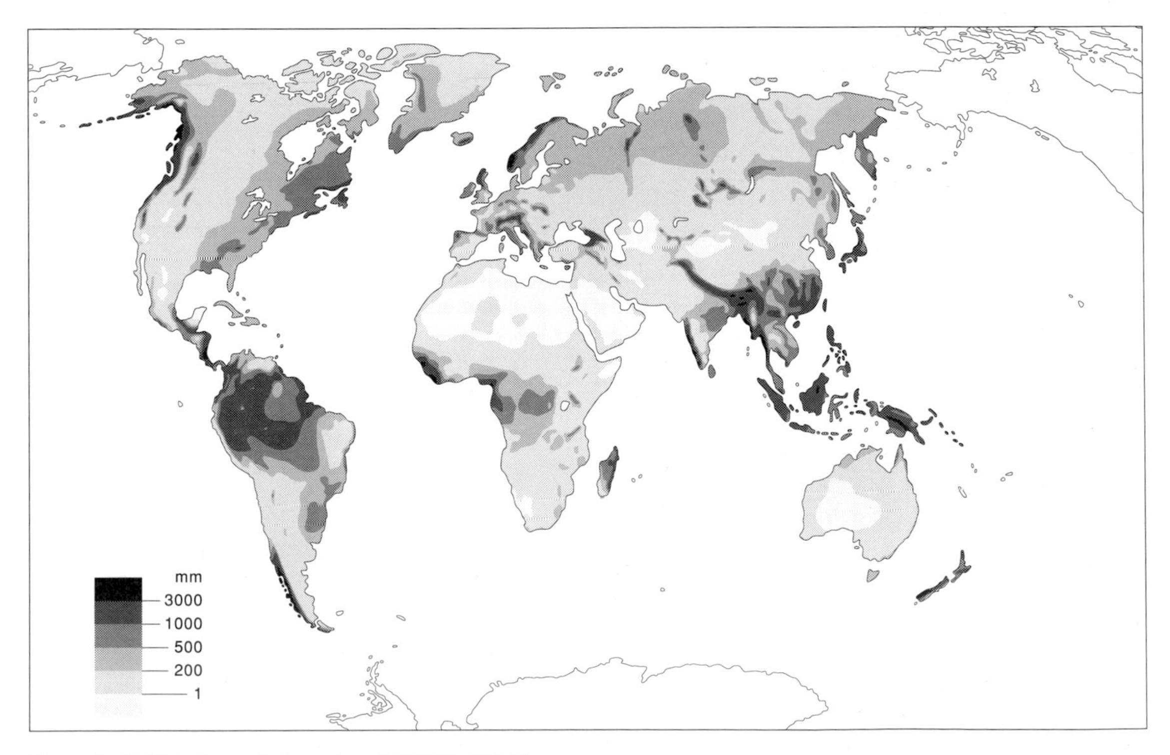

Figure 2.17 Global runoff. Based on UNESCO (1978)

the world's largest river, the Amazon, in 1963–64 resulted in a doubling of previous estimates with a 10 per cent probable error. Such is the dominant position of the Amazon that there are only nine rivers in the world with actual discharges greater than that probable error of 17 500 m^3s^{-1}.

There are also large areas of the world, such as South-east Asia, with virtually no discharge measurements. Making allowances for these deficiencies, Nace (1970) managed to raise his hydrometric estimate to just 770 000 m^3s^{-1}. Extrapolating an empirical power relationship between cumulative discharge and cumulative drainage area, he raised his estimate to just 5 per cent below the climatological estimate, but unfortunately this relationship holds only for larger basins.

Another possibility is that significant amounts of water actually drain directly into the ocean from groundwater, bypassing the rivers. Estimates for this **runout** have ranged from zero up to 12 000 km^3a^{-1}. Nace (1970) made a very hypothetical calculation which suggested that global runout is insignificant, at only 7000 m^3s^{-1} (221 km^3a^{-1}) or less than 1 per cent of river runoff. However, a more recent estimate of 12 000 $km^3\ a^{-1}$ (Speidel and Agnew 1988) is globally significant.

The pattern of global runoff outlined shows two broad zones of high runoff, in the tropics and in mid-latitudes, especially in windward coastal areas. These broadly coincide with patterns of precipitation. However, because of high evaporation in the tropics, the most extensive zones of very high discharge are not there, but on the mid-latitude margins of the eastern Pacific in both hemispheres, where moisture-laden westerly winds, frontal systems and orography combine. Hence, the UNESCO map indicates a water balance in the upper Amazon basin of:

$$RO = P - E$$
$$2000 = 3200 - 1200$$

compared with $2800 = 3200 - 400$ mm

on the coast of British Columbia.

2.5 Regional water balances and resources

Many of the problems of securing sufficient water for human needs stem from the uneven distribution of water resources and from the fact that the distribution of human society is rather different, being determined by a multitude of cultural, political and resource

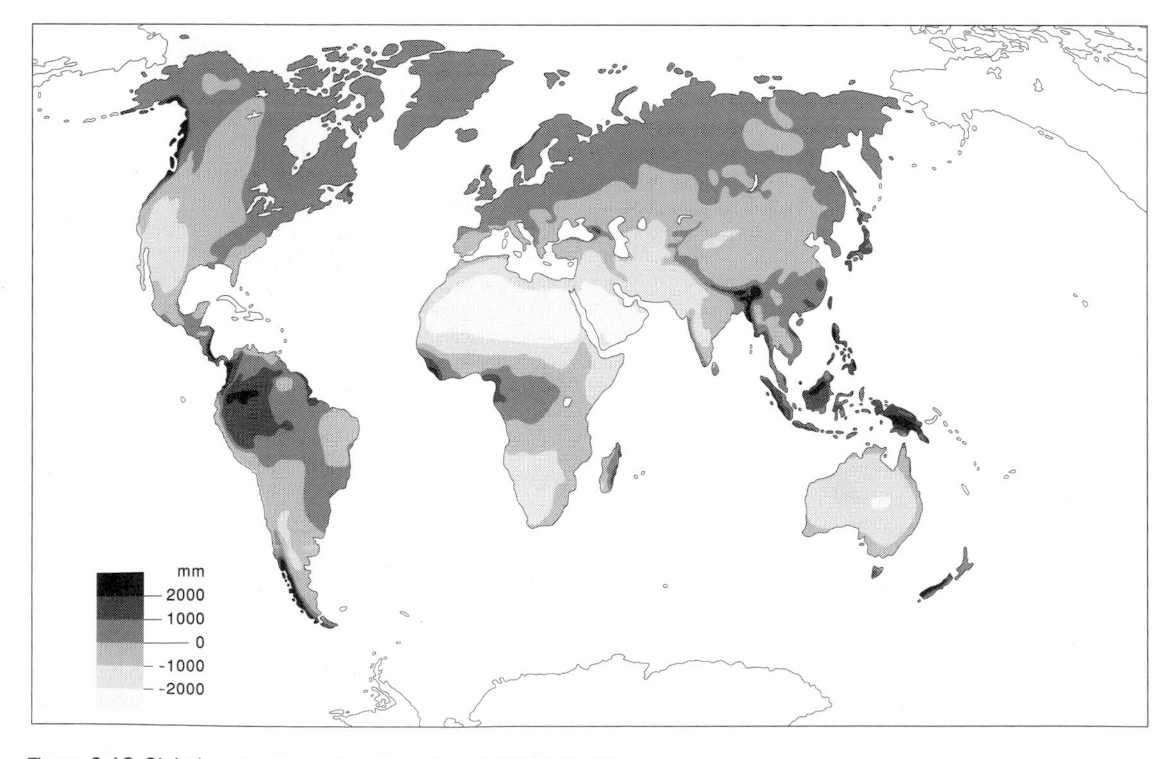

Figure 2.18 Global water resources. Based on UNESCO (1978)

criteria. Two-thirds of humanity live in dryland environments, while the bulk of annual freshwater surplus lies in the relatively uninhabited equatorial rainforests and the bulk of long-term freshwater storage lies in the ice caps.

The UNESCO map of river resources (Figure 2.18) demonstrates the marginal nature of surface supplies in much of North America, Mediterranean Europe, across the Eurasian land mass from the Black Sea to Beijing, and critically across most of the Indian subcontinent.

The use of potential resources is also complicated by the time distribution of runoff. Only one-third of total runoff is stable or **baseflow** runoff. The remainder is storm runoff, which is less frequent and highly variable and generally requires regulation by reservoirs to convert it into a reliable resource. Table 2.5 shows that this problem is particularly acute in Asia, where periods of intense rainfall, like the monsoons, cause 74 per cent of runoff to occur in flood flows. Consequently, Asia has the largest programme of regulation.

Hence the generous global mean river surplus of 10 965 m^3 a^{-1} per capita hides the fact that Asia and Europe have little more than 4000 m^3 a^{-1} per capita and under half of this is stable flow (Table 2.5). Asia has almost the largest stable water resources, but ironically has the least in per capita terms. In stark contrast, Australasia is the least well endowed in terms of total natural resources and has the lowest amount of artificial regulation of rivers, yet has the largest per capita resources. North America has barely twice the natural runoff of Europe, yet supplemented by one of the largest reservoir programmes in the world this yields nearly four times the resources per capita. Nevertheless, it is only by continuing this vast dam building programme that North America has maintained per capita resources over the last 25 years. In Africa, where dam building reached a peak before 1970, per capita stable resources have fallen.

Africa still has 1.5 times the per capita stable resources of Europe, but statistics aggregated at this scale can be misleading. Water shortages are the bane of the continent. The area of surplus runoff covers barely one-seventh of the continent. This type of imbalance is repeated at regional and national scales throughout the world and has resulted in a plethora of planned and implemented schemes for long-distance water transfer (Chapter 9).

The water balances of the oceans are also important, because they influence ocean currents and hence sea surface temperatures. The Atlantic and Indian Oceans have a negative water balance (Table 2.6) and therefore receive water from the oceans with a positive water balance, the Pacific and the Arctic. This results in a significant flow from the Pacific into the Atlantic around Cape Horn.

The water surplus in the Arctic is critical for the water balances of Europe. It accrues because of low evaporation losses and large volumes of freshwater runoff from major rivers like the Mackenzie, Lena, Ob and Yenisei. This gives the Arctic Ocean a tendency to flow outwards, especially into the Atlantic, so limiting the penetration of warm waters and preserving polar snow and ice surfaces. Deflection of the Gulf Stream by the Coriolis effect and the shape of the eastern seaboard of North America tend to cause this outflow to concentrate and to form the Labrador Current. The relative strengths of the Gulf Stream–North Atlantic Drift and the Labrador Current control sea temperatures off Newfoundland and thence evaporative input into mid-latitude depressions. This has a major influence on rainfall receipts in Western Europe, and the UK Meteorological Office has focused attention on sea temperatures off Newfoundland for medium- to long-range weather forecasting in Britain.

2.6 Hydrological effects of climate change

The present status of the global hydrological cycle is merely a snapshot in a long continuum of change. The geological record embodies periods when the balance of the cycle was very different, from the Carboniferous swamps to the Triassic deserts. Changes in the amount and distribution of land masses have wrought fundamental shifts in the cycle. The sensitivity of the hydrological cycle is illustrated by the fact that a 10 per cent change in global continental precipitation requires only a 2 per cent change in oceanic evaporation. The creation of new deserts requires an even smaller change, 0.2 per cent, in the overall cycle (Kayane 1996).

Many factors alter the nature and distribution of world climates, from astronomical movements and changes in solar output to volcanic eruptions. From Table 2.7 it is clear that some of these factors operate on a reasonably regular cyclical basis, while others, like continental drift, do not. Over the period of the Quaternary, which began approximately 2.5 Ma ago with the first of the series of Pleistocene glaciations, it seems likely that oscillations in the Earth's orbit and attitude to the Sun, the **Milankovich oscillations,** have acted as triggers to glaciation and deglaciation by altering the intensity and timing of the seasons (Figure 2.19). At least 17 similar periods of glaciation have

Table 2.5 Regional water resources

Continent	Natural riverflow ($km^3 a^{-1}$)			Regulated runoff ($km^3 a^{-1}$)	Total stable runoff ($km^3 a^{-1}$)	Percentage increase in stable runoff	Total	Per capita runoff ($m^3 a^{-1}$)		
	Total	Stable baseflow							Stable	
		Total	Per cent	1970 (1992)	1970 (1992)	1970 (1992)	1970 (1992)	1970	1992 without extra capacity	1992 with extra capacity
Europe	3100	1125	36	200 (312)	1325 (1437)	18 (28)	4850 (4334)	2100	1852	2009
Asia	13 190	3440	26	560 (1198)	4000 (4638)	16 (35)	6466 (4210)	1960	1276	1481
Africa	4225	1500	36	400 (564)	1900 (2064)	27 (38)	12 250 (6536)	5500	2939	3193
N America	5950	1900	32	500 (1115)	2400 (3015)	26 (59)	19 100 (14 280)	7640	5760	7236
S America	10 380	3740	36	160 (4135)	3900 (7875)	4 (110)	56 100 (35 791)	21 100	13 447	27 154
Australasia	1965	465	24	35 (273)	500 (738)	7.5 (59)	109 000 (74 274)	27 800	18 899	27 895
TOTAL	38 830	12 170	31	1855 (7597)	14 025 (19 767)	15 (62)	10 965 (7442)	3955	2688	3789

Source: Original 1970 estimates according to Lvovich (1970, 1977). Estimates for the early 1990s are based upon increases in maximum reservoir capacity reported in Gleick (1993), according to International Water Power and Dam Construction (1992), and recent population figures. These estimates should be treated with caution. Not all reservoirs are listed and many are not operated primarily for water supply. Insufficient data are available on operating practices.

Table 2.6 Water balances of the oceans and continents (mm a^{-1})

(a) Oceans

	Evaporation	Precipitation	Runoff receipts	Net balance
Atlantic	1240	890	230	–120
Pacific	1320	1330	70	80
Indian	1320	1170	80	–70
World oceans	1140	1260	120	0*

*Arctic Ocean in positive balance

(b) Continents

	Evaporation	Precipitation	Runoff
Europe	390	640	250
Asia	310	600	290
Africa	430	690	260
N America	320	660	340
S America	700	1630	930
Australasia	420	470	5
Continents	420	730	310

Source: Based on Budyko (1970)

occurred in the Earth's history back to 2.5 Ga ago, probably with similar causes.

However, the exact mechanisms by which the Milankovich oscillations are translated into glacial–interglacial cycles are still in dispute. This is probably partly because climatic change involves crossing a variety of thresholds in different parts of the Earth–atmosphere system, including oceanic circulation (Broecker 1989; Rind and Chandler 1991). Moreover, the Milankovich theory does not explain the long periods when the Earth was not locked into a glacial–interglacial cycle. The Earth appears to have spent only about 3 per cent of its existence in glacial phases, yet there is no reason to suspect major shifts in the Milankovich oscillations. **Continental drift** is probably the main cause, especially when it brings a large land mass into the polar regions to allow the accumulation of land ice. In the Quaternary, we have had an Antarctic land mass and a land-locked Arctic Ocean, which is probably unique in Earth history. The land-locking minimizes the penetration of warm ocean currents into the Arctic, fortified by an oceanic water balance surplus, and aids ice development. If the North American plate continues to move away from Europe at a rate of 20 mm a^{-1}, this protected Arctic should effectively disappear in 50 Ma. This could then release the world from the threat of a future glaciation for a considerable time.

Another factor in the onset of glacial cycles may have been mountain-building or **orogeny**. Uplift of the Antarctic mountains and the Himalayas–Tibetan Plateau occurred in the late Tertiary. These could have created more snowfall. A recent theory also suggests that large volumes of cold meltwater in the Asian highlands could have caused cooling by dissolving excessive amounts of atmospheric CO_2, which then remained locked up for an extended period in marine sediments (Raymo and Ruddiman 1992), although initial computer simulations clearly overestimate this effect.

While the northern mid-latitudes were experiencing glacial periods, the intertropical belt experienced periods of greater aridity. These appear to have been linked with a general weakening of Hadley cell activity and of the associated tropical monsoons (Kutzbach 1983; Street-Perrott and Roberts 1983). In contrast, the demise of the last glaciation around 10 ka BP was marked by a **pluvial** period, a distinct rejuvenation of monsoonal rains and expansion of the northern hemispheric Hadley cell, associated with the conjunction of the maximum tilt of the Earth towards the Sun and perihelion occurring around the June solstice (Figure 2.19).

Table 2.7 Factors causing global climate change (arranged in astronomical space)

Typical period of operation	Factors	Process and effects
Years–millennia	PLANETARY ORBITS	gravitational pulls affect – solar activity* – cosmic radiation* – volcanic activity affecting radiation receipts and atmospheric opacity/reflectivity
Decades	SUN'S INTERNAL PROCESSES	affect solar output of electromagnetic and particulate (solar wind)* radiation
20 000–100 000 years	EARTH'S ORBIT	Milankovich mechanisms–distance from Sun, angle of polar axis and seasonality affect distribution and intensity of solar radiation
Hours–decades	EARTH'S ATMOSPHERIC COMPOSITION	affects transmission, radiation and reflection of electromagnetic radiation
Days–millennia	EARTH'S SURFACE PROPERTIES	affect radiation and heat balance, evaporation, precipitation and wind movement
Years–millions of years	EARTH'S CORE AND MANTLE	movements (plate tectonics) affect – volcanic emission of dust and gas – land–sea distribution – mountain building fluctuations in geomagnetic fields affect – cosmic radiation* distribution/intensity – ocean currents (seawater weakly magnetic)

*Charged particles from the Sun and Outer Space increase NO_x gases in atmosphere affecting 'greenhouse' properties.

2.6.1 The last 10 000 years

Greater detail is emerging on postglacial or Holocene climatic changes. Table 2.8 summarizes the main climatic shifts in Europe. The table links hydrological and human impacts to the shifts in airflow patterns between colder drier periods dominated by polar and continental air and warmer wetter periods dominated by strong maritime influence from the Atlantic.

Figure 2.20 shows a reconstruction of temperature and hydrological regimes undertaken by Lamb (1977). It indicates increased rainfall and generally higher runoff during warmer periods, when the hydrological cycle is accelerated. During the warmest postglacial period around 7 ka BP, the Hypsithermal, the UK temperature was probably 1.5°C higher than now and rainfall 10–15 per cent higher. Riverflow is estimated at 16 per cent higher. In the next major warm period, the Early Mediaeval Optimum, around AD 1100, temperatures are estimated to have been *c.* 1.0°C higher and rainfall 3 per cent higher.

One cause of increased rainfall in Europe must be higher evaporation from a warmer and largely ice-free North Atlantic, but it is also very dependent upon the exact tracks of depressions, i.e. the latitude and intensity of the Ferrel jetstream and the position of the Rossby waves. Stronger zonal flow in the westerlies brings a stronger oceanic influence into Western Europe, as indicated by the terms Atlantic and subAtlantic for the milder, wetter periods *c.* 5–7 ka BP and post 900 BC. These periods were interspersed with colder drier periods dominated by polar air masses, when the westerlies were weaker, less zonally concentrated and focused on more southerly latitudes, i.e. the Boreal and subBoreal periods of 7–9 and 3–5 ka BP.

Work by Lamb (1977) on the **Little Ice Age** (LIA) in

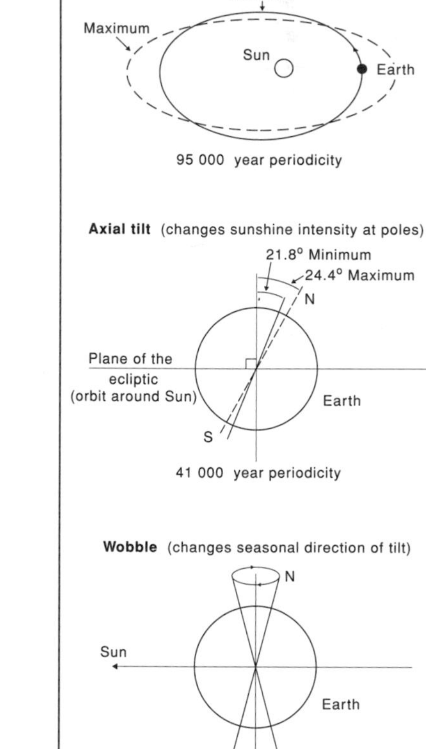

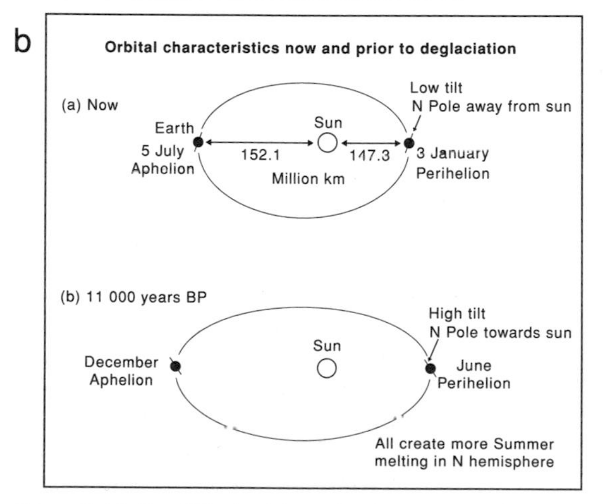

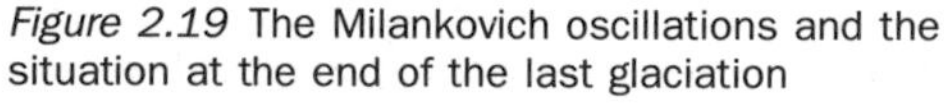
Figure 2.19 The Milankovich oscillations and the situation at the end of the last glaciation

Europe, 1550–1800, has revealed another interesting aspect of regional climate. Increased strength in the westerlies is associated with increased wavelengths in the fixed Rossby waves. The LIA was marked by a reduction in zonal westerlies and a westward shift in the European Rossby trough to a position centred over London bringing predominantly polar air masses to Britain. In contrast, in the stronger westerlies of the preceding optimum the Rossby trough was located in Eastern Europe as now, bringing a ridge with milder more southwesterly air streams over Western Europe. The strength of the westerlies is partly determined by meridional temperature gradients and partly by latitude as the spin effect (relative vorticity) increases towards the poles.

Mid-latitude climates, dominated by a varying mixture of tropical and polar air masses, are therefore very sensitive to shifts in wind systems that are only minor in terms of the General Circulation of the atmosphere. This makes predicting climatic change and its hydrological effects particularly difficult in such regions. Fortunately, this is also the zone in which most climatic research has been carried out and where the new high spatial resolution windows in the latest **General Circulation Models (GCMs)** are being run first (Box 2.1).

2.6.2 Predicting the next 100 years

For more than a decade, climate models have been predicting that global warming will cause some significant changes in rainfall patterns. The evidence that an enhanced greenhouse effect will affect the Earth in the coming century is still largely theoretical. Observational evidence indicates a reasonably clear rise in global mean temperature of 0.5°C this century, an overall rise in mean sea level of about 10 mm per decade, and most conclusively of all a rise in concentrations of the major greenhouse gas, CO_2, from 315 ppm when measurements began in 1958 to 360 ppm in 1990. But there is no proof that the temperature rise is linked to rising CO_2 levels; in fact in the period immediately following the Second World War, when fossil fuel burning should have raised CO_2 levels markedly, the world temperature was either falling or static. The evidence from sea level rise is even less clear. It reflects a combination of factors, of which increased glacial melting and thermal expansion of ocean water are just two. Tectonic land movement at monitoring stations is a major complicating factor and the newly planned global monitoring network, GLOSS, should provide a sounder database (Pugh 1990).

Table 2.8 Hydrological, cultural and agricultural impacts associated with postglacial climatic fluctuations in Europe and the Middle East

Years (approx.)	Dominant airflow & climatic phase		Key climatic features and hydrological impacts	Cultural events	Agricultural impacts
	Polar and continental	Tropical and maritime			
14000–8000 BC	Old Dryas Younger Dryas	Allerod	Series of sudden temperature oscillations mark end of last glaciation.	Cave dwellings.	
8000–6000 BC	Boreal		Rapid warming. Mountain glaciers melt and deep depressions enter the Middle East creating a humid environment.	Settlements in Iranian foothills.	Gathering of wild grains in well-watered Kurdistan. Agriculture invented.
5000–3500 BC		Atlantic (Hypsi-thermal)	Temperatures 1.5°C above present. Demise of Scandinavian ice sheet (5000) & Laurentide ice sheet later. Desertification in Middle East. Subtropical High N and monsoonal rains penetrate into Sahara.	Rise of urbanism in floodplains and dominance of states with dependable water supplies.	Invention of irrigated agriculture in Middle East. Agriculture later spreads W & N into Europe in favourable climate.
3500–1000 BC	subBoreal		Drying of Sahara (3500–2800). Subtropical High S, Nile flow reduced – droughts of 22nd, 20th & 18th centuries. Possibly severe floods in Mesopotamia linked with S shift in mid-latitude jetstream.	Israelites roaming in drying landscape, enslavement of pastoralists (2100–1700 BC). Demise of Old & Middle Kingdoms of Egypt linked to Nile droughts?	Agriculture widely adopted in Europe as cooling may have caused impoverishment of hunting resources.
900 BC – AD 300		subAtlantic	Cool & wet in N. Regrowth of peat bogs, Russian forests expand southwards. AD – warmer & sea level rise. Wetter Mediterranean with more depression rainfall.	Southward tribal migrations in Europe, BC.	Wetter Mediterranean supports agricultural base for Greece and Rome. Vine & olive cultivation expands N.
AD 400–800	'Dark Age recession'		Droughts in C Asia. Severe North Sea floods. Readvance of Alpine glaciers.	Westward tribal migrations beginning in Asia topple Rome. AngloSaxon migrations from borders of North Sea.	
AD 800–1200		Early Mediaeval Optimum	Temperatures 1°C above present. Drier climate in NW Europe.	Viking explorations. Colonization of Greenland (+2–4°C)	Vine *c*. 5 degrees N.
1350–1820	Little Ice Age		Temperatures 1°C below present. Cool and wet in W Europe. Readvance of Alpine glaciers. Cooling spreads W from China, where jetstream wave No. 1 extends furthest S.	Enfeebled populace, plague & depopulation. Westward migrations culminate in fall of Byzantium (1453). Thames Frost Fairs. French Revolution (1789).	Retreat of cultivation limits. Crop failures.
1850+		'Global warming'	Retreat of Alpine glaciers. Sea level rise. Global temperature + 0.5°C. Warmer sea surface temperatures in N Atlantic and increased dominance of W winds. More rainfall in NW Europe.		Crops introduced from tropical America (maize & potato) helped by nineteenth-century warming. Twentieth-century productivity partly due to warming & CO_2 fertilization.

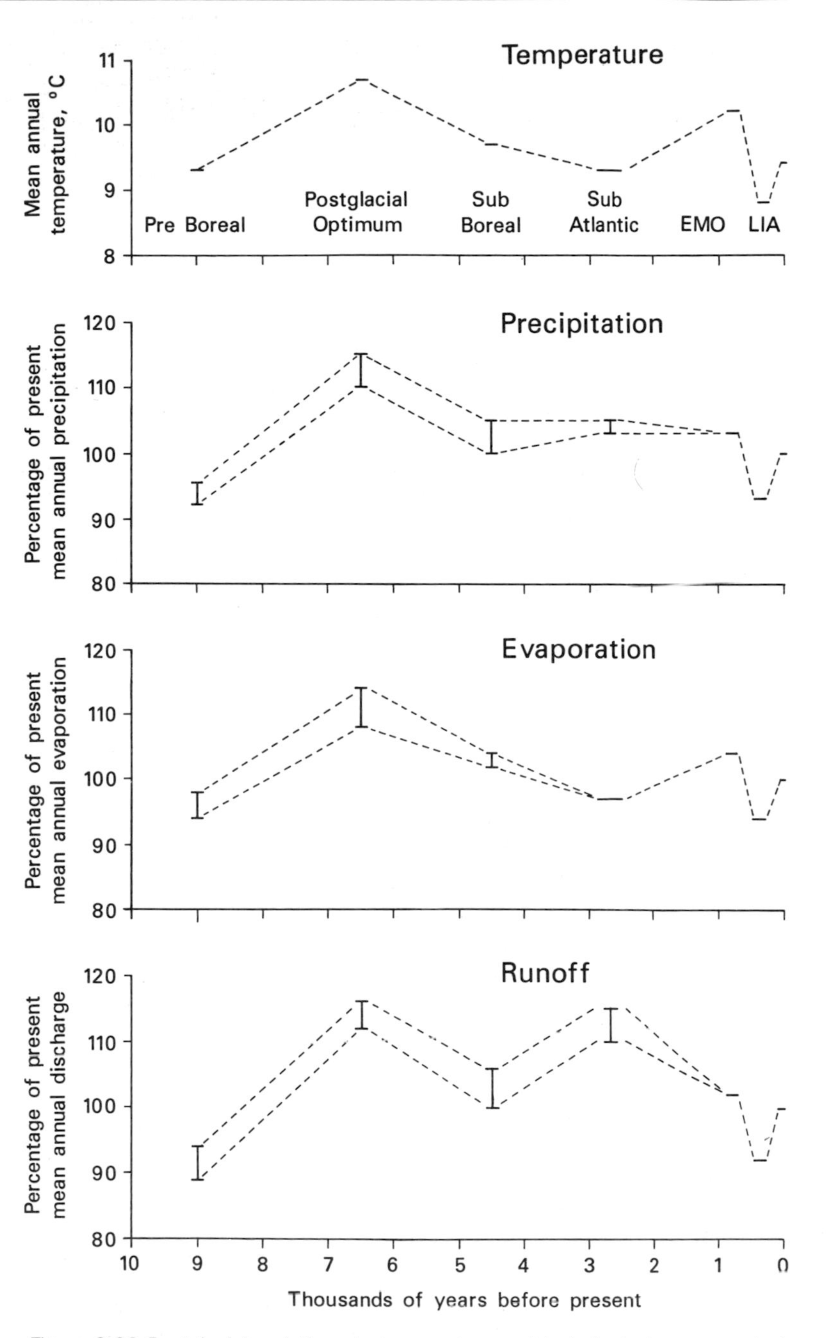

Figure 2.20 Postglacial variations in temperature and hydrological components in England and Wales, based on the calculations of Lamb (1977)

Nevertheless, when the CO_2 curve is extrapolated it suggests that concentrations will reach double the estimated pre-industrial baseline of 290 ppm by 2030–2050. The strongest evidence for global warming is that all the GCMs currently operational predict that this increase will cause marked warming with an average rise of about 2°C, increasing to perhaps 7°C in polar regions (largely through the demise of snowcover) and reducing to less than 1°C in equatorial regions. Although there has been a small but discernible trend towards less marked effects as these GCMs have been refined during the 1990s (e.g. Houghton *et al.* 1990,

1992), the predicted rates of change are still 4–5 times any that have occurred during the twentieth century (Wigley and Raper 1992).

These predictions are based on the effective doubling of CO_2 concentrations (Box 2.1). It is conceivable that if other gases that act as more effective greenhouse gases increase significantly, the date of effective doubling could be sooner. Perhaps of greatest concern is methane, CH_4, which is 30 times more effective than CO_2 and for which a major source is irrigated agriculture. Methane concentrations are increasing at 1 per cent a^{-1}, twice the rate of CO_2, and seem to be less controllable in principle. At the current state of technology, the burgeoning world population cannot feed itself without expanding irrigation, whereas it could survive by switching to alternative energy sources. Moreover, point emissions are more controllable than nonpoint, agricultural sources.

Box 2.1 Global Climate Models

In order to assess the reliability of hydrological predictions derived from climatic models, it is essential to appreciate some key aspects of these models.

The **General Circulation Models** used in weather forecasting need only consider meteorological variables, applying the laws of physics to the movement and internal alteration of air bodies. The need for medium-range forecasting, 5–10 days ahead, has required the additional consideration of radiative, conductive and material exchange with the surface. Predicting global warming scenarios requires even more information on surface variables, ideally incorporating realistic models of ocean circulation and the dynamics of the cryosphere and biosphere (Figure 2.21). The

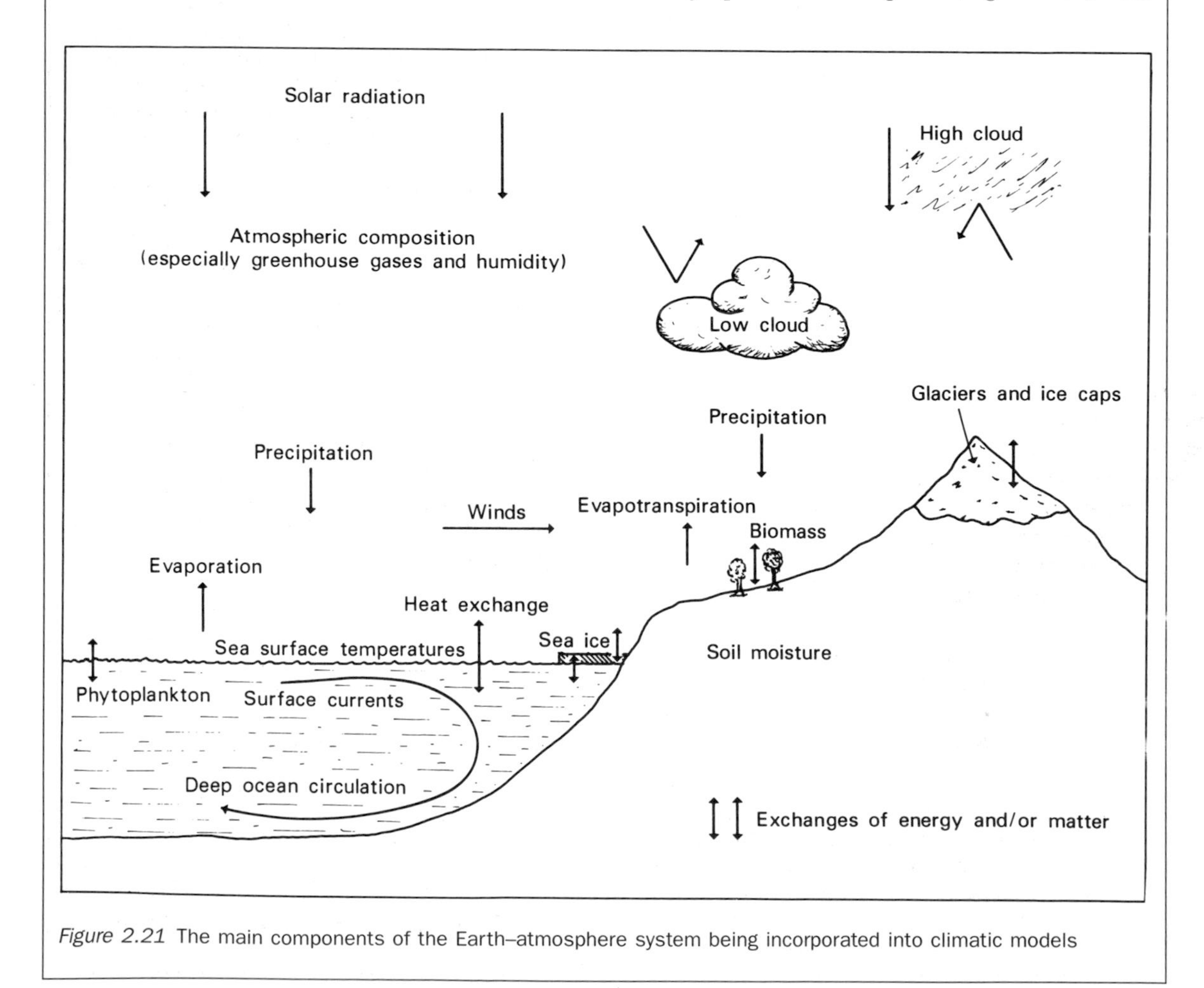

Figure 2.21 The main components of the Earth–atmosphere system being incorporated into climatic models

Box 2.1 (cont.)

respiration and population dynamics of marine phytoplankton are now thought to be crucial to CO_2 balance in the atmosphere, yet these are poorly monitored and understood.

GCMs have generally treated the ocean as a largely passive slab of water about 50 m deep. Second-generation GCMs are truly **Global-scale Climate Models** incorporating environmental interaction submodels, especially oceanic circulation in **Coupled Atmosphere–Ocean GCMs.** These should provide much more reliable information on water budgets. Indeed, one of the more discouraging features of GCM predictions has been the wide differences between models in predicted rainfall patterns.

There are a number of other limitations that need to be appreciated:

1. Available computer capacity and operating time still limit **spatial resolution**, the density of points for which predicted values are output. Spacing in the UK Meteorological Office (UKMO) model was 150 km or more, which was too crude to give usable intra-national water balance estimates. The new Hadley Centre coupled model (UKHI) improves this by meshing a detailed 'window' over Western Europe, with points at 50 km, into the coarser global model (Hadley Centre 1993).
2. Similar factors have limited the use of **transient runs**, which more realistically model the gradual adaptation of the system to a steady build-up in CO_2 as opposed to **equilibrium** or 'shock runs', in which the model is simply presented with a double-CO_2 scenario. Only a handful of transient runs have been reported (e.g. Murphy and Mitchell 1995). Yet biospheric adaptations could turn out to be noticeably different.
3. Many atmospheric and environmental subsystems are very sensitive and yet our knowledge of them is inadequate through lack of data or theory. An increase of just 3 per cent in average global low cloud cover (increasing planetary albedo) could negate a 2°C forcing by greenhouse gases, yet this is close to the error margin of current satellite surveillance. In contrast, a 3 per cent increase in the water vapour content of the air would add 0.5°C to greenhouse warming. The balance is fine. Will higher vapour content in a warmer atmosphere be counterbalanced by more low cloud? Will there be an increase in the *proportion* of high cirriform cloud, in which absorption of global infrared (IR) outweighs reflection of solar shortwave, which could *add* to global warming?

Computer experiments at the UK Meteorological Office have demonstrated the importance of cloud physics, especially the extent of cold-cloud processes. Differing assumptions about ice content caused a 2°C difference in mean global temperatures, 2.7–4.5°C (Mitchell 1991). Ice crystals are more efficient infrared absorbers, but they also tend to fall out of clouds with weak updraughts faster than water droplets, thereby truncating the Bergeron process and maintaining a more liquid cloud (Rowntree 1990). It is difficult to acquire observational data on internal temperatures and crystal densities on a global scale.

The dust content of the air is also important and seems to be increasing. Dust too is ambivalent in its effect. Dark industrial soot and smoke can enhance absorption, but light dust, especially from agriculture in semiarid regions, seems to be most on the increase, with opposite effects. Dust may also interact with water vapour as shown in Figure 2.22.

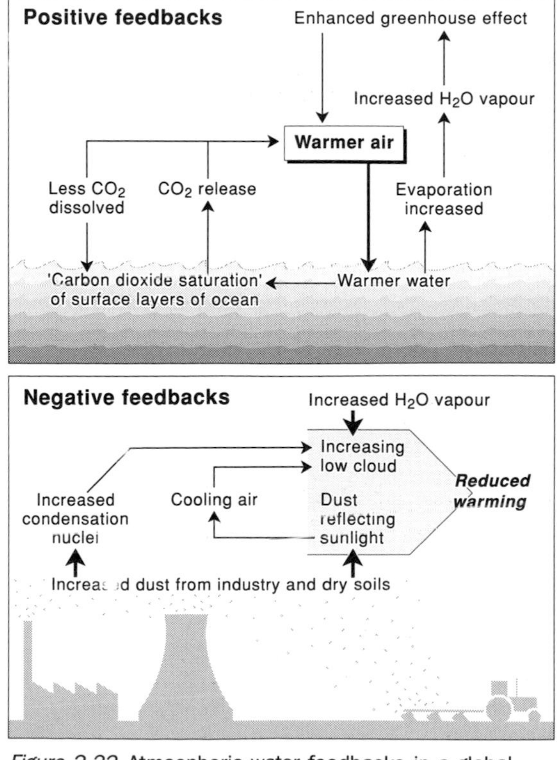

Figure 2.22 Atmospheric water feedbacks in a global warming scenario

Box 2.1 (cont.)

Dust could be a major 'anti-greenhouse' factor. So may phytoplankton, which absorb CO_2 in photosynthesis and are expected to flourish in a CO_2-rich world. But phytoplankton response could be reduced by genetic damage from extra solar UVB radiation, if the shielding effect of the ozone layer is further reduced (Hardy and Gucinski 1989). Other mechanisms may counter global warming and also be self-correcting, e.g. condensation and precipitation processes will limit both vapour and cloud build-up (but also wash out aerosols!). The exact balance is extremely difficult to compute.

4 Political, social, economic and technological changes which are impossible to foresee can have marked effects. The oil crisis of 1973 reduced the annual increment of CO_2 output from fossil fuels from 4.5 per cent to 2.5 per cent a^{-1}. Current best estimates of the timing of effects assume that the 1987 Montreal Protocol on CFCs and its 1990 supplement will be fully implemented. However, continued decimation of the carbon sink in the tropical rainforests will hasten events.

Historical analogues and the validation of GCMs
The only way to be confident about GCM predictions is to test their ability to 'hind-cast' past climatic events. The 1993 UK Hadley Centre model was validated against observed climatic data for the first 90 years of this century, and tuned to improve the 'fit', before being run in 'forecast' mode. Some improvements may be achieved by comparing results from different GCMs, especially by comparing the effects of different simplifications, the '**parameterization**', of physical processes.

Another approach is to run simulations of past climates like the Cretaceous warm period 220–65 Ma BP, which was 10–20°C warmer, the Pliocene Optimum 4–3 Ma BP, which was 4°C warmer, or the Hypsithermal, 9–6 ka BP, 1–2°C warmer (Bach 1989). Budyko (1989) suggested that these climatic analogues provide a *better* method of prediction because of the difficulties with GCMs, but this ignores both the rapid progress with GCMs and the inexact nature of the signal in palaeoclimatic reconstructions (Huntley 1990).

There are similar problems with *reconstructing palaeohydrological environments* to use as analogues for global warming, despite considerable refinements in recent years in the techniques used for reconstructing flood histories (Starkel *et al.* 1990; Gregory *et al.* 1995) as discussed in section 4.2.3. In any case, climatic history rarely repeats itself exactly. Perhaps the most important difficulty is finding a close enough analogue to the state of the regional atmospheric circulation system. Although Newson and Lewin (1990) suggested that the flood-prone period around the demise of the Little Ice Age might be a possible analogue for global warming events in Europe, the climatic system is in a very different state now from that 150–200 years ago, when the stationary Rossby trough was beginning to migrate east from Western Europe, the Ferrel jetstream was beginning to shift north from a markedly lower latitude, and sea surface temperatures to the west of Europe were decidedly lower (Lamb 1977). Analogues therefore provide valuable benchmarks for validation rather than substitutes for GCMs (Jones 1996).

Moreover, as global climate models, rather than simply atmospheric models, the new GCMs are bound to develop next century into powerful tools offering a viable method of conducting a wide range of global environmental experiments by computer. They will provide the highest level of synthesis of environmental science and enable us to conduct experiments that cannot be undertaken in the real world. One of the first uses will be **sensitivity analyses** designed to determine the sensitivity of variables, such as water supply, to global warming. This should stimulate better modelling of other economic and environmental systems.

The current status of predictions is taken up in Chapter 11.

Conclusions

The global hydrological cycle is the most effective integrator in our environment. It either affects or is affected by almost every process of exchange on Earth. The extent and complexity of these interactions underline its central importance. They also mean that it is still extremely difficult to assess, model and predict. Rapid developments in Coupled General Circulation Models stimulated by the need to predict the effects of global warming are set to reveal a lot more about the sensitivity and operation of the global hydrological system.

Further reading

Meteorological aspects are well covered in:

Ahrens C D 1991 *Meteorology today*, 4th edn, New York, West Publishing Co: 576pp.

Barry R G and R J Chorley 1992 *Atmosphere, weather and climate*. London, Routledge: 392pp.

Mason B J 1962 *Clouds, rain and rainmaking*. Cambridge, Cambridge University Press. A classic and readable introduction.

Present and past states of the hydrological cycle are discussed in:

Street-Perrott A, M Beran and R Radcliffe (eds) 1983 *Variations in the global water budget*. Dordrecht, Reidel: 518pp.

Solomon S J, M Beran and W Hogg (eds) 1987 *The influence of climate change and climatic variability on the hydrologic regime and water resources*. Wallingford, International Association of Hydrological Sciences Pub. No. 168: 640pp.

UNESCO 1978 *World water balance and water resources of the Earth*. Paris, UNESCO: 663pp. World maps and detailed explanatory book.

Discussion topics

1. Outline the main factors that control the intensity and spatial distribution of any one hydrological component.
2. Consider the differences in the fluxes and geography of the hydrological cycle between an ice age climate and the present day.
3. Discuss the importance of accurate assessments of hydrological fluxes and an adequate understanding of exchange processes for the sound exploitation of water resources.

CHAPTER 3

Hydrological processes within the river basin

Topics covered

3.1 Evaporation and transpiration losses

3.2 Infiltration and soil moisture

3.3 Runoff-generating processes

3.4 Groundwater

The drainage basin provides the natural framework for the generation of riverflow. As such, it is widely recognized as the natural unit for water management and for the scientific study of hydrological response. In most cases, the topographic divide that forms the periphery of the basin acts as an impermeable boundary, dividing the land surface into discrete units for generating runoff. Within each unit, runoff response is controlled by the general water balance:

$$\text{Runoff} = \text{Precipitation} - \text{Evaporation} \pm \Delta\text{Storage}$$

where evaporation includes transpiration from plants and ΔStorage is the change in the amount of water stored in the rocks and soil. Each of the components of the water balance is affected by certain properties of the land surface, but those which are land based are affected most. As the residual product of the water balance, runoff integrates the effects of topography, vegetation, bedrock and surface materials on the other components, and so it tends to be the most sensitive and the most variable.

The drainage basin has been recognized as the essential unit of study since the earliest field experiments. For practical reasons most basins selected for detailed process studies have been relatively small (e.g. 0.1–10.0 km^2). These have provided a wealth of information on the processes of runoff generation and the effects of surface properties on 'hillslope hydrology', i.e. the pathways of water loss or drainage from the land. Over the last decade or so, they have begun to provide equally detailed information on water quality processes.

There are still many problems in extrapolating and integrating the information gained from small basins into hydrological models for larger catchment areas (Chapter 6). However, process studies in small basins are proving valuable as a basis for pollution control, environmental impact assessment (EIA) and land-use planning.

Figure 3.1 is a schematic representation of the drainage basin system. The succeeding sections will analyse its major components.

3.1 Evaporation and transpiration losses

On land, Dalton's Law really only describes **potential evaporation** (PE), i.e. the evaporation rate given an unrestricted supply of water (section 2.2). More often than not soil surfaces have a restricted water supply. The **actual evaporation** (AE) depends on: (1) the depth to the **water table** (section 3.4); (2) the size, density, orientation and connectivity of pores within the soil, which control the height to which water may rise above the water table by electrical attraction to the walls of fine tubes or **capillaries**; and (3) the **local heat budget**, which is far more variable on land than water because of differing slope angles and orientation (or

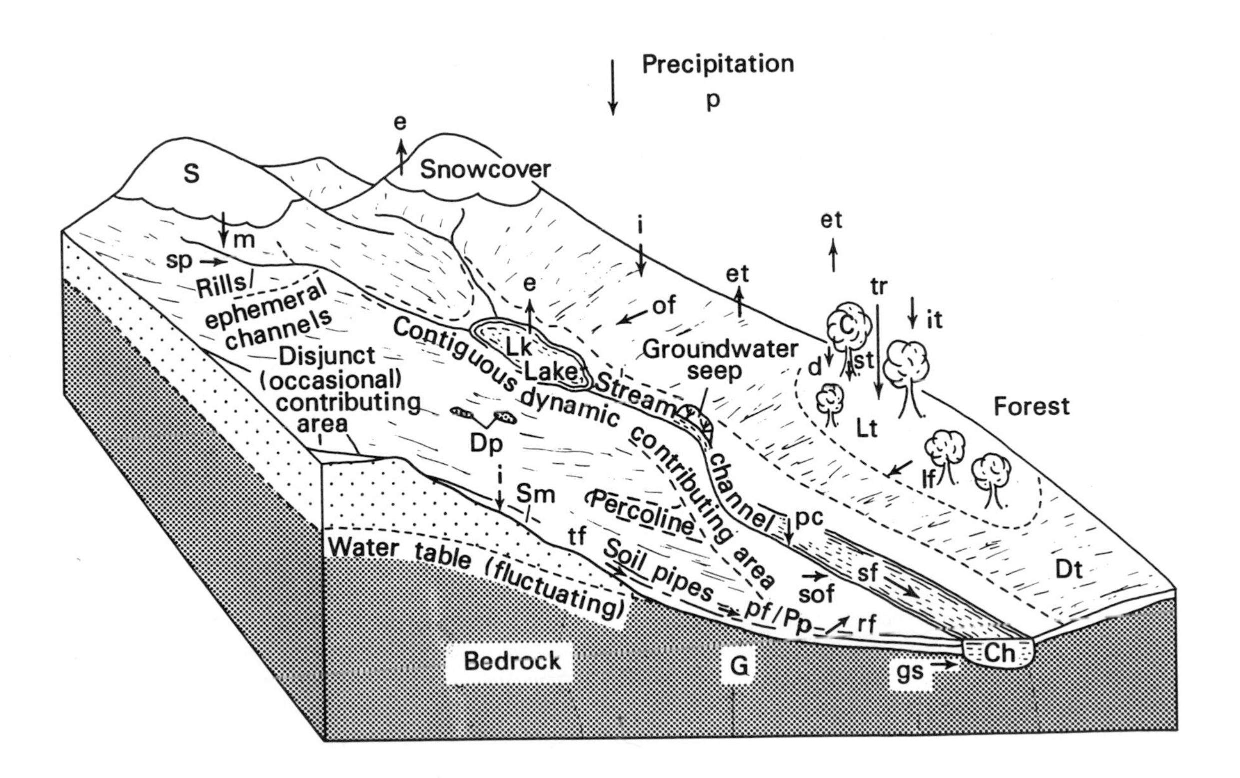

Figure 3.1 The drainage basin system. *Storages*: C, canopy; Ch, channel; Dp, depression; Dt, surface detention; G, groundwater; Lk, lake; Lt, litter; Pp, pipe; S, snow and ice; Sm, soil moisture. *Processes and flows*: d, drip; e, evaporation; et, evapotranspiration; gs, groundwater seepage; i, infiltration; it, interception; lf, litterflow; m, meltwater; of, overland flow; p, precipitation; pc, channel precipitation; pf, pipeflow; rf, return flow; sf, streamflow; sof, saturation overland flow; sp, spring flow; st, stemflow; tf, throughflow; tr, throughfall. Contributing areas contribute storm runoff.

Table 3.1 Evapotranspiration rates from different vegetation

A. Relative rates

	Larch, birch, Douglas fir	Spruce	Maquis scrub	Beech	Pine	Pasture	Bare earth
Ratio to bare earth	9	8.5	8	7.5	5	3	1

B. Sample measurements (mm a^{-1})

UK (Yorkshire)		Holland (Castricom)		UK (Berkshire)		Germany (Harz Mountains)	
Sitka spruce	801	Oak, beech, alder	500	Scots pine	679	Norway spruce	579
Grass, irrigated	456	Australian pine	655	Open water	597	Grass	521
Grass, non-irrigated	416	Open water	611				

Sources: Baumgartner (1967), Schachori *et al.* (1967), Geiger (1957), and others

'aspect'), differing soil properties, such as albedo and local drainage status, and vegetation shading. Considered over a longer period of time, AE is also affected by the pattern of precipitation, its timing, amount and phase. Thus, at a given location, the same amount of rain falling in a number of frequent light storms is likely to remain at or near the surface and available for evaporation for longer than if it fell in one heavy storm. Again, snowfall would have lower vaporization losses.

Transpiration is controlled by more variables, because it depends upon many dynamic aspects of plant activity. (1) Different plant species can transpire very different amounts, depending upon the nature of the evaporative openings in the leaves, the **stomata**, especially their size, density and location or exposure (Table 3.1). (2) *Season* determines whether deciduous plants have leaves and along with (3) *time of day* alters radiation balances, photosynthesis and growth rates, and stomatal activity. Transpiration at night is negligible, only 5–10 per cent of the daytime rate, which is driven by solar heating and photosynthesis. (4) The *stage of growth* of the vegetation, since plants use more water during periods of active growth or building of biomass, and when root systems have reached maximum spread and efficiency. For a group of plants, transpiration rates tend to reach a maximum when they completely cover the ground. (5) *Meteorological factors* also affect the opening of stomata, for example stomata tend to close in high winds to reduce potentially damaging water loss. In contrast, in high humidities plants may continue to lose water by exuding it as liquid in order to maintain sap movement. (6) *Soil properties* also affect the ease of abstraction of water by the roots, as illustrated in Figure 3.2.

Transpiration is notoriously difficult to measure or estimate on its own, because of the scale of the process and its dynamic variations. Consequently, evaporation and transpiration are generally lumped together as **evapotranspiration** (ET).

3.1.1 Calculating surface losses

As process theory has become more sophisticated, there has been a tendency for the formulae used to calculate losses to incorporate ever more detailed representations of vaporizational processes.

Temperature indices These represent the lowest level of sophistication. In early formulae, evaporation itself was often not calculated as a separate entity, e.g. Thornthwaite's P/E ratio in Table 3.2, eq. 1.

Nevertheless, a handful of these formulae have found widespread acceptance and are still used, notably Thornthwaite's 1949 formula for monthly PE, the Blaney and Criddle PET formula and to some extent the Turc equation (Table 3.2, eqs 2 to 4). These use varying combinations of annual, monthly and maximum monthly temperatures.

C Warren Thornthwaite and his colleague John Mather played a major role in developing practical theory, based on experimental work begun in the 1940s in America. The 1949 formula (Table 3.2, eq. 2) was

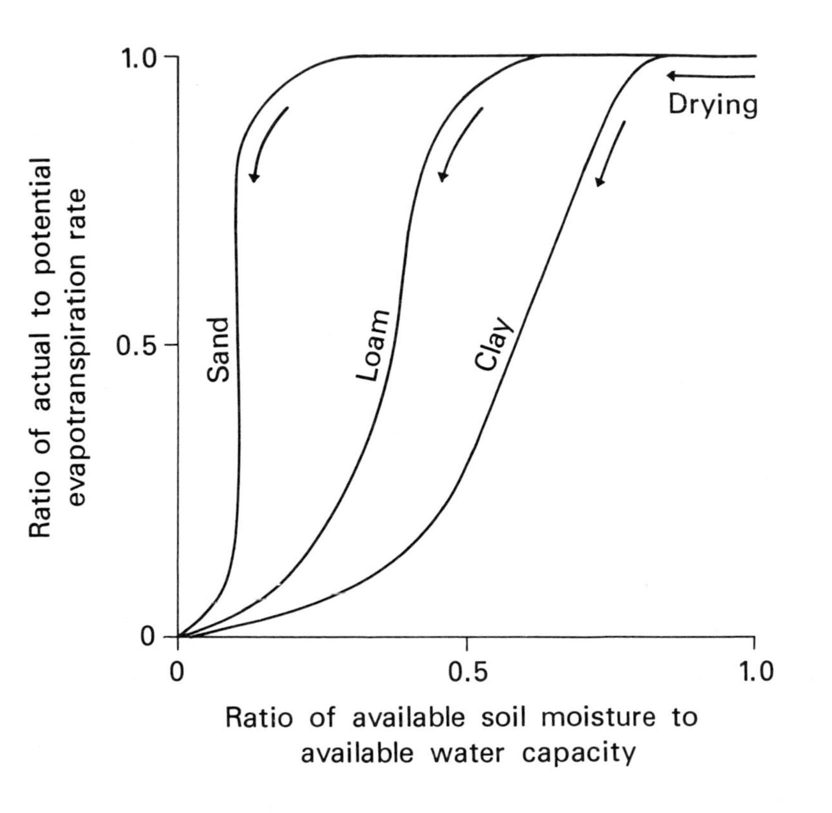

Figure 3.2 The graph compares the rates of reduction in evapotranspiration as the soil dries out (right to left) for sand, loam and clay, quantified as a proportion of the total water-holding capacity of the root zone, the part of the soil that contains water available to plants ('the available water capacity'). In sand, plants can extract water at full potential rates until near their wilting point, but in clay the supply is restricted by smaller soil pores so that uptake and transpiration rates can rapidly fall below potential much earlier. Based on Dunne and Leopold (1978)

based on statistical relationships established between climatic data and 'evaporimeters' developed to enable direct measurement of evaporation. It was the first temperature-based formula to be applied to short-term, monthly evaporation (Thornthwaite and Mather 1955). It was successfully tested against the much more complex and process-based Penman formula (*v.i.*) in the Thames basin by Ward (1971).

However, some of the physical relations in the formula are open to question. For example: (1) the variable I is calculated from the *annual sum* of monthly mean temperatures, yet it is used to obtain evaporation estimates for *individual months*; and (2) the formula implies that there will be no evaporation in a month when the mean temperature is 0°C or below, yet a *mean* of 0°C normally indicates that some days have experienced temperatures above that, when evaporation is likely to have occurred. In general, the formula does not work so well in regions that receive air masses with strongly contrasting relative humidities, i.e. where advected humidity affects vapour uptake, since Thornthwaite ignored humidity effects. Tests have shown a tendency to overestimate losses by 20–30 per cent or worse in high latitudes and also in West Africa, where the area of 'no runoff' (negative water balance) based on the formula has been found to extend some 350 km or more into areas of known riverflow. The problem may be due to the fact that net radiation balances are not uniquely related to the daylight factor, d, which is used to estimate available energy levels.

The Blaney–Criddle formula is a development of this formula that incorporates an extra variable for transpiration, K (Table 3.2, eq. 4). It has been adopted by the USDA Soil Conservation Service to estimate consumptive use of water by irrigated crops, as an aid to apportioning water supply to irrigation schemes.

Table 3.2 Selected formulae for evaporation and evapotranspiration*

Source and date	Equation	Explanation
Temperature based		
1. Thornthwaite – 1931	$\frac{P}{E} = 11.5\left\{\frac{P}{(\overline{T}-10)}\right\}^{10/9}$	**annual precipitation/evaporation ratio** annual precipitation (P), evaporation (E) and mean annual temperature ($\overline{T}$)
2. Thornthwaite – 1949	$E = 1.6d\left\{\frac{10\bar{t}}{I}\right\}^{a}$	**monthly evaporation** (cm) d = total monthly daylight hours/360 (360 hours in 30-day month) $\bar{t}$ = mean monthly temperature $I = \sum^{12} i$ where $i = (\bar{t})^{1.514}$ $a = bI^3 + cI^2 + dI + 0.49$ where b, c and d are empirical coefficients
3. Turc – 1954	$E = \frac{P}{\sqrt{0.9 + \left\{\frac{P}{L}\right\}^2}}$	**annual evaporation** (mm) $L = 300 + 25T + 0.05T^3$ (Also a 10-day version)
4. Blaney and Criddle – 1950	$ET = (0.142\bar{t} + 1.095)(\bar{t} + 17.8)Kd_{ann}$	**monthly potential evapotranspiration** (cm) K is empirical crop factor d_{ann} = monthly daylight hours as fraction of annual
Mass transfer 5. Dalton – 1802	$E = f(u)(e_{sat} - e_{act})$	**evaporation** $f(u)$ = function of windspeed e_{sat} = svp at temperature e_{act} = actual vapour pressure
6. Thornthwaite and Holzman – 1939	$E_0 = \frac{0.623\kappa^2\rho(e_1 - e_2)(u_2 - u_1)}{p\left\{\ln\frac{z_2}{z_1}\right\}^2}$	**evaporation rate** (cm s^{-1}) E_0 = evaporation from open water e_1 and e_2, u_1 and u_2 are vapour pressure and windspeed at heights z_1 and z_2 p = atmospheric pressure (hPa) κ = von Karman's index of turbulent transfer ≈ 0.4 ρ = density of water Valid only in neutral atmosphere, not when lapse rates high (unstable air)
7. Pasquill – 1949	$E_0 = \frac{p\kappa^2(e_1 - e_2)(u_2 - u_1)}{\ln\left\{\frac{z_2 - z_0}{z_1 - z_0}\right\}^2}$	**evaporation rate** as eq. 6 z_0 is zero plane displacement Valid as eq. 6

Table 3.2 (cont.)

Source and date	Equation	Explanation
Energy balance 8. Basic equations	$Q_{in} = Q_{insol}(1-\alpha) - Q_{back} + Q_{advect}$ $Q_{out} = Q_{heat} + Q_{evap} + Q_{stored}$ $Q_{evap} = \rho\lambda E_0$ $\therefore E_0 = \frac{Q_{evap}}{\rho\lambda}$ *or* $E_0 = \frac{Q_{insol}(1-\alpha) - Q_{back} + Q_{advect} - Q_{heat} - Q_{stored}}{\rho\lambda}$	**evaporation rate** (mm s^{-1}) variables as above, plus Q is energy in = incoming insol = insolation back = longwave losses advect = advected heat out = outgoing heat = sensible heat stored = stored heat λ = latent heat of vaporization α = albedo
Combined formulae 9. Penman – 1948/1963	$E_0 = \frac{\frac{\Delta}{\gamma}H + f(u)(e_{sat} - e_{act})}{\frac{\Delta}{\gamma} + 1}$	**evaporation rate** (mm day^{-1}) H = net radiation balance in mm water equivalent Δ = rate of change of svp with temperature γ = psychrometric constant
10. Penman and Monteith – 1965	$ET = \frac{\Delta H + \frac{\rho c_p(e_{sat} + e_{act})}{r_{sfc}}}{\lambda\{\Delta + \frac{\gamma(r_{sfc} + r_{aero})}{r_{aero}}\}}$	**evapotranspiration rate** (mm day^{-1}) as eq. 9 r_{sfc} = surface resistance r_{aero} = aerodynamic roughness c_p = specific heat of water

* For further information see text.

K has been tabulated from empirical observations of crop type and stage of growth by the SCS (Dunne and Leopold 1978).

Despite the relative success of some temperature-based formulae, it should be remembered that: (1) temperature is only one factor in Dalton's Law, and not always the dominant one; and (2) the link between temperature and evaporation is not fixed. There tends to be a *hysteresis loop*, whereby early in the season there is more evaporation at a given air temperature, because cold, dry air is being warmed up, creating larger saturation deficits. In the autumn, air is being cooled, leading to higher RHs, which inhibit evaporation. This is true for both actual and potential ET, but probably more so for actual ET (AET) because soils are wetter early in the season so that rates are nearer potential levels, whereas soils are drier in the autumn.

Mass transfer equations These describe and use the physical processes outlined in Dalton's Law (Table 3.2, eq. 5; section 2.2). This approach has also been termed *aerodynamic* or *saturation deficit*, although each of these titles strictly refers to only one part of the formula.

Many adaptations have sought to make an operational formula out of Dalton's Law by comparing observed and calculated evaporation and statistically deriving empirical coefficients for the aerodynamic term. The Waite formula, developed at the Waite Agricultural Research Station in Australia, is one example. In this, the aerodynamic coefficient, $f(u)$ in Dalton's Law, varies between 200 and 350, depending on site characteristics such as roughness and exposure to wind.

Explicit representation of the physical processes of vapour dispersion was achieved in the Thornthwaite–Holzman and Pasquill formulae (Table 3.2, eqs 6 and 7). These use measurements of windspeed, u, and vapour pressure, e, at two heights above the surface in order to estimate *vertical windshear* and *humidity gradients*. Differences in horizontal wind velocity at different heights (the windshear) generate vertical turbulent eddies, which cause mechanical dispersion of water vapour, while vertical humidity gradients control molecular diffusion of the vapour. The formulae incorporated advances in the theory of vapour diffusion. Pasquill's equation and the version of Thornthwaite–Holzman in Table 3.2 introduced von Karman's constant as an index of turbulent transfer. Pasquill used the **zero plane displacement**, which is the height in a vegetation cover at which windspeed falls to zero, as indicating the point at which the vertical gradients in temperature and humidity stop. It is also called **roughness length**. Measurements suggest this may be as little as one-tenth or as much as two-thirds of the height of the vegetation. In theory, it can be extrapolated from the windspeed measurements by assuming an exponential increase in windspeed with height. This assumption may not be justified, however. It applies in a neutral atmosphere, but not in situations with strong convective instability or in a stable temperature inversion. More recent adaptations have used different versions depending upon atmospheric stability on the day in question, for example, in detailed studies of glacier mass balance.

The Pasquill and Thornthwaite–Holzman formulae have been used mainly for research purposes, since they require specific instrumental installations which are not normally found at meteorological stations. Thornthwaite's subsequent retreat from the high ground of scientific theory in 1949 therefore proved a well-judged practical decision.

The energy budget approach This is based on the premiss that all surplus energy available from the net heat balance is used up in evaporation. The two basic equations in Table 3.2 (eq. 8) describe the quantity of heat entering the local environment (Q_{in}), which consists of the local radiation balance plus heat advected by warm air or water entering the locality, and the subsequent use of that heat (Q_{out}) as sensible heat to warm the air (Q_{heat}), latent heat used in evaporation (Q_{evap}) and the residual heat stored in the soil or water body (Figure 3.3). By assuming that the system is in equilibrium, so that $Q_{in} = Q_{out}$, and by reorganizing the equation to leave Q_{evap} as the only unknown, an equation is obtained to predict the quantity of heat available for evaporation:

$$Q_{evap} = Q_{insol}\,(1-\alpha) - Q_{back} + Q_{advect} - (Q_{heat} + Q_{stored})$$

The actual amount of water evaporated (mm^3) is then simply Q_{evap} (J) divided by the latent heat of vaporization, L ($J\ kg^{-1}$), times the density of water, ρ($kg\ mm^{-3}$), for each mm^2 of surface.

The approach has been very successful in assessing evaporative losses from reservoirs, for which Bruce and Clark (1980) provide a full discussion. To use the approach for water bodies, it is necessary to calculate radiation balance, if possible by directly measuring incoming solar shortwave (Q_{insol}), albedo (α) and outgoing terrestrial longwave radiation (Q_{back}), and to calculate changes in air and water temperatures, in order to estimate the other terms. This can require a considerable amount of data collection and instrumentation.

A complication peculiar to water bodies is the development of a layered structure in certain

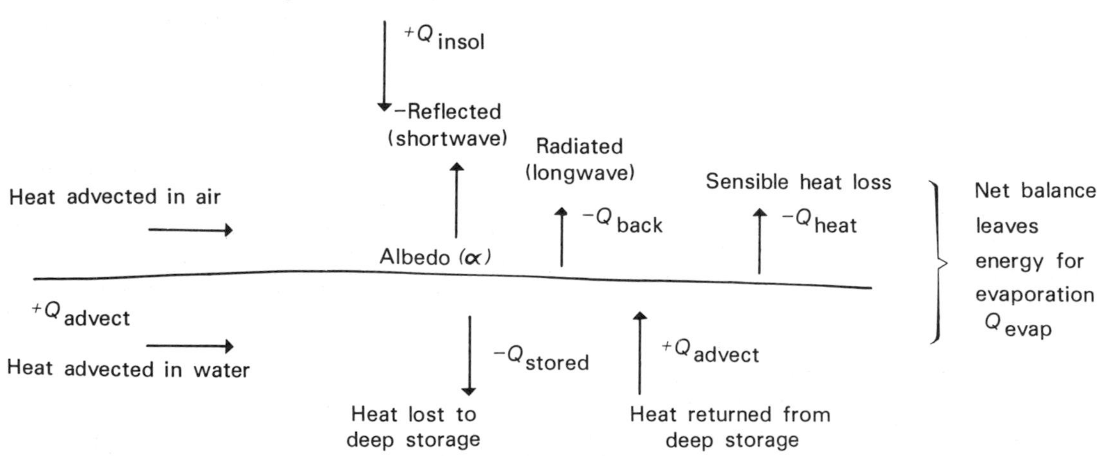

Figure 3.3 Components of the surface heat budget. Symbols as in text and Table 3.2

circumstances. In general, warmer water entering from a stream ($+Q_{advect}$) will be less dense and spread out over the lake surface making the heat readily available for evaporation, but colder water ($-Q_{advect}$) will usually sink and therefore not reduce evaporation rates immediately. In addition, at around 4°C, the maximum density of water, convective overturning will take place within the lake as surface water that has cooled to 4°C sinks and causes warmer water to upwell. This creates an inverse layering in the lake, with the cooling surface actually remaining warmer than the bottom water, and it prolongs evaporation at higher rates than would be suggested by the mean heat budget of the lake. These complications can be monitored by installing a vertical string of temperature probes (thermocouples or thermistors) in the lake.

Subsurface temperature profiles are also needed on land surfaces to estimate stored and advected terms. However, the central assumption of the energy budget approach, that a linear relationship exists between surplus energy and the depth of water evaporated, is invalidated in soils when water supply becomes restricted. The approach is therefore estimating potential not actual evaporation.

Similarly, in order to estimate sensible heat exchange (Q_{heat}) properly, measurements are needed of the temperature profile above the surface and estimates of atmospheric stability, as in Pasquill's formula. The approach can, however, be simplified to use readily available meteorological data by utilizing the **Bowen Ratio**, β

$$\beta = \frac{\text{sensible heat}}{\text{latent heat}}$$

which can be calculated from measurements of temperature and relative humidity at screen height and at the surface. A simplified approach to an energy budget estimate would then be

$$e = \frac{(\text{net radiation})}{\rho L} \; \frac{1}{(1 - \beta)}$$

as outlined by Linsley *et al.* (1982).

Even so, the energy budget approach does not attempt to trace the mechanisms of water transfer.

Combined formulae A major step forward on both the scientific and the practical fronts was taken by H C Penman in 1948. He set out to produce a method that considered all the significant factors in the evaporation process and at the same time would require a minimum of specific data collection and where necessary could utilize existing meteorological instrumentation. He began by considering evaporation from open water and gradually refined his formulae over three decades in order to better represent the processes of transpiration. Refinements in the representation of transpiration and of evaporation from intercepted water have also been added by Monteith and by the British Meteorological Office.

In fact, the great strengths of Penman's approach have been its *adaptability* and its *pragmatic use of available data*. The adaptability is founded on the basic scientific veracity of the original approach. Its worldwide popularity is largely due to Penman's provision of alternative methods for calculating the essential parameters, based on his appreciation of lack of instrumentation in many parts of the world.

Hence while the net radiation budget is ideally estimated from direct measurements using long- and shortwave radiometers, the more commonly available measurement of 'sunshine hours' can be substituted. Penman provided two formulae which use the hours of bright sunshine to calculate: (1) net incoming shortwave radiation, along with data on the latitude, season and surface albedo; and (2) longwave losses, incorporating measured saturation vapour pressure and mean air temperature. Shaw (1994) and Ward and Robinson (1990) discuss these in detail and Rodda *et al.* (1976) summarize alternative data inputs.

In essence, Penman combined the energy budget and mass transfer approaches. He used the energy budget method to calculate the energy available for evaporation, but he simplified it by ignoring the effects of storage and advection. This is reasonable since heat flux through the surface has been shown to amount to only a few per cent of total incoming energy, at least for a soil surface. The approach has been successfully adapted by the US National Weather Service (NWS) for use on larger, natural lakes (Kohler *et al.* 1955) despite the complications of heat distribution in lakes discussed above.

By eliminating the stored and advected heat terms, he removed two of the more difficult to measure components. If all surplus radiative energy, Q_{net}, is assumed to be divided only between energy used for evaporation, Q_{evap}, and energy that warms the air, Q_{heat}, then:

$$Q_{evap} = Q_{net} - Q_{heat}$$

Since hydrologists require calculations of evaporation, E_o, in millimetres of water, rather than energy, these terms must be divided by the amount of energy consumed in evaporating 1 mm of water, i.e. ρL, to give new terms, respectively H and K, in millimetre equivalents:

$$E_0 = H - K$$

Sensible heat transfer, K, depends on two factors: the difference in temperature between the evaporating surface and the air, which controls conduction and convection, and the windspeed, which controls the turbulent transfer of heat. Penman therefore used the **mass transfer** approach to calculate sensible heat exchange. Hence:

$$K = \gamma f(u)(T_{surface} - T_{air})$$

in which the aerodynamic coefficient, $f(u)$, uses windspeed measured at 2 m and is calculated from a formula based on Penman's experimental measurements. γ is the 'psychrometric' or 'hygrometric constant', which takes different values depending on the temperature units and the method of ventilation or 'aspiration' of the wet and dry bulb thermometers in the 'psychrometer'.

If it is assumed that gradients in temperature and humidity between the surface and the level of a standard meteorological measuring screen are small and similar, it is possible to dispense with the awkward measurement of temperature and svp gradients from the surface upwards, which are required by normal mass transfer techniques, and just use standard screen measurements. Similarly, if it is assumed that air turbulence affects both temperature and humidity alike, the same $f(u)$ can be applied to both. Finally, it is possible to use the theoretical relationship between svp and air temperature to calculate the effect of warming the air on its vapour holding capacity (Figure 2.9). The rate of change of svp with T, de/dT, is represented by Δ in the subsequent equation: it represents the efficiency with which the surplus energy can be utilized in evaporation and it increases with temperature:

$$E_0 = \frac{\Delta H = \gamma\{f(u)(e_{sat} + e_{air})\}}{\Delta + \gamma}$$

Shaw (1994) describes the derivation of the formula in detail. Penman later simplified this formula by dividing through by γ and tabulating the changing values of Δ/γ with temperature (Table 3.2, eq. 9).

This approach represents the processes involved reasonably accurately, although Penman derived the windspeed effect, $f(u)$, by empirical comparison of calculated E_0 with field measurements from evaporation tanks. Van Bavel (1966) refined this aerodynamic function by using diffusion theory along the lines of Pasquill, but using only the inferred gradient in windspeed between 2 m and z_0, thus obviating the need for measurements at more than one height. The improved formula is regarded as very good for daily potential evaporation and acceptable for hourly estimates, given good radiation measurements. The basic Penman formula is normally only applied to periods of five days or more, although good net radiation and windspeed measurements can enable it to be used for shorter periods.

Despite the simplifications, the general approach is a significant improvement upon the two separate approaches it incorporates. The addition of the net radiation balance of the surface to the mass transfer approach allows evaporation to be driven by surplus energy at the surface. This frees the mass transfer approach from the assumption of equilibrium between water and air temperatures and so makes the formula more applicable to smaller water bodies. At the same time, it frees the energy budget approach from the constraints of instrumentation, for example, by using principles of vertical heat transfer and some empirical means of estimating energy budgets.

Most subsequent refinements of Penman's formula for E_0 have been concerned with adapting it to calculate **evapotranspiration.** Penman himself began with a purely empirical approach, comparing calculated E_0 with ET losses from a well-watered plot covered in short-cropped grass at Rothamsted Research Station, near London. For this purpose, he defined **potential evapotranspiration** (*PET*) as 'evaporation from an extended surface of short green crop, actively growing, completely shading the ground, of uniform height and not short of water'. His aim was to hold biological and pedological factors constant and thus to establish a notional potential rate of loss that changes only with meteorological factors.

Collecting data from similar plots in a wide variety of climates from Europe to the humid tropics, Penman produced the empirical formula:

$$PET = f E_0$$

where f is a seasonal correction factor which covers the effects of differing insolation intensity, day length, stomatal response and geometry. He still recommended local empirical confirmation wherever possible, because of the large variation that factors such as age and species can cause.

His subsequent work concentrated on developing a more process-based formula for PET that would be more generally applicable, introducing factors covering the effects of stomatal opening and the resistance of the plant membranes to vapour diffusion. Experimental work by Monteith provided a further valuable refinement of the biological parameters which has enabled PET calculations to be extended to a variety of crops. Monteith distinguished two scales of canopy resistance affecting processes: (1) **aerodynamic resistance** or roughness, r_{aero}, which generates turbulent diffusion and is considered in the *f*(*u*) term; and (2) **stomatal resistance**, r_{sfc}, which controls molecular diffusion (Table 3.2, eq. 10).

At present, however, the refinements are of more value to agricultural research than to hydrology. Data on stomatal resistance are largely lacking and experiments at the UK Institute of Hydrology (IH) have shown how sensitive it is to a complex mixture of meteorological and botanical factors (Stewart 1988). Monteith (1985) noted that his formula has mainly been used 'in reverse', in order to estimate surface resistance rather than to calculate *PET*. There is also considerable spatial variability in evapotranspiration processes at scales appropriate to considering stomatal effects, as the British Meteteorological Office has acknowledged in adopting the Penman–Monteith formula as the basis for its national computerized system, MORECS, the Meteorological Office Rainfall and Evaporation Calculation System (section 5.3.2).

3.1.2 Interception

Interception is now known to explain much of the variability in *ET* between plant species or associations, once thought to be due to different transpiration rates. Leaves and branches intercept precipitation before it reaches the ground and hold a certain amount where it is more exposed to direct evaporation. The process has been quite accurately dubbed 'the washing-line effect'. Intercepted water has a high surface area to volume ratio and is held higher in the air, where windspeeds are greater. Hence, evaporation from a wet forest canopy can be much higher than would be estimated from the radiant energy balance alone.

Research at UK IH Plynlimon catchments has given valuable insight into the processes involved (e.g. Calder 1986, 1990; Hudson 1988). The paired catchment experiments, which compare neighbouring forested and moorland basins, have indicated that the bulk of the 20 per cent or so difference in annual runoff between the two basins can be attributed to interception losses in the forest catchment. This suggests that afforestation could *increase* evapotranspiration losses by up to 50 to 100 per cent in parts of upland Britain (Newson 1981). In practice this may mean *net* losses of only 30 per cent on 3000 mm of annual rainfall, and a recent re-evaluation at Plynlimon suggests that the effect may be getting less, perhaps due to acid rain damage or ageing of the trees (Hudson and Gilman 1993). Nevertheless, the results confirm the higher water losses from forests found in Law's early experiments, although Law (1956) was not aware of the importance of interception. And they carry important consequences for land-use policy and environmental impact assessment (EIA), especially in light of plans by the UK Forestry Commission to plant 2.5 million ha between 1980 and 2000, mainly in the uplands, which are the prime water resource areas.

Two broad categories of factor affect interception losses: botanical characteristics and meteorological variables.

1. *Vegetational characteristics*, such as the shape, size, roughness, flexibility (turgidity) and orientation of leaves, total biomass and deciduous or non-deciduous habit, control the **crown capacity**, which is the maximum amount that can be stored in the canopy.

2. *Rainfall intensity–duration* characteristics control the speed with which crown capacity is reached and may even result in higher losses if the intensity is so low that evaporation losses take place before crown capacity is reached. Evaporation losses are greatest from long-duration, light rain or drizzle. Long durations tend to decrease losses even in heavy rain, but the effects decrease exponentially as rainfall amounts increase. *Rainfall frequency* can also have an effect because losses tend to be greater at the start of storms.

3. *Phase of precipitation*: snowfall is more prone to interception and trees can reduce ground level snow receipts by 80 per cent or more. However, evaporation is much lower from snow and in the end losses tend to be greater from intercepted rainfall.

4. *Windspeed and turbulence* affect the supply of unsaturated air and heat transfer to the canopy and the removal of vapour from the canopy zone. Hence, greater exposure to the wind will tend to increase the evaporation. This tendency may be partly offset where increased shaking by the wind reduces crown capacity, for example at the edge of a woodland.

5. *Radiant energy balance* affects the temperature of the leaves, the intercepted water and the air around the canopy. Forest canopies tend to have higher net

energy balance than grassland because of their lower albedo, around 14 per cent compared with 20 per cent for grassland.

Calculating interception losses The most common general formula for calculating losses on a *storm basis* was proposed by Robert Horton in 1919:

$$\text{Interception loss} = \text{Crown capacity} + (E.d.V)$$

where d is storm duration and V is the ratio of the surface area of the vegetation to its projected area on the horizontal ground surface. The crown capacity determines how much water can be held on the vegetation, while the term in brackets determines the amount of evaporation from that water *during* a storm in mm mm^{-2} of ground. It is assumed that all intercepted water is eventually evaporated after the storm. Merriam introduced a sophistication in 1960 in which crown capacity is treated as an exponential function of rainfall amount in order to reflect greater penetration by rainfall in heavier storms.

More practical formulae for calculating *annual losses* on a catchment basis have been developed by Calder and colleagues at the IH. For a forest,

$$E = PET + f(\alpha P - \omega PET) \text{ mm a}^{-1}$$

where f is the fraction of the catchment area covered by forest, α is the average fraction of annual rainfall, P, intercepted (35–40 per cent for areas of the UK with over 1000 mm a^{-1}) and ω is the fraction of the year when the canopy is wet ($\cong$0.000 12P). PET is estimated by the Penman formula.

For shorter periods, more attention must be given to the physical processes. Rutter *et al.* (1971) produced a physically based model that can be used for individual storms. This explicitly considers the rate of drip drainage from the leaves, which is estimated by an exponential function of storage capacity proportional to the depth of water stored during the actual storm. The Rutter model has formed the basis for much subsequent development, though an alternative approach adopted by Gash (1979) takes a more simplified view of drip processes.

Research at the IH has also investigated interception by heathers and indicated canopy storage or crown capacities almost 50 per cent as large as mature trees, e.g. 2.7 mm compared with 6.1–7.6 mm, although actual transpiration losses were 50 per cent less than for grass. Calder (1985) presents a variant of the tree interception formula to take account of transpiration rates from heathers that are half those suggested by Penman's formula. He has since produced a general formula for daily and seasonal interception by forest, heather and grass using a variant of Merriam's exponential function (Calder 1988; Newson and Calder 1989).

Spatial variations The limited number of environments that have been instrumented for process studies reveal considerable variations in interception rates. One of the greatest contrasts is naturally between deciduous and evergreen woodland, 16 per cent in Figure 3.4. Interestingly, the examples in Figure 3.4 suggest that this contrast can exist irrespective of season. It suggests that even during the growing season coniferous needles are more efficient interceptors. This is partly because deciduous leaves bend more under weight and are shaken more by the wind. The difference may remain similar in winter because leaf-drop also occurs in evergreens and because rainstorms are heavier and more prolonged.

The overall physiognomy of plants is also important. Coniferous trees tend to have branches that bend downwards away from the trunk, whereas deciduous trees have a more typically 'dendritic' branching mode. Figure 3.4 shows that the contrasts occur predominantly in the amount of **stemflow** concentrated by the trees' own surface drainage systems. There is less difference in **throughfall**, the precipitation that has avoided interception or dripped off the canopy.

Data from subtropical forest in Brazil suggests that canopy interception may account for only half of total interception losses there. Absorbent bark, mosses and other plants living on the trees can prevent 40 per cent of stemflow from ever reaching the ground. On this evidence tropical and subtropical rainforests have the highest interception rates in the world. The more rapid recycling from this to the atmosphere must contribute to local rainfall production.

In extending their modelling studies to tropical rainforest in Java, Calder and the IH team have concluded that total evaporative losses can best be estimated there by assuming that all surplus radiant energy is used in evaporation. The net annual radiation balance is therefore a usable surrogate for the energy equivalent of annual ET, given evaporation and transpiration that are unlimited by water deficiency (Newson and Calder 1989).

Horizontal interception or 'occult precipitation' In certain circumstances plant, rock or soil surfaces can collect water directly from the air, bypassing conventional precipitation processes. This may be a very

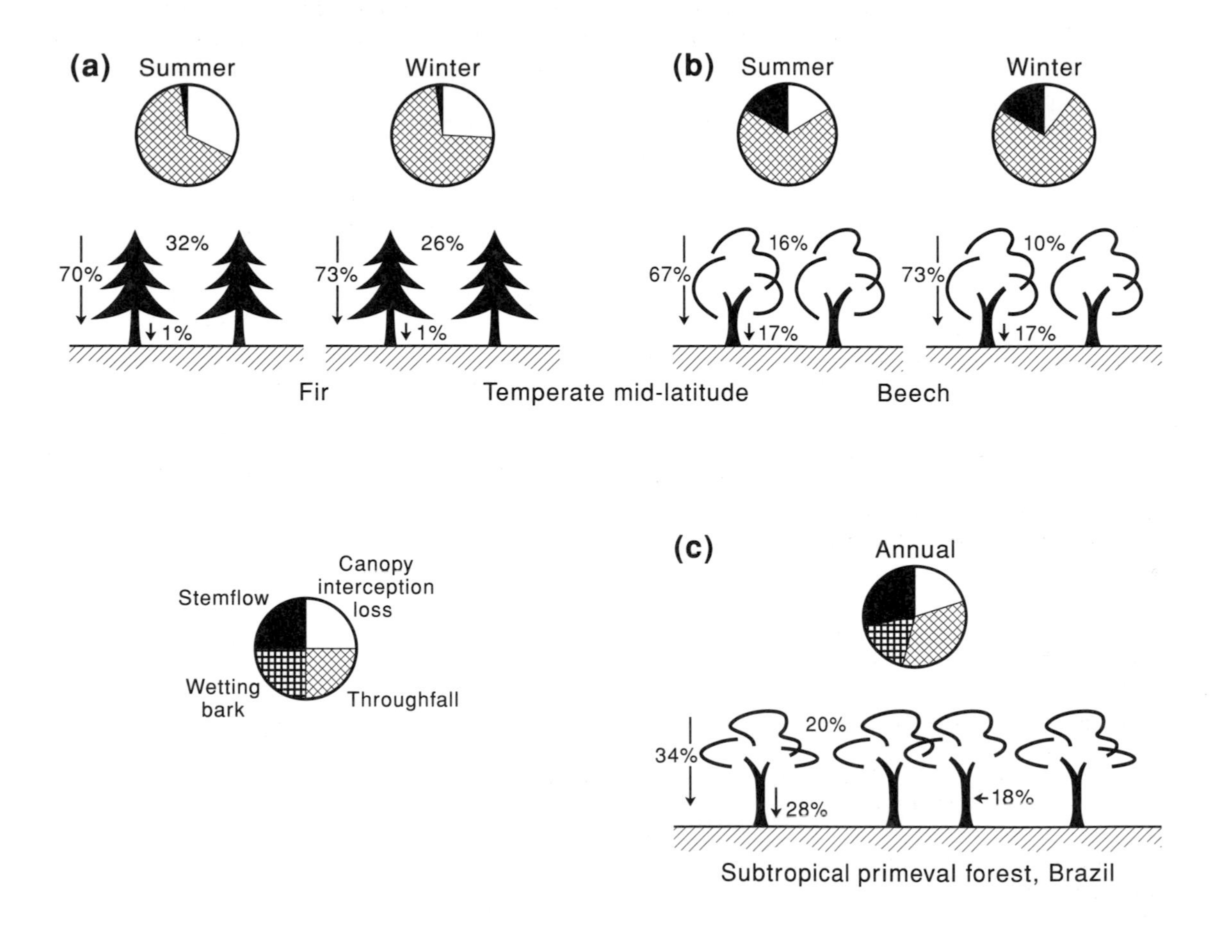

Figure 3.4 Interception, throughfall and stemflow in a variety of forest environments. Based on data in Geiger (1957) and other sources

important source of supply in some ecosystems. Measurements of **dew** collection on Table Mountain, Cape Town, have indicated total annual water receipts of 3300 mm compared with conventional rainfall measurements of less than 2000 mm (Nagel 1956). Whole ecosystems can survive entirely without rain by intercepting fog or low cloud, as in the Namib Desert, where beetles collect fog droplets by raising their forewings.

Yet, despite recognition in classic microclimatic texts like Kittredge (1948) and Geiger (1957), little research has been carried out in this area. There are clearly two main processes involved: (1) *impaction* of cloud droplets on obstructing surfaces; and (2) *radiational cooling* of surfaces at night causing condensation of water vapour. The Israelis have made practical use of the latter for decades by piling stones around fruit trees to stimulate dew. Occasional suggestions have been made that fog drip might actually contribute to a small rise in stream discharge (Gurnell 1983).

Forested mountainsides that rise through the normal local condensation levels may well receive substantial amounts from this 'occult precipitation', but very few quantitative evaluations have been conducted. Even so, the detailed measurements of forest water balance undertaken in Britain by the IH do indicate that the net effect of afforestation in the UK is to reduce receipts. Quite apart from any effects on water balance, forests may also have important effects on the *quality* of intercepted water, which will be considered under acid rain in section 8.1.

3.2 Infiltration and soil moisture

The infiltration of water into the soil is perhaps the most pivotal process within the drainage basin. Most agricultural crops and natural plant species depend upon infiltrated water. Most non-infiltrated water is either quickly evaporated or else drains rapidly into streams, arguably providing the bulk of storm runoff (section 3.3). In general, the process of infiltration 'condemns' water to a role of movement some two orders of magnitude or more slower than non-infiltrated water.

It is possible to write an equation for the water balance of a block of soil thus (Figure 3.5):

$$\Delta SMC = f + c - d - ET + v + TFLO_{in} - TFLO_{out}$$

in which ΔSMC is the change in soil moisture content, f the infiltration rate, c the rate of addition of water raised by capillary attraction from the water table, d the rate of drainage to the water table, ET evapotranspiration losses, v changes in vapour content (small enough to ignore), and $TFLO$ is lateral seepage (throughflow). Apart from the throughflow terms (discussed in section 3.3.4), this is essentially as defined by J R Philip when he propounded his influential theory of infiltration.

Interest in infiltration processes was first stimulated by R E Horton's theory of runoff (section 3.3.1), which identified its crucial role in dividing rainfall between storm runoff and groundwater seepage. According to Horton's theory, the hydrologically *effective rainfall* is all the water that remains on the surface after infiltration and any losses in surface detention, interception and evaporation (Figure 3.6). Despite some erosion of their key position by proponents of throughflow processes (sections 3.3.4 and 3.3.5), infiltration processes remain a major component in physically based runoff models (section 6.3) and the refinement of infiltration models continues. Much recent research has concentrated on the spatial variability of infiltration processes, especially on the effects of larger voids or macropores.

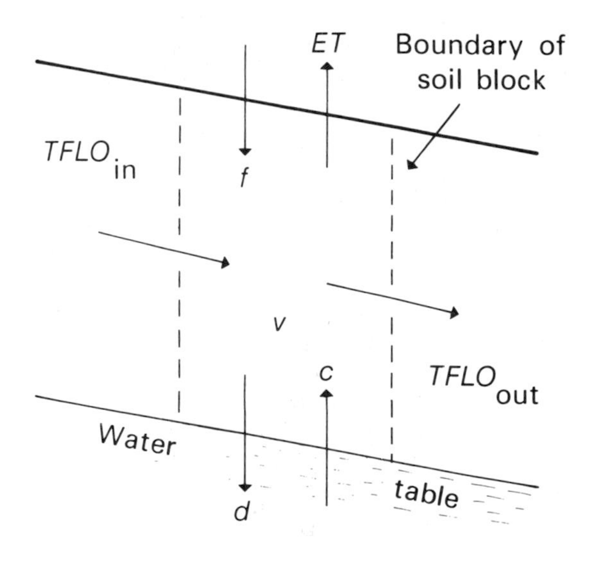

Figure 3.5 Water balance components for a block of soil

It is important to distinguish between infiltration rate and infiltration capacity. An **infiltration rate** is the volumetric rate of seepage into the upper surface of the soil, expressed as mm^3 per mm^2 of surface (i.e. mm) over a period of time, e.g. mm s^{-1}. Rates are highly variable, depending on the soil and its current water content; typically higher in a dry soil and reducing during rainfall. In contrast, the **infiltration capacity** is more stable and may be treated as a constant parameter of a given soil type. It is the maximum rate at which a given soil can absorb water *when it is in a specified condition*. The most meaningful definition is the **saturated infiltration capacity** (Figure 3.6), which Horton termed the limiting infiltration capacity, 'the relatively low and steady infiltration rate in a soil surface free from cracks and major holes, that has been wetted to saturation point for long enough to permit full swelling of clays and colloids and full adjustment of soil structure to a stable saturated state'. This definition has a lot in common with Penman's definition of *PET* in that it aims to produce an easily measured, stable and reproducible value, while making simplifying assumptions (in this case about cracks) that may limit its applicability in some circumstances.

Saturated infiltration capacities vary with soil properties, vegetation, soil faunal activity, land management, slope, geomorphic location and even temperatures. Soils with **massive** structure, with no cracks or substantial voids, are least porous, whereas **blocky** or **prismatic** structure favours infiltration along vertical cracks or pores (Figure 3.7). **Rainpans** or **splashpans**, formed by raindrops dispersing soil aggregates and washing the silts and clays into the soil pores, have very low infiltration capacities unless they crack (Römkens *et al.* 1990). The acidity or base status of the soil can also affect infiltration. Neutral to moderately alkaline soils tend to have the best **crumb** structure (Figure 3.7), as a result of the binding action of calcium and magnesium. The better drainage and nutrient status in these soils also attract soil fauna and

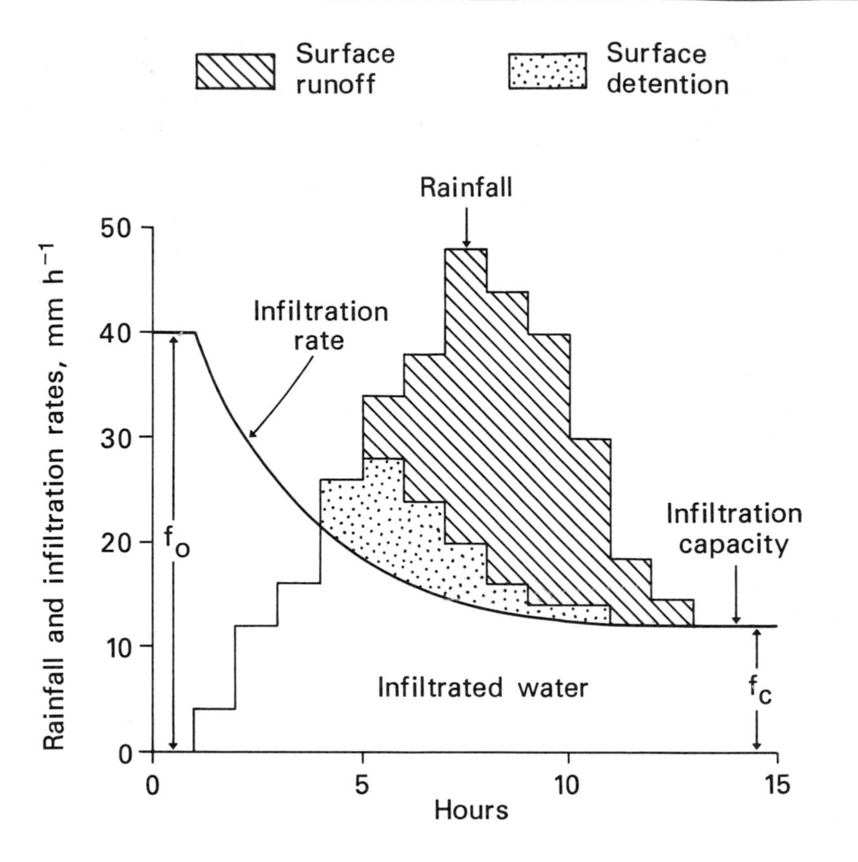

Figure 3.6 The role of saturated infiltration capacity and Horton's 'effective rainfall' in generating surface runoff

micro-organisms that create or maintain open pores and can excrete colloids that bind the soil particles. Earthworms are extremely important in this respect, especially the large *Lumbricus terrestris* (Bonell *et al.* 1984). Conversely, low pH and low temperatures can result in peat accumulation. Infiltration measurements on peat generally indicate very low capacity, e.g. 0.0001 mm s^{-1} (Jones 1990), but, like clay, peat is **histic**, i.e. it expands and contracts quite noticeably during wetting or drying, causing cracking. In the presence of substantial cracking the common definition of saturated infiltration capacity, as given by Horton, becomes somewhat irrelevant, since most water is likely to enter via the cracks.

Vegetation can also increase infiltration by creating cracks around the stem or trunk, and rootholes, or simply by protecting the soil surface from rainsplash and sun-baking. Even infilled dead stump holes, **pedotubules**, can still have higher infiltration capacity (Gaiser 1952; Brewer and Sleeman 1963). By increasing surface roughness and holding up surface water drainage, vegetation and forest litter also increase the total amount of infiltration, and by transpiring and drying the soil it will tend to increase initial infiltration rates. In general, the effects of vegetation on infiltration capacities far outweigh the effect of interception losses (Figure 3.8).

Both soil and vegetation can be sensitive to land management practices. Different crops and agricultural practices can create greater variability in infiltration capacities than differing natural vegetation. Ploughing is intended to improve infiltration and drainage, although some soils may develop impermeable **plough-pans** due to the washing down, or **eluviation**, and accumulation of fine particles just beneath plough depth. Heavy machinery, tramping animals, and over-exposure of bare soil to heavy rain cause compacted surfaces. Classic data from Nebraska show capacities 450 per cent higher under alfalfa (lucerne) than in bare, naturally compacted earth in the same silt loam soil. Data presented by Musgrave and Holtan (1964) show that silt loams with high organic content have up to 500

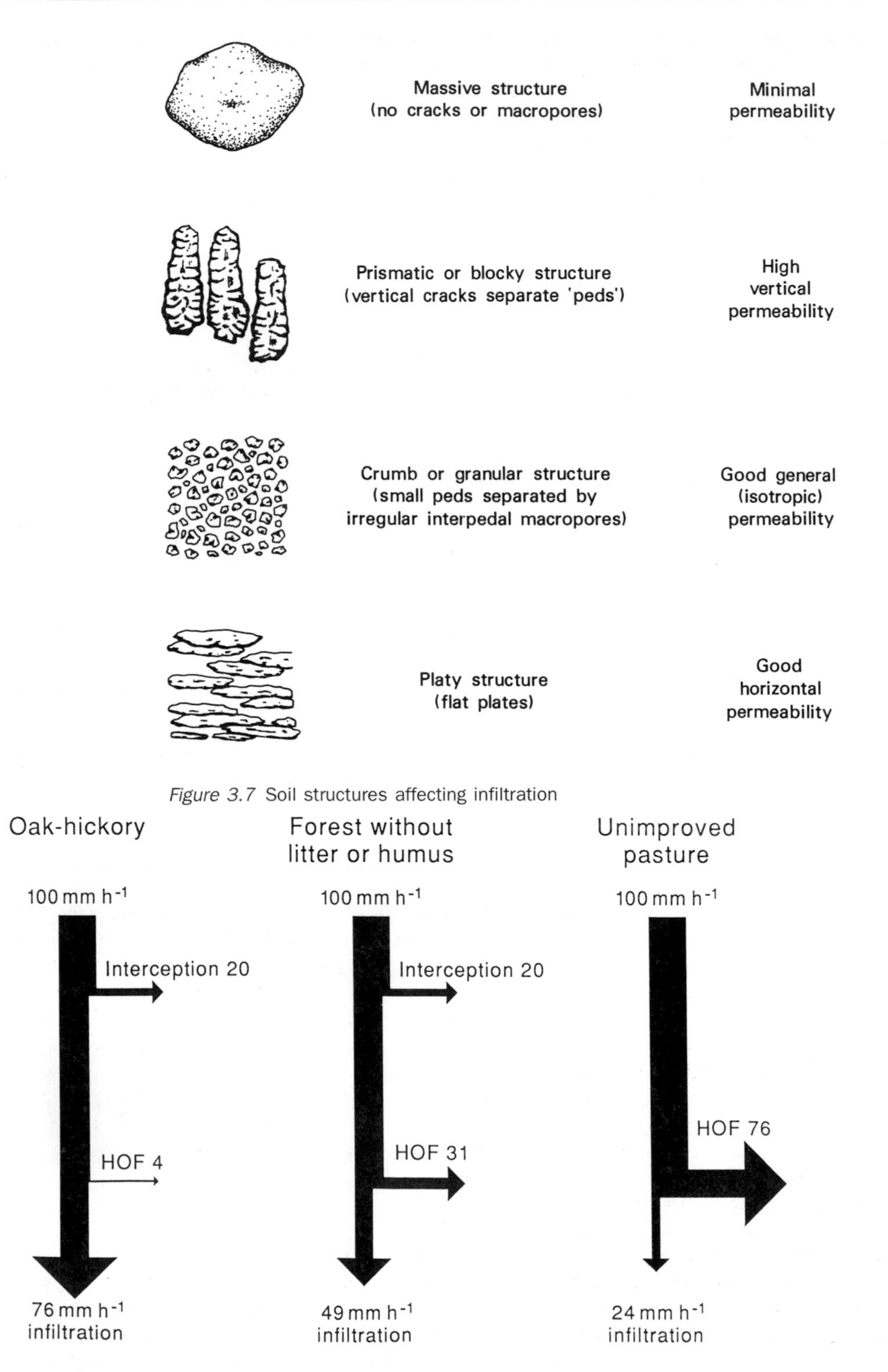

Figure 3.7 Soil structures affecting infiltration

Figure 3.8 The effect of vegetation on infiltration capacity and runoff. HOF is Hortonian overland flow (section 3.3.1). Based on data from Miller (1977)

Table 3.3 Examples of saturation infiltration capacities for different vegetation cover, soils and agricultural management (mm h^{-1})

Soils, season, etc.	Agricultural treatment		Source
1. Interaction between soil type and agricultural treatment			
	Bluegrass pasture	Cornfield	
Silt loam with high organic content	27.3	6.8	Midwest USA (Musgrave and Holtan 1964)
Silt loam with low organic content	8.3	6.5	
2. Seasonal effects of mulching			
	Cornfield, unmulched	Cornfield, mulched	
June	22	42	Midwest USA (McCalla and Army 1961)
October	6	39	
3. Effects of grazing on different soil types			
	Rangeland, grazed	Rangeland, ungrazed	
Coarse soils on sandstone	31–33	42–55	Colorado (Lusby *et al.* 1963)
Fine soils on shale	14–21	14–19	
4. Effects of afforestation			
	Old unimproved pasture	Old oak–hickory forest	
	43	76	SE USA (Lull 1964)

per cent higher infiltration capacities under bluegrass pasture than under corn, compared with only a 200 per cent difference when organic content is low (Table 3.3). Mulching with straw can reduce soil compaction in arable fields. In the new zero-tillage methods of cultivation, which are designed to reduce soil erosion, mulching can be essential for maintaining infiltration rates.

Slope and topographic location affect both surface and subsurface drainage. However, these effects tend to get confounded with those of soil type, since soil drainage is an important factor in the development of different soil types. It affects chemical processes by controlling aeration as well as the eluviation and deposition of finer mineral and organic material. More important than the actual angle of the slope is its geomorphic context. Hence, for a given slope angle, infiltration rates are likely to be higher on convex or 'shedding' slopes than in concave or 'receiving' sites, because of better soil drainage at shedding sites. This is due partly to higher hydraulic gradients within the soil and partly to coarser grained soils in shedding sites. This generalization is supported by the observed patterns of saturation overland flow, which are indicative of zero infiltration capacity (section 3.3.2). Thin or crusted soil or peat can, however, create soils of low permeability on upper slopes.

3.2.1 Infiltration theories

The processes of infiltration are extremely complex, often involving changes in soil structure as well as water content and gradients in water pressure within the soil. No attempt to quantify them can consider all aspects and there is a range of models available that make varying degrees of simplifying assumptions.

Horton's model describes the commonly observed pattern of decaying infiltration rates shown in Figure 3.6 by fitting a smooth mathematical curve of exponential decline, without considering the processes involved:

$$f_t = f_c + (f_0 - f_c)e^{-kt}$$

where f_t is the infiltration rate at a given time t from the start of the storm, f_c the saturated infiltration capacity, f_0 the initial rate (at time $t = 0$) and the final part is the exponential weighting in which k can be regarded as a constant for a particular soil. In effect, this parameter k subsumes the total effect of all the physical processes of soil packing, swelling, pore blocking and cessation of microfaunal activity that reduce infiltration rates as the soil becomes wetter. Horton (1933) assumed that the soil is freely drained, that water is ponded on the surface and that there is a constant intensity of rainfall.

Green and Ampt devised a formula in 1911 that gave more consideration to the processes. Their formula has been found to apply very well to simple sands with a narrow range of pore sizes and has enjoyed some popularity among computer modellers in recent years. It identifies two separate components, conduction and diffusion, which are driven respectively by gravitational force and by capillary attraction caused by surface tension between the water and the walls of the finer soil pores. **Conductivity flow** (A) is seen as a steady rate of flow caused by a potential gradient, i.e. a gradient in potential energy between the wet or flooded surface and the drier soil beneath. **Diffusivity flow** (B) occurs at the wetting front as it advances into the soil (Figure 3.9). Here the water diffuses through the drier soil. The formula is:

$$f_t = A + B/S_t$$

where S_t is the volume of water stored in the depth of soil saturated by infiltrating water at time t. A and B are constants for a particular soil in a given moisture state.

A more realistic model for normal soils would include a provision for a variety of pore sizes. The Green–Ampt formulation applies to smaller pores, but capillary forces are not important in larger pores.

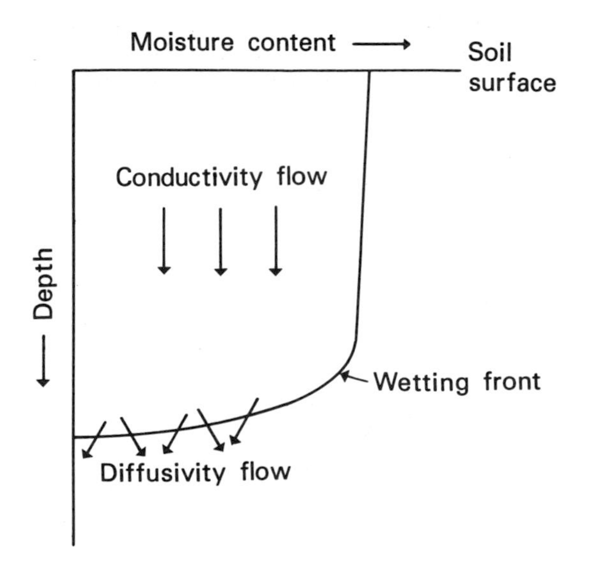

Figure 3.9 Conductivity and diffusivity flow during infiltration

Kirkby (1985) has shown how it is possible to define weighted values of A and B which take into account the proportion of pore space in capillary and non-capillary pores.

The classic formulation produced by J R Philip in the 1950s extended the Green–Ampt approach by allowing for a deceleration of diffusivity flow over time as the wetting front approaches the base of the soil or the water table, the zone of long-term saturation. The basic Philip equation is:

$$f_t = A + (B/2)t^{-1/2}$$

Philip's model offers a better prediction of infiltration rates than Horton's at the start of a storm or for brief periods of infiltration. For longer periods, however, he found it necessary to employ a more complicated approach using the Richards equation, which is based on Darcy's Law of seepage (section 3.3.4). Useful discussions of this approach are provided by Childs (1967), Knapp (1978) and Kirkby (1985).

Despite continuing advances in infiltration models, the application of current models is limited by two problems: (1) accounting for the spatial variability in infiltration capacity; and (2) errors and uncertainties elsewhere in modelling the basin system. Infiltration capacities can vary considerably within a basin, even over very short distances. Values have been shown to vary by over three orders of magnitude in just 130 m, even when measurement points have been selected to avoid obvious cracks (Jones 1990).

Beven and Germann (1982) drew special attention to the effects of macropores, which tend to cause water to bypass the soil matrix. They proposed that both infiltration and subsequent seepage should be viewed as occurring in two largely separate 'domains', the **matrix domain** and the **macropore domain** (Figure 3.10). Four types of macropore can be distinguished: cracks and planar voids, cylindrical holes (mainly biotic), vughs (between these two types) and water-sculpted soil pipes (section 3.3.4). The resulting spatial variability in infiltration has important implications for both modelling and monitoring (Chapters 5 and 6).

Some tests have indicated that there is very little to choose between the available infiltration models when they are used to calculate a single infiltration value for a basin within a runoff prediction model (Peschke and Kutile 1982). This implies either that some account needs to be taken of spatial variability in infiltration or that improvements are needed in the description of processes further down the cascading system of runoff, or both.

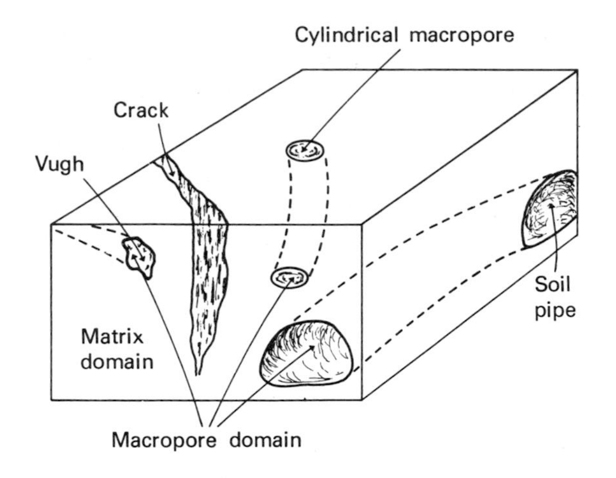

Figure 3.10 Macropores and matrix infiltration domains. After Beven and Germann (1982)

3.2.2 Soil moisture

The amount of water stored in the soil prior to a rainstorm tends to be an important factor in the stormflow response of a basin. Wetter soils encourage more overland flow or direct runoff, and the larger the zone of soils at or near saturation in hollows and along streambanks the quicker will be the stream's response to rainfall and the greater the likelihood of flooding.

Considering its importance, soil water has been remarkably poorly understood in hydrological terms until quite recently. Traditionally, the zone of soil water has been seen as predominantly a **zone of aeration**, i.e. of unsaturated soil with pores that normally contain more air than water. Input has been taken as due to local infiltration and losses as due to evapotranspiration or vertical percolation to a **zone of saturation**, permanently resident at some depth commonly assumed to be in bedrock or parent material rather than the soil. The upper surface of this zone of saturation is commonly called the **water table** and seen as intersecting the streambanks and supplying baseflow to the stream, a process known as **effluent seepage**. Normally the **capillary fringe**, a narrow zone in which water can rise by capillary attraction above this water table, is the only part of the soil seen to contain any of this groundwater (section 3.4), and then only adjacent to the stream (Figure 3.11).

In reality, the water table can exist more or less permanently within the soil at low-lying sites, as proven by what soil scientists call groundwater gley soils, bluish anaerobic soils in a state of chemical reduction. Perhaps more importantly, layers (horizons) with reduced permeability within the soil or at its base can create **perched water tables,** as evidenced by 'surface water gley' soils. This saturation of the soil does not have to last long enough to have a permanent effect on soil properties for it to be hydrologically significant. Horizons saturated only for the period of storm drainage are the main source of throughflow discharge to the stream (Figure 3.11).

In general use, the term **soil moisture** refers to the water content of the zone of aeration and essentially relates to storage, whereas movement, especially lateral drainage, is better considered as **throughflow**. It is also worth while looking carefully at two terms commonly supposed to apply to critical thresholds in soil moisture content: the field capacity and the wilting point. Each has been regarded as constant for a particular soil, although this is really an oversimplification.

Field capacity was defined by Horton (1933) as the maximum water content that a soil can hold in a free draining situation. Hence, it is argued that field capacity must be reached before drainage occurs out of the soil column towards the water table. However, there is generally some vertical drainage loss well before field capacity is reached and no marked change in the rate of drainage when it is reached. This is because soils normally contain some macropores which conduct gravitational drainage within them relatively freely, largely irrespective of the moisture content of the surrounding matrix. Gravitational flow may be lateral as well as vertical and it may be in the form of unsaturated flow (section 3.3.4). The concept is probably more valid for drainage following saturation, where a marked reduction in the release of water often

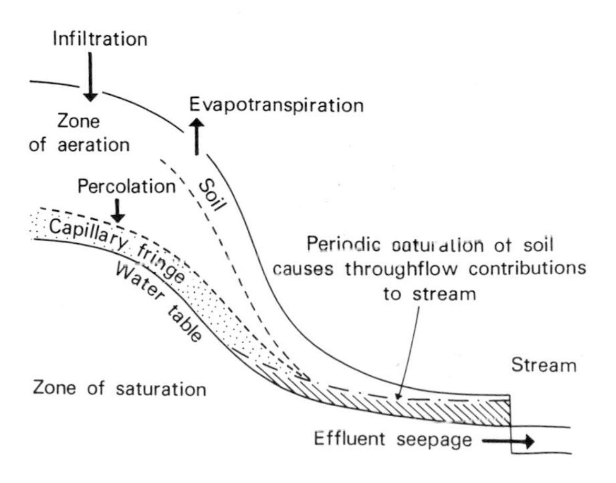

Figure 3.11 Zones of soil water storage and storm response

occurs. Even so, field capacities vary over time, depending on temperatures and biotic activity.

The concept of field capacity is particularly important as the source of the parameter **soil moisture deficit (smd)**, which is the amount needed to raise the moisture content to capacity. This is commonly calculated as a means of assessing irrigation requirements and is used in some riverflow models in order to estimate the yield of direct runoff from the hillslopes. Despite its theoretical limitations, it has generally proved to be of practical value. Nevertheless, obtaining a spatially representative value for either field capacity or soil moisture deficit can be as problematic as for infiltration capacity.

Wilting point stands at the other end of the water content spectrum. It is the moisture content at which plants exhibit permanent wilting, i.e. the suction needed to remove the remaining water from the soil is too great for the plants to exert. This will, nevertheless, differ from species to species. In any case, it is not important hydrologically, except perhaps as an indicator of the onset of desiccation cracking in some soils, which will dramatically increase their macropore space.

Moisture-holding forces Soils generally consist of three components – a skeleton of more or less coarse mineral particles, clays and colloids finer than 2 μm in diameter that coat and bind the skeleton, and an assortment of living and dead biotic matter. The clays and colloids give *structure* to the soil, creating aggregates of particles and larger inter-aggregate drainage spaces. Together with the organic matter, they also retain moisture.

Moisture is held by means of absorptive, adsorptive, osmotic and capillary forces. **Absorption** involves uptake of moisture into a body. Clay minerals absorb water into their crystalline structure, causing expansion: this water is then generally unavailable to plants and is lost only by evaporation. In contrast, **adsorption** involves a thin film of water molecules on the surface of particles held either by weak electrical charges or by the chemical transfer or sharing of electrons. **Osmotic forces** occur where water is drawn into the soil aggregates by a saline solution. By and large, these are only significant in alkaline or saline soils. In contrast, **capillary forces** are more widespread and significant. They tend to dominate adsorptive forces in wetter soils. Capillary forces are due to surface tension in fine pores and their strength increases as the radius of pores gets smaller.

For hydrological purposes, capillary forces are the main forces and the total force needed to remove water from the soil has been called simply the **capillary potential**. Perhaps a better term for this is **matric potential**, i.e. the total suction force that must be exerted in order to remove the water from the soil matrix. Matric potential increases as soil moisture content gets less. Because it rises rapidly at low water contents, it is normally measured on a logarithmic scale, the pF scale (Figure 3.12). If the suction potential is viewed as a negative pressure equal to the force exerted by a column of water x cm tall resting on 1 cm^2, then the pF is $\log_{10}$ of the height of that column in centimetres.

A saturated soil matrix, with only the macropores drained, would have a p$F = 0$. Drier soils have a p$F > 0$ and are said to be in *tension*. A freely drained soil at field capacity will have a p$F = 2$. In contrast, in a fully saturated soil, where all pores are filled with water, there is a positive water *pressure* and the pF, somewhat confusingly, is by definition negative. In practice, the exact relationship between water content and pF varies from soil to soil, depending on the specific mix of moisture-holding forces and the porosity, i.e. the size and number of pores (Figure 3.12).

The hydrologist is generally interested in soil moisture for two reasons: (1) because of its importance in providing an environmental setting for the transformation of precipitation into runoff; and (2) as a potential contributor to runoff itself. The patterns and gradients in soil moisture are important controls in the generation of both surface and subsurface runoff, as will be explored in section 3.3.

3.3 Runoff-generating processes

Streamflow is generated from a variety of sources. Conventionally, groundwater held within the bedrock is the main source of low flows, while surface flow and possibly shallow subsurface flows are the main sources of stormflow. Both surface and shallow subsurface sources are now known to consist of a range of processes from infiltration-excess overland flow, saturation overland flow and rillflow to saturated and unsaturated diffuse throughflow, macropore flow and pipeflow (Figure 3.1), as described in subsequent subsections. Groundwater will be discussed in section 3.4.

These processes occur in differing strengths and combinations from one basin to another, depending on climate, bedrock, relief, soils, vegetation and land use. Indeed, the variety has been said to be 'bewildering' (Whipkey and Kirkby 1978).

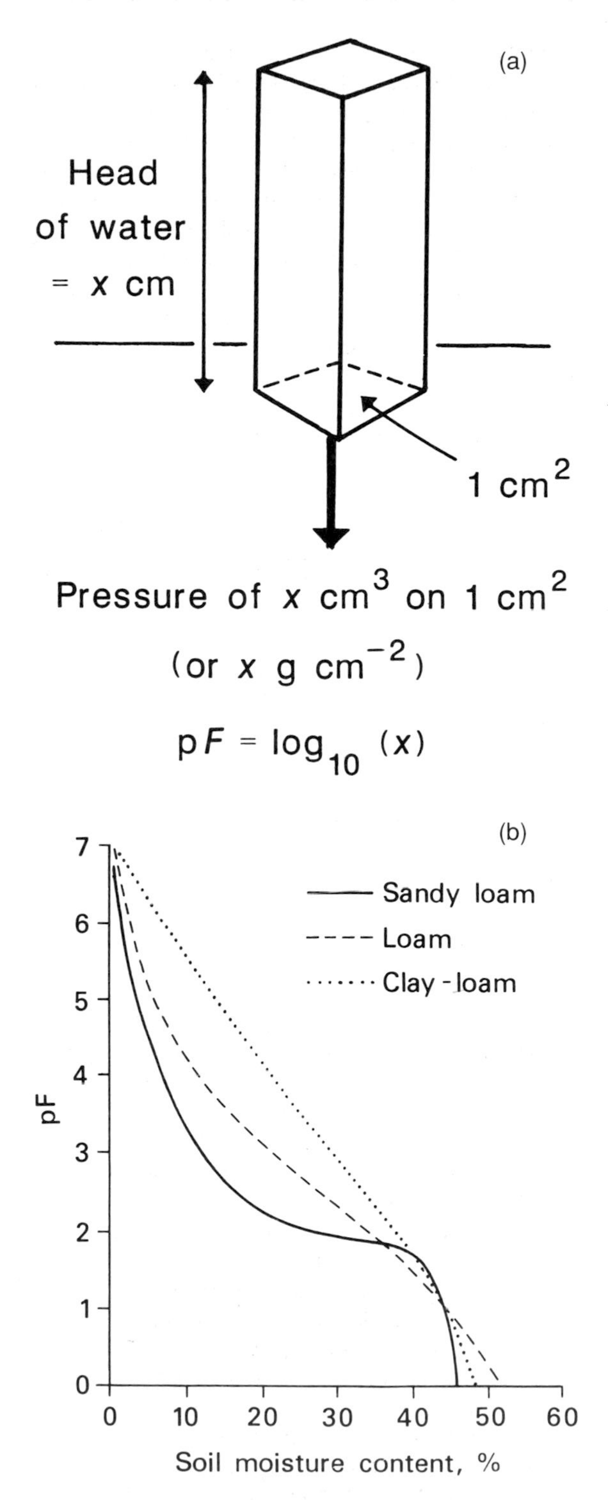

Figure 3.12 Matric potential and the p*F* scale: (a) explanation; (b) examples of response in selected soils

Considerable progress has been made in determining the environmental factors that control the geographic distribution of individual processes. Yet there are still major gaps, especially in relation to subsurface processes and the ability to predict the mix of processes in many environments.

3.3.1 Infiltration-excess overland flow

The first scientific theory of storm runoff was published in 1933 when the American hydrologist Robert E Horton proposed his theory of **infiltration-excess overland flow**. The theory seemed to offer a scientific basis for one of the most influential 'black box' models of river runoff, Sherman's (1932) unit hydrograph (section 6.1.2).

Horton proposed that the process of infiltration irrevocably divides precipitation receipts into **overland flow**, which fails to infiltrate the soil and flows rapidly over the surface to generate storm runoff or **quickflow** in the streams, and infiltrated water, which is either lost via evapotranspiration or percolates down to a body of **groundwater**, and then seeps slowly downhill to sustain low flow or **baseflow** in the streams (Figure 3.13).

Water fails to infiltrate the soil when rainfall intensities (or snowmelt rates) exceed the infiltration rate. The term **delayed Hortonian overland flow** has been used for the case where this occurs only late in the storm due to changes in soil structure caused by wetting or raindrop impact (Gerits *et al.* 1990). However, it is normal for the final saturated infiltration rates to be less than initial dry rates in most soils. Initially, such water will pond up on uneven surfaces until the depth is sufficient for ponds to merge and **detention storage** is succeeded by downslope overland flow. Horton envisaged that this flow commences uniformly over the whole of a basin and that it gradually increases in depth downslope as flow concentrates the water in the valleys.

He later proposed that this increasing depth would lead to a point at which non-erosive laminar flow is replaced by erosive turbulent flow, and that this would cause the initiation of rills, gullies and eventually stream channels (Horton 1945). The supposed point of this transformation, Horton's x_c **distance**, has been highly influential in geomorphological theory. Despite some intriguing laboratory evidence for such rill initiation (Moss *et al.* 1982), however, there are numerous inhomogeneities on natural hillslopes and competing processes that tend to make the x_c distance more hypothetical than real (Jones 1987a).

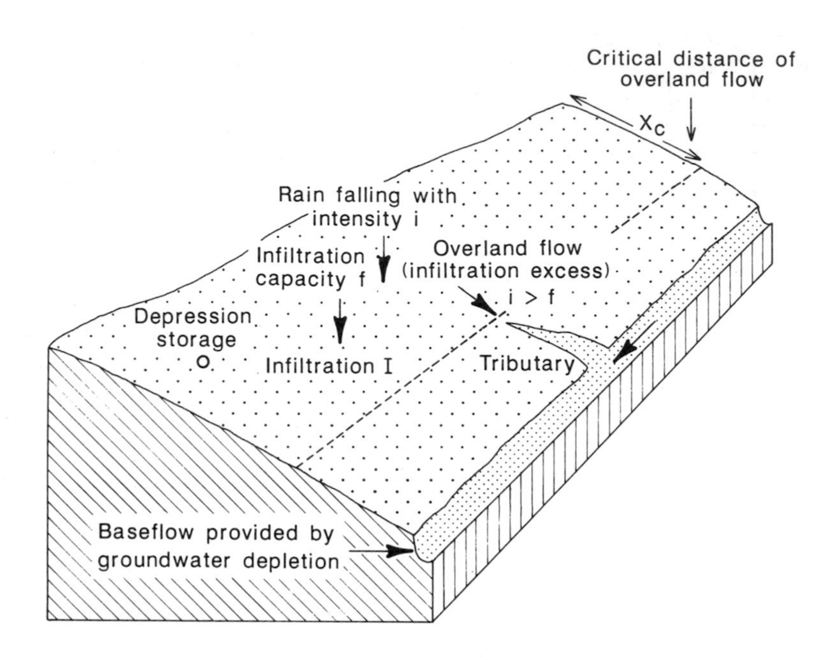

Figure 3.13 Horton's model of storm runoff

Horton's theory of runoff generation is best viewed as an end-member in a series of possible combinations of sources of discharge. A troublesome fact that has been reported throughout this century is that widespread overland flow is rarely observed in humid vegetated catchments. One explanation is that rainfall intensities are commonly less than the infiltration capacities of vegetated soils. It is also likely to occur instantaneously over a whole basin only if the area is small and has uniform soil and vegetation throughout. Thin soils, low infiltration capacities and sparse vegetation favour Hortonian overland flow.

It has been widely observed on bare badland hillslopes, for example, by Schumm (1956) in New Jersey, Bryan *et al.* (1978) in Alberta and Yair and Lavee (1985) in Israel. However, as Yair and Lavee (1985) observe, the relatively small size of convective rain cells in arid areas tends to limit surface runoff to only part of the catchment and sometimes not even the whole of a single slope. Moreover, both Bryan *et al.* (1978) and Yair and Lavee (1985) present experimental field data that indicate the importance of subsurface runoff. Arid lands are not the sole preserve of Hortonian overland flow as was believed until recently.

Horton's theory, especially as a justification for the unit hydrograph method, has formed the basis for many computer models of stormflow generation (Chapter 6). It has also been a valuable stimulus to research on infiltration processes and the mechanics of overland flow and soil erosion.

3.3.2 Saturation overland flow

Cook (1946) suggested that overland flow can be created by subsurface flow that resurges on to the surface. This **return flow** occurs over only *part* of the catchment, typically on lower slopes where soil water seepage has developed higher soil moisture levels prior to the storm and where groundwater levels may be nearest the surface. Field experiments by Kirkham (1947) confirmed the process.

It was not, however, until the 1960s that Horton's model began to be seriously challenged. The challenge came largely from forest hydrologists in the USA, who as a group had fairly consistently doubted the generality of Horton's theory. Experiments on subsurface stormflow at the US Forest Service Experiment Station at Ashville, North Carolina, by John D Hewlett were eventually very influential in proving the importance of return flow (Hewlett 1961; Hewlett and Hibbert 1967; Hewlett and Nutter 1970).

At the same time, the concept of a **partial contributing area** was formalized by Betson (1964) and Betson and Marius (1969), and the Tennessee Valley Authority incorporated it into a hydrological model. The TVA recognized that areas contributing storm runoff are not only partial, but dynamic. **Dynamic contributing areas** (DCAs) expand during a storm as the saturated area adjacent to the stream channels is fed by seepage from upslope (Figure 3.14). When this supply is cut off after the storm, the DCA contracts again. Moreover, the areas and dynamics vary from storm to storm, depending on pre-storm moisture distribution and storm intensity, hence Hewlett's term **variable source areas**.

Kirkby and Chorley (1967) used the term **saturation overland flow** to describe surface stormflow from the DCA, which may be comprised of return flow and/or direct precipitation that cannot infiltrate the saturated soil surface because the ponding effect of water already in the soil has created a near-zero infiltration capacity (Figure 3.14). They suggested that most overland flow is saturation flow. Field experiments by Dunne at the USDA Sleepers River watershed in Vermont finally established that saturation overland flow can be the dominant source of storm response in the stream (Dunne and Black 1970).

Debate has continued as to the relative contribution of subsurface sources either to saturation overland flow or direct to the channel. In practical terms, saturation overland flow may produce a storm response or **streamflow hydrograph** indistinguishable from Hortonian overland flow and, provided the response changes approximately linearly with the size of storm, unit hydrograph techniques may model it satisfactorily (section 6.1.2). Even so, the dependence of saturation flow upon topography and upon pre-storm conditions that are spatially and temporally variable offers the distinct probability that responses will be nonlinear. In addition, the pathway of return flow through the soil is likely to alter the solute content of the runoff, e.g. the take-up of nitrates from agricultural fertilizers, which may have important effects on water quality (section 8.2.2).

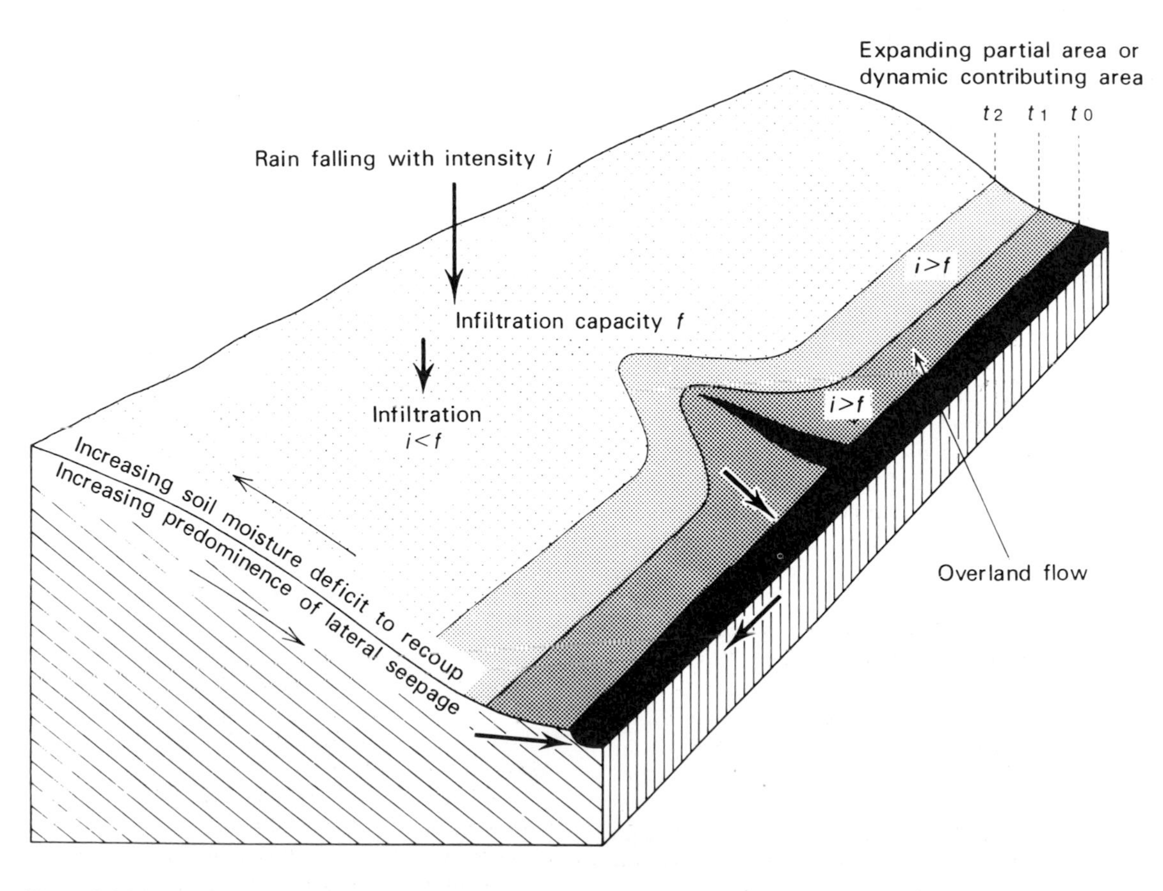

Figure 3.14 Hewlett's model of storm runoff

3.3.3 Spatial patterns of overland flow

Horton's (1945) model led him to define a **mean distance of overland flow** (MDOF) on simple geometric grounds. Since the average distance between streams is given by dividing the basin area (A_b) by the total length of streams $\Sigma(L_s)$, overland flow which covers the whole basin must travel half this distance between crestlines and channels:

$$\text{MDOF} = \frac{1}{2}\frac{A_b}{\Sigma L_s} = \frac{1}{2D_d}$$

where D_d is the drainage density. The MDOF has been used frequently in American models for predicting streamflow (section 6.1).

However, equations derived by Kirkby (1978), which cover both Hortonian and saturation overland flow generation, indicate that Horton's formula is only likely to be valid in a dryland climate with impermeable surfaces. As annual rainfall, vegetation cover, infiltration capacity and soil storage capacity increase, so the pattern of overland flow becomes increasingly sensitive to topography. Shallow swales or hollows often carry most saturation overland flow on vegetated hillslopes (Figures 3.15 and 3.19b).

Kirkby and others have developed the concept of the ***a/s* index**, in which the accumulation of water by any hillslope drainage process can be related to *a*, the area drained per unit contour length (Figure 3.16), and the gradient of the slope, *s*. Points in the landscape on which flow converges have larger areas upslope that contribute flow to them, so that where the contours form a hollow more water drains across every metre unit of contour than elsewhere. Conversely, steeper slopes mean less accumulation.

The most important factor determining the sensitivity of overland flow to topography is probably the ratio between the mean rainfall per day and the mean infiltration rate. Figure 3.17 shows how areas where there is greater sensitivity to Kirkby's topographic index have higher infiltration rates and lower rainfall/infiltration ratios. In the semiarid environment of Arizona, overland flow is insensitive to topography and typically Hortonian, because infiltration rates are near zero and rainfall is infrequent but intense. It is more sensitive in the humid UK and eastern USA, where initiation tends to require convergent flow. The convergence is predominantly in subsurface seepage, which reduces infiltration rates and may add return flow. Figure 3.17 suggests that in the lower rainfall areas overland flow will occur for only a day or more a year in definite hollows (*a/s* > 1.0 km). Kirkby also showed that the number of days with overland flow is controlled by the ratio of the storage capacity of the soil to the mean rainfall per rainday. Typical storage capacities under forest (100 mm), grass (40 mm) and on bare ground (10 mm) help to explain the relative infrequency of overland flow under denser vegetation.

In general, therefore, the spatial pattern of overland flow generation in humid climates is controlled by **soil**

Figure 3.15 A seepage line, the site of linear surface flow during heavy storms

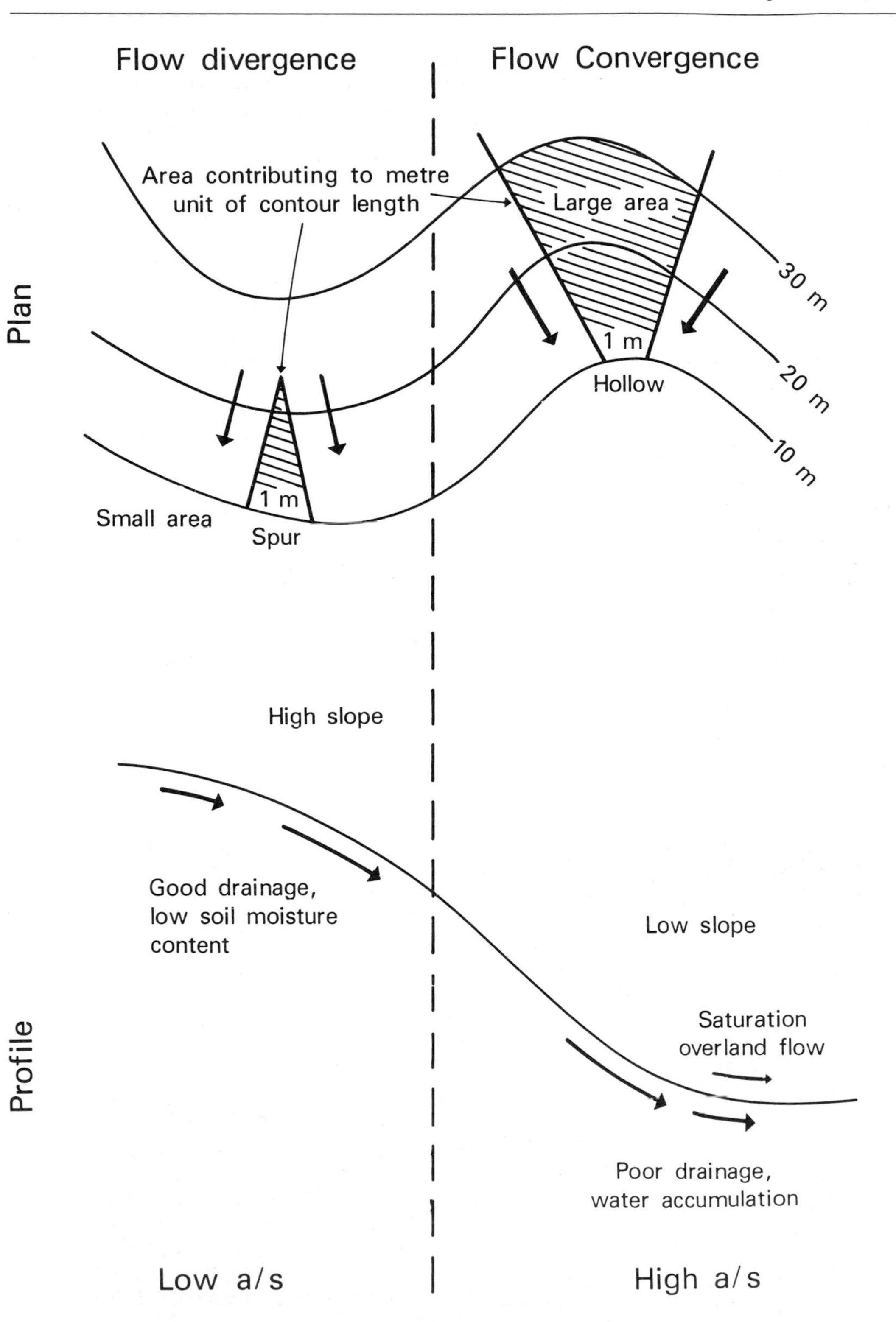

Figure 3.16 How topography concentrates flow: the derivation of the area drained over slope, a/s, index

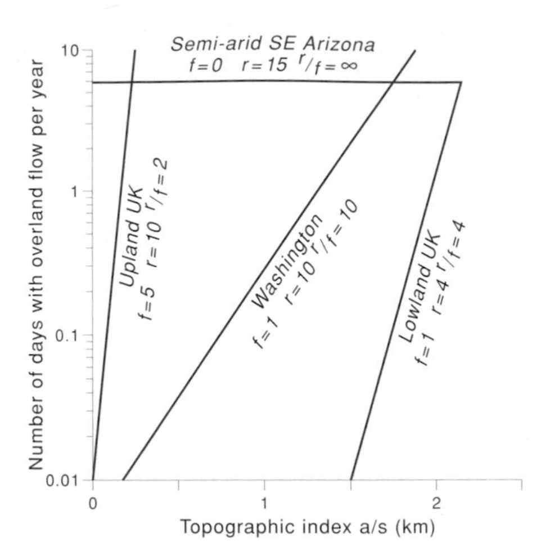

Figure 3.17 How the occurrence of overland flow is affected by topography in four climatic regions. Based on Kirkby (1978). (f = infiltration capacity r = mean rainfall per rainday

moisture distribution. Field studies by Anderson *et al.* (1984) have shown that temporal patterns are affected similarly, with hollows exhibiting the greatest variability in response and the longest recession tails (Figure 3.18).

Even in drylands, however, field experiments have shown that flow may occur on only part of the hillslope. Emmett's (1978) observations suggest that the classic uniform sheet of Hortonian overland flow is more the exception than the rule (Figure 3.19a). In addition, Yair and Lavee (1985) observed that frequent cracks in the desert crust create very discontinuous overland flow. This can cause drier surfaces and less overland flow near the base of the slope, the exact opposite of the situation in humid lands. They concluded that the *physical and chemical properties of the soil* are the most important controls in arid climates.

3.3.4 Throughflow processes

Throughflow is the predominantly lateral movement of water down a hillslope within the soil profile. It largely creates the patterns of soil moisture that are the primary controls on the generation of overland flow in humid lands. The term throughflow is more or less synonymous with the older term **interflow**, which is still used, though decreasingly, in the USA. However, it was introduced by Kirkby and Chorley (1967) in order to refer to the specific processes of lateral soil water movement, and to avoid confusion with the 'interflow' of hydrograph analysis, which may cover a variety of actual sources (section 6.1.2).

Throughflow processes can be divided into slow **diffuse flow** and rapid **macropore flow**. Diffuse flow passes through the fine pores of the soil matrix, and is easier to model than macropore flow, which follows a range of discrete passageways through the soil. It can be considered to be a response to a gradient in potential energy, known as a **head**. The total soil water potential, Φ, is effectively the sum of the gravitational potential, ψ_g, due to the elevation of that point above a reference datum (e.g. the bottom of the slope), and the water

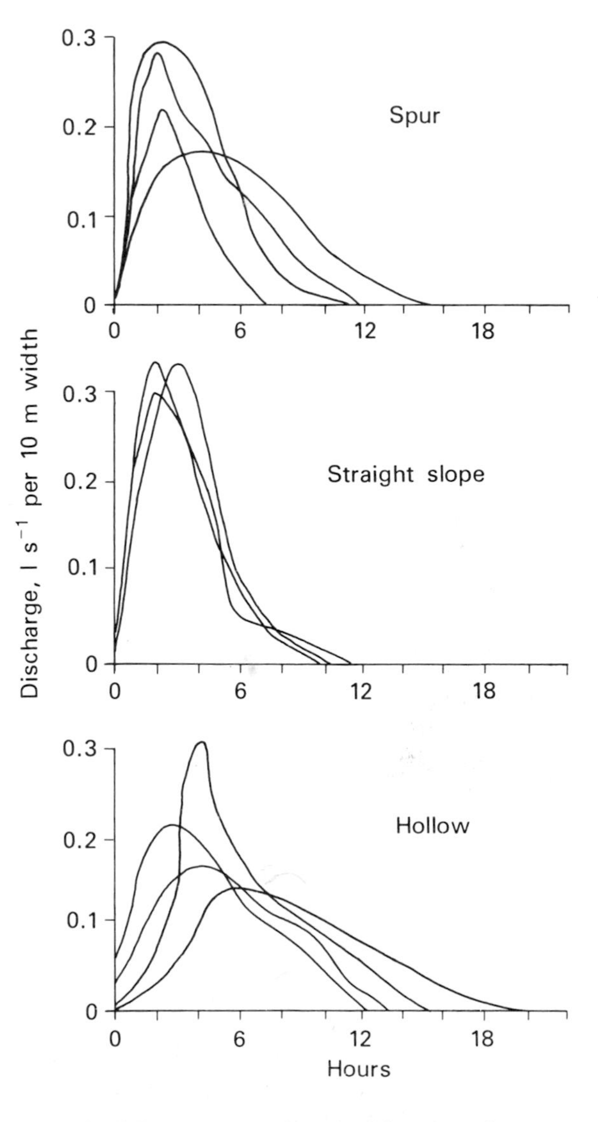

Figure 3.18 The response of overland flow to surface form during storms. Based on observations by Anderson *et al.* (1984)

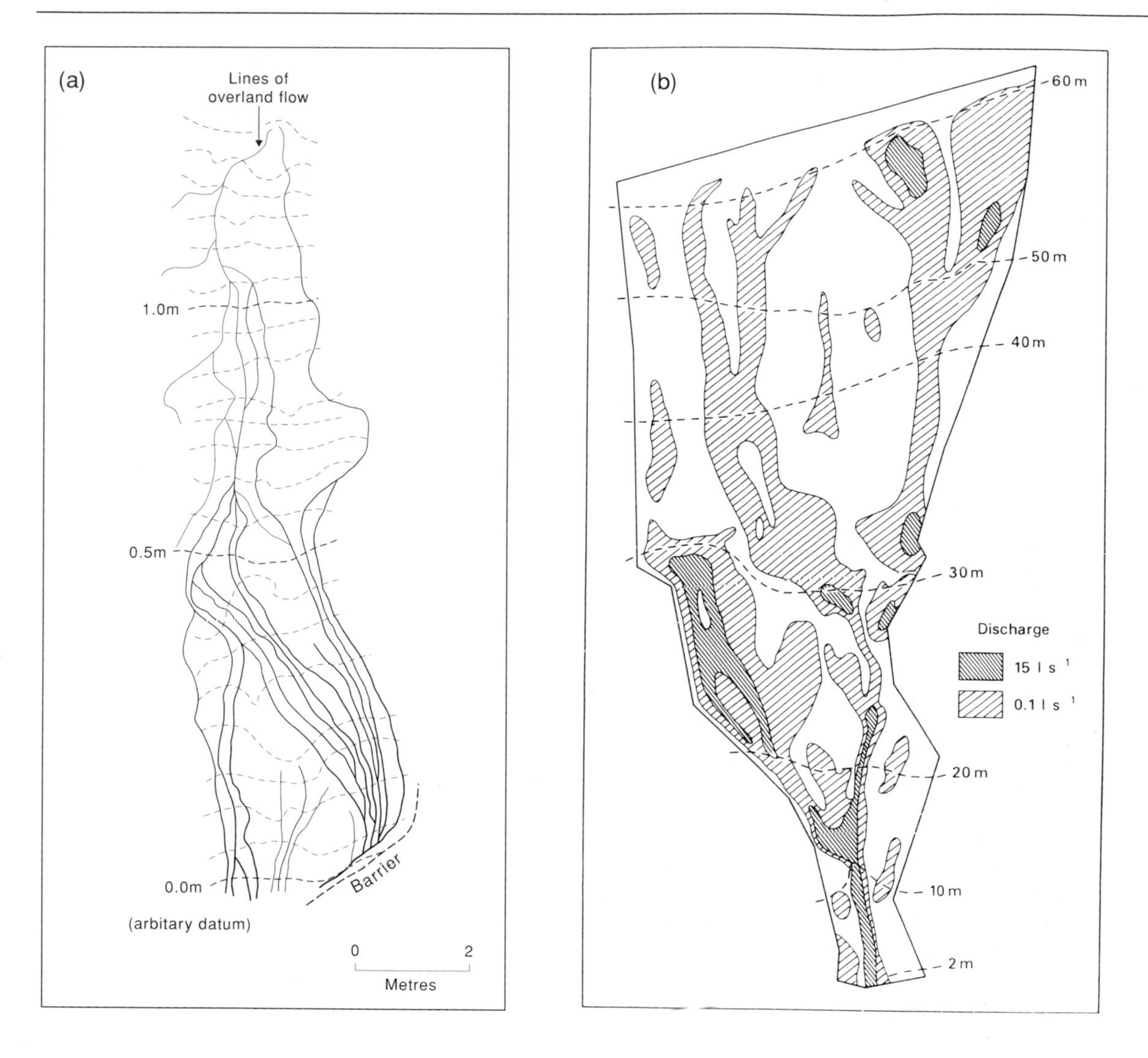

Figure 3.19 Patterns of overland flow: (a) Hortonian overland flow observed by Emmett (1978); (b) Saturation overland flow observed by McCaig (1984)

pressure potential, ψ_p, due to the water content of the soil:

$$\Phi = \Psi_g + \Psi_p$$

Soil water tends to migrate from areas of higher to areas of lower total potential. For lateral flow down a hillside, the gravitational potential dominates, so that water will move from a site at a higher elevation to a lower one even though both sites have the same water pressure potential. Hence at the basin scale, sites at lower elevations tend to be water-receiving sites and topography controls the processes of soil water redistribution. At a given site, however, it is quite possible for differences in water pressure potential to exceed the effects of the relatively small differences in elevation within the soil profile. Hence, water may move upwards within the profile if the surface is considerably drier than lower down, i.e. the water pressure potential is more negative at the surface.

It makes sense to consider both positive water pressure, in a saturated soil, and negative pressure or tension, in an unsaturated soil, as aspects of a continuum of water contents. However, it is well established in usage that we refer to the **piezometric potential** in the saturated state and to the **matric potential** (a negative potential) in the unsaturated state, which is rather confusing and partly arises from different methods of measurement (sections 3.2.2 and 5.5.1).

The movement of diffuse throughflow can be described by one of the oldest 'laws' in hydrology, **Darcy's Law**, which states that the discharge of water, Q, through a saturated cross-section of soil of area, a, is given by:

$$Q = kia$$

where the velocity is the product of k, the hydraulic conductivity of the soil, and i, the gradient in total soil water potential in the direction of flow through the cross-section. The **hydraulic conductivity** is the normal rate of transmission of water through the soil at a certain state of wetness. Darcy's Law was originally applied to flow in saturated soil and the saturated hydraulic conductivity is still the parameter normally used in the equation. However, Richards (1931) and others have extended the equation to cover flow in unsaturated soils, by allowing the hydraulic conductivity to vary with the water content. The nonlinear partial differential equations in the Richards formula are difficult to solve for anything other than a homogeneous soil on a flat slope. However, they are increasingly used in computer models to develop numerical solutions. The so-called **Richards Equation** includes provision for reciprocal changes in gradients, conductivities and flows over time, for example, during a storm or the subsequent drainage (cf. Ward and Robinson 1990), whereas Darcy's Law strictly applies to a fixed, saturated or '*steady state*'.

Spatial variability and the effects of topography
Darcy's Law and its extensions assume that the soil is uniform. It is possible to apply separate equations to different soil horizons, but it is still assumed that each horizon is laterally uniform and that flow occurs through a uniformly porous matrix like a sponge rather than through recognizable channels. This continues to serve as a very valuable and practical simplification.

However, most soils exhibit greater variability. Macropores create preferred flow routes in the majority of soils (Germann 1990). Recent research in Holland has also found **fingered flow**, i.e. flow in 'fingers' 10–40 cm wide, within the soil matrix in soils ranging from clay and peat to even homogeneous, wettable sands. This is proving important for studying pollution transport (section 8.1.2).

Moreover, the thicknesses and properties of the horizons vary laterally, creating an ordered sequence of

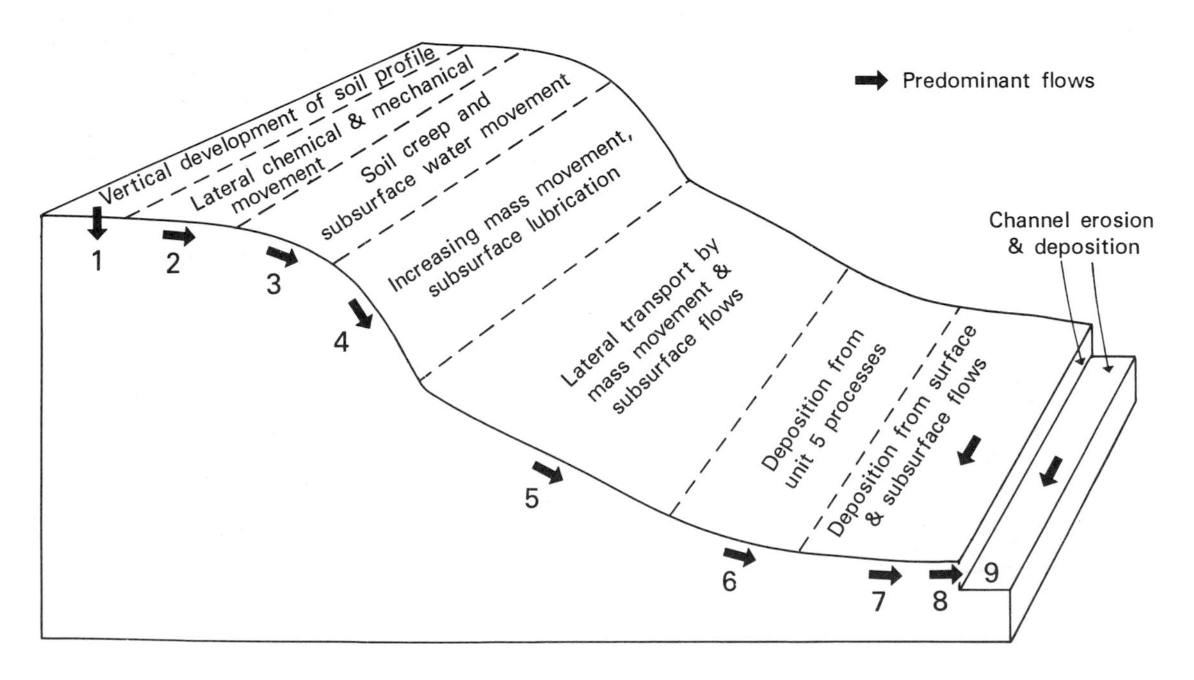

Figure 3.20 Interrelationships between soil water movement, soil-forming processes and geomorphic location as expressed in the NULM model by Conacher and Dalrymple (1977)

related but different soils known as the **soil catena**. In Figure 3.20 the catena concept is related to the processes of water movement that both follow and create soil variability in Conacher and Dalrymple's (1977) Nine Unit Landsurface Model (NULM). These downslope trends occur within the framework of a variegated three-dimensional landscape, in which 'hollows' or valleys cause convergent flow and crests or spurs cause divergent flow.

Thus a high *a/s* ratio should indicate the areas with convergent throughflow, highest soil moisture levels and greatest chance of saturation overland flow (Carson and Kirkby 1972). Anderson and Burt (1978) first presented instrumental proof of this relationship, and showed how the area of saturated soil in a hollow expanded and contracted during a storm in close correspondence with stream discharge (Figure 3.21). Modifications, particularly logarithmic transformations, of the original *a/s* formula have since been proposed to fit the observed spatial patterns better (Burt and Butcher 1986; Burt 1992).

Subsurface topography may also be responsible for flow convergence. In fact, water can be concentrated by the surface topography of any interface with lower hydraulic conductivity, whether a soil horizon or bedrock. Bunting (1961) introduced the term **percolines** to refer to lines of deeper soil running down the hillside like infilled channel networks (Figure 3.22). These may be invisible at the surface, but the concentration of flow eventually tends to erode slight surface hollows or **swales** and to transform their downstream parts into **seepage lines**, which frequently develop saturation overland flow.

Macropore flow Most soils contain a variety of pores that are much larger than the matrix 'micropores' envisaged in Darcy's Law or in more recent models of stochastic flow in porous media (Scheidegger 1974). These meso- and macropores have numerous causes, from simple gaps in the packing of the aggregates of soil particles to desiccation cracking and microfaunal activity. Pores with diameters greater than about 30 μm and that are also connected to similar or larger pores can materially assist drainage. They can also allow sufficient velocity of flow (1.0 mm s^{-1} and more) to cause erosion, enlarging and developing integrated, three-dimensional micro-drainage networks within the soil (Jones 1987a). Macropores that show the first signs of erosional smoothing have been given the rather unattractive name of **vughs**.

There is no universally accepted definition of a macropore. Hydrologically, the key characteristics are the ability to drain the soil rapidly and, possibly, to develop erosive turbulent flow. Beven and Germann (1982) defined them as voids in which the water is not held up by capillary forces. This would place a lower limit of about 4 mm diameter, but others have suggested values as low as 60 μm. Although standard definitions are based on size or capillary tension, the role of macropores as drainage channels clearly also depends upon their length, continuity and connectivity.

Modelling and monitoring macropore flow is very much a research frontier at present. There is increasing interest in macropore flow as a source of nitrate leachate from agricultural land and as an aggravating factor in the drainage of acid rain (Chapter 8). Germann (1990) has made the first valuable attempts to model

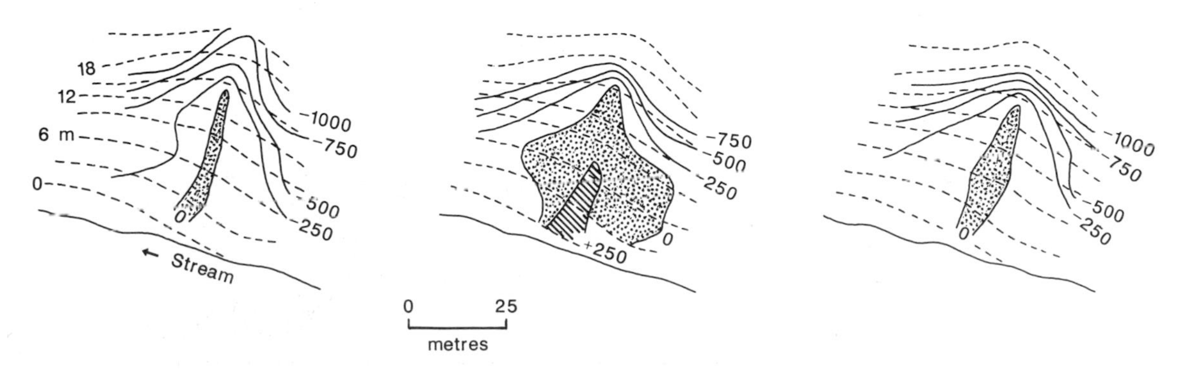

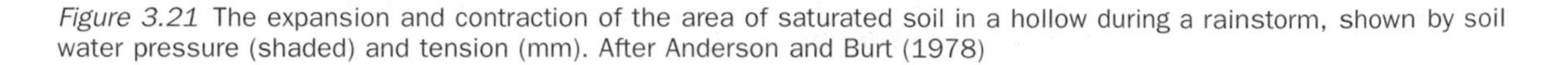
Figure 3.21 The expansion and contraction of the area of saturated soil in a hollow during a rainstorm, shown by soil water pressure (shaded) and tension (mm). After Anderson and Burt (1978)

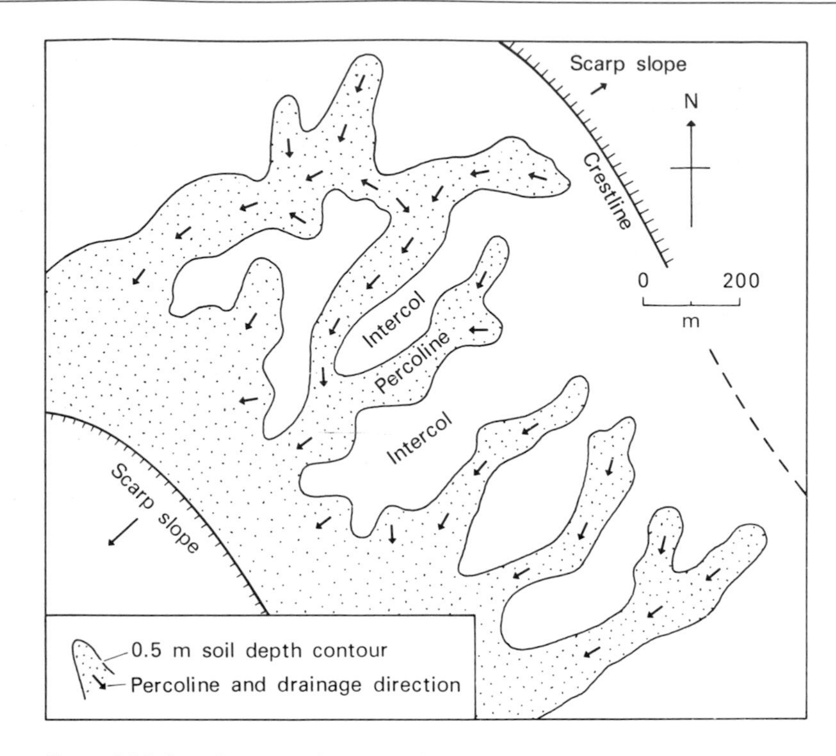

Figure 3.22 Percoline networks, lines of deeper soil that resemble channel networks. After Bunting (1961) *The role of seepage moisture in soil formation, slope development and stream initiation.* By permission of the American Journal of Science

flow in macropore systems, but the intense variability of nature at this scale presents quite formidable problems for mathematical analysis.

Pipeflow Occasionally, extremely large macropores develop as a result of hydraulic pressure or of hydraulic enlargement of other macropores, which are effectively

Figure 3.23 A set of soil pipes

natural water pipes (Figure 3.23). These **soil pipes** can form extensive channel networks that can have a major influence on hillslope drainage and upon streamflow, in a wide range of environments from semiarid to perhumid (Jones 1971, 1990). Sizes range from a few millimetres to over a metre in diameter, and pipeflow velocities average six times faster than overland flow. Pipeflow has been found to contribute 50 per cent of streamflow in one headwater catchment in Britain with a particularly dense network of pipes (4.4 km of pipe within 23 ha). Higher percentages may occur when moderate rainstorms fall on moderately wet catchments. In contrast, overland flow is likely to contribute more in severe storms and on wetter soils, whereas in drier situations sources near to the stream, like riparian seepage areas, contribute more (Jones 1987b). These pipes seem to receive water partly through rapid infiltration, via cracks and holes, and partly from older groundwater. The height of the **phreatic surface**, or level of saturation in the soil, is critical to their response: many pipes operate only when the phreatic surface rises to their level, others flow perennially because they intersect the general level of groundwater.

One of the most significant aspects of piping is its effect upon contributing areas (Jones 1997b). As Figure 3.24 shows, pipes can link remote areas of a catchment to the stream, and these linkages themselves vary according to which pipes, and possibly linking swales, are activated in a given storm. We should no longer think of the DCA as comprising a simple expansion into contiguous areas (section 3.3.2). Research suggests that pipes are common in the wetter parts of upland Britain (Jones *et al.* in press) and that they could have a significant effect in reducing the chemical buffering of acid rain by cutting contact time with the mineral soil.

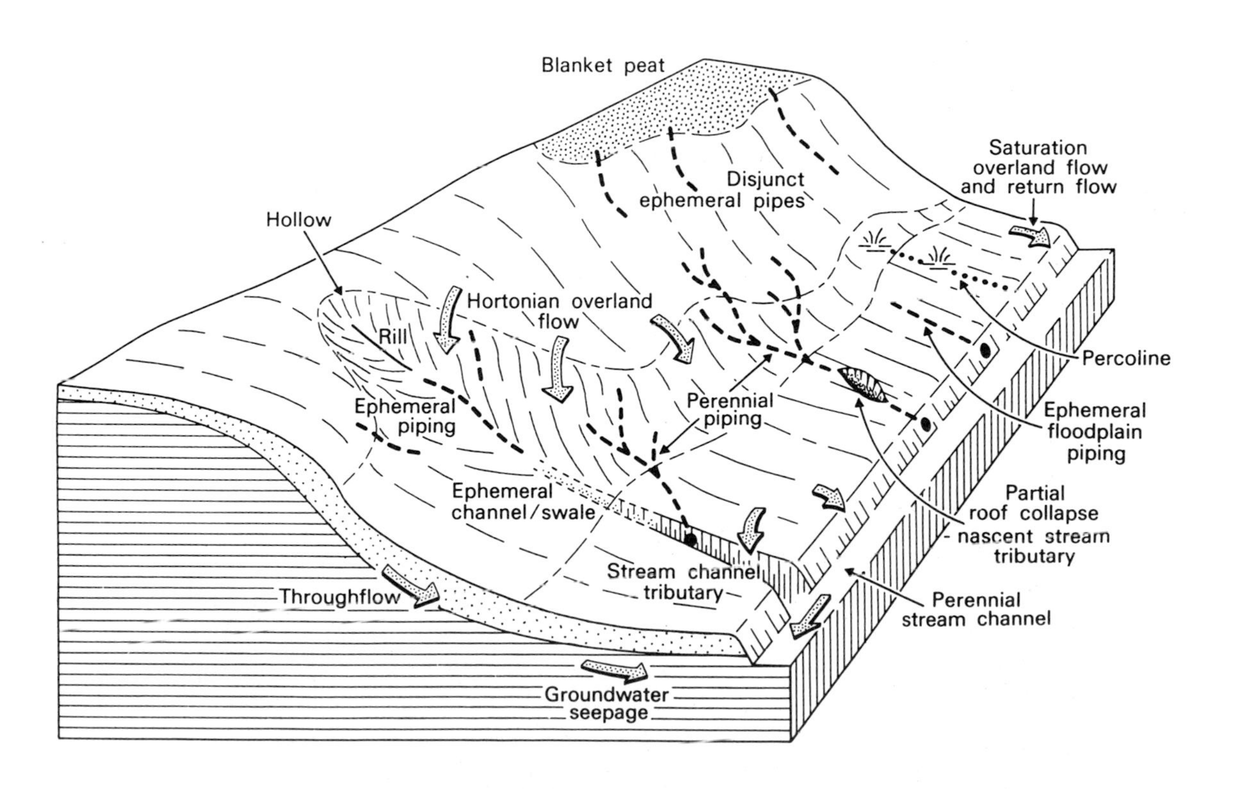

Figure 3.24 The full range of hillslope drainage processes feeding stormflow to the stream

Box 3.1 Comparison of stormflow-generating processes

It is now clear that we must replace the single process and spatial homogeneity of Horton's theory with a multi-process system operating in heterogeneous space. Saturation overland flow is more common in humid landscapes than Hortonian. It is spatially restricted and generally dependent upon the pattern of soil moisture produced by throughflow. Throughflow itself can contribute significant amounts of runoff, either by macropore flow or by surfacing (exfiltrating), especially in swales. Shallow swales and deeper 'dry valleys' offer a mechanism by which the surface streamflow network can expand beyond the strict channel network and tap slower moving throughflow (Figure 3.25).

The relative efficacy of hillslope processes as sources of stream runoff depends on the velocities,

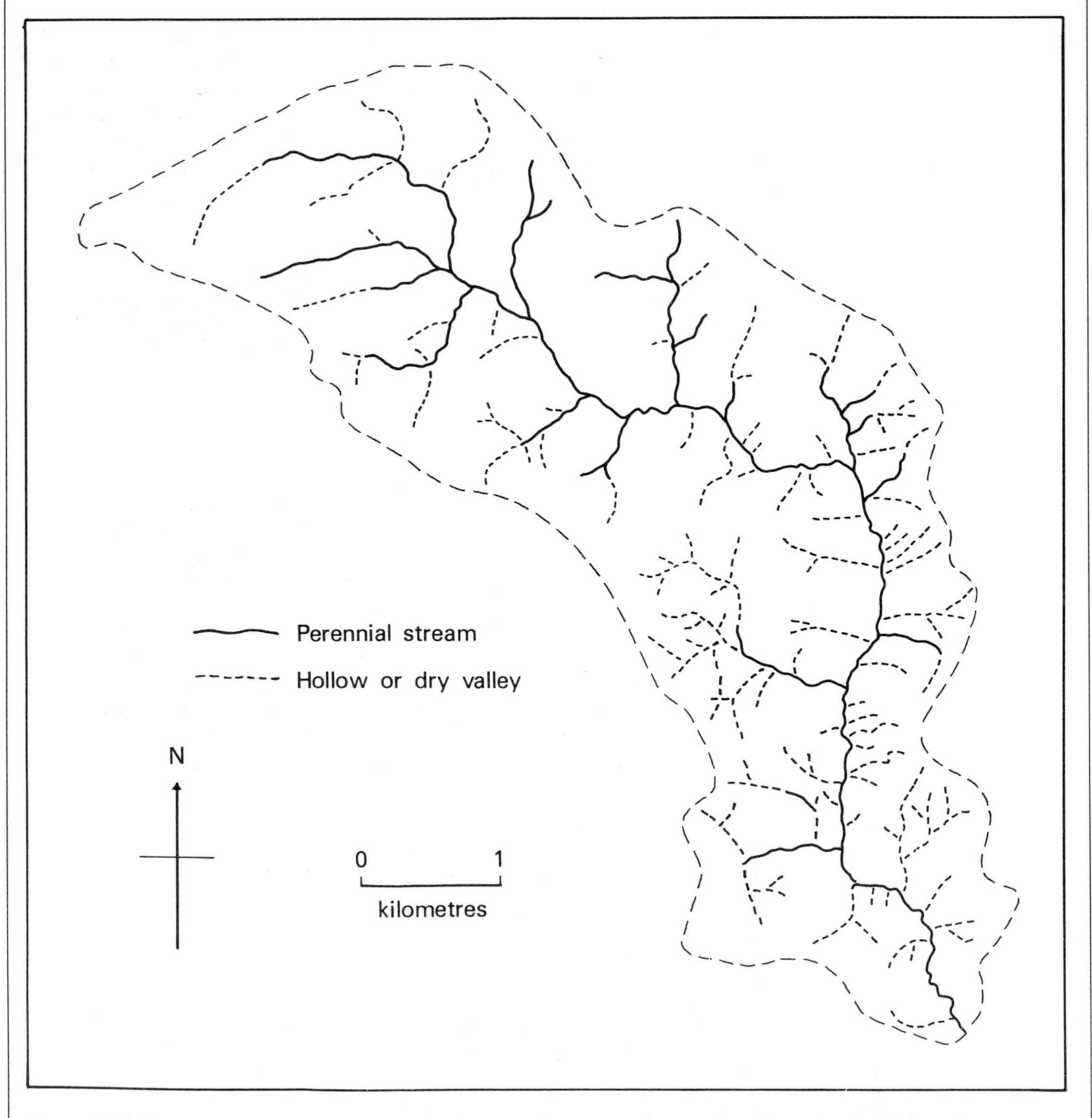

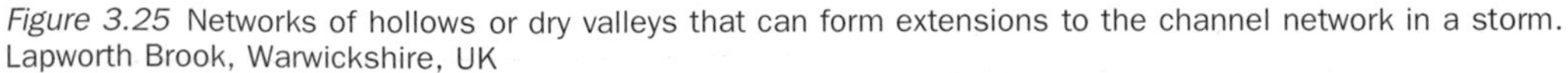
Figure 3.25 Networks of hollows or dry valleys that can form extensions to the channel network in a storm. Lapworth Brook, Warwickshire, UK

Box 3.1 (cont.)

volumes and time delays in their response to storms. In *velocity* terms, the average rank order is pipeflow (300 mm s^{-1}), overland flow (50 mm s^{-1}), macropore flow (3 mm s^{-1}), and finally diffuse matrix throughflow (0.65 mm s^{-1}). These rates suggest that overland flow would take less than 1 hour to reach the stream channel in the average British basin, whereas diffuse throughflow would take over 100 hours and therefore arrive too late to contribute to the stormflow hydrograph. However, field observations in Britain suggest that overland flow is more likely to be saturated than Hortonian (therefore spatially restricted), and partly fed by throughflow. Even where conventional infiltration measurements suggest that Hortonian flow will be common, in practice large macropores and pipes can cause subsurface flow to dominate (Jones 1990).

Nevertheless, *discharge* is the product of velocity and cross-sectional area, and rankings can change significantly when the overall size of each drainage pathway is considered. Hence, overland flow tends to have a much larger cross-sectional area than pipeflow, which compensates for the lower velocities. Rankings can change from storm to storm and within a storm, as the total areas and depths of the different types of flow fluctuate. And they vary from basin to basin as the mixture of flow routes and frequencies of activation of those routes is affected by soil properties, topography and climate. Kirkby's (1978) calculations show why overland flow is so rare in lowland Britain, given the relatively low annual rainfall and low rainfall intensities (Figure 3.17), and why throughflow is important in upland Britain, because of higher and more frequent rainfall and higher effective infiltration rates. Table 3.4 looks at the way different hillslope processes depend on topography, based on area drained, a, and discharge, q, per metre of contour.

Table 3.4 Discharge in hillslope drainage processes related to topography

Hortonian overland flow	Throughflow (saturated)	Saturation overland flow
$q_{hof} = a(R_i - f)$	$q_{tf} = a(f - q_0)$	$q_{sof} = a.R_i + q_{tf}^*$

Discharge per metre width of slope, q, in each process is related to a, the area drained per metre width, rainfall intensity, R_i, infiltration rate, f, leakage to groundwater, q_0. q_{tf}^* is the portion of throughflow contributing return flow.

Finally, Figure 3.26 and Table 3.5 relate observed responses to the size of drainage basin. The table confirms the efficiency of overland flow as a source of streamflow, but it also shows the potency of pipeflow in small areas. The **runoff coefficient** is the ratio of storm runoff to rainfall.

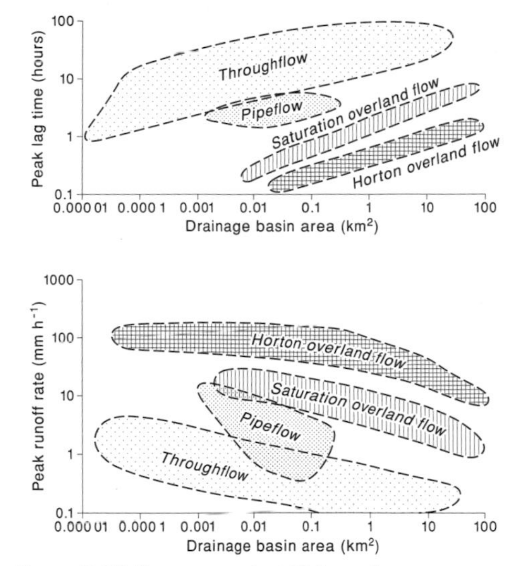

Figure 3.26 Responses in hillslope flow processes related to basin area. Based on Dunne (1978), Kirkby (1985), Burt (1992) and Jones (1997b)

Table 3.5 Typical responses for hillslope drainage processes

Basin area (km^2)	Peak runoff rate (mm h^{-1})			Peak lag times (h)			Runoff coefficient		
	HOF	TF	PF	HOF	TF	PF	HOF	TF	PF
10	24	(0.1)	(0.3)	0.7	(25.3)	(9.6)	0.18	(0.1)	(0.11)
1	36	0.15	(0.7)	0.4	14.3	(6.7)	0.21	(0.1)	(0.17)
0.1	54	0.2	1.6	0.3	8.0	4.6	0.25	0.1	0.25
0.01	81	0.3	3.6	0.2	4.6	3.2	0.29	0.1	0.38
0.001	122	0.4	8.4	0.1	2.6	2.2	0.35	0.1	0.56

HOF, Hortonian overland flow; TF, throughflow; PF, pipeflow. Brackets indicate extrapolation beyond the range of data. Overland flow and throughflow calculations based on data from Dunne (1978) and Jones (1997a).

3.3.5 Snowmelt processes

Snowmelt processes fall into two groups: *melting* and *drainage*. Older studies and most modelling work up to the 1980s (section 6.5) concentrated on the melting processes. But as more is understood about the drainage processes, many parallels are appearing between snowpack processes and drainage through the soil.

The rates and pattern of snowmelt are controlled by four major factors: (1) the environmental energy balance; (2) the presence of anisotropy and spatial heterogeneity in snow depth, density, porosity and permeability; (3) processes of crystal metamorphosis; and (4) the development of an isothermal profile. The **energy balance** of the snowpack is particularly important at the time of melt, when radiative heating and advected heat from warm air and rainfall can initiate melting. The heat content of the rain is generally quite low in comparison with the heat needed to melt a cold snowpack, but when the snowpack is 'ripe' through general warming rainfall can tip the balance (section 6.5.2). The rain also carries heat down through the snowpack to affect the parts not readily warmed by the slow conduction of heat through the dry snow. Herein lies the most important difference between snowmelt drainage and soil throughflow: erosion is predominantly thermal rather than mechanical and drainage rapidly destroys its own pathways.

Energy balance history may also be important, especially in long-term snowpacks. Together with the *history of snowfall* (its crystal form and amount), temperature fluctuations and periods of melting during the life of the snowpack can affect the crystalline structure of the pack and the growth of impermeable ice layers. These two factors are the principal causes of vertical and horizontal differences in permeability (Marsh and Woo 1984).

Crystal metamorphosis occurs under the influence of temperature and pressure changes. Snow crystals may be randomly broken and eroded during drifting, but once they are deposited in their final resting place they undergo systematic metamorphosis, which begins with crystal breakdown and ends by creating new crystals. Under **destructive metamorphism** dry snow crystals break down into rounded snow grains. Some break-up may occur under the weight of the new snow above, but the key process in grain production is vapour migration from convex surfaces and sublimation on to concave surfaces to achieve surface–free energy equilibrium (Sommerfield and Lachapelle 1970). This involves very localized vapour movements, which occur when the snow is a uniform temperature. Hence it is also termed **equitemperature metamorphism** (Jones and Stein 1990).

Constructive metamorphism involves vapour migration on a larger scale, which follows a temperature gradient and sublimates on to the colder snow. This **temperature-gradient metamorphism** produces larger, faceted bead-like crystals known as **depth hoar**. Metamorphism is much faster in snow approaching the melting point. The rationalization of the pore space that results, especially where larger crystals are formed, tends to increase the hydraulic conductivity of the snowpack.

There may also be important consequences for meltwater quality (Jones and Stein 1990). Crystals tend to expel pollutants from their crystal structure as they grow, so that they remain on the crystal surfaces. Metamorphism should cause more rapid leaching of pollutants from these surfaces and greater risk of acid flushes (section 8.1).

Further **melt metamorphism** occurs in wet snow when the temperature is near 0°C, the liquid water content is high and it is said to be 'ripe'. This tends to begin in the lower part of the pack, which is shielded from the cold atmosphere by the insulating cover of the upper snowpack, and progresses towards the surface until the whole pack is *isothermal* around the melting point. Melting and refreezing causes **grain clusters** to form, which further improve the conductivity of the snow until it becomes saturated **slush** and the clusters break up again. At this stage, refreezing and compression under the overburden of snow can produce **firnification** (section 2.1.2) and **ice layers**.

The snow is denser when it is ripe, with water equivalents of 30–60 per cent against 20 per cent or less in the original snow, which encourages thermal transmission and more rapid melting. Heat penetration may also be aided earlier on in the process by ventilation tubes formed in dry snow when gusts suck air out of the snowpack. Melting usually begins in the top few centimetres and meltwater percolates downwards until it reaches a less permeable horizon, ice layer or the frozen or saturated soil surface. Compressed vegetation on the surface can add a 'thatched roof' effect. Perched water tables (section 3.4) may develop within the profile, especially in areas where the topography tends to cause water to accumulate. In snow, water accumulation and permeability are more closely related than in soil, because drainage waters can so rapidly modify the permeability of the snowpack, as in the process of **fingering** described by Wankiewicz (1979). Figure 3.27 illustrates some of these processes.

Melt piping can exploit the more permeable layers.

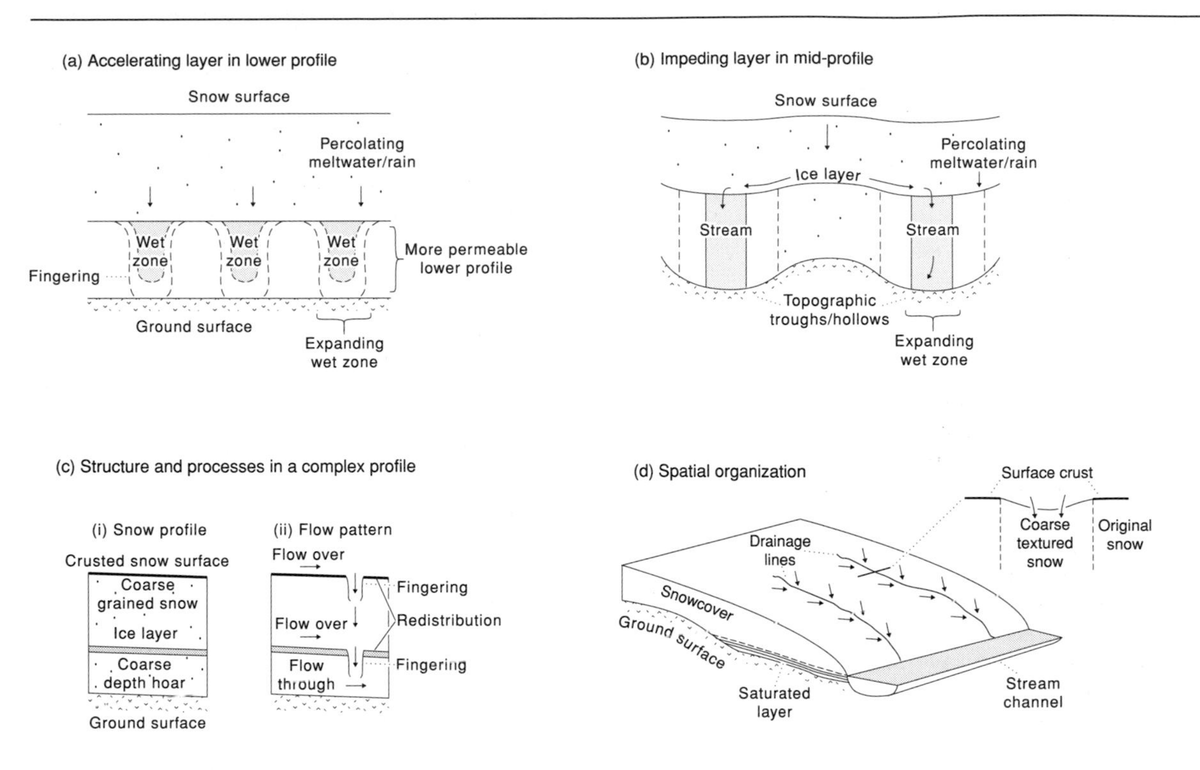

Figure 3.27 Processes of snowmelt drainage. After Wankiewicz (1979)

And as the snow clears, meltwater from the snow-free slopes can drain under the remaining snowpack through melt tunnels.

Sub-snow drainage processes vary widely from place to place. In the Alaskan tundra, Hortonian overland flow beneath the snowpack tends to dominate, especially over permafrost (permanently frozen ground), but under birch–aspen forest in permafrost-free areas more meltwater infiltrates the soil and stream response is slower. Dunne and Leopold (1978) report a similar situation in the Canadian Shield, with Hortonian overland flow dominant in the north but insignificant under hardwoods in the south. Studies by Roberge and Plamondon (1987) at Lac Laflamme in southern Quebec showed that soil pipes typically transmit one-fifth of meltwater there, but towards the end of the melt period this may rise to three-quarters. Soil pipes are also important carriers of meltwater in the Albertan badlands (Bryan and Campbell 1986). In Vermont, Dunne and Black (1971) found meltwater drainage evenly split between above- and below-surface processes.

3.4 Groundwater

Groundwater sustains low flow discharges in perennial streams. It is also commonly the largest store of water within a drainage basin. Both of these roles make it important as a source of water supply, and increasingly a matter of concern as groundwater pollution and in some regions 'groundwater mining' are becoming more common.

It can be defined as a body of water that totally fills the pore space in a rock. The rock may be hard 'consolidated' rock or loose 'unconsolidated' rock. The definition is wide enough to encompass the soil, in which 'groundwater' merges with 'saturated throughflow', but in practice it is useful to try to distinguish groundwater as a generally larger, deeper and more slowly reacting body principally located in bedrock or drift materials, such as river alluvium, rather than in the organic soil.

Despite the impression often given by classical theory (see below), it is wrong to imagine groundwater as a ubiquitous or continuous feature in every drainage basin. There are many areas even in humid regions where it is difficult to identify a true body of bedrock groundwater, because the bedrock is too impermeable. This is true of many areas of igneous or metamorphic rocks. Here, groundwater may only exist to any great extent in pockets of unconsolidated deposits like river gravels. Most large groundwater bodies are held in sedimentary rocks, such as sandstone or limestone. Even in many such permeable rocks, the dominant influence of fissures or cave systems on water movement can make the idea of a unitary body of water occupying the interstices of the rock matrix inappropriate. There is a close analogy here with matrix flow and macropore flow in soils (see page 85). As in the latter case, nature offers a complete spectrum of flow patterns from totally canalized to totally diffuse flow through the rocks. The art of the hydrogeologist or groundwater hydrologist is to generalize at the most appropriate level for each case.

The basic vocabulary of hydrogeology recognizes a variety of types of groundwater storage (Figure 3.28). The **water table** or **phreatic surface** is the upper limit of saturation in an *unconfined* body of groundwater (as in soil). It separates the **vadose zone** above, which is normally unsaturated, from the saturated **phreatic zone** below. The term 'water table' is unfortunate. It rarely resembles a table and in homogeneous bedrock it more normally takes the form of a rather subdued version of the surface topography. At the water table, water stands at atmospheric pressure. Rocks holding this water are called **aquifers**. Rocks that hold and transmit very little water are called **aquicludes**. **Aquitards** lie somewhere in between, but all these terms are relative. One hydrologist's aquiclude may be another's aquitard, recognizing that its slow permeability is hydrologically significant. Aquifers held between aquicludes are said to be *confined*, whereas those bounded by more permeable aquitards are called *leaky*. Confined aquifers may also be called **artesian**, although the term can be used in various ways. The classic **artesian spring** or **well** is one in which pressure in the confined water body forces water upwards; in effect, the **hydrostatic level** of the water, i.e. the phreatic surface as it would exist if the water were free to establish equilibrium with atmospheric pressure, lies above the ground surface. Such locations are particularly advantageous for human settlement. A prime example is the London Basin, where water drains through the chalk aquifer from the surrounding hills down into the centre of the synclinal basin confined beneath the London Clay aquiclude. It is, however, possible to get a similar effect at the base of steep slopes or geological folds even in unconfined aquifers.

In reality, geological structures and the permeability of the rocks tend to vary over an area. Hence, aquicludes may be more permeable locally or aquifers less efficient. **Perched water tables** can form above an aquiclude that is of very limited extent or at times when infiltration rates temporarily exceed the permeability of

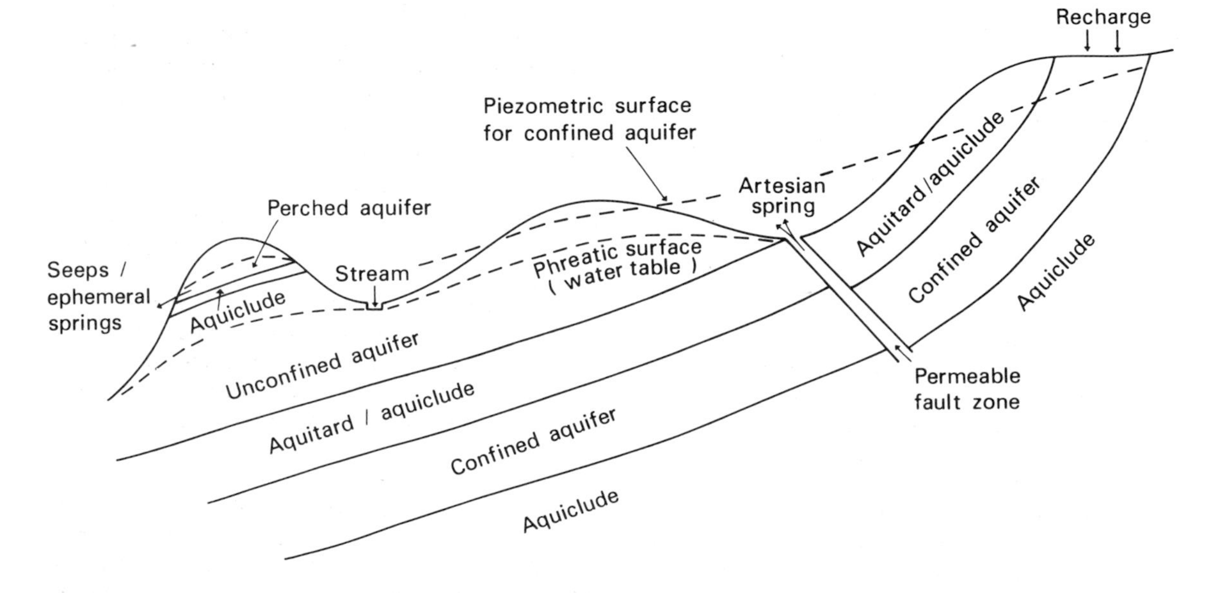

Figure 3.28 Forms of groundwater storage

an aquitard. They are called 'perched' because they stand above the main groundwater body.

From the water resources viewpoint, a good aquifer should possess three key properties: (1) it should be large; (2) it should have a moderate to high drainable porosity, in order to hold enough *extractable* water in the voids; and (3) it should have sufficient permeability to allow easy abstraction of the water. A number of parameters can be used to determine the efficiency of yield. Although the porosity determines the total amount of water that can be held, the amount that is extractable is determined by the **storage coefficient** or **storativity.** This is the amount of water that will raise or lower the pressure head by one unit (1 m), i.e. that will alter the pressure within the water body by an amount equivalent to 1 m^3 of water pressing on each 1 m^2 of aquifer. It is measured in volume per unit area per unit change in head (m^3 m^{-2} m^{-1}). The **specific storage** is the volume released per unit volume of saturated aquifer for a unit reduction in hydraulic head in m^3 m^{-3} m^{-1} (which may be quoted simply as m^{-1}). In unconfined aquifers, where this causes the water table to fall, storativity is known as the **specific yield.** Storativity is more complicated in confined aquifers, where it can partly depend on the compressibility of the aquifer, since compaction will squeeze water out. The related measure of **specific capacity** is used for water supply wells and is the ratio of the discharge achieved by pumping to the amount of fall in the water table, in m^3 day^{-1} m^{-1}.

Permeability is determined by the **coefficient of permeability**, the k in Darcy's Law, which is strictly measured in m day^{-1} at 15.6°C, since the viscosity of water varies with temperature. But the rate of transmission is commonly measured by the **transmissivity** or **transmissibility**, T. This is the rate of flow per unit width of aquifer under unit hydraulic gradient, for the whole depth of the aquifer, y, i.e.

$$T = k.y$$

which has the units m^2 day^{-1} (or m^3 m^{-1} width day^{-1}). Multiplied by the total width of the aquifer this gives the discharge in familiar m^3 day^{-1}. Transmissivity is highly variable, even within a specific rock stratum, and it is difficult to generalize. Figures quoted by Dunne (1978) suggest moderate to high values for unconsolidated rocks (*c.* 300 to 37 000 m^3 m^{-1} day^{-1}), low to moderate in sandstones (70–400), and low in metamorphic rock. Yet limestones can show great variability, even within the same strata (e.g. 60–17 000).

In addition to being variable from place to place, or **inhomogeneous**, rocks may also display distinct differences in permeability according to the *direction* of flow. Such rocks are said to be **anisotropic**.

Factors controlling groundwater flow Because groundwater flow and the geological structures that affect it remain mostly unseen, flow patterns are normally inferred from observations of wells, which give an approximate indication of water table level, or piezometers, which measure pressure head (Chapter 5). The height of the water surface above an appropriate datum level gives the **hydrostatic head**, as in soils (section 3.3.4). Interpolating between the observation points gives the **potentiometric** or **piezometric surface**. This is the prime control on groundwater flow: the distribution of potential energy within the water body.

Figure 3.29 shows a simple model of groundwater flow, based upon the **piezometric gradients** within this surface. Topographic valleys cause a corresponding dip in the potentiometric surface, aided by drainage from springs, seeps and channels in the valleys. On a broader scale, gradients may be enhanced by higher rainfall receipts in the uplands. Contours joining points of equal pressure head are called **equipotential lines**. Flow will take place perpendicular to these lines of equipotential, i.e. down the piezometric gradient. These flow lines or **streamlines** tend to flow vertically downwards on crests or groundwater 'divides' and upwards in the valleys. Over much of the hillslope, flow will be approximately parallel to the ground surface, centred on a midline that separates the realms of downward and upward components. Where the potentiometric surface intersects the ground surface, exfiltration or surface seepage will occur.

The zone of downward movement is the area of **recharge**. In reality, zones of recharge may occur in a variety of topographic locations. In arid lands, the main recharge areas are often the beds of large ephemeral rivers. This is a case of **influent seepage**, as opposed to the usual **effluent seepage** in which groundwater seeps *into* the stream. It creates a 'groundwater mound' beneath the beds, which is a rare instance of the water table opposing rather than following the topographic contours.

In practice, the potentiometric surface is also a product of varying transmissivity within the bedrock, so that zones of greater permeability will encourage faster 'drawdown' in the surface (Figure 3.30). Such zones often coincide with topographic valleys, perhaps

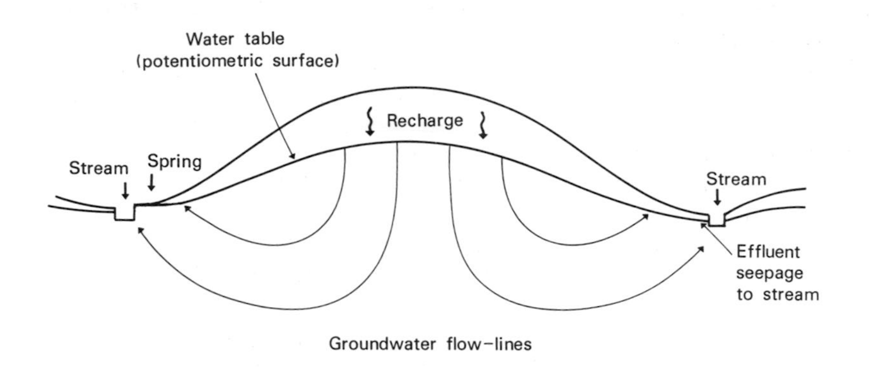

Figure 3.29 Classic model of groundwater flow in an unconfined aquifer

because the valleys have been created by erosion along lines of weakness or because the topographic concentration of streamlines creates more erosion of subsurface voids (Dunne 1980; Jones 1987a).

Flow nets like that in Figure 3.30 form the basis for modelling, which will be discussed in section 5.5.2 and more generally in Chapter 6. It is, however, pertinent to point out here that the discharge of baseflow, Q_b, entering a stream per unit length of channel may be calculated as:

$$Q_b = fx$$

assuming a uniform infiltration rate, f, across the distance, x, between the channel and the groundwater

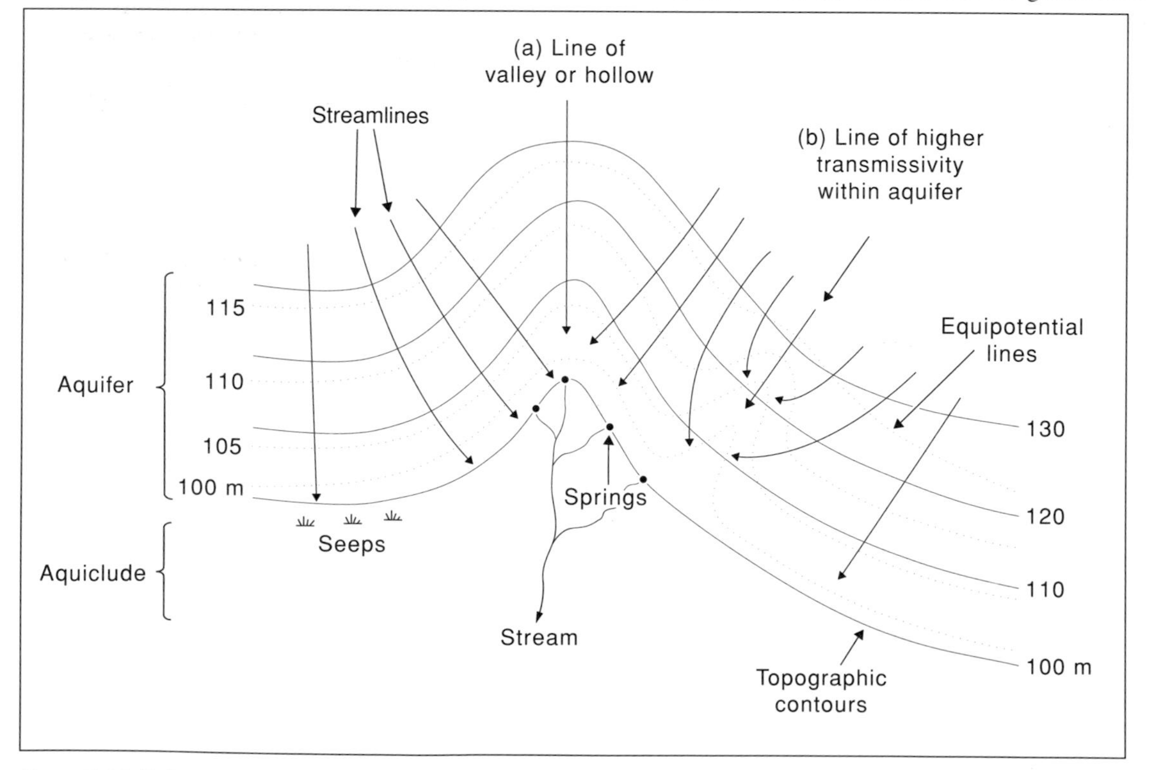

Figure 3.30 Hydraulic gradients control groundwater flow in streamlines perpendicular to lines of equal potential energy – the equipotential lines. Dips in the piezometric surface may be related to topographic valleys (point a) or to zones of higher transmissivity (point b)

divide, and an isotropic, homogeneous and unconfined aquifer. This simple solution is reasonable for situations in which groundwater flow can be regarded as being in a 'steady state', i.e. inflow balances outflow and the potentiometric gradient remains the same. More complicated circumstances require more elaborate modelling, for example, where equilibrium is disturbed by storm events or where seepage is not spatially uniform. Groundwater spring networks can clearly disturb the pattern.

Conclusion

Major advances in our understanding of hydrological processes at the basin or subbasin scale are still being made. The world is nothing like as simple as it appeared 50 years ago and there are considerable geographical differences in the dominant processes. Process studies continue to inform modellers and to improve prediction and control. Later chapters will show how they are proving especially valuable in predicting the impact of environmental change and pollution.

Further reading

Good coverage of hydrological processes can be found in:

Anderson M G and T P Burt (eds) 1985 *Hydrological forecasting*. Chichester, Wiley. Contains a few useful chapters on processes.

Anderson M G and T P Burt (eds) 1990 *Process studies in hillslope hydrology*. Chichester, Wiley: 539pp. Too detailed for most introductory purposes, but contains some good chapters.

Conacher A and J B Dalrymple 1977 The nine unit landsurface model: an approach to pedogeomorphic research. *Geoderma*, 18(1/2): 1–154. This offers a unique integrating approach to hydropedogeomorphic processes on hillslopes, supporting general theory with detailed observations of soils.

Dunne T and L B Leopold 1978 *Water in environmental planning*. San Francisco, Freeman: 818pp. Ageing, but clear and readable.

Kirkby M J (ed.) 1978 *Hillslope hydrology*. Chichester, Wiley: 389pp. Ageing but still a readable introduction.

An in-depth coverage of groundwater processes is provided by:

Fetter C W 1994 *Applied hydrogeology*, 3rd edn, New York, Macmillan: 691pp.

Discussion topics

1. Consider the variety of processes generating streamflow and discuss the factors that may cause one or another to become more dominant.
2. Outline the similarities and dissimilarities between flow through snow and soil.
3. Why is it important to study evaporation processes if most practical estimates of evaporation losses are calculated using statistical formulae?
4. Discuss the proposition that spatial heterogeneity is not just a complicating factor but the very key to runoff generation.

CHAPTER 4

Floods, droughts and magnitude–frequency relations

Topics covered

Extremes in hydrology take many forms, from extremes of magnitude to extremes of length or both. By definition, an extreme event must be rare and unusual, but there is no sudden break or gap between extreme events and normal events; the normal grades into the extreme. Essentially, all hydrological events are part of a continuous series.

This does not mean that some extreme events cannot be so different from normal experience that they appear to be a completely different type of phenomenon. Mandelbrot and Wallis (1968) coined the memorable terms Noah events and Joseph events to describe some of these. A Noah event, recalling the legend of Noah's ark, is proof that extreme events can be very extreme indeed. Hydrologists may console themselves with the notion of the 'probable maximum precipitation' (see page 103), but a wise statistician would not rule anything out. We hardly know the limits of present climates, let alone of past or future climates. On the other hand, a Joseph event, recalling Joseph's predictions to Pharoah of seven drought years and more, proves that persistent events can be very persistent indeed. In fact, Joseph must have been a keen observer of nature; it was not until the 1950s that H E Hurst formally drew the attention of modern hydrology to the patterns of 'persistence' in the flow of the River Nile, the so-called 'Hurst Effect' (section 4.3).

For flood protection it is normally the extremeness of *individual events* that is of importance and concern. In the case of droughts and low flows, it is the extremeness of the *sequence of events* that normally matters, for example, for calculating the storage capacities that are required for reservoirs to maintain a reliable water supply throughout those periods. Different extremes are critical for different purposes and a variety of thresholds may be defined in terms of severity and lengths of events that are important in different ways. Nevertheless, when we look at the raw statistics of natural events without making 'value judgements' as to what is critical for what, we find continuous distributions from the very small to the extremely large, the short to the long and from the very common to the extremely rare. Even Noah and Joseph events can be brought within the fold of frequency distributions that encompass more common events.

The fact that floods and droughts are different from common events by degree rather than kind is of tremendous importance for understanding and predicting their occurrence. It means that we can use observations of everyday events to learn more about them. This idea was used to great practical effect by the statistician E J Gumbel, who developed the highly influential concept of **extreme value analysis**. At the heart of Gumbel's theory lies the notion that the magnitude and frequency of extreme events can be estimated from observing the magnitude–frequency distribution of more ordinary and not so severe events. Based on this notion, he developed theories for both floods and droughts. According to Gumbel, the frequencies of rare flood or drought events can be estimated by fitting theoretical statistical distributions to as large a collection of peak flow, low flow or rainfall records as possible (section 4.2).

4.1 The nature and causes of floods and droughts

As we discuss the nature and causes of floods and droughts, it is necessary to begin by considering their definition. Both are remarkably difficult to define. This is partly because in nature there are no real breaks in flows and partly because the thresholds that are selected are defined in terms of human requirements, which vary.

4.1.1 Floods

As commonly understood, a flood occurs when a river overtops or 'bursts' its banks and flows across the floodplain. To the hydrologist, however, this is just a flow which exceeds a particular level; albeit an extremely important one in nature and for human occupancy of the floodplain. This **bankfull level** is also only one of many levels or thresholds that could be selected as being critical for some particular purpose. Engineers, planners and insurers are often more concerned with floods that exceed some other, higher level that is more critical to flood defences, and they seek to predict how often that higher level will be exceeded. For agricultural purposes, the *depth* of flooding may not determine the extent of damage as much as the *length* of the inundation or the season.

In practice, the bankfull level is often difficult to define, not least because the height of the banks varies along the river. Similarly, the depth of flow is only one component of the discharge. The same bank height may not accommodate the same discharge, since the discharge that a channel can contain also depends on the width and slope of the channel. As a result, the same discharge may inundate some stretches of the floodplain and not others.

It is scientifically better to define a flood in terms of the *discharge*. Hence, hydrologists can refer to 'flood flows' when no real 'flooding' has occurred. The highest flow in any one year is called the 'annual flood' whether it 'bursts' the banks or not. In this sense, the term is used synonymously with **peak discharges**.

In rivers, floods and low flows are usually expressions of the temporal variability in rainfall or snowmelt interacting with river basin characteristics. These characteristics can be broadly divided into: (1) overall basin form; (2) hillslope properties; and (3) channel network properties (Figure 4.1). Both hillslope and channel properties can be further subdivided into the general form of hillslopes and drainage networks, on the one hand, and specific properties of materials, surface or channel roughness and channel cross-sections on the other. The most important interactive processes are: (1) the relationship between the rates of supply and infiltration of water; and (2) the antecedent conditions determined by the recent history of precipitation or melt. These mainly relate to processes on the hillslopes.

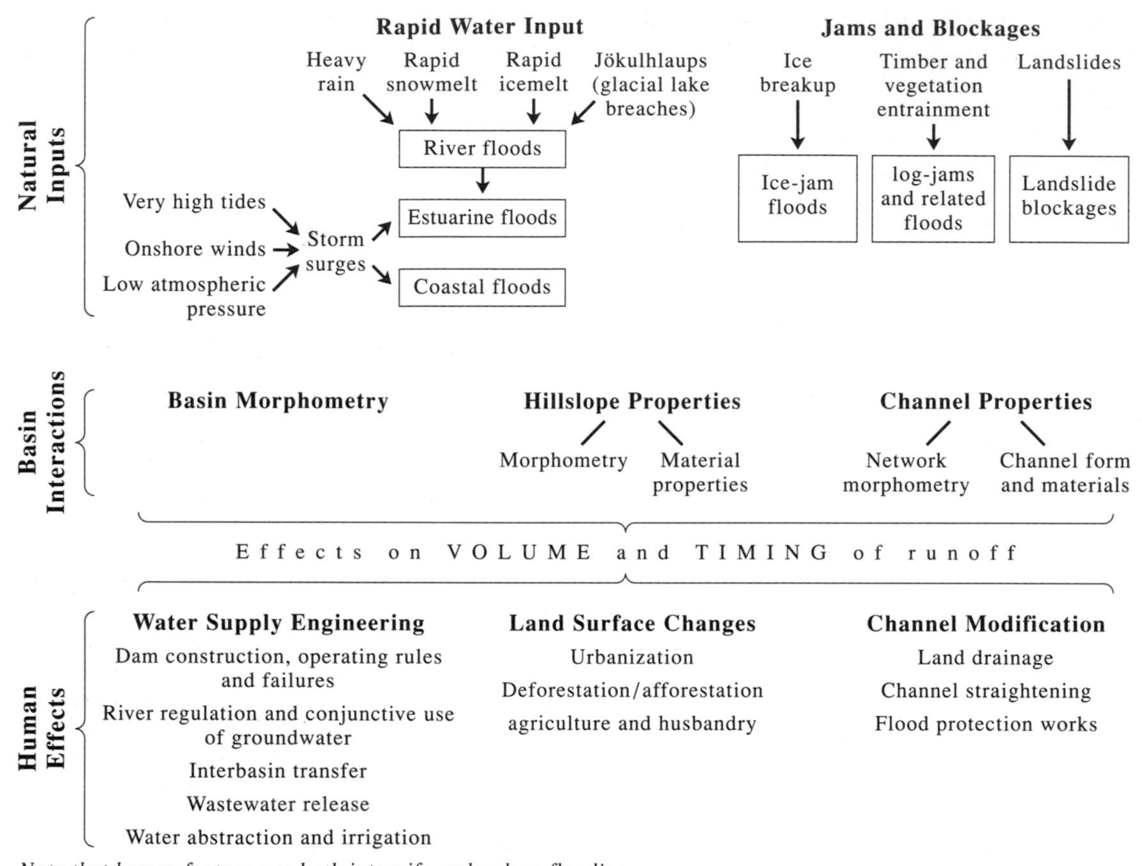

Figure 4.1 Major factors initiating and modifying floods

The two features that most determine whether a flood occurs are: (1) the volume of direct surface or near-surface runoff; and (2) the uniformity of runoff times from different parts of the basin. The more uniform the response and travel times, the greater the likelihood of the riverflow building up into a high peak flow. In basins that have such uniformity, this may create flood flows out of runoff volumes that in others would pass as a more subdued event. In basins with soils and rocks that absorb more of the storm- or meltwater, river response is likely to be even more subdued owing to lack of volume.

Each of these two key features is typically a complex product of basin characteristics and their differing interactions with rainfall or snowmelt inputs. The properties of hillslope materials, especially infiltration capacities and lateral permeabilities in the topsoil, largely determine runoff volumes but in conjunction with antecedent soil moisture conditions. As we saw in section 3.3.4, hillslope planform can also be important in controlling the pattern of redistribution for the remnant water from antecedent rainfall events, so that it controls the amount of saturation overland flow. For all of these factors, their effects on the speed of runoff is as important as the actual creation of the runoff. Indeed, speed of drainage is a critical determinant of whether runoff arrives at a certain point in the channel in time to add to peak flow or to be regarded as 'storm runoff' at all (see section 6.1.2). High slopes and drainage densities also contribute to speed of runoff. Similarly, uniform response is largely a product of a compact basin form and a compact, dendritic channel network, but it may also depend on hillslope planform and patterns of saturation overland flow. Further consideration will be given to these factors when we discuss modelling in Chapter 6 (Box 6.1).

River flooding may also result from the sudden release of water from dams or lakes, such as jökulhlaups (section 6.5.4), or from natural damming by landslides, log jams or ice jams (Figure 4.1). In addition, some flooding can be found in valley bottomlands that is not due to overbank flow at all, but caused by groundwater levels rising through the fluvial deposits underlying the floodplain. This flooding can actually precede the arrival of floodwaters in the river. There are other types of floods, such as coastal flooding caused by storm surges (as in the Rhine Delta in 1953; section 2.6.2 and Box 10.1) and estuarine floods, which may be caused by high river discharges meeting a high tide.

In Chapter 7, we will review the very significant effects that human modification of hillslopes and channels can have on the risk of flooding. But for the moment it is appropriate to discuss just one further natural source of river flooding that is limited to cold regions.

Ice-jam floods Ice jams are of particular concern because they are almost an annual occurrence and a major source of flooding in many parts of Canada, Russia, Alaska and Scandinavia. Ice-jam floods have been particularly problematic in southern Ontario, which is one of the most densely populated areas of Canada, and where the climate ensures that all rivers freeze up for at least a quarter of the year, from early January to late March or April. In urban areas like Belleville, special channels have been constructed which are designed to break up the ice floes into smaller pieces and which have smooth concrete banks to prevent jamming. Most flooding is caused by ice floes grounding on the banks and packing against other floes, rather like log jams. Indeed, most current theories of ice jamming are based on work with log jams, undertaken largely by the logging industry. During the 1980s, the Canadian National Water Research Institute (NWRI) at Burlington, Ontario, carried out an extensive programme of field monitoring on the Thames and Grand Rivers together with physical simulation experiments in laboratory flumes (e.g. Beltaos 1981).

Ice jams tend to be an integral part of spring break-up on the rivers. Break-up commonly begins in upstream reaches when snowmelt runoff causes stream stage to rise. Normally, the river ice cover is still frozen to the banks and only breaks free once the snowmelt discharge has exceeded the maximum that can be transmitted under the ice. This is called the **maximum stable freeze-up stage** and is usually equivalent to the water level when the ice cover formed. A further rise in stream stage is normally required to set the broken ice in motion. Break-up then generally proceeds downstream, although the middle reaches may already have alternating open and covered sections. Once an ice jam occurs it can have a positive feedback effect on ice movement by raising water levels still further and so enabling more ice floes to float free and to add to the jamming.

The NWRI has used field measurements of ice surface roughness in models based on standard jamming theory in order to predict jamming effects. Larger jams tend to contain ice floes that have higher initial surface roughness, which creates internal friction and helps to hold the jam in position. They also tend to occur in the narrower sections of the channel. Internal friction is increased markedly when ice floes freeze together, which can happen early in the melt season. This makes it more difficult to predict ice-jam behaviour. Difficulties can arise again when the floes begin to melt and the internal friction is suddenly reduced. This can release a wave of floodwater downstream.

Not all flooding is during the melt season. Floods can be caused by the build-up of slush ice during the freeze-up period while river discharge is still fairly high. This is caused by **frazil ice** (crystal flocs entrained in the flow) jamming up against obstructions or stationary ice and freezing. At the same time, **anchor ice** on the bed of the channel can increase bed roughness, and frazil ice will increase the viscosity. Measurements on the Beauharnois Canal near Montreal have shown that frazil ice can thus reduce the conveyance capacity of the channel by 40 per cent. On the Niagara River, anchor ice and dense frazil ice can retard the flow by more than 850 $m^3 s^{-1}$. These two effects may initiate flooding as the cross-sectional area of flow increases to accommodate the incoming discharge. Such floods are more difficult to predict than floe jams.

Along the shores of the Great Lakes in southern Ontario flooding can also be caused by snowmelt runoff meeting a frozen lake and spreading out over the lake ice and surrounding land. An early and not particularly environmentally friendly solution to this problem was to sprinkle carbon black over the lake ice to induce melting. A better alternative might be to create artificial leads along the shoreline by mechanical means.

4.1.2 Droughts and low flows

Perhaps even more than with floods, droughts are a matter of definition. It is a question not only of trying to draw a line in a continuum of different degrees of water shortage, but also of a wider basis for definitions, because droughts tend to cover a larger area than floods and have a greater range of impacts. Perhaps the most common basis for definition is meteorological, which is generally taken as shortage of rainfall, but hydrologists and agriculturalists are also concerned with net water balance, and botanists may be concerned with the quality and therefore the physiological value of the water available. Hydrologists and water resource engineers are also concerned with longer histories of meteorological shortfalls than meteorologists, because they are dealing with the responses of throughput within often complex catchment and reservoir systems. A hydrological drought is best expressed as a protracted period of **low flows**, which may result not simply from the current meteorological drought, but also from lack of recharge of groundwater stores or lack of reservoir refilling over a much longer period.

In defining a **meteorological drought** it is particularly important to take account of differences between climates. The old British Rainfall Organization's definitions of an **absolute drought** as more than 15 consecutive days with less than 0.25 mm on any day, or a **partial drought** as at least 29 consecutive days with a *mean* daily rainfall of less than 0.25 mm, were designed for the British climate. They would be totally inappropriate for a drier climate. Drought must be distinguished from the general aridity of a climate, which can be defined as the long-term ratio between annual precipitation and evapotranspiration.

Hence, it is best to define drought in terms of deviations from the average rainfall, water balance or riverflows in a certain climate. Herbst *et al.* (1966) defined **drought intensity** in terms of rainfall as:

$$\text{Drought intensity} = \frac{\sum_{t=1}^{t=d}[(EP_t - \overline{P_t}) - \overline{D_t}]}{\sum_{t=1}^{t=D}\overline{D_t}}$$

where $\overline{P_t}$ is the long-term mean precipitation for each month, t, and $\overline{D_t}$, the monthly mean 'deficit' for that year, which is an index based on the difference between actual and long-term average precipitation for each month: $\overline{D_t} = [\sum_{t=1}^{t=12}(\overline{P_t} - P_t)]/12$. The effective precipitation, EP_t, is the precipitation during each month plus a weighted function of the difference between the rainfall and the monthly mean rainfall for each previous month. Each summation, Σ, is over all the actual drought months, $t = 1$ to d. Droughts are also generally seen as prolonged and widespread, and **drought severity** can be defined as the drought intensity multiplied by the duration of the drought.

Agricultural droughts may be defined in relation to crop tolerances, and evapotranspiration losses are likely to be an important consideration. The US NWS defines an agricultural drought as one that causes 'at least partial crop failure'. Water balance calculations, as outlined in section 5.3.2, can be used to estimate soil moisture status as a preliminary to defining a drought. The US SCS (1970) provided a practical instruction manual to help agriculturalists to use the Blaney–Criddle evapotranspiration formula to estimate soil moisture levels and calculate irrigation requirements.

In contrast, a **physiological drought** refers to the condition of plants that suffer from an excess of salty water, often on poorly drained irrigated land. In this case, the problem is lack of physiologically usable quality rather than quantity.

For a **hydrological drought**, the actual flow in the rivers is of most concern. Engineers concerned with maintaining water supply or with the dilution of waste effluents must consider the *length* of the period of low flows as well as the extremeness of flows *below* a certain level. Flows may be low because of a lack of precipitation in the preceding wet period(s) or winter(s) as well as low rainfall during the current summer or dry period. The more the riverflows depend on groundwater effluence from aquifers, the longer may be the 'memory' within the system.

Again, as with flood flows, the definition of a low flow can vary according to the application. And it may not be simply a question of defining low flows according to fixed discharges which are critical for different purposes. Definitions commonly employ frequency and duration, rather than absolute discharge. The UK Low Flow Studies Report (IH 1980) produced equations to predict the discharge for low flows of 10-day duration that occur no more than 5 per cent of the time (section 6.1.1). In this case, a low flow is defined not by its size or magnitude but by the frequency with which it occurs as a sustained flow for a 10-day period. A similar approach was taken by Hindley (1973) who defined 'dry weather flow' as flow during the driest 7-day period. Note that this approach implicitly measures the extreme flows in terms of normal flows, i.e. flows that occur more of the time. The methods by which low flows are thus defined will be described when we discuss the estimation of frequencies in section 4.2.2.

From the discussion so far, it should be clear that 'droughts' are primarily generated by aberrations in the weather. This may be due to the failure of atmospheric circulation patterns to follow their normal course, for example depression tracks are diverted or monsoons fail, or else it may result from cooler temperatures, particularly sea surface temperatures (SSTs), which generate less evaporation and less convective activity. Drought in the Sahel has been linked to lower SSTs in the tropical Atlantic, which may be due to the strength of oceanic upwelling. Sahel droughts are also linked with the Pacific El Niño current (sections 2.2.2 and 4.3) and with a failure of the ITCZ to penetrate as far north as normal so that the West African monsoon does not reach the northern interior. The penetration of the ITCZ appears to be limited by a strengthened and extended high-pressure system over the Sahara. According to one theory (Bryson 1973), increased dust entrained in the air from desiccated and devegetated surfaces causes cooling aloft and sinking, stable air that suppresses rain production. In the Indian subcontinent, the monsoon rains are highly variable from year to year, and droughts can occur either as a result of late arrival of the monsoon or of more frequent or extended breaks in the monsoon rains, which occur after the eastward passage of each active convective wave in the circulation (Barry and Chorley 1992).

In Europe, drought is often associated with a prolonged 'blocking anticyclone', a zone of high atmospheric pressure that becomes virtually stationary and wards off the advances of rain-bearing depressions. The western European droughts of 1976 and 1984 were caused by blocking anticyclones in the area. When these anticyclones finally gave way, the restored westerly winds brought in an unusual sequence of intense depressions, which 'broke' the droughts spectacularly. The exact causes of blocking anticyclones are still in question, but they can be related to a quasi-regular sequence of oscillations in the westerlies known as the **index cycle**. Normally, the winds oscillate over a period of 3 to 8 weeks between a strongly 'zonal' flow, paralleling the zones of latitude, and more 'meridional' flow, in which the amplitude of the Rossby waves is increased and much of the flow is in a north–south sense following the meridians of longitude (Barry and Chorley 1992). Towards the end of the cycle, the airflow can exhibit chronic meandering, which cuts some of the flow off in islands of high- and low-pressure. If the cycle sticks in this phase, a prolonged period of blocking will occur. The sudden change in the weather at the end of the blocking phase is associated with an equally sudden return to strong west–east airflow. The exact effects of the blocking anticyclone on an area depend on its position relative to that area: areas on the fringes can receive more rainfall.

Low flows are, of course, not just a product of the weather but also of local basin characteristics. Rivers fed by substantial aquifers can maintain flow levels longer than others during a meteorological drought. All the catchment characteristics that encourage flood flows also increase the risk of low flows, e.g. higher slopes and lower infiltration capacities, except, that is, for instances in which floods may result from heavy rain adding to river levels that have been raised by high levels of effluent groundwater.

4.2 Estimating the magnitude and frequency of extreme events

Estimates of the maximum size of event likely to occur in a certain period are required for many reasons, from protection against natural disasters to planning water supply systems. The earliest approaches tended to concentrate on estimating the magnitudes of supposedly maximum-sized events. Later approaches followed Gumbel's lead in recognizing that no ultimate maximum can be defined and that frequency is as important as magnitude. Some of the most recent work in practical hydrology is again stepping back from the concept of probability and planning on the basis of a notional 'worst scenario', for reasons that will be explained in section 4.2.2.

4.2.1 Estimating the maximum magnitude of storm events

The simplest approach is to use the worst events on record. This naturally assumes that the records contain at least one sufficiently extreme event. This can rarely be relied upon at an individual locality. However, a number of authorities have established **'envelope curves'** for heavy rainfall or floods based on regional or global collations. Unlike the usual method of fitting curves which seeks a middle way through the observed values, envelope curves, as the name implies, are fitted so that they enclose the maximum values. Foster (1948) and Todd (1970) published compilations of extreme rainfalls from around the world which seem to show a simple relationship between depth and duration. As Figure 4.2 shows, the method is very sensitive to the length of records and perhaps to changing climates. In

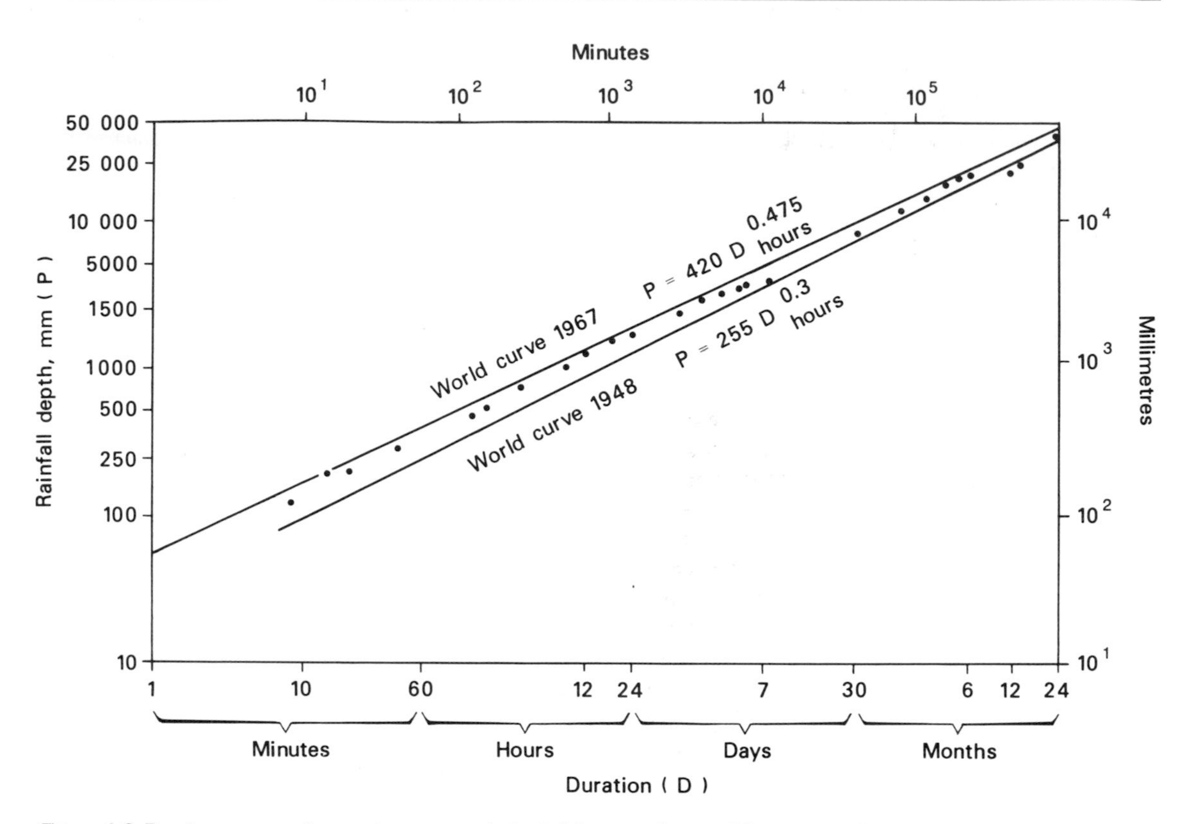

Figure 4.2 Envelope curves for maximum recorded rainfall, according to different compilations

this case, the longer data set suggested that the envelope curve should be raised by a factor of about 1.5.

An envelope curve relating peak discharge to drainage area was used in the UK Institution of Civil Engineers' first attempt to establish maximum floods in the UK in 1933, which was updated (ICE 1960) after the disastrous floods at Lynmouth in August 1952 breached the curve (Figure 4.3). These curves were the source of the first empirical flood flow 'models' (section 6.1).

Recognizing the limitations of using only events that have occurred in the records, techniques were developed that make some use of rainfall or runoff theory. These have tended to combine a degree of hydrometeorological theory on rainfall production, and perhaps runoff, with pragmatic methods for utilizing records of severe storms from similar areas. A number of these techniques are still in use and providing good service.

Standard Project Storm (SPS) Developed by the Miami Conservancy District in the 1920s, the technique takes severe but not extreme, catastrophic storms recorded in the general vicinity or within the same hydrometeorological region, and shifts their position so that they produce the maximum discharge. It may also be called the Standard Project Flood. It established the important principle of **storm transposition**, whereby a storm, or more usually a composite of the four or five worst storms, is modified to take account of the meteorological effect of differences in altitude and relief. The US National Weather Service has issued guidelines for the transposition of storms, which emphasize that: (1) storms should only be selected from within the same 'homogeneous rainfall region'; and (2) any adjustments should be limited to 'slight' differences in topography between areas. The WMO (1986a) offered a choice between a complex physical approach for transposition and a simpler one based on average rainfall-elevation gradients.

The SPS does not aim for 'the worst possible scenario'; merely for relocating storms that are 'reasonably' characteristic of extremes in the region. It is therefore a fairly conservative estimate of extreme events that can be used for small-scale projects such as farm dams, but it is neither sufficiently reliable nor

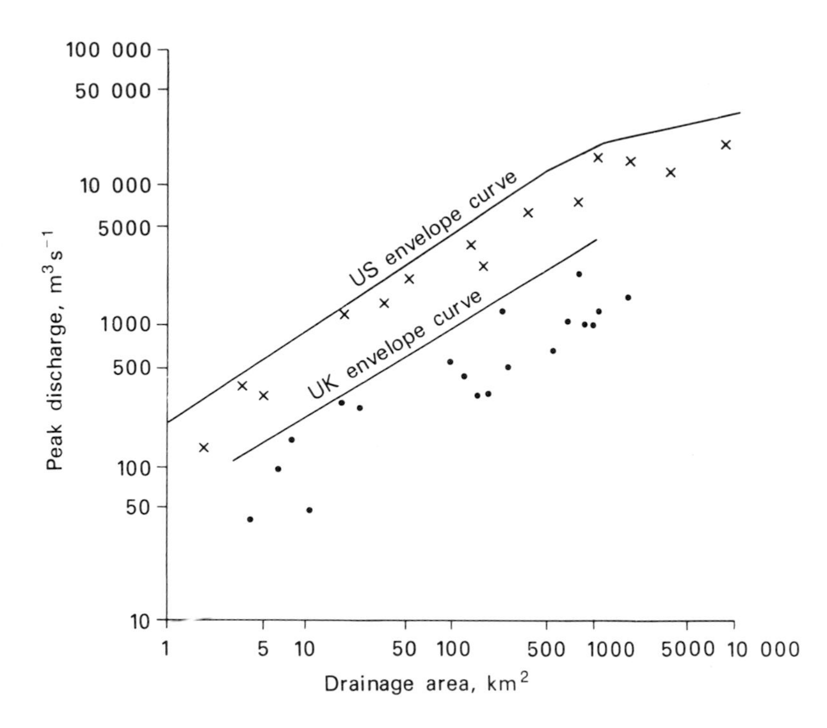

Figure 4.3 Envelope curves for maximum flood discharges after Shaw 1994

extreme enough to be used in projects involving possible danger to human life. As such, SPS figures tend to be about half of those given by the next method to be discussed, the Probable Maximum Precipitation (PMP); the exact proportion varies slightly from one geographical region to another. The continued value of the SPS as an economical design criterion for minor works is emphasized by the fact that the US Bureau of Reclamation has published maps of the USA showing the coefficients needed to convert the US NWS maps of PMP into the SPS.

Probable Maximum Precipitation (PMP) This technique was developed in the 1940s by the US Weather Bureau (now the National Weather Service), which published charts, nomograms and regionalized coefficients to simplify computation (USWB 1947). It is widely used by the US Corps of Engineers and other US water agencies and continues to be adopted or adapted for use in many parts of the world.

The concept takes the severest storms already experienced in the general region and then uses meteorological theory to maximize rainfall production within them. This additional maximization process utilizes air mass properties, the synoptic situation at the time of the recorded storms, and information on the area itself, such as size, regional location and topography. The amount of water vapour entering the storm cell is estimated from observations of moisture content (from temperature and dew point measurements), and the amount of air inflow (from measurements of windspeed or barometric pressure gradient and the estimated depth of the inflowing layer) and used to estimate the 'effective precipitable water content'. This is calculated for the maximum recorded storm and for the situation in which inflow and condensation rates are set at a theoretical maximum. The calculations can be based on direct observations or on estimates made from the synoptic situation and the season.

The final estimate of the PMP is given by:

$$\text{PMP} = \text{Observed maximum precipitation} \times \frac{\text{Maximum possible water content in clouds}}{\text{Water content in observed storm}}$$

This can be done for storms of different lengths or 'durations', e.g. 6, 24 or 72 hours, over different size areas and at different seasons. In all, it is a rather complex method and is best handled by a trained meteorologist. Details are given in WMO (1986a).

Where no large storms have been recorded, it may be possible to 'transpose' a storm from a similar area, as in the calculation of the SPS (WMO 1986a). The difference, however, is that the PMP requires the most extreme storms to be transposed and then proceeds to make them even more extreme.

The hydrologist may then use the PMP to estimate a **Maximum Probable Flood (MPF)**. This can be achieved by taking this maximum precipitation as input to a rainfall–runoff model or a catchment unit hydrograph, as described in sections 6.3 and 6.1.2.

Quite clearly, the use of the word 'probable' does not imply any quantifiable level of probability. Any hope of making a probability estimate has been destroyed by applying a theoretical maximizing procedure to the observed storms. It is really an estimate of what seems 'likely' to be the worst possible event in a certain area at a particular time of year, assuming that the proportion of rainfall to water content in that event remains in line with what has already been observed in the worst storms in the region. The accuracy of the estimate is also limited by the amount of meteorological data available and by the simplified treatment of meteorological processes. Miller (1977) describes it as a 'convenient fiction'; it is nevertheless one of considerable practical value.

Hershfield (1961) tested a simple statistical version of PMP, which overcomes the common problem of lack of sufficient meteorological data by sidestepping any attempt at physical modelling altogether. This was based on a general formula developed by Ven te Chow (1951) for the frequency analysis of hydrological events, which assumes a Normal distribution of magnitudes. It takes the PMP to be some large multiple of the standard deviation of annual maximum storm rainfalls, σ, above the average storm:

$$\text{PMP} = \bar{x} + k\sigma$$

In statistical terminology, k is a 'standard score' or z-score value. Calibrating this equation with records from thousands of raingauges, Hershfield obtained a coefficient of $k = 15$. Later tests suggested that this was too high in heavy rainfall areas and too low in low rainfall areas (Hershfield 1975). In other words, there is a greater range between extreme and average storm rainfalls in areas with low rainfall. Dhar *et al.* (1975) confirmed the value of this simplified approach and outlined the possibility of regionalizing the k values for different rainfall regions. In line with his aim of enabling PMPs to be calculated where no data are available other than standard rainfall measurements, Hershfield also proposed that the same formula could be applied to *daily* rainfall, since more data are collected on a daily basis than on autographic gauges or data loggers that would allow individual storms to be identified.

It has been claimed that the estimates are comparable with the complex US NWS procedure. Although Ward (1975) criticized it as a 'rough and ready' technique, it has found acceptance, especially where data or local expertise are insufficient for the full PMP approach. It is, however, advisable to use more than one method of estimation where important engineering projects are involved. It should also be remembered that Hershfield's method is based on raingauge data and therefore predicts values *at a gauge*. Depth–area–duration curves are needed to spread the PMP over an area, or perhaps over time, if the time base is changed (Shaw 1994).

A snowfall PMP based on physical theory is used in Canada. This makes it possible to maximize a series of snowstorms and then add up the results through a winter to obtain an estimate of the maximum end-of-season snowcover. This is then fed into a snowmelt model. Where snowmelt and rainfall mix to form a major flood hazard, it is possible to combine this type of estimate of maximum spring snowmelt with a standard PMP, which could be established specifically for the main type of heavy rainstorm occurring during the melt season.

The original US NWS PMP method has been widely accepted as a criterion for flood protection in urbanized catchments, where a high degree of security is needed. It is, nevertheless, open to criticism from two diametrically opposite points of view. By using only storms that have occurred in the records as a starting point, it is quite likely that these do not adequately represent the worst that can be expected. Even though they are then put through the maximization procedure, the point of taking real events as a starting point is to constrain the maximized storm within apparently reasonable bounds. Another viewpoint is that the PMP is too extreme even so, and that it requires too much expense to accommodate it in designs. This is why the SPS is still used in non-critical areas, and why engineers have so often turned to techniques that offer a measure of the *likelihood* of events of a certain magnitude as a guideline in cost–benefit decisions.

4.2.2 Estimating frequencies

When the costs of engineering are balanced against the resulting benefits, it is often decided that it is only economically worth while to make provisions to be 98 per cent or 99 per cent sure, say, that the flood control works will withstand the worst storm events or that a water supply reservoir system will be able to continue to supply sufficient water throughout the worst drought. It is usually stated that the works are designed to withstand the 1 in 50 year event (98 per cent) or the 1 in 100 year event (99 per cent certainty). The lower figure is commonly accepted for major works, but where human life is at risk 1:100 is specified. The higher figure may also be specified in major public supply systems, such as the River Dee Regulation Scheme, which combines a flood protection role for agricultural land with public water supply for Chester and Wrexham. By comparison, far greater probabilities of 'exceedences' are acceptable for urban or road drainage; perhaps only 1 in 2 years for drains on minor country roads. Hydrologists tend to view the risks in terms of the **return period** or **recurrence interval**, which is the reciprocal of the frequency: an event with a frequency of 1 in 100 has a return period of 100 years.

These risks are normally assessed by analysing past records of rainfall or runoff. The approach has been called 'actuarial', because it is based on the same premiss as the traditional approach of an actuary or insurance risk assessor, i.e. that the past record is the key to future performance. Long records are required in order to estimate the risks of the most extreme events. Even for relatively frequent events, the longer the record the better the risks can be assessed. Unfortunately, the shortness or lack of records is a common constraint. Even in Britain, there were considerably fewer discharge stations before the 1960s. In dryland regions, the assessor may be lucky to have more than 10 years of runoff records. Rainfall records are more common, with a reasonable coverage for 100 years in Britain. Hence the proliferation of techniques to reconstruct riverflows from rainfall records discussed in section 6.2.

Past records may be used in two main ways: (1) as they stand, or (2) as the basis for extrapolation. Neither of these approaches is essentially dependent upon computer analysis or requires much calculation, although certain forms of modern computer modelling do owe their origins to the same philosophy and can also be used to extrapolate frequencies (sections 4.2.3 and 6.2).

Duration curves and cumulative flows In the first type of approach, streamflow records can be used to construct **flow duration curves** like that in Figure 4.4, which can show the amount of time flows fall above or below a certain discharge during the period of record.

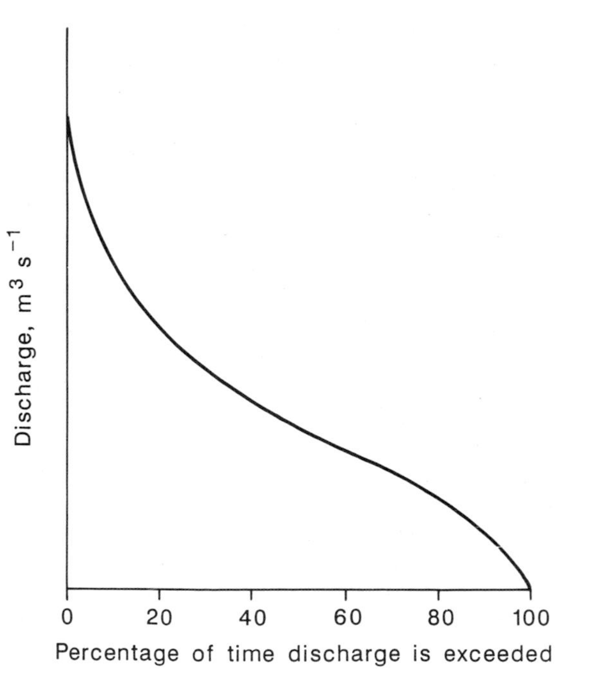

Figure 4.4 A simple flow duration curve

The flow duration curves for different rivers can be more easily compared if $Q/\overline{Q}$ is plotted, standardizing each flow, Q, in terms of average flow in each river. Pirt and Simpson (1982) standardized the curves by dividing through by basin area and mean annual effective precipitation, in order to quantify the effect of lithology and soils on low flows. They found that different master curves could be established for each soil type in the UK Flood Studies Report classification (section 6.1.1), so that selecting the curve appropriate to the hydrological soil type in an ungauged basin and multiplying by drainage area and effective precipitation would enable low flow risks to be estimated without any need for flow records.

Flow duration curves provide a first estimate of the overall risk of low flows, but tell us nothing about the sequence of flows. For designing water supply reservoirs, the length of sustained sequences of low flows is very important. In 1883, Rippl developed a method which has found longstanding and widespread acceptance called **mass curve** or **Rippl analysis**. The method selects the sequences of lowest flows in the historical record and calculates cumulative inflows for the *worst historic drought* sequence. The result is then used to calculate the storage needed to guarantee supplies during periods when inflows fall below estimated

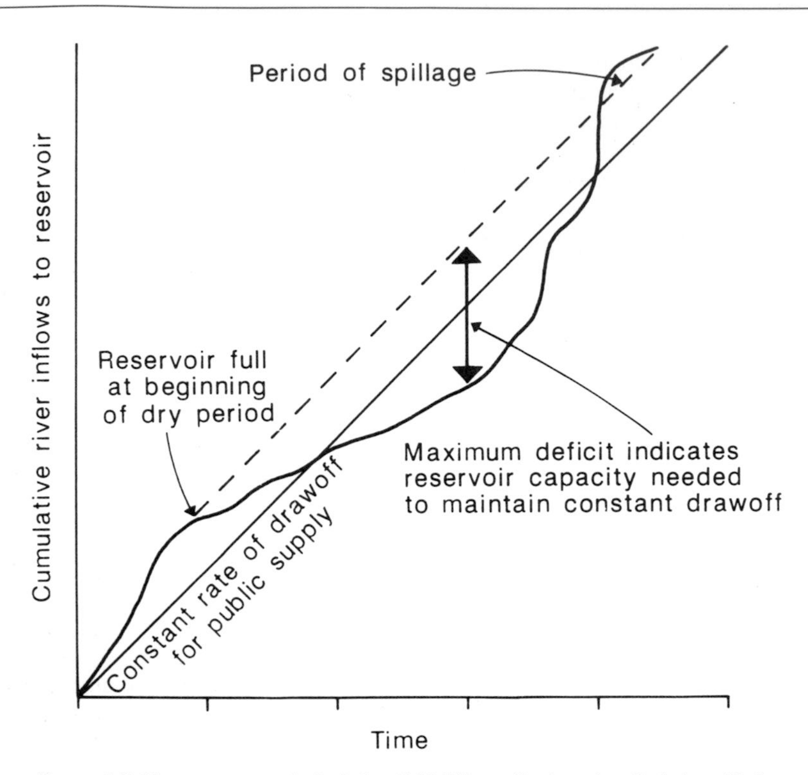

Figure 4.5 Mass curve analysis (after E M Wilson *Engineering Hydrology* (3rd edn). Macmillan Press Ltd)

demand. As Figure 4.5 illustrates, the curves are Rippl by name and ripple by nature.

Extreme value analysis The approach pioneered by Gumbel (1958) is founded on the theory that the frequency with which certain events have occurred in the records is a good guide not only to their own future occurrence but also for *lesser and greater events.* Extrapolation and interpolation are usually achieved by fitting a straight line through the data points, which are plotted according to magnitude and frequency. Special graph paper is available for these plots. In order to achieve a staight line, the plots are taken to be plots of cumulative frequency and the frequency axis is graduated unevenly to match the cumulative frequency distribution of a model statistical distribution or **probability density function (pdf)**, such as the Normal distribution (cf. Figure 4.10). However, no single statistical distribution has been found that fits all data, so a number of different graph papers exist for analysing extreme events.

Again, while it is easy to see where to plot an event on the magnitude scale, it is less clear how to plot it on the frequency axis, especially since this is really *cumulative frequency*. The most obvious definition of frequency, F, for annual floods is

$$F = \frac{\text{Rank order of event } (m)}{\text{Number of years } (n)}$$

where the largest event is given a rank of one. Thus, the 10th largest event in 100 years has a probability of 1/10 or 0.10 and a return period ($1/F$) of 10 years. However, each graduation on a cumulative frequency graph relates to the frequency of events that are not simply equal to but *equal to or greater than* the indicated magnitude: a point that is often forgotten when interpreting the predictions. In addition, this simple definition is not the best estimator when records are short and the sample of extreme events is likely to be somewhat unrepresentative. Hence, at least seven less obvious definitions have been proposed for plotting the frequency, which all tend to increase n and give m less weight. The most popular is the Weibull formula:

$$F = \frac{m}{(n+1)}$$

The analysis begins by selecting either: (1) the highest rainfall or discharge in each year; (2) all rainfalls or discharges above a certain threshold; or (3) a mixture of the two which takes the same number of

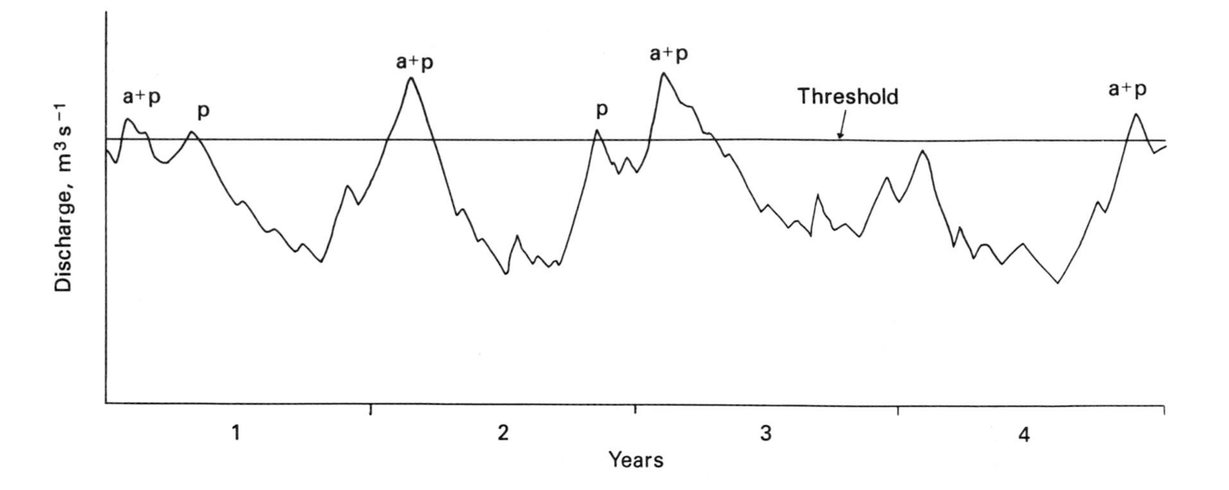

Figure 4.6 Selecting events for the annual (a) and partial (p) duration series. See text for explanation

peaks as the number of years, but selects them irrespective of yearly boundaries (Figure 4.6). All of these methods filter out the most common events from the **full or complete series**, which are of no use for the study of extremes. The choice of method depends on the purpose of the analysis and has to balance the advantages and disadvantages of each; none is a perfect answer.

The first option is called the **annual maximum series** and is the best option for ensuring the assumption that all events are independent, which is required by the statistical analysis, is not broken. Events close together may not be independent random occurrences, especially when the analysis is made using daily values (which are more widely available than values archived on a storm-by-storm basis). High flows on one day tend to be followed by similar flows on the next day, so the rarer the chance of two close 'events' both being selected for the analysis the lesser will be the chance of breaking the fundamental statistical assumption. If the year is defined as the Water Year (beginning when the annual groundwater recharge cycle starts) rather than the calendar year, then the chances of breaking this assumption can be reduced further. On the other hand, this method of selection means that there is a high probability that the second highest flow in some years will be higher than the highest in other years, yet it is omitted from the analysis.

The second option, known as the **partial duration series** or more explicitly the 'peaks over threshold series', answers this latter criticism, but increases the chance of breaking the assumption of independence. This method may also select more events for the analysis, which increases its statistical reliability and may be a crucial consideration if records are short. The UK Flood Studies Report concluded that this method is a particularly good one for estimating the mean annual flood from records as short as 3 to 10 years (NERC 1975). It is also the better method for assessing the frequency of fairly frequent events, whereas the slightly simpler annual maximum series is normally preferred for low frequency events like the 'hundred year flood', i.e. 1 in 100 years. For events with return periods of under 2 years, the partial duration series gives significantly shorter intervals, the difference increasing to more than 50 per cent at the annual level.

The third option, the **annual exceedence series**, which combines both criteria is not commonly used. It could be seen as a valuable compromise on theoretical grounds, but in practice there is so little difference between the other two methods in the estimates of return periods greater than 10 years that there is hardly any reason to introduce the extra complication.

Once the appropriate data series has been selected, the analyst usually seeks to fit a straight line through the data that will enable them to be extrapolated to events which are more extreme than any on record. Finding a statistical distribution that will enable a simple straight line summary of the data can be a matter of trial and error. There are no firm guidelines as to which is best. In broad terms, these distributions can be divided into Gaussian and non-Gaussian models, named after the nineteenth-century mathematician Karl Friedrich Gauss. As Figure 4.7 shows, a true Gaussian or 'Normal' model is symmetrical. Varying degrees of asymmetry or 'skewness' may be caused by natural

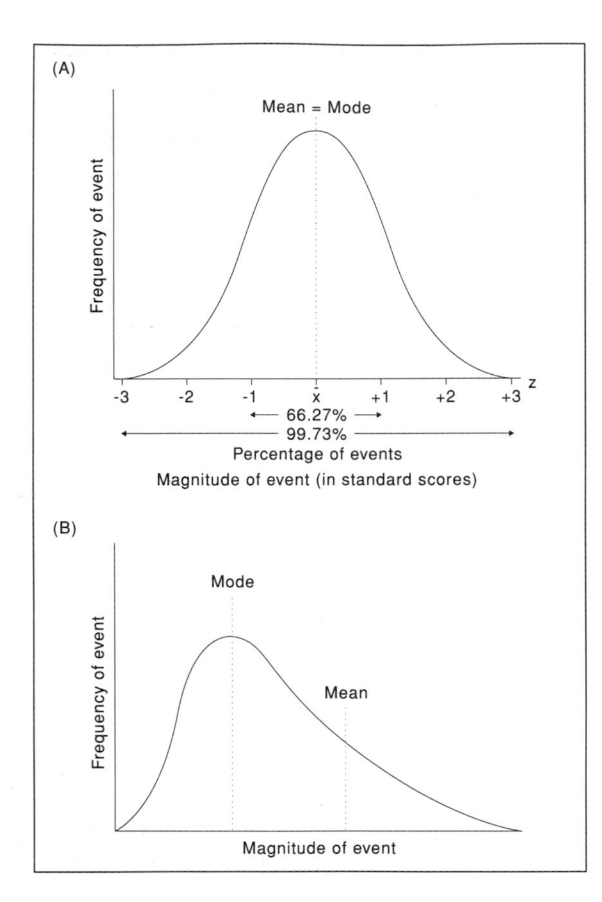

Figure 4.7 Gaussian and non-Gaussian distributions of events: (a) Gaussian or Normal distributions are 'bell-shaped' and centred on the mean; (b) non-Gaussian distributions in hydrology are 'skewed' with a long 'tail' of high-magnitude events, which causes the mean to fall some way above the peak frequency.
(Note that in Figures 4.8 to 4.11 magnitude is on the vertical axis and the horizontal axis is *cumulative* frequency, unlike here)

deviations from the simple assumptions of the Normal model, or they may be due to a few extreme events turning up in a comparatively short record.

It is possible to undertake the same exercise purely by computing. Chow (1951) proposed a simple formula (identical to that described for Hershfield's PMP in section 4.2.1) that is based on the Normal distribution and a 'frequency factor', k, which is synonymous with the 'normal deviate', z, found in statistical tables. The US Water Resources Council has published frequency factors for a popular skewed distribution, Pearson Type III (Viessman and Lewis, 1996).

Gaussian models In the 1940s, the statistician E J Gumbel investigated a number of distributions related to the basic **Normal distribution**. A very large collection of independent, random events should approximate a Normal or Gaussian distribution. In practice, hydrological series tend not to be totally random and the data sets are not very large, so alternatives are needed. Rainfall series often tend to be more Normal than riverflow series. In part this is because they are the result of one set of 'random' processes (the rain-producing processes), whereas riverflow is also the result of river basin properties, particularly how wet the soils are at the beginning of the storm. Gumbel invented his own distributions and used them as the bases for special graph papers to facilitate plotting.

Figures 4.8 to 4.11 illustrate four different model distributions, beginning with the Normal distribution as portrayed on Normal probability paper. Any Normally distributed series will plot as a straight line on this paper. Note that the graduations on the paper expand both ends of the graph to achieve this effect. The figures at the top of the plot in Figure 4.8 indicate the probabilities of daily rainfalls being at least as large as those on the vertical axis. Hence, the straight line fitted to the data suggests that rainfalls of 100 mm or more have a probability of 0.02, i.e. 2 per cent. The figures on the bottom axis translate this into frequencies; daily rainfalls with a probability of 2 per cent should occur twice in 100 years, i.e. they have an average return period of 50 years.

Runoff data tend to be more skewed than rainfall data, perhaps because they combine the effects of more than one set of 'random' processes, comprising both meteorological and topographic effects. The proportion of extreme rare (low frequency) events tends to be disproportionately large compared with a simple Normal distribution. Such a distribution may be 'tamed' by transforming the scale of magnitudes. In Figure 4.9 a logarithmic transform has successfully converted such a distribution into a Normal distribution by contracting the upper end of the magnitude scale. This straight line **log-Normal** graph is easier to extrapolate than the exponential curve of non-transformed data.

The next two Figures, 4.10 and 4.11, show two of Gumbel's own distributions. Figure 4.10 takes the same data as Figure 4.9. It suggests that the data for the Tana River in Kenya are fitted slightly less well by a **Gumbel Type I** distribution than on the log-Normal graph in Figure 4.9. The slight hint of an upwards curvature in Figure 4.9 might suggest trying a logarithmic transform of the magnitude scale.

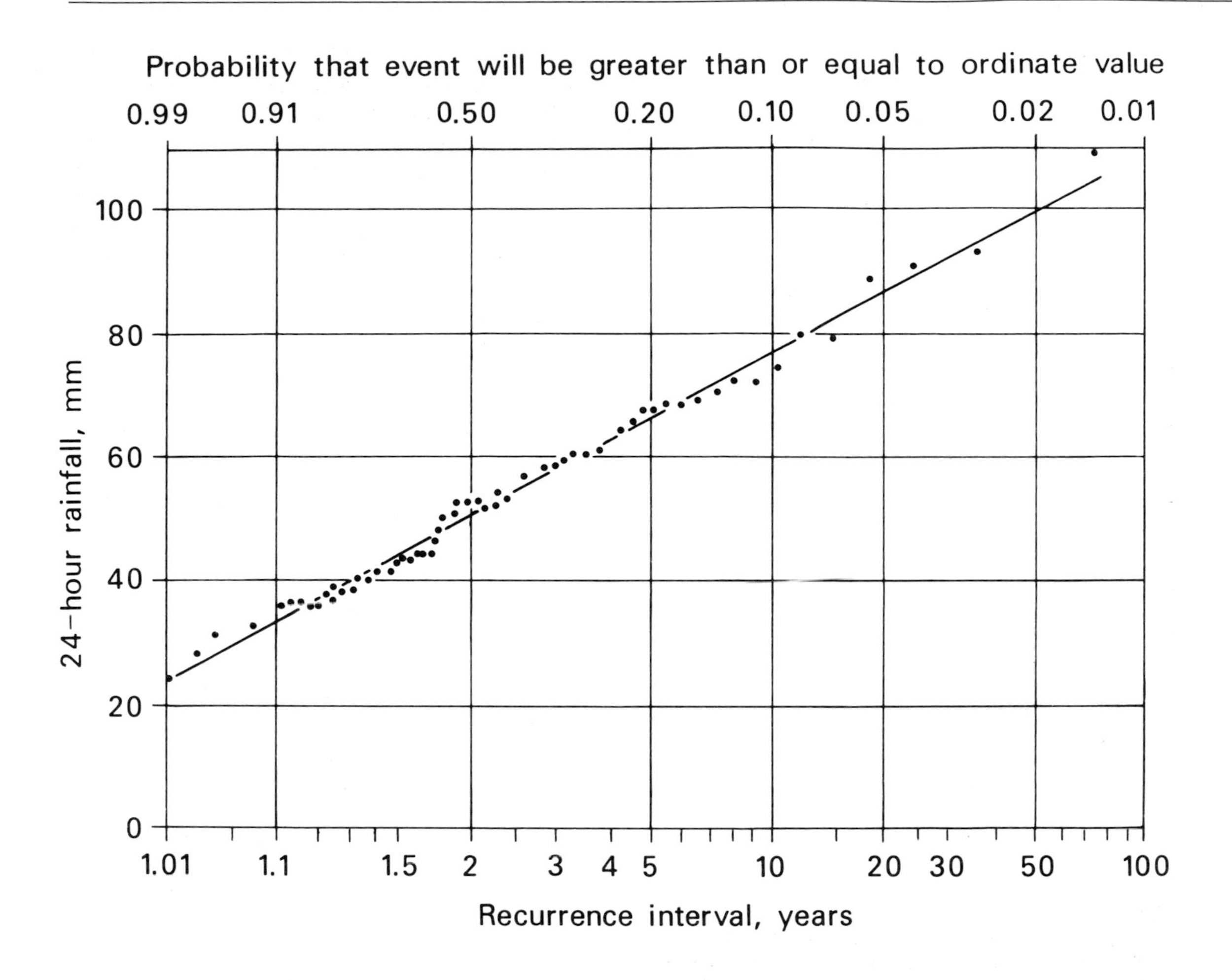

Figure 4.8 Normal probability paper used to plot hydrological events. Daily rainfall annual maximum series, Buffalo, New York, 1891–1961 (after Dunne and Leopold 1978, from US NWS data)

Gumbel's Type III distribution does just this. But as Figure 4.11 shows, in this particular instance the result is worse; the upwards curvature was too slight and has now been overcorrected.

Non-Gaussian models Sometimes the skewness in the data is too marked for a logarithmic transformation to reduce it sufficiently (Figure 4.7). In other words, there is evidence of rare events of quite exceptional magnitude, far in excess of what would be expected in a Normal distribution. Alternative, 'non-Gaussian' statistical distributions may be fitted to these cases.

The family of distributions known as **gamma distributions** is often used because it not only covers very skewed distributions but it is also very adaptable to different degrees of skewness. It may also bear some similarity to the physical processes involved in the generation of riverflow. One way in which a gamma distribution may arise is through the combination of two or more random series of occasional events spread over a continuous period of time. This could be a combination of random rainfall events and random antecedent soil moisture conditions, which between them control the magnitude of the flood. Another way of looking at a gamma distribution is as random events that occur with shifting frequencies over time. This could be due to the same physical argument, i.e. superimposed random series, or it could be due to the inherent shifting nature of the climatic system, or both. Gamma distributions probably fit flood series so well because they can cover the multiple strands of variability in nature.

The **Pearson Type III** distribution is a very popular gamma model with its own graph paper available. For

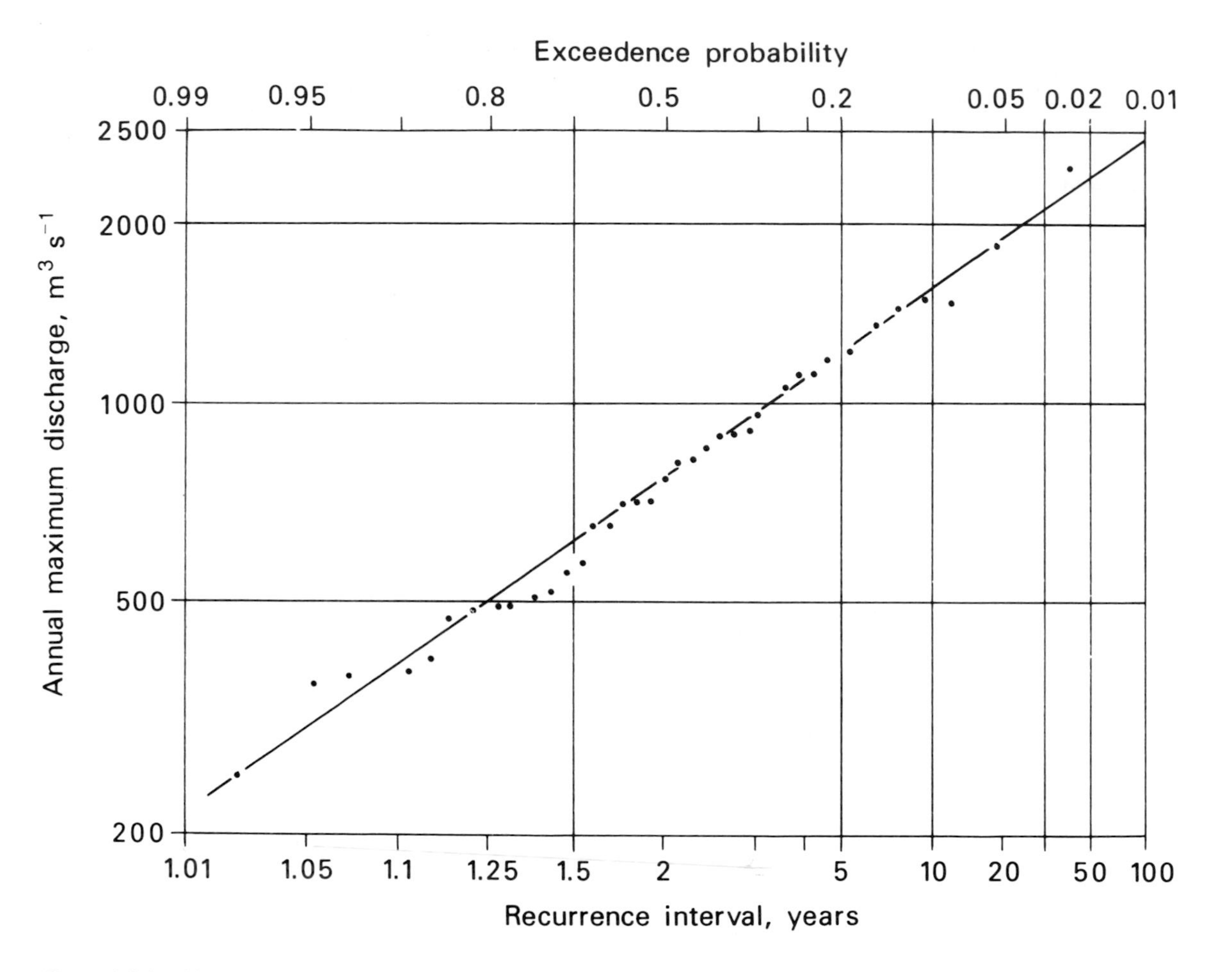

Figure 4.9 Log-Normal probability graphs of annual maximum daily discharge for the Tana River at Garissa, Kenya 1934–70 (after Dunne and Leopold 1978, data from Ministry of Water Development, Nairobi)

even greater skewness, there is **log-Pearson Type III,** where log-discharge is plotted. This latter distribution is often found to fit annual flood series very well and the US Water Resources Council has issued maps of generalized skew coefficients for the logarithms of annual flow maxima in the United States.

Problems with return periods Despite the widespread and successful use of these statistical approaches to estimating the risk of extreme events, there are a number of problems that need to be appreciated. At the very least, these require caution in interpreting the meaning and reliability of the estimates. At worst, some practising hydrologists claim that they completely undermine their value, although this seems to be a rather extreme judgement.

Most problems arise from one of three sources:

1. These are only statistical, not deterministic, estimates, so that a 50-year flood does not just occur rigidly at 50-year intervals. Rather, the return period is a measure of its *average* frequency: in any individual instance the next '50-year' flood could follow any number of years after the last. The probability of getting two 50-year floods in consecutive years is extremely low, but it exists: nothing can be ruled out entirely from a statistical point of view.
2. The estimate is inevitably based on inadequate data. The shorter the record, the more doubt there must be about estimating rare events or selecting the 'right' model to fit. Worse than this, because

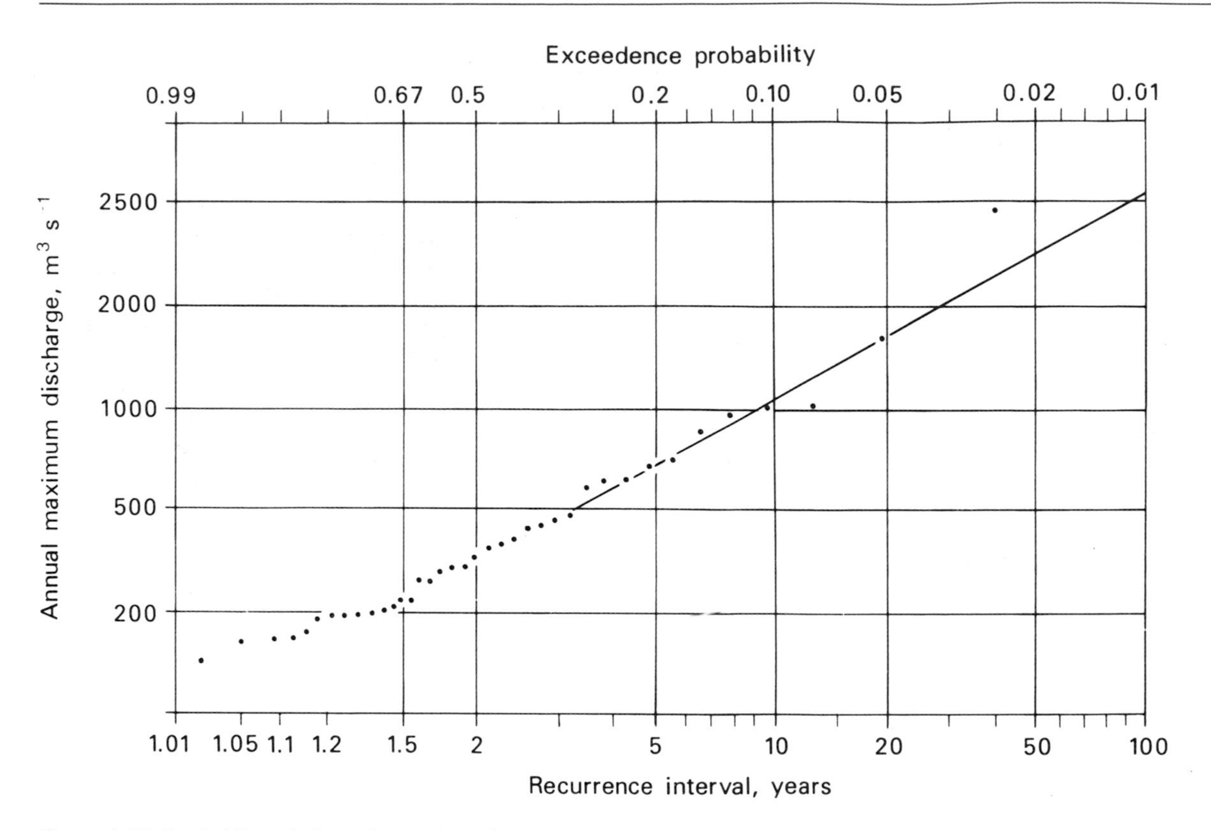

Figure 4.10 Gumbel Type I plots of annual maximum discharges for the Tana River, Kenya (after Dunne and Leopold, 1978)

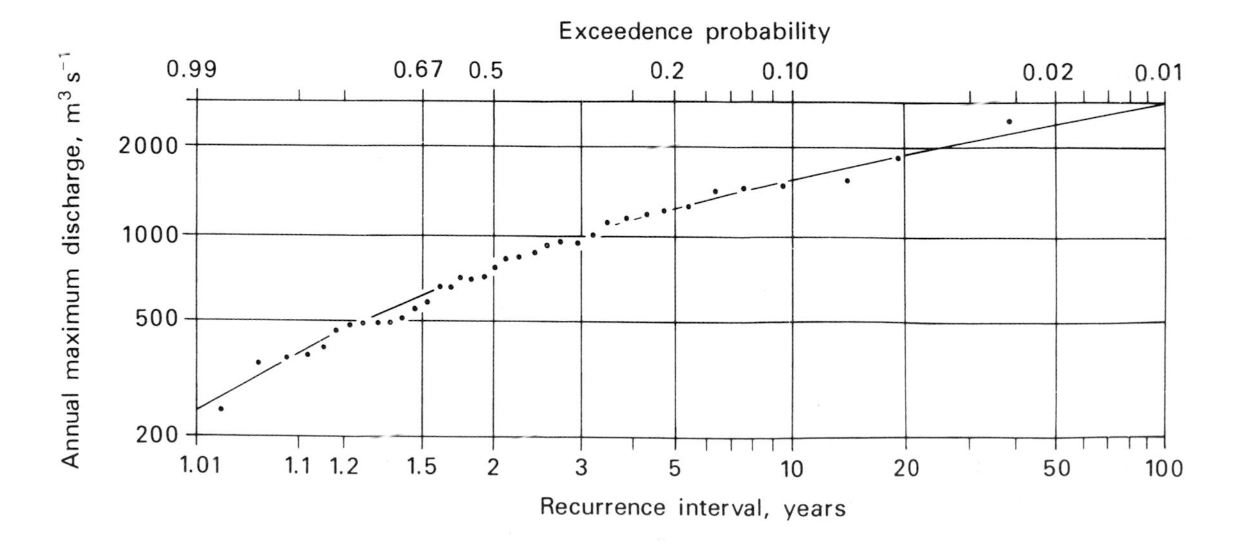

Figure 4.11 Gumbel Type III plot of annual maximum discharges for the Tana River, Kenya (after Dunne and Leopold, 1978)

50-year floods, or any others, can pop up any time, in practice we are never sure that, say, a 20-year record does not contain a 50-year flood. This seems obvious: 50-year floods must turn up somewhere and therefore in some shorter records. The problem lies in the implicit assumption of the technique that they do not. If we have only 20 years of records and the largest flood in that record is really a 50-year flood (that would be recognized as such if we had 50 or 100 years of record), the technique assumes that it is the 20-year flood.

3. To add to problem 2, climate is forever fluctuating and shifting. Any record is only a finite sample in a sea of change. The Colorado Compact was very unfortunately based on records covering some unusually wet years (section 4.3). Fears of the effects of global warming suggest that recent records will not be a good guide to future risks (section 11.5). Even without this extra artificially induced change, Figure 4.12 illustrates the wide range in estimates that can be obtained by taking records from different periods even within this century. Each of these periods approximates to the WMO guideline of basing climatological estimates on 30 years of records, yet the 50-year daily rainfall in Oxford is twice as large when estimated from the wettest period as it is from the driest.

   This result might conventionally be put down to 'climatic variability'. However, it might also be explained by a few rarer events turning up, or not turning up, in individual 25-year periods. In other words, it is the weather not the climate that is displaying variability; the events are merely samples from the set of probabilities within the current climate. It can be shown that for perfectly random annual events there is only a 64 per cent chance that any particular 30-year sample will contain a storm with a return period of 30 years or more. This means that there is quite a high risk (a 36 per cent chance) of the sample not containing anything as large as the 30-year event. At the same time, we find that there is a 26 per cent probability that an event with a return period of 100 years or more will turn up in any 30-year sample. Bruce and Clark (1980) give a clear discussion of these calculations.

These observations raise the question of what is the current climate. Water resources engineers must plan

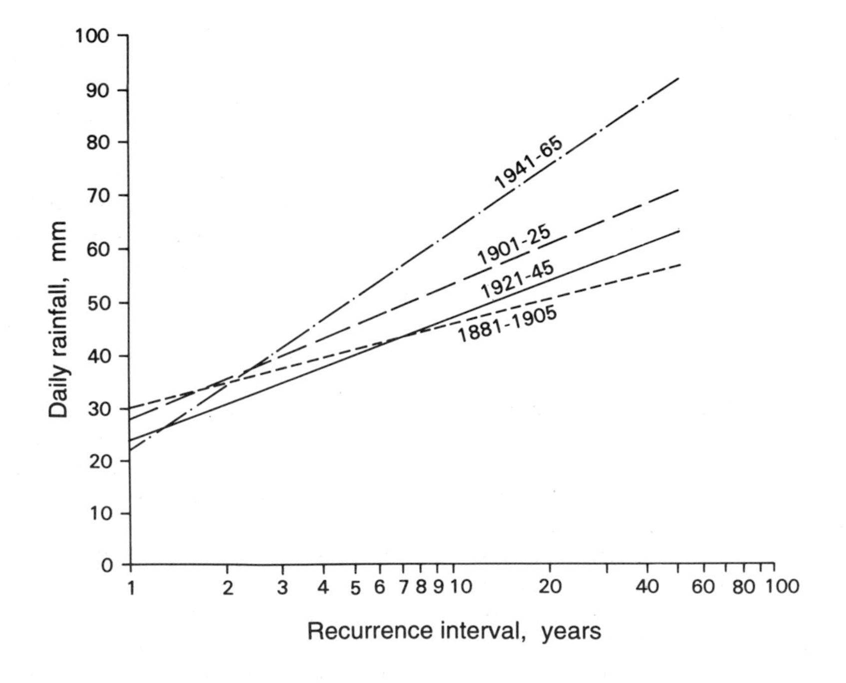

Figure 4.12 Estimates of return periods or recurrent intervals for maximum daily rainfall at Oxford, England, derived from different periods of record. Based on Rodda (1970)

for 50- to 100-year events. We may even find it useful to estimate the likelihood of floods or droughts that have return periods counted in centuries (e.g. Doornkamp *et al.* 1980). Logically, there must be some finite risk of such extreme events occurring 'now', under the present climate. Yet climate certainly changes on the scale of centuries and seems to do so even over decades. So, although, according to standard statistical theory, longer records should enable us to establish the risks of extreme events better, the climate is a 'non-stationary series', forever changing, and the observations taken at the end of a long record could be from a climate with a set of event probabilities that is significantly different from that at the beginning. Perhaps we are here confronting the natural limits to our means of estimating risk. Future improvements in climate modelling might push forward the frontier, but in the end how predictable is the climate (section 11.5; Gleick 1988)?

This problem is compounded by the fact that 'Gumbel analysis' in general also implicitly assumes that the terrestrial environment remains unchanged. There is no allowance for the effects of environmental changes such as major land drainage or afforestation.

Operational yields Disillusionment with the concept and methods of return period analysis led Jack and Lambert (1992) to propose the technique of **operational yields**. This aims to determine the yield that can be expected from a reservoir during the 'worst possible drought sequence'. The worst possible drought is an artificial creation rather like the PMF. The computed operational yield is based on inflows under the worst drought scenario *and* on manipulation of the quantity of water supplied to consumers using all the legal and technical means available.

Figure 4.13 illustrates the technique. The worst possible drought is constructed by selecting the worst summer drought on record and extending it to mid-November, whereas in a normal year in Britain heavy rains during September–October should end the drought. The Supercalc spreadsheet model now in use with Welsh Water adds a variable 'supply factor', which begins the year at 1.0, climbs to 1.2 in spring when consumers are beginning to increase demand at the first signs of a drought, and is then progressively reined in. The checks on supply follow the normal progression of operational decisions, in which the operators will not know how long the drought is to last, so they begin with calls for voluntary economies and end by applying legally binding 'drought order' restrictions. Jack and Lambert (1992) have proposed that one further option might be to reduce compensation waters early in the year. There are, indeed, grounds for believing that in some cases established 'minimum acceptable flows' released by reservoirs can be too generous at certain times of year, and should be flexible. Extreme care is needed, however, to safeguard the river environment, especially wildlife.

The operational yield is proving a valuable tool in the management of existing reservoirs, and is now used throughout Welsh Water with the exception of the Dee Regulation Scheme (section 9.3), where legislation specifies fixed return period criteria.

4.2.3 Extending the record and improving predictions

Since one of the major factors that limits the applicability of extreme value analysis is shortness of records, numerous attempts have been made to artificially extend records. Early, pre-computer age approaches included the station–year and double mass curve methods for extending rainfall records. The **station–year** method would hold that if there are 10 years of record at 10 rainfall stations then this can be treated as 10 × 10 = 100 years of records. This is, of course, a rather dangerous assumption, first, because long-term rainfall can vary so much from site to site, and, second, because in the short term it may not! If long-term rainfall differs significantly from station to station, then the station records cannot be regarded as samples from a single population. But if storm rainfall records have a high degree of inter-station correlation, then they are not independent samples; they are likely to be recording the same storms, which would then be given double (or more) weight in the extended 'record'. These problems can be minimized by careful inspection and selection of the data.

The **double mass curve** method is illustrated in Figure 4.14. This is not to be confused with Rippl's mass curve described in the last section. The objective here is to establish a relationship between a certain gauge site where the record has ceased or been otherwise interrupted by a change in the method or location of measurement, and nearby sites with uninterrupted records. Plotting cumulative rainfall for the interrupted site against cumulative mean totals for the other sites should allow an adjustment factor to be calculated and applied to reconstruct data for the interrupted site.

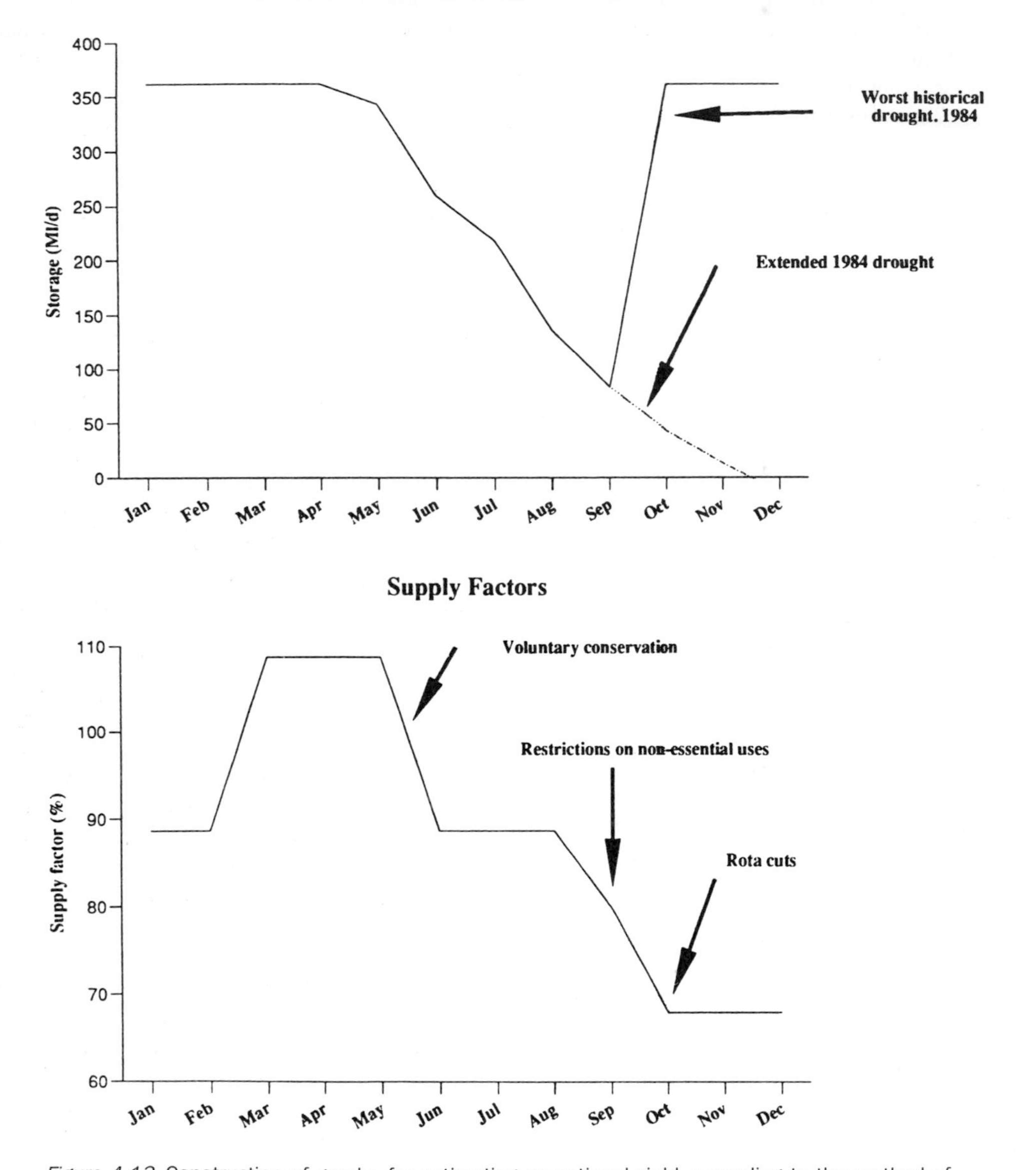

Figure 4.13 Construction of graphs for estimating operational yield, according to the method of Jack and Lambert (1992)

This is really a graphical case of correlation or multiple correlation, and the ease of statistical analysis on modern computers has led to many different examples of records being extended by correlation. A common example is where rainfall records are used to extend a short riverflow record by establishing a correlation for the period of parallel measurements.

Palaeohydrologic reconstructions Riverflow and rainfall may also be reconstructed by correlations with a wide variety of predictor variables. The IAHS International Commission for Surface Waters has collated source material on suitable hydrological proxy data to aid in these reconstructions (Liebscher 1987). Jones *et al.* (1984) undertook reconstruction of past riverflow in

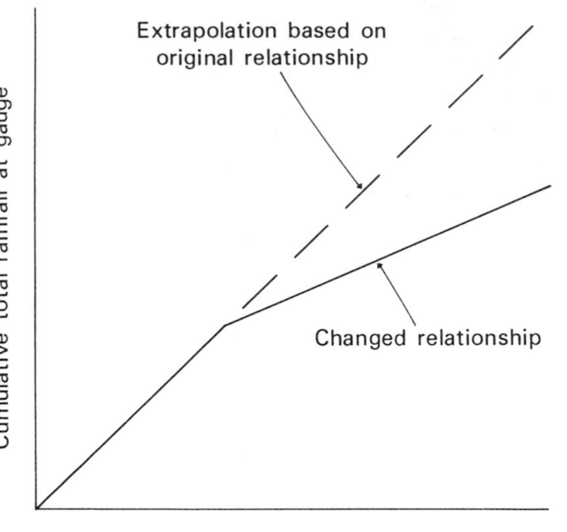

Figure 4.14 The double mass curve method of extending a record or filling in missing data based upon correlation with surrounding stations

Britain using tree-ring analyses, in which variations in the width of annual growth rings were correlated with current rainfall. They then used the historic tree-ring record derived from old trees and building timbers to reconstruct past rainfall. Schadler (1989) reconstructed a 450-year sequence for the Rhine using regression, based on modern rainfall and temperature. He applied the regression equation to a set of proxy weather data that had also been reconstructed.

The WMO World Climate Programme is also encouraging further research into palaeohydrologic analogues, which might be used to estimate flood magnitude–frequency curves for global warming scenarios. An approach which is being actively developed by fluvial geomorphologists is to use geomorphic evidence of past floods to reconstruct magnitude–frequency curves directly. These tend to use deposits left by past floods and take some representative measure of the size of clasts, or individual large boulders, to estimate flood flow velocities. For some severe events scour marks, wash limits or trim lines in lichens growing on rock surfaces may be available to indicate the depth of flow.

The methods are clearly incapable of reconstructing full flow sequences, but this is not necessary. Just the frequencies of the most severe events might be useful information in its own right. However, there are a number of considerations that are of more concern:

1. The depositional record is often only partial, as later floods have erased the evidence of earlier floods.
2. The techniques for estimating palaeovelocity and palaeodischarge from deposits still require considerable improvement (Thornes and Gregory 1990). Maizels (1983) gives a detailed review of the technical problems. There are still many sediment–flow relationships that are not fully understood, even for present rivers. Hydraulic equations tend either to make important assumptions that may not be relied upon or to require field data that cannot be obtained. Natural streams are extremely variable, for example in the spatial and temporal distribution of velocities, in bed roughness and in sediment properties and supply. Hence, the fundamental assumption that flow was steady and uniform, which is generally needed to apply these hydraulic equations to palaeohydraulic reconstruction, is unlikely to be valid (Maizels 1983). There is also no guarantee that the remaining portion of the flood deposit is a representative one: there could be an argument for supposing that it may have remained untouched because it has always been in a part of the valley that is away from the main flow, even perhaps in the flood that deposited it. Similarly, it may be very difficult to determine whether the deposit represents one single flood or a number, especially without any datable material within the deposit, and the question then must be asked as to what velocity or discharge is being estimated. Equally, in estimating peak discharge the cross-sectional area of flow is needed, but this is often more difficult to reconstruct than peak velocity. In the final analysis, geomorphic estimates of palaeovelocity and palaeodischarge can still only be regarded as 'first approximations'.
3. Finally, the fluvial system contains many complex interactions and while climate may be the 'driving force' there is a considerable 'cultural blur' in the history of European and many other rivers (Newson and Lewin 1990), which can make it difficult to distinguish between changes in flood frequency that are climatically induced and those that are due to human activity. Often the changes are a mixture of the two. An excellent example of this problem is provided by the British River Severn. Rodda *et al.*

(1978) explained a slight increase in median flows at Bewdley over the period 1851–1975 as due to land-use changes and, in the last decade, to river regulation. Howe *et al.* (1967) suggested that afforestation in the upland reaches had effectively increased drainage density and so increased flood peaks. Yet Higgs (1990) has concluded that much of the change can be attributed to changes in the location and frequency of heavy rainstorms.

Stochastic synthesis A common modern approach to extending records is to use computer techniques for 'stochastic synthesis' – this is described in section 6.2. These can create an artificially long synthetic record, which preserves the statistical properties of the original shorter record. This is achieved by randomly sampling events from the same statistical distribution that fits the events in the record. It provides an infinitely extendable 'record' and may produce *rare events* which do not appear in the instrumental records, but which may be expected to occur in a large sample taken from the fitted statistical distribution. In effect, the distribution provides a statistically preserved 'current climate' in which these rare events may materialize. It also randomly resequences the order of events so that it produces *sequences* that have not appeared in the instrumental record. This enables a fuller exploration of the probabilities of sequences, which is particularly important in estimating the risks of extended droughts. Time series synthesis can also help to take climatic change into account by adding in a 'trend' factor (section 6.2).

Over the last 20 years, synthetic techniques have found widespread acceptance as a means of estimating risk, but they too fall short of being a panacea. Most techniques suffer, like 'Gumbel analysis' in general, from the implicit assumption that neither the climate nor the terrestrial environment will change significantly in the future and has not done during the period of synthesis. A better way of accounting for the effects of such changes is to use 'simulation' models, which are discussed fully in Chapter 6. Simulation models take more account of the physical processes involved. Blackie and Eeles (1985) describe the use of a 'lumped' IH rainfall–runoff simulation model to extend flow records.

4.3 Patterns, cycles and teleconnections

Most of the techniques outlined so far tend to treat floods or droughts as random events in a stationary series. Yet climate and riverflow are clearly non-stationary and follow trends and cycles. One of the classic and most critical misjudgements in the history of hydrometric assessment proves the point. The 1922 Colorado Compact apportioned water rights on the Colorado River on the basis of records taken during what turned out to be an exceptionally wet period. The consequences may have hampered development over a large portion of the southwestern USA ever since. In the years 1896–1930 the average flow at Lee's Ferry was nearly 21 billion m^3, whereas the 1931–65 average was only 16 billion m^3; nearly a 25 per cent reduction. The Compact was drawn up when the 10-year moving mean was at its highest.

The problem was not caused by a short record. Quite simply, the record was collected during a period that has not proven representative of the rest of the twentieth century. This kind of problem can be caused by the **Hurst effect**, or **persistence**, with periods of similar years clustered together. If running means, like 5-year averages, are plotted through these clusters, cycles tend to appear with each run of below average years followed by a run of wetter years (Figure 4.15a). This means that long-term variability is much greater than might appear from a record collected during just one half of the cycle; and it makes the dangers of predicting on the basis of short records even worse. It can be a problem even for modern stochastic synthesis, despite Markov, ARIMA and time series models that do take persistence into account (section 6.2), because they can only take account of any persistence they find *within* the record, not of what happened beyond the end of the record.

Cycles in rainfall and riverflow have been identified in the USA, the CIS and many other areas, often with a variety of wavelengths of cycle superimposed. Russian scientists largely pioneered this work, and identified particularly strong cycles of 2 and 11 years in Soviet rivers. In some cases regional patterns have been observed, in which certain cycles are dominant in certain regions. Some interesting long-distance correlations have been observed in which regions as far away as South America and Asia seem to vary in unison, either positively or negatively. They may switch from wetter than average to drier than average together or out of phase. These **teleconnections** are both fascinating and controversial. We do not yet know

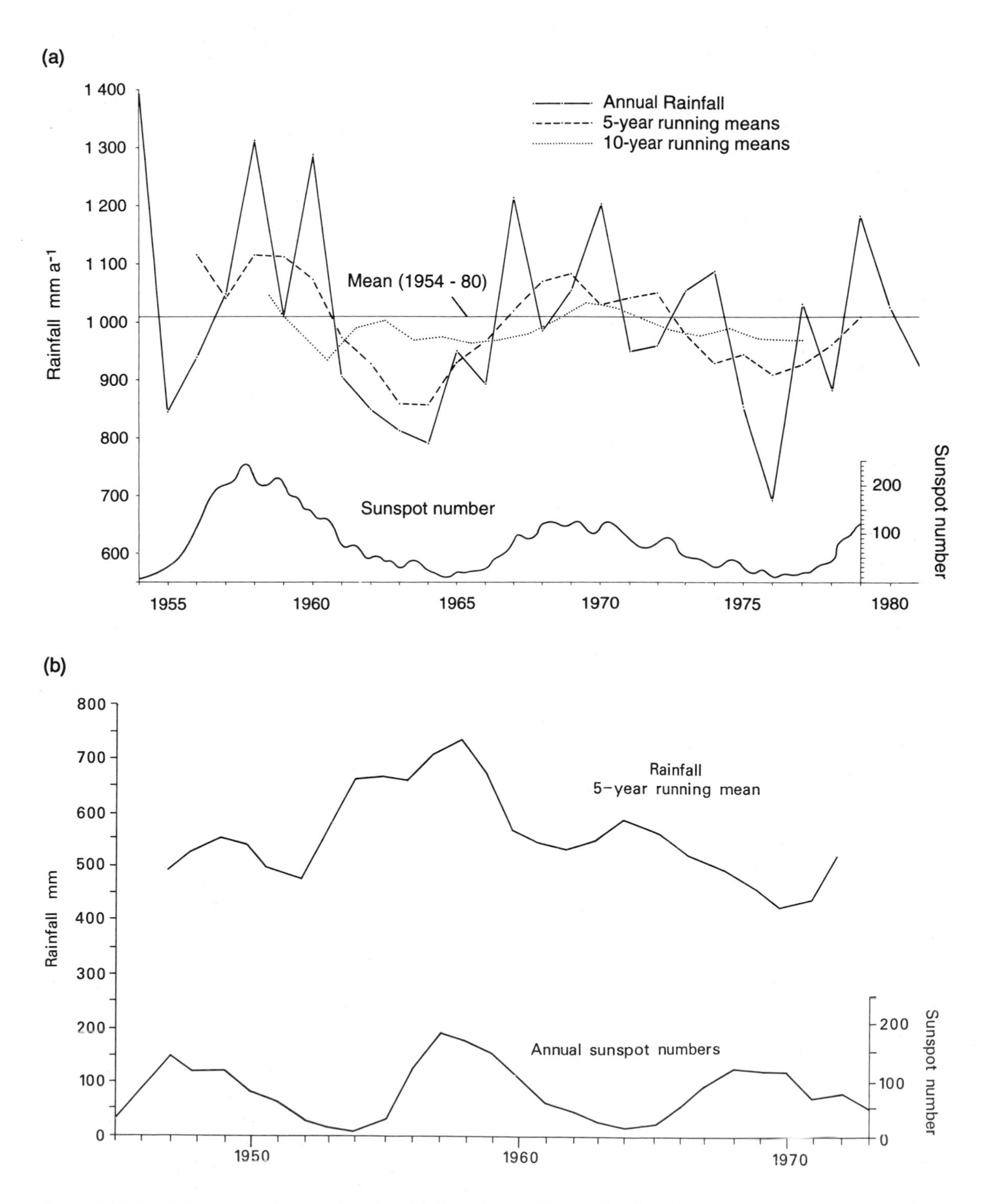

Figure 4.15 Rainfall cycles and sunspot cycles: (a) Aberystwyth, UK, showing the effect of smoothing rainfall records by running means; (b) Lahore, Pakistan, showing positive correlation up to the 1960s

whether they are real effects worth investigating, or simply statistical freaks. Most of them seem very unlikely to be real from what we currently know about atmospheric circulation patterns. But our knowledge is rapidly growing in this field. Not long ago the Southern Oscillation (section 2.2.2) was considered to be of interest for the tropics but not for mid-latitudes. The accompanying reversal of the ocean current, El Niño, carries warmer water to the west coast of South America, causing heavy rains there and a matching drought in tropical Africa and Australia. However, recent evidence connects ENSO events with river discharge and daily precipitation in the southeastern USA, California and the Pacific Northwest (Kahya and Dracup 1994; Woolhiser *et al.* 1993). Following an ENSO event, which brings warm water and more rain to the eastern tropical Pacific, the Rossby waves in the mid-latitude westerlies increase in amplitude between the central Pacific and the western Atlantic. This **Pacific/North American (PNA)** effect causes sea level pressure anomalies in the eastern North Atlantic (Hamilton 1988). The Hadley Centre high-resolution GCM for Europe (section 2.6) has been tested for its reproduction of ENSO events (Hadley Centre 1993).

The last decade has actually seen a marked increase in interest in causal links between climatological cycles and runoff cycles, and in the forcing factors behind the climatological cycles. Frequent attempts have been made to link **sunspot cycles** and rainfall. Sunspots are darker areas of the Sun's surface that follow quasi-regular cycles of increasing in size and number and then decreasing again over periods of about 9–17 years, with an average cyclical period of around 11 years. The 'sunspot number' is actually a statistic based on their size and number. Many of these correlations have turned out to last for a while, then mysteriously disappear or suddenly to flip sign from a positive to a negative correlation; high rainfall years associated with high sunspot numbers for a while, then suddenly associated with low sunspot numbers. For decades, many scientists have pointed to this and the continuing lack of a totally accepted physical explanation for the connection as grounds for disregarding sunspots. But correlations do continue to turn up. The almost total absence of sunspots during the peak years of the Little Ice Age between 1645 and 1715, known as the Maunder Minimum, seems to confirm a broad link between solar activity and global temperature and rainfall patterns. Rodda *et al.* (1978) found links between sunspot and rainfall/runoff cycles in more recent instrumental records in Britain.

Figure 4.15 shows two plots of sunspots and rainfall for Lahore, Pakistan, and Aberystwyth, Wales, which illustrate both the relationships and the problems. The Lahore record shows a sudden reversal in the correlation that is coincident with a change from a high sunspot peak to a lower amplitude oscillation. However, this apparently damning inconsistency, like many other similar cases, *might* be explained by slight shifts in storm tracks or frequencies associated with some changes in energy input, perhaps occasionally crossing some critical threshold level. In fact, there is a double sunspot cycle of 22 years, the **Hale cycle**, which seems to be reflected in some hydrometric records, that typically comprises a higher peak and a lower peak. The incidence of drought in the western USA over the last 300 years has been correlated with the Hale cycle: drought areas have been most extensive 2–5 years after Hale minima. The Hale cycle is related to a cyclical reversal in the magnetic polarity between pairs of sunspots (in which the right-hand spot switches from north to south and back in 22 years) and to similar cyclic activity in solar flares, fountains of plasma issuing from the Sun's surface, which are a major source of the 'solar wind'. The solar wind is a fluctuating stream of charged particles, some of which get ensnared by the Earth's magnetic field and cause the northern lights. The solar wind is associated with severe magnetic storms which occur around the Earth at approximately 11-year intervals and cause intense heating in the upper atmosphere. Perhaps more significantly for weather patterns, when the solar wind is strong, it provides a screen that reduces the penetration of cosmic rays into the atmosphere. When it is weak, more cosmic rays enter the ionosphere, causing intense heating and the production of more nitrogen oxides, NO_x, in the upper atmosphere. NO_x may well be the key link in the chain, because it absorbs electromagnetic radiation from the Sun and so reduces energy receipts at ground level. It could therefore influence evaporation rates, temperature profiles and rainfall production (sections 2.2 and 2.3).

Venne and Dartt (1990) appear to have made progress in improving the fit between sunspot cycles and rainfall patterns by adding in the Quasi-Biennial Oscillation. The **Quasi-Biennial Oscillation (QBO)** occurs in the middle and upper stratosphere, and involves a shift from north to south of the equator and back again in the position of the transition from westerly flow in the winter hemisphere to easterly flow in the summer hemisphere. It has an average periodicity of 2.2 years. Thus, the equatorial stratosphere changes from an easterly to a westerly flow and back every two years. Van Loon and Labitzke (1988) have shown that westerly winds in the North Atlantic region are stronger during sunspot maxima than during sunspot minima under the west phase of the QBO but weaker under the east phase. This is related to accompanying fluctuations in the patterns of atmospheric pressure in the upper tropophere and of sudden warmings in the polar stratosphere caused by air descending towards the tropopause. Tinsley (1988) found a link between the average latitude of winter storm tracks crossing the North Atlantic north of 50° N, the phase of the QBO, and the solar flux in the 107 mm band, which is correlated with sunspot activity. Tracks shift over a range of 6° of latitude during these cycles. Could the 2-year component in Russian riverflow already mentioned be related to such cycles?

Yet another oscillation, the **North Atlantic Oscillation (NAO)**, was identified by Walker in 1924, whereby stronger westerlies cross the North Atlantic in years when temperatures are above average in Greenland and below average in northern Europe. The NAO phase can be defined by sea level pressure anomalies between the Azores and Iceland. There are indications that this may affect rainfall in western Europe (Rogers and van Loon 1979). Indeed, both the tracks and the intensities of depressions crossing the North Atlantic are likely to be influenced by an amalgam of ENSO/PNA, QBO, NAO and, perhaps, short- to medium-term solar effects.

Other longer cycles also exist in solar activity, some of which may find expression in rainfall/runoff patterns. But, if it is difficult to prove such links for short term cycles, it is even more so for cycles such as the 80–90-year Gleissberg cycle, over which the amplitude of the sunspot cycles seems to gradually increase and then rapidly fall off. There also appears to be a 180-year cycle in solar activity that might be linked to the tidal effects on the surface of the Sun caused by the alignment of the larger planets, such as Jupiter and Venus (Gribbin and Plagemann 1977). Links have been made between climatic patterns and the 18–19-year lunar cycle, whose tidal effects also affect atmospheric pressure (Currie and O'Brian 1992).

Ultimately, the Earth–atmosphere system is an *open system*, and while external astronomical influences may not be of primary importance for daily weather forecasting, their role in climatic cycles seems highly likely. More scientific proof is still needed for most of these external effects, as Folland *et al.* (1992) point out. Nevertheless, some interesting links have been found, such as that between bombardments of meteors and an increase in cyclonic activity in the North Atlantic 5 days later.

Conclusion

If the present author were allowed to dabble in prediction or at least to express a hope, it would be that hydrologists in the foreseeable future might be able to improve upon the static approach of current extreme event analysis by placing the period of available records within a wider (if not global) framework of spatial and temporal patterns in climatic variation. This should enable a judgement as to how representative that record may be, whether it is likely to have been taken during unusually wet or dry years, and perhaps open the way to adjustments in average discharges and the predicted probabilities of extreme events.

Further reading

Good introductions to most of these methods can be found in:

Dunne T and L B Leopold 1978 *Water in environmental planning*. San Francisco, Freeman: 818pp.

Gregory K J, L Starkel and V R Baker 1995 *Global continental palaeohydrology*. Chichester, Wiley: 352pp. Offers details of recent progress in palaeohydrology.

Shaw E M 1994 *Hydrology in practice*, 3rd edn. London, Chapman and Hall: 569pp.

Wilson E M 1983 *Engineering hydrology*, 3rd edn. London, Macmillan: 309pp.

WMO 1994a *Guide to hydrological practices*. Geneva, WMO. Detailed descriptions.

Discussion topics

1. Discuss the problems caused by short records and changing climates for estimating the risks of extreme events.
2. Consider the relative merits and different uses of actuarial assessment versus methods that try to consider processes.
3. Are teleconnections merely an illusory distraction?
4. Should sunspot cycles be taken seriously in climatic forecasting?

CHAPTER 5

Monitoring and assessing processes

Topics covered

5.1 World trends, satellites and hydrological networks

5.2 Measuring precipitation

5.3 Measuring evaporation and evapotranspiration

5.4 Measuring runoff

5.5 Measuring water in the ground

Monitoring is the foundation stone for sound development and protection, for both water resources and the environment. It may not in the past have been the most exciting aspect of hydrology, but its fundamental position is unassailable. A knowledge of what aspects can be measured and how well they are measured is vital to a full understanding of hydrological processes and to any application of that knowledge. Moreover, improvements in theory and technology over the last two decades have made it a much more interesting topic.

5.1 World trends, satellites and hydrological networks

5.1.1 Broad trends

Recent decades have seen increasing attention to the costs of running hydrological monitoring schemes and the accuracy of assessment. The high costs of maintaining ground-based monitoring stations have led to widespread rationalization of station networks. Much of this rationalization has been ad hoc. Lack of funds, new priorities, social re-evaluation of cost–benefit or simple lack of interest have all contributed to dramatic reductions in the numbers of raingauging sites in many former colonial states in the Developing World since the early 1950s, especially in tropical Africa, Indonesia and the West Indies. In Dominica, barely 25 per cent of the former raingauges now exist. Despite some growth in the number of rainfall stations in Latin America during the early 1980s, the WMO/UNESCO (1991) survey shows a continuing worldwide decline since (Figure 5.1). Fortunately, some of this decline is scientifically planned, particularly in Europe and North America, where new technology and theory have often been adopted along with a careful assessment of the required and attainable accuracies.

The situation is somewhat better for river discharge and water quality stations (Figure 5.1). Here a general increase in monitoring sites during the 1980s can be linked to increasing concern for the human and environmental impact of river pollution, and the efforts of the WHO International Drinking Water Supply and Sanitation Decade. Even so, there is clear evidence that the rate of increase is on the decline throughout most of the world, long before the density of monitoring stations has reached optimal levels.

The 1990s has seen the launch of a new initiative by the WMO to stem this decline and to provide timely, and if possible real-time, quality-controlled data at about 1000 stations worldwide to strengthen the global programme of water resource assessment. The **World Hydrological Cycle Observing System (WHYCOS)** uses satellite communications to link strategically selected new and existing stations reporting water levels, discharges, water quality and meteorological variables (Figure 11.2).

Box 5.1 looks at the fundamental questions that need to be asked when monitoring systems are being set up.

5.1.2 New technologies: remote sensing and telemetry

Over the last 25 years, automation, telemetry and remote sensing have revolutionized data collection. Gauges can be linked via land telephone line or via radio to headquarters. **'Intelligent' gauges** can preprocess the data collected to store or transmit only a summary, e.g. mean, maximum and minimum in a given period. Computers can be programmed to interrogate these stations at regular intervals and to update data banks and modelling programs, as on the River Dee (section 9.3). VHF/UHF radio is essentially line-of-sight communication, so it remains local or 'basin-scale' unless repeater stations are used. Alternatively, **communications satellites** or **observational satellites** can act as the repeater station and so extend the coverage to regional or even world scale. The satellites receive data from **data collection platforms (DCPs)**, which vary from complete automatic weather stations and multi-parameter water quality stations to perhaps a single gauge, each using a low-power radio transmitter. Data collected by the European Space Agency's (ESA) Meteosat 1 is retrieved by the European Space Operations Centre at Darmstadt in Germany, processed within a few minutes and retransmitted to users by radio via Meteosat 2. The American Geostationary Operational Environmental Satellite (GOES) Data Collection System (DCS) is offered free to agencies like the USGS and by 1987 25 per cent of the USGS water-monitoring network reported via GOES.

A novel form of telemetry is offered by **meteor burst** (or scatter) technology. Every day some 10 billion meteors enter the atmosphere. These create two to three ionized trails every minute that can reflect or reradiate VHF signals between ground stations up to 2000 km apart. The USDA SNOTEL network is one of the largest using this technology, with over 500 stations recording daily snow accumulation, precipitation and air temperature. Each trail lasts only a few seconds at most, so messages have to be brief, but greater use is now possible with high-speed computer processing and perhaps some preprocessing by intelligent DCPs before transmission.

The primary role of observational satellites is to provide a spatial image of surface features, such as land use, topography and soil moisture, and of

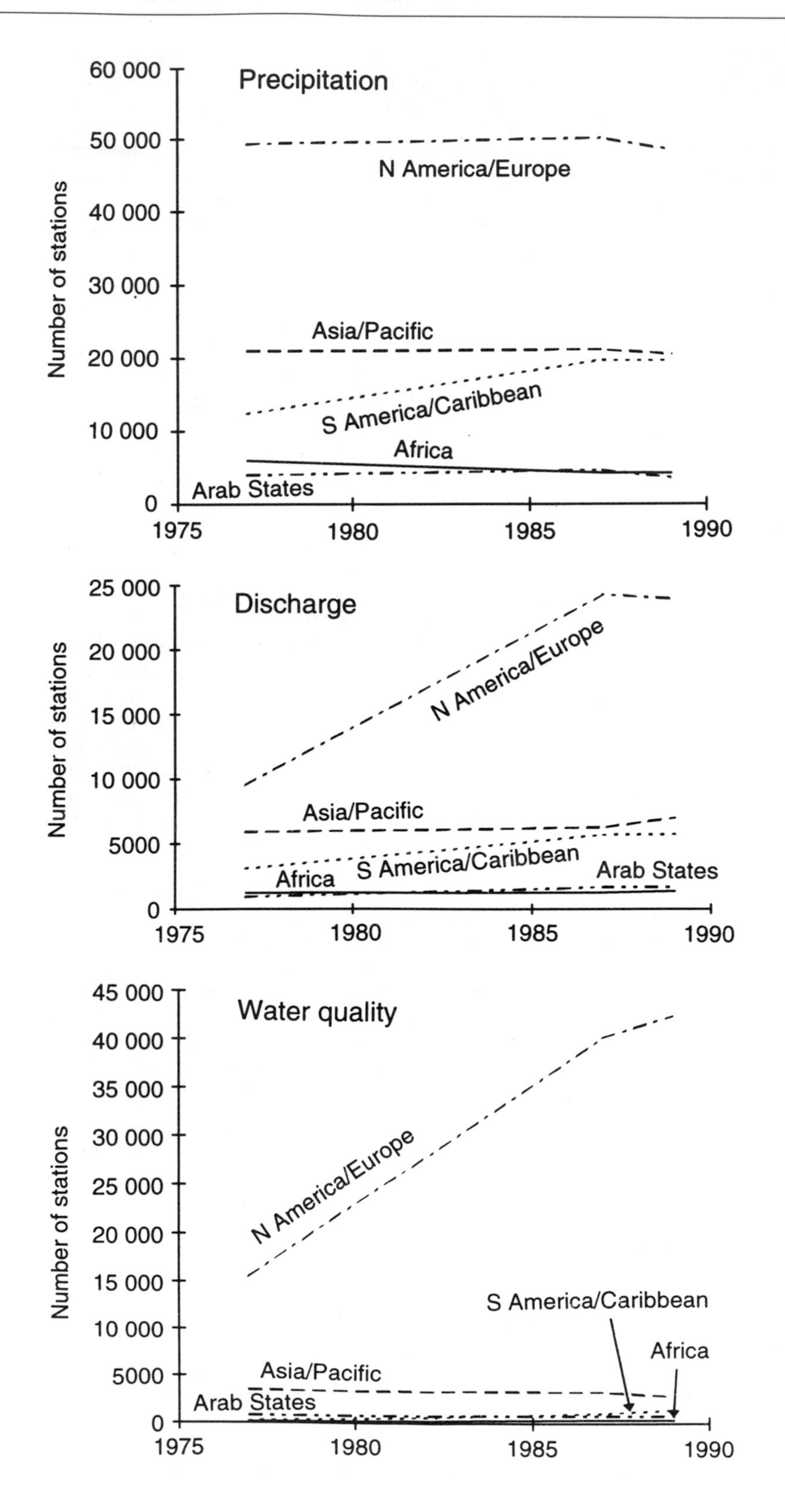

Figure 5.1 Global trends in precipitation, riverflow and water quality stations, according to WMO (1991)

Box 5.1 Network design philosophies

For all questions of **hydrometry**, that is monitoring the fluxes or exchanges of water, three types of problem need to be addressed:

1. The type of monitoring instrument to be used
2. The number and location of those instruments
3. The frequency and length of measurement.

For many questions there is a fourth type of problem:

4. How to estimate exchanges over a wide area of land on the basis of measurements taken at just a few locations.

A variety of principles have been proposed for the theoretical design of hydrometric networks, and the range of approaches is continually being augmented. The earliest and still the most common approach has been to combine a general understanding of hydrological processes with the practicalities of where and how frequently it is feasible to take measurements. Nemec and Askew (1986) called this the **basic pragmatic** philosophy. Other designers have undertaken revisions of the original network based on **data analysis**, often reducing gauge densities in the light of measured variability, as described in section 5.2.3. The most recent additions have been applications of **statistical theory**, which may be used both to design and to rationalize networks. Kriging and Kalman filters are examples of statistical techniques which can improve spatial extrapolation or time-averaged areal estimates of mean precipitation, respectively, and at the same time facilitate the assessment of gauging requirements (Bastin *et al.* 1984; Bras and Colon 1978; section 5.2.4).

Despite all the attention and particularly the stimulus from the WMO over the last 40 years, there is still no all-embracing theory for network design, nor perhaps is there likely to be one in the foreseeable future. The WMO has produced widely accepted guidelines, for example on *minimum densities* for raingauges (Table 5.1), in WMO (1994a), but these are essentially pragmatic rather than theoretical. They are based on estimated natural variabilities and average data requirements.

Part of the lack of uniformity in criteria for the design of networks may be put down to continually developing philosophies. However, other important elements are: the variety of purposes for establishing the network; changing data requirements; changing technology; and changing economic circumstances. Data requirements for assessing *water resources* may be different from those for *hydrological forecasting*. Indeed, Nemec and Askew (1986) referred to design principles themselves as 'time series', ever-changing just like the data. Economic criteria are bound to become more important in design decisions in the future, but how these can be formalized may be problematic. It is often pointed out that it is difficult to assign a monetary value to a human life lost in a flood (section 9.1.2).

Networks frequently need to be redesigned to accommodate new techniques of data collection or of rainfall–runoff modelling. An example of the effects of changing technology is the replacement of integrated rainfall–runoff stations on the River Niger linked by radio with a superior system linked by satellite. The WMO design for an integrated network in the Amazon basin is an example of the feedback between network design and data interpolation. Resources for building new stations were limited. The final plan involved an evenly distributed but low-density network of rainfall–streamflow stations, combined with better interpolation. Extrapolation and interpolation between stations was carried out on a grid square basis using equations for calculating precipitation, evaporation and runoff that were derived from relationships established by detailed monitoring in representative physiographic regions.

Table 5.1 WMO-recommended minimum densities for rainfall, runoff and evaporation gauging stations in square kilometres per gauge

Climate	Precipitation			Streamflow			Evaporation
	Mountain	Plain	Coast	Mountain	Plain	Coast	
Humid	250	575	900	1000	1875	2750	50 000
Arid		10 000			20 000		100 000

Source: WMO (1994a).

meteorological indicators, such as clouds (which can indicate wind directions and strength as well as precipitation potential), surface temperatures or ocean waves (again indicating wind properties). Meteorological satellites may also provide vertical profiles of atmospheric properties. Observational satellites range from **geostationary** meteorological satellites whose orbital speed matches that of the Earth, to **Sun-synchronous** satellites that cross the Earth in near-polar orbit, taking about 100 minutes to circle the globe. Geostationary satellites orbit the equator and give continuous monitoring of a selected portion of the globe, even at night by recording infrared radiation. Each satellite can cover about 40 per cent of the globe, a 'full Earth disc'. A set of five is now in place providing complete coverage, although the view is badly distorted in high latitudes.

Polar-orbiting satellites like NASA's Landsat series only return to exactly the same location every 18 days and the coverage follows a linear 'footprint' on the surface of the globe which wraps around the Earth rather like rolling string around a ball. As the orbital tracks cross in high latitudes they more often return *almost* to the same location. This can improve the usable frequency of imaging and data collection from DCPs, as well as offering the opportunity for some 'pseudo-stereoscopic' viewing above 70° N or S. A single Landsat image covers 34 225 km^2. The low frequency of coverage from polar-orbiting satellites is one of the main detractors from their value, especially for monitoring flood events.

Unfortunately, there are still no plans to launch a specifically hydrological satellite, although the ESA did sponsor a review of hydrological requirements (Barrett and Herschy 1986). The hydrological value of a satellite image depends very much on the size of the smallest features that can be seen. In this, the polar-orbiting satellites, which orbit at lower altitudes, score better. The smallest resolvable feature or **pixel** (picture cell) visible from a geostationary meteorological satellite is typically a kilometre or more across. This is called its **resolution**: it is essentially the area of the surface within view of the scanning sensor during the recording of a single reflectance value. Most of the Landsat series have had resolutions of around 80 m, although the Thematic Mapper reduced this to 30 m and the French SPOT (Système Probatoire d'Observation de la Terre) has reduced it to 10 m (Table 5.2). Somewhat enigmatically, linear features like a river that are rather smaller than the resolution may still be seen. This is because of the way they influence the reflected radiation over a series of adjacent pixels in a recognizable pattern. It is possible to resolve linear seepage lines on hillslopes with SPOT imagery.

Another key property of satellite imagery is the spectral coverage of the radiometer channels. The number of channels, the central wavelength of each channel and the range of wavelengths covered in each channel all affect the potential value to hydrology. One of the most valuable channels for hydrology is channel 7 of the Landsat Multispectral Scanner (MSS), which covers the near-infrared (IR) (Table 5.2). Differences in vegetation cover and photosynthetic activity in general are most apparent in the *reflected near-IR*, as also is the extent of surface water or snowcover. Surface and cloud-top temperatures are best indicated by longer wavelength, thermal IR, which is *emitted* from the Earth. More sophisticated work may involve combining more than one of these **spectral slices** (images based on a single sliver from the radiation spectrum) in order to distinguish different features by establishing their **spectral signatures**, i.e. typical patterns of reflection in the various wavebands.

Satellite imagery lends itself to a range of handling procedures, from manual interpretation to fully automatic analysis, although little use has yet been made of the potential for a fully automated route inputting the results of computer-processed image analysis directly into process-based hydrological simulation models (section 6.3.2). One of the most long-established techniques is **nephanalysis**, in which a meteorologist carries out manual interpretation of the cloud cover for use in rainfall estimates (section 5.2.5). The other end of the range is represented by fully automatic or *unsupervised* computer analysis. This uses one of a variety of classification algorithms to produce a map of the 'scene', classifying areas into groups that are statistically recognizable as having different spectral signatures. Perhaps greatest use to date has been made of an intermediate procedure known as *supervised*, in which the classification categories are determined from the outset by a human operator. The operator will indicate the class and location of sites that have been visited and classified on the ground, i.e. the *ground truth*. The program then attempts to place each pixel into one of the ground truth categories according to its spectral signature. These procedures are most commonly used to map land use, vegetation cover and surface water areas, which can provide background data for hydrological modelling.

For more complex phenomena, like soil moisture and snow properties, it is possible to calibrate the satellite

Table 5.2 The key specifications and hydrological uses of major satellites

Space platform	Launch schedule	Main sensor of use	Wavelengths	Resolution	Frequency	Main hydrological application
GOES series (USA)	1975–1999 (11)*	Visible and infrared spin-scan radiometer (VISSR)	Visible, thermal IR	0.8, 6.9 km	19 min	Cloud meteorology
Meteosat series (ESA)	1977–1995 (7)	Imaging radiometer	Visible, IR, thermal IR	2.4, 5 km	30 min	Cloud meteorology
Meteor (USSR/CIS)	1969–1994 (58)	Imaging radiometer	Visible, thermal IR	1–2 km	12 h	Cloud meteorology
GMS series (Japan)	1977–1994 (5)	VISSR	Visible, thermal IR	1.25, 5 km	30 min	Cloud meteorology
Insat series (India)	1982–1997 (8)	Very high resolution radiometer (VHRR)	Visible, thermal IR	2.75, 11 km (2 km 1992+)	30 min	Cloud meteorology
NOAA Tiros-N series (USA)	1978–1999 (18)	Advanced very high resolution radiometer (AVHRR)	Visible, IR, thermal IR	1.1 km	12 h	Cloud meteorology Surface hydrology Surface features
Landsat series (USA)	1972–1997 (3) 1982–1997 (5)	Multispectral scanner (MSS) Thematic mapper (TM)	Visible, IR, thermal IR	79 m, 237 m 30 m, 120 m	18 days 16 days	Surface hydrology Land use, topography
SPOT (France)	1986–1999 (5)	High resolution visible (HRV)	Visible, IR, panchromatic	20 m, 10 m	3–4 days	Surface hydrology land use, topography
ERS (ESA)	1991–1994 (2)	Synthetic aperture radar (SAR)	Microwave	26 m	3 days	Snow and ice cover, soil moisture
MOS (Japan)	1987–1991 (3)		Visible, IR	50 m	17 days	Surface hydrology land use, topography
DMSP (USA)	1966–1995 (30)	Scanning microwave radiometer (SMR)	Microwave	5–15 km	12 h	Rainfall, snow water equivalent, ice
Radarsat (Canada)	1994–2002 (3)	SAR	Microwave	10 m	16 days	Snowcover

*Number launched or currently planned in brackets.

imagery with more detailed aircraft remote sensing, which is commonly taken at the same time, i.e. underflying the satellite. Balek *et al.* (1986) described using a model aircraft flying at 200 m for this purpose. More elaborate airborne systems are provided by IR thermal linescanners and the Daedalus multispectral scanner with 10 visible to near-IR channels and one thermal. These give high resolution, but are costly and are only flown for special research purposes.

Indeed, one of the main detractions at the present time is that very few completely operational systems are in place for the automatic processing of remote sensing for hydrological applications. Current use is especially restricted by the lack of easy access to image-processing equipment, lack of trained personnel and lack of specifically hydrological software. Future developments also seem to be threatened by the escalating costs of obtaining computer-compatible tapes (CCTs) or photographic products from agencies that are now being run on a more commercial footing.

Some of the most significant developments in the near future are likely to come from deploying **micro-wave** technology. Ground-based weather radar is now well established for precipitation measurement (section 5.2.5). **Radar** is an *active* system based on the return or backscattering of radiation emitted from an artificial source. However, one of the key recommendations made by the European Association of Remote Sensing Laboratories' (EARSeL) Working Group on Hydrology and Water Management was that *passive* **microwave** imagery (measuring *natural* environmental emissions in the microwave wavelengths) should be available from civilian polar-orbiting satellites to improve precipitation estimates (Table 5.3). A second recommendation was that future satellite-borne **synthetic aperture radar** should be designed to cover the spectral range required for monitoring snowcover and melting processes.

These technological developments not only provide better coverage of spatial patterns, but they also call for repeated re-evaluation of existing conventional networks.

5.2 Measuring precipitation

Precipitation is an areal input. Traditional technology has provided only point measurements, along with numerous techniques for deriving areal estimates. Although modern technology is tending to supplant the traditional point gauge and provide areal values directly, the gauge is still the ultimate reference and for some purposes may be an adequate indicator on its own.

5.2.1 Gauges for rainfall and snowfall

Ideally, the type of instrument used to measure precipitation should be determined by the use that is to be made of the records. Measurements of total daily or monthly precipitation may be adequate for assessing water resources, whereas continuous records are necessary whenever rainfall intensity is important, for example in flood warning. In practice, the choice is commonly determined by what is available or by the cost of instrumentation and of observers to collect the data. Most instrumental records are taken by 'standard' gauges that have to be emptied into graduated measuring bottles at regular intervals. Such gauges are normally emptied daily, although at full 'synoptic' weather stations some gauges are measured every six hours, while at more remote sites larger gauges may be installed to hold a whole month's precipitation. The fact that the most common measurements are daily totals recorded regularly at 0900 GMT without regard to the timing and length of precipitation events can be problematic for the hydrologist. The problems are illustrated by a storm at Aberystwyth, Wales, on the night of 29/30 December, 1986, and part of the following day, which caused a flood on the River Rheidol. The rainfall at the official meteorological station is recorded in two successive daily totals. Hydrologists concerned to estimate the frequency of the flood storm find that if the total 64 mm fell over a 24-hour period it has a frequency of 1 in 20 years. If, however, the rain all fell within 12 hours, which is probable, its frequency is only 1 in 50 years. Since the rain fell during a holiday period when the autographic gauge was inoperative, there is no way of telling which is nearer the truth.

Recording or autographic gauges cost more in capital outlay but require less effort in data collection. They also give extra information on the time of occurrence, duration and intensity of precipitation. A continuous record is kept either by a 'tipping bucket', a large weighing device, or by a siphon or tilting siphon that empties the collector when it is full (Figure 5.2). Older, autographic versions recorded on to paper charts. While these have the great advantage of providing a visual impression of events, the quantitative information is difficult to abstract from them. The UK Flood Studies programme speeded up the process by photographing the original charts and digitizing the

Table 5.3 Required remote sensing specifications for hydrological applications and current provisions from satellites and Space Shuttle. Minimum and ideal specifications are based on EARSel, and WMO Commission for Hydrology reports with additions in brackets

Hydrological parameter	Spatial resolution		Frequency of coverage		Accuracy		Suitable sensors and space platforms*									Main deficiencies at present
	Minimum acceptable	Ideal	Minimum acceptable	Ideal	Minimum acceptable	Ideal	Geostationary satellites	Space shuttle		Polar-orbiting satellites						
							Imaging radiometers (e.g. VISSR)	LFC survey camera	SIR	AVHRR	MSS	TM and HRV	SAR	SMR	OCM	
Precipitation	10 km	**1 km**	1 month	1 h	30%	**20%**	cloud cover and temperature			Cloud cover and temper-ature				liquid water content, rain rates		SMR in USAF-DMSP, but restricted civilian availability globally
Snowcover	10 km	1 km	1 month	24 h	100 mm	50 mm	area	area	✓	✓	✓	✓	wet snow	water equi-valent, depth		SMR as above. SAR on, defunct Seasat (1978), then not available till ERS-1 (1991), JERS-1 (1992) and Radarsat (1994)
Ice cover	1 km	25 m	7 days	24 h	20%	10%	✓	✓		✓	✓	river and lake ice	river and lake ice	sea ice		As above
Glacier area	500 m	25 m	10 years	2 years	5%	2%	✓	✓		✓	✓	✓	✓			As above
Surface water area	100 m	30 m	7 days	**24 h**	5%	**3%**	area, flash floods	✓		major floods	✓	✓	✓			Improved frequency generally needed for flood coverage, though SPOT-1 (1986) HRV gave 2.5-day repeat off nadir
Aquifer area	1 km	100 m	5 years	3 years	30 m	10 m			✓		✓	✓				
Evaporation	10 km	1 km	10 days	24 h	30%	20%	drought			drought	drought	drought				
Soil moisture	(100 m)	**(10 m)**	(1 month)	(24 h)	(20%)	(10%)			✓				✓			SIR and SAR resolutions are 30 and 26 m respectively. Shuttle programme resumed 1993 after 5-year gap. Launches occasional, coverage limited.
Water turbidity	300 m	100 m	24 h	6 h	50%	20%						3			✓	OCM not available till EOS-COLOR in 1998
Topography: Drainage area	100 m	20 m	10 years		1%	0.5%		✓✓	✓		✓✓	✓✓			✓	As above. Landsat and SPOT adequate
Land use & vegetation	(50 m)	(20 m)	(5 years)					✓✓	✓		✓✓	✓✓			✓	

Figures in bold are *not* generally achieved by current systems. Figures in brackets are estimated by the author. Ticks indicate suitability. Special applications are indicated in full.
*For abbreviations see Table 5.2, except for long-focus camera (LFC), shuttle-imaging radar (SIR) and ocean-colour monitor (OCM).

(a)

(b)

Figure 5.2 Inside three types of automatic raingauge: (a) tipping bucket – each half of the 'bucket' collects an amount of water equivalent to, for example, 1 or 0.5 mm of rain over the surface area of the orifice before it tips, activating a reed switch, which sends an electrical pulse to a logger; (b) tilting siphon – the charting pen floats up as rainwater fills the chamber, which tips when full returning the pen to the bottom.

Figure 5.2 (cont.) (c) A totalizing gauge that weighs both rainfall and melted snow – it records on punch-tape and can be interrogated by telephone link (cover removed)

photographs on a digitizing table, which produces a stream of computer-readable *x* and *y* coordinates of the plots. Another design of autographic gauge records on punch-tape that can be fed directly into a computer (Figure 5.2c). However, the most efficient modern method records directly in electronic digital format on to tape, disk or solid state memory, which is then physically collected or electronically downloaded.

Traditional raingauges and snowgauges suffer from a number of technical problems, including undermeasurement due to air turbulence around the gauge, evaporation losses prior to measuring, and losses due to 'wetting' (see page 133). The most important of these is **turbulence**. Figure 5.3 shows the extent to which standard gauges undermeasure as windspeed and therefore turbulence increase. The graph suggests that in winds of 22 m s^{-1} (80 km h^{-1}) the typical gauge catches barely 50 per cent of the true rainfall and only 25 per cent of true snowfall. Snow is more vulnerable than rain because of the less aerodynamic shape of typical snowflakes and because of its lower density. Measurements at UKIH indicate an 18 per cent undermeasurement of total precipitation with standard British gauges at the exposed site of Carreg Wen, Plynlimon, in the Welsh mountains, compared with only 8 per cent at Wallingford in the English lowlands. Rodda (1967) reported 12 per cent undermeasurement of snow even at Wallingford. The author's own measurements on an exposed coastal site at Aberystwyth in West Wales show an average undermeasurement of 14 per cent in rainfall alone. They also suggest an increase in undermeasurement from 6 per cent in winds of 2 m s^{-1} to 21 per cent at 7 m s^{-1}, which is in line with Rodda and Smith's (1986) estimates based on average values at 17 long-term sites in Britain. These results have significant implications for the accuracy of assessed water resources in the wetter regions of Britain, and imply that the gradient in precipitation receipts between northwest and southeast is greater than estimated (cf. section 11.5.2). Fortunately, the error is 'on the right side' and there are more resources than is apparent!

Gauge designers have devoted considerable attention to ways of minimizing the effect of turbulence. The most popular solution has been to fit a *shield* on to the gauge, such as the Alter, the similar Tretyakov, which is popular in Russia, or the Nipher (Figure 5.4). Atmospheric Environment Canada uses a particularly effective design of snowgauge in which the Nipher shield is integrated with the gauge (Figures 5.4 and 5.5). This has long been regarded as the best snowgauge available. However, even these can perform badly in certain circumstances, particularly in windier regions and on exposed sites. Atmospheric Environment Canada has tested versions of the Wyoming gauge in the windy, treeless landscapes of the Northwest Territories. This uses a double ring of permeable fencing to becalm the air around the gauge (Figure 5.6). The UK Institute of Hydrology developed a new form of shielding for research in Scotland.

Windspeed and turbulence increase the higher the gauge stands above the ground, but, unfortunately, there is no international standard height for gauges. The standard UK and Commonwealth daily raingauge is set at 305 mm above the ground surface. This is a compromise between reducing the turbulence problem and minimizing the opposing problem of rain splashing up and into the gauge from the ground surface. However, autographic gauges can stand at a variety of

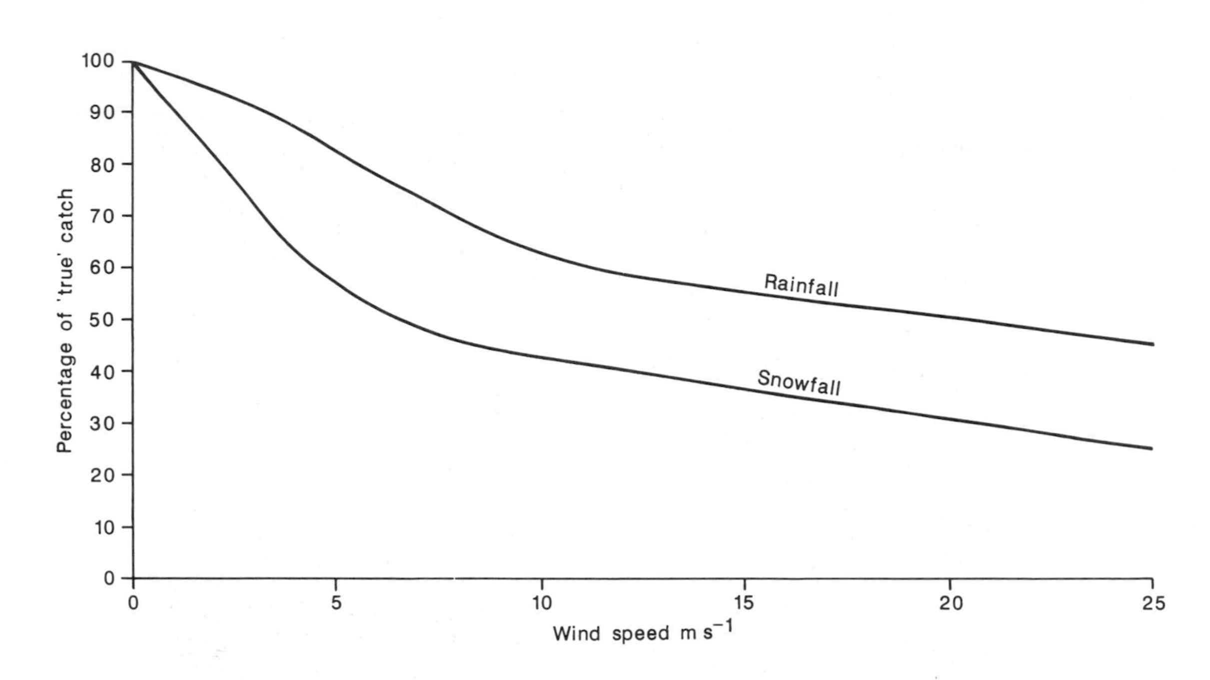

Figure 5.3 Undermeasurement of rain and snowfall in relation to windspeed in standard, unshielded gauges. After Bruce and Clark (1981)

(a)

Figure 5.4 Shields and the problem of turbulence: (a) Alter; (b) Nipher; (c) Canadian Nipher gauge; (d) windfields around a gauge

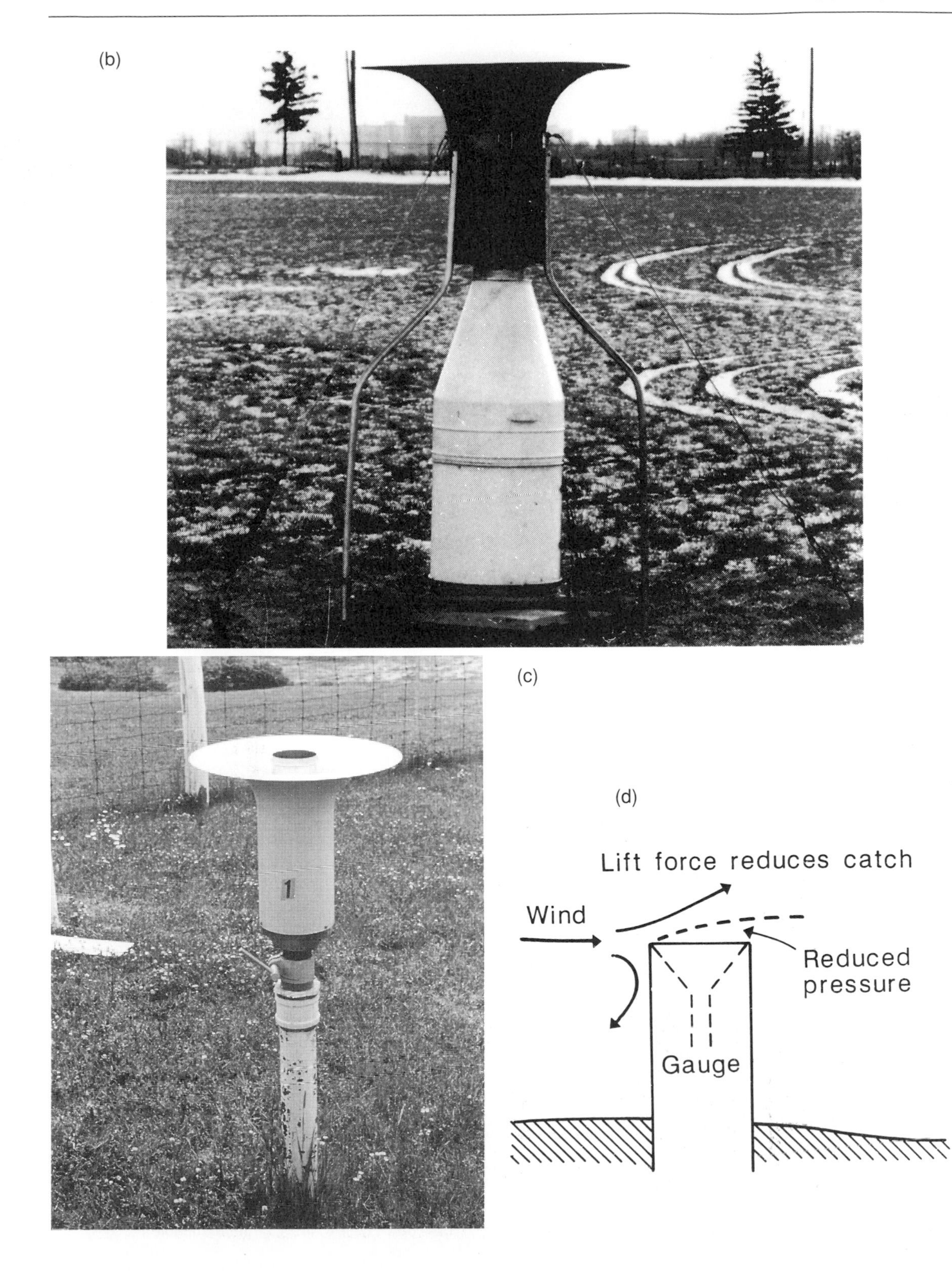
(b)
(c)
(d)
1
Lift force reduces catch
Wind
Reduced
pressure
Gauge

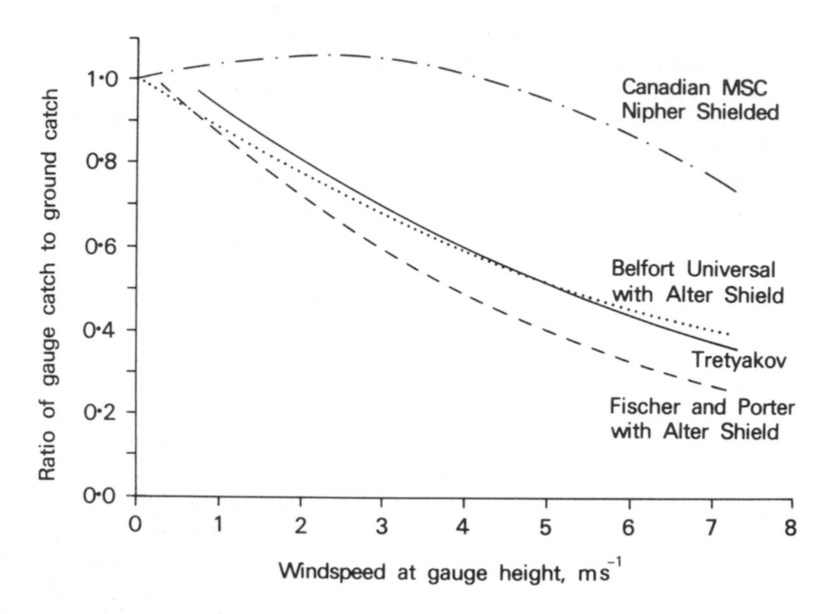

Figure 5.5 Performance of shielded gauges

Figure 5.6 An experimental shielded gauge for snowfall. The Wyoming gauge has two concentric permeable rings of fencing or netting to slow down high winds – an impermeable barrier would cause turbulence in its lee

heights. Greater elevation is needed where snowfall is common, in order to allow for snow accumulating on the ground and to avoid drifting snow entering the gauge. But snowfall catch is also reduced as a result of the extra elevation. The Russian and CIS standard gauge is mounted so that its orifice stands 2 m above the ground. As a consequence, the UNESCO (1978) atlas of the world water balance estimated that precipitation can be undermeasured by up to 60 per cent in parts of Central Asia. The authors suggested ways of redressing this by using multiple regression equations relating snowfall to temperature and relative humidity; but one is tempted to ask why continue to use instruments with such a large measuring error! The problem is exacerbated when rainfall is measured in gauges set at elevations considered necessary for snowfall measurement, as with the Russian gauges and with all-purpose gauges like the Fischer–Porter totalizing gauge in Figure 5.2c, which is commonly used in North America.

An alternative solution to the turbulence problem is to lower the gauge, so that the orifice is as near to the ground surface as possible. The UK Meteorological Office standard for windy areas is the **turf wall** (Figure 5.7a), which effectively reduces windspeed across the gauge close to that very near the ground while maintaining the 305 mm standard elevation at the gauge in order to minimize insplash. The Meteorological Office has continued to use this model, despite encouraging results from experiments with 'anti-splash' devices set around lower gauges. The UK Institute of Hydrology built upon the Meteorological Office experiments and produced a highly successful **ground-level raingauge** that is set in a 305-mm-deep pit and the orifice is surrounded by an anti-splash grid which offers virtually no horizontal surface that would allow insplash (Figure 5.7b). This model has become an unofficial 'standard' for hydrological research in Britain, but it is not likely to displace the Meteorological Office model as a meteorological standard simply because one of the prime tenets of record keeping is to maintain as long a sequence as possible based on the same instrument and the same site. Ultimately, the sequence *will* be broken by revolution or evolution of one sort or another, but the resolve is to reduce such upheavals to a minimum in order to provide the best estimates of trends and risks (section 5.2.3).

A quite different category of problem can be introduced by the rules of operation. This is most manifest in the case of snowfall measurements at synoptic stations in climates prone to frequent light snowfall. Here, very frequent measurement combined with low snowfall intensities can cause more instances where the amount of new snowfall is less than the measurable minimum, so that it gets recorded only as a 'trace' (less than 0.125 mm). Calculations based on Nipher gauge observations at the McGill Sub-Arctic Research Station in northern Quebec have suggested that up to 50 mm a year (10 per cent of the local receipts) might go unmeasured as a result, whereas they may have become measurable if they had been allowed to accumulate for longer. Losses due to **wetting**, that is, water left clinging to the inlet funnel and collecting bottle during measurement, are also increased by over-frequent measurement, whether it be rain or snow, which must be gently melted in the collector to obtain its **water equivalent**. Tests on Russian gauges indicate losses of up to 0.25 mm per measurement due to wetting; even in Canadian gauges 0.15 mm can be lost each time. Undermeasurement can also result from exactly the opposite of this problem. When measurements are too infrequent and snowfall intensities are too high, or the snowflakes stick together, **bridging** can occur, which clogs the gauge and prevents further collection.

A partial, but spurious, counterbalance to this undermeasurement can occur when snowgauges catch **blowing snow**. Blowing snow has been lifted up from the ground to above gauge height by the wind, and it is therefore recycled rather than new precipitation. Observers' manuals instruct staff to disregard snow that has been received under a cloudless sky, to try to minimize this overmeasurement. Even so, the problem is rather more complex. There is no sure way of determining how much recycled snow has been caught when clouds may be precipitating at the same time as high winds are blowing up a snowstorm from the ground. At Knob Lake in Central Labrador–Ungava about 40 per cent of all days with snow falling also have blowing snow. Even the 'clear sky' rule is not foolproof: in polar and subpolar regions, snowfall can occur without clouds (section 2.3.1).

In addition to these technical problems, traditional gauges can also suffer from being sited in **unrepresentative locations**. The McGill station offers a good illustration of this. The synoptic Nipher gauge was sited at the airstrip near the outlet of the Knob Lake research basin (Figure 5.12). Much of the basin is tree-covered with large expanses of open tundra downstream and on the adjacent ridges. Thus, while the Nipher is in a windswept site, much of the basin is sheltered and receives a net horizontal transfer of snow by blowing and **drifting** (defined as transfer *below* the

Figure 5.7 Ground 'shields': (a) the UK turf wall; (b) the UK IH ground-level gauge

height of the gauge orifice) from the surrounding area. Comparisons between accumulated winter snowfall at the gauge and areal snowcover surveys at the end of winter have suggested that the gauge could be 'undermeasuring' by up to 40 per cent a year. This very high figure is probably due more to lack of a representative site than to aerodynamic problems with the gauge itself. It emphasizes the problems of using snowgauges for hydrological purposes.

5.2.2 Assessing snow on the ground

Snow has two unique and very important hydrological properties: it does not infiltrate the soil and it is only of hydrological interest when it melts. These mean that the hydrologist is not normally interested in snowfall intensities or in the timing and duration of snowfall. Consequently, measurements need not be very frequent and continuous measurement is not generally required. This highlights one of the many conflicts between standard meteorological procedure and the requirements of hydrology, and means that some of the under-measurement problems described in the last section could be significantly reduced if measurements were being taken solely for hydrological purposes.

These properties also mean that the hydrologist has the opportunity of a 'second bite of the cherry' by measuring the snow lying on the ground, the **snowcover**. Not only does measuring snow on the ground overcome the technical problems of collecting as it is falling, it can also be more appropriate. For predicting meltwater volumes, the amount, timing and spatial distribution of snowfall are not as important as the final distribution of snow on the ground after it has been reworked by blowing, drifting, metamorphosis, and episodes of melting and refreezing.

Two broad approaches have been adopted for measuring snow on the ground, either (1) following the meteorological tradition of measurements at a point location, or (2) snow surveys covering a wider area. The longstanding meteorological standard has been to use a **snow board**, a square piece of wood or concrete slab. Snow falling on these boards is usually measured daily. The procedure consists either of taking a representative number of measurements of the depth of new snow or, if there is sufficient snow, of taking complete vertical cores of snow by pressing a cylinder into the surface. After the measurement is taken, the board is swept clear. Since it is water equivalent not snow depth that is required, cores have to be melted and depths have to be converted. Herein lies a major weakness of the method. In the UK, a factor of 1:12 is used, that is, 12 mm of snow would be converted to 1 mm of water equivalent. In Canada, the factor is 1:10. Moreover, that factor was derived from experiments at Toronto and it is a reasonable assumption that the mix of crystal types falling in southern Ontario, and therefore the density of snow lying, is not entirely representative of the whole of the second-largest country in the world, extending as it does from the intense cold of the Arctic to the mild, damp coastlands of British Columbia. Unfortunately, this method is the only standard for measurement in the UK and even coring is only done at a small number of special 'Snow Survey' sites. Indeed, the most common type of snow record is '**days with snow lying**', which is not really a

Figure 5.8 Measuring new snowfall: a photoelectric 'snowdrop' gauge at McGill University, Montreal

measurement but is based on a purely qualitative visual assessment of the area around a station at 0900 GMT. Even in Canada, about 80 per cent of the 2500 Atmospheric Environment Service stations rely solely upon the snow board.

More sophisticated approaches to point measurements rely on either the weight of the snow or the attenuation of radiation passing through the snow. The 'snowdrop' design in Figure 5.8 is based on the attenuation of a beam of light between the ground and the sensor as measured from a photoelectric cell. But light is rapidly attenuated in snow, and at remote sites that cannot be cleared, where snowfall is heavier or where an undisturbed snowpack is required, radioactive isotopes such as cobalt-60 may be used.

The accumulation of the snowpack can also be monitored by devices that weigh the snow, and with less concern about environmental safety than with radio-isotopes. These can be divided into snow plates and snow pillows. **Snow pillows** are set on a firm horizontal surface and are filled with a fluid, commonly an antifreeze mix (Figure 5.9a). The pressure exerted on the pillow by the accumulating snow can be measured either with floating level recorders or via pressure transducers that convert the pressure to an electrical current. The pillow itself must be non-stretchable: strong artificial fibres or even encasements with sliding steel segments have been used. Most pillows require a fairly deep snowcover to operate effectively, although USDA SNOTEL stainless steel pillows are only 30 mm thick. There is also the possibility that the contact between snow lying over the pillow and adjacent snow may affect the recorded weight, especially if there are ice lenses forming a link. Otherwise, the weight of the snow is directly proportional to the water equivalent. The USDA obtains 90 per cent response and 10 per cent accuracy in transmission using clusters of three to four 1.8 m^2 pillows and meteor burst telemetry.

Snow plates measure shallower snowcovers better, because they are flat or slightly concave rather than convex and sit flush with the ground surface. The plate rests on some weighing system, which may be like a pillow or a tyre inner tube. Again, it is important that this is essentially non-stretchable if a water level recorder is to register the changes in pressure accurately. One of the problems in regions that receive shallow and often very ephemeral snowcover is that rain may occur along with the snowfall or that the snow melts on contact, so that no weight is recorded, and yet at the same time any raingauges will tend to undermeasure the snowfall element. The snow plate in Figure 5.9b offers a universal measuring device for rainfall, snowfall, snowcover and snowmelt which is well suited to this situation (Waring and Jones 1980). This is achieved by a fibreglass dish that slopes very slightly towards a central drainage hole connected to a tipping bucket. Rain and meltwater give a tipping bucket record, while snow is recorded by the weighing device. Even if snow melts on contact, the water will be collected and undermeasurement of the snowfall is minimized by the ground-level position and a wiremesh screen, which simulates the surface roughness of the surrounding vegetation.

For snowmelt forecasting, however, wetness, crystalline state and layering are also important (sections 3.3.5 and 6.5.3). These mainly require manual surveys, for example, by digging snow pits. Wetness can be measured by inserting a 'snow fork', which uses radiowaves to measure the dielectric properties of wet snow, into the snow (Tiuri and Sihvola 1986).

Snow survey Unfortunately, the fact that snow does not infiltrate the ground also means that it is available for reworking and redistribution by the wind. This increases the spatial variability of snowcover, and therefore snowmelt, compared with the original precipitation (Figure 5.10). Consequently, more attention needs to be given to spatial patterns.

A **snow course** is a set of measurement stations laid out so as to cover a representative cross-section of accumulation sites. To economize on measuring effort, it is accessible and short. Sometimes more than one snow course is needed to encompass the variety within a basin. Normally measured only weekly, or only during the melt period, snow depth is measured on marker stakes and water equivalent is obtained by weighing snow cores. The course in Figure 5.10 covers a convenient set of representative sites in an area where drifting is important and forest vegetation cover dominates the pattern of snow accumulation. Other basins have less easily defined patterns and less correlation between snow depth and water equivalent (Figure 5.11).

Larger-scale **snow traverses** and **end-of-season basin surveys** require considerably more effort, but they are one of the best ways of estimating the potential volume of meltwater. The basin in Figure 5.12 contains a large area of lake (24 per cent of the basin), where significant amounts of snow get flooded and frozen into the lake ice cover. This '**white ice**' can absorb 25–30 per cent of the annual snowfall and not release the meltwater until after the land-based melt process is nearly complete (Jones 1969). The amount of white ice

(a)

(b)

Figure 5.9 Monitoring the developing snowpack: (a) a snow pillow; (b) a snow plate designed to measure snowfall and snowmelt

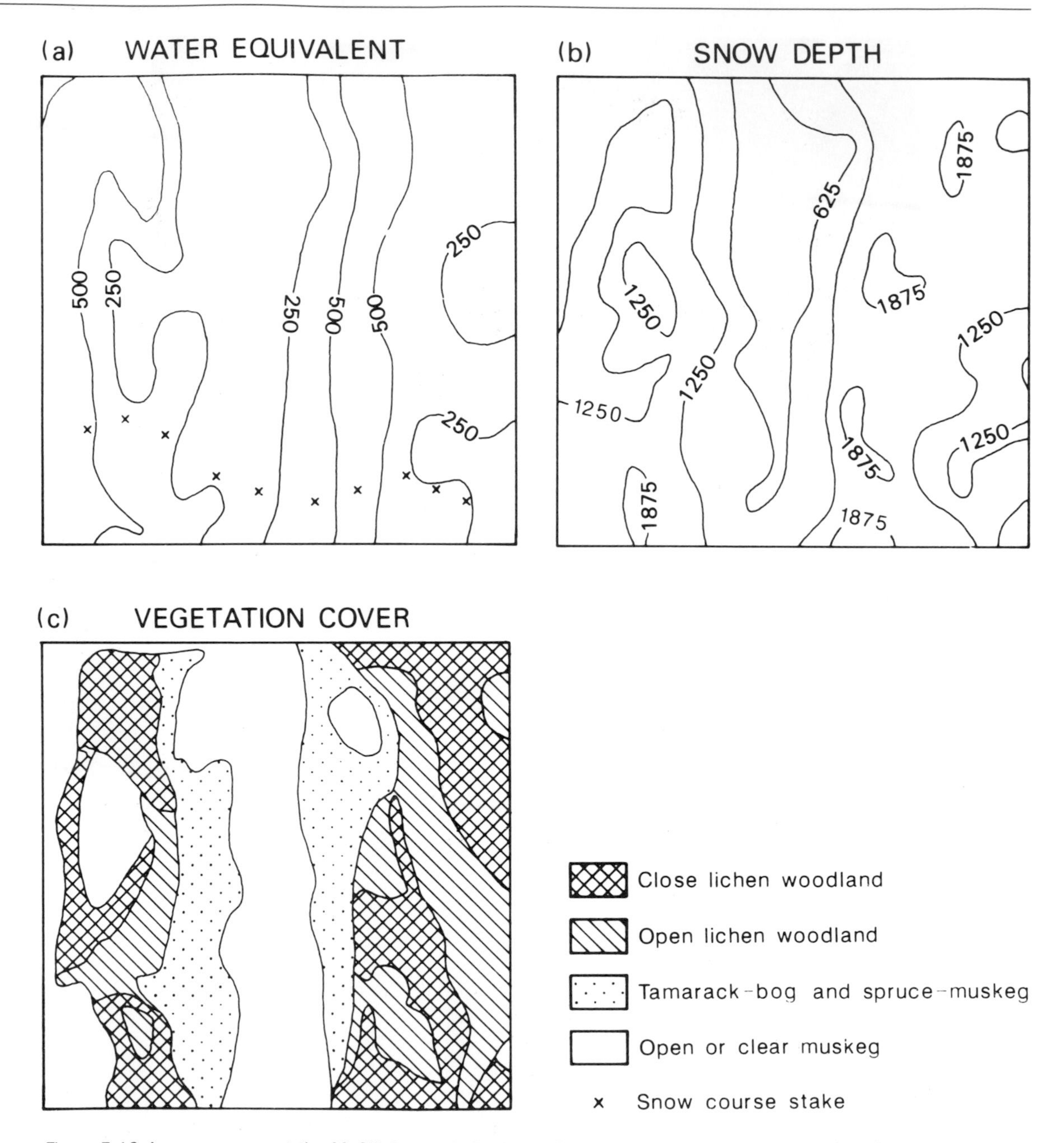

Figure 5.10 A snow course at the McGill Subarctic Research Station, which samples a representative range of vegetation cover types within a convenient distance. Mapping around the snow course confirms the close relationship between snow depth (mm), water equivalent and vegetation cover. Regular snow course measuring sites are marked by 'x'

therefore must also be surveyed (Adams 1981). In some cases, aerial or satellite surveillance may be linked with basin surveys or snow courses to monitor changes in the snowcovered area during the melt period (section 6.5.3).

A novel form of basin survey was pioneered in Britain in the 1970s by the Institute of Hydrology using **ground-based photogrammetry**. The essence of the technique is to produce a detailed contour map of the hillside before and after snowfall. The difference between the two gives the depth of snow. Ground reconnaissance was needed to relate snow depth and

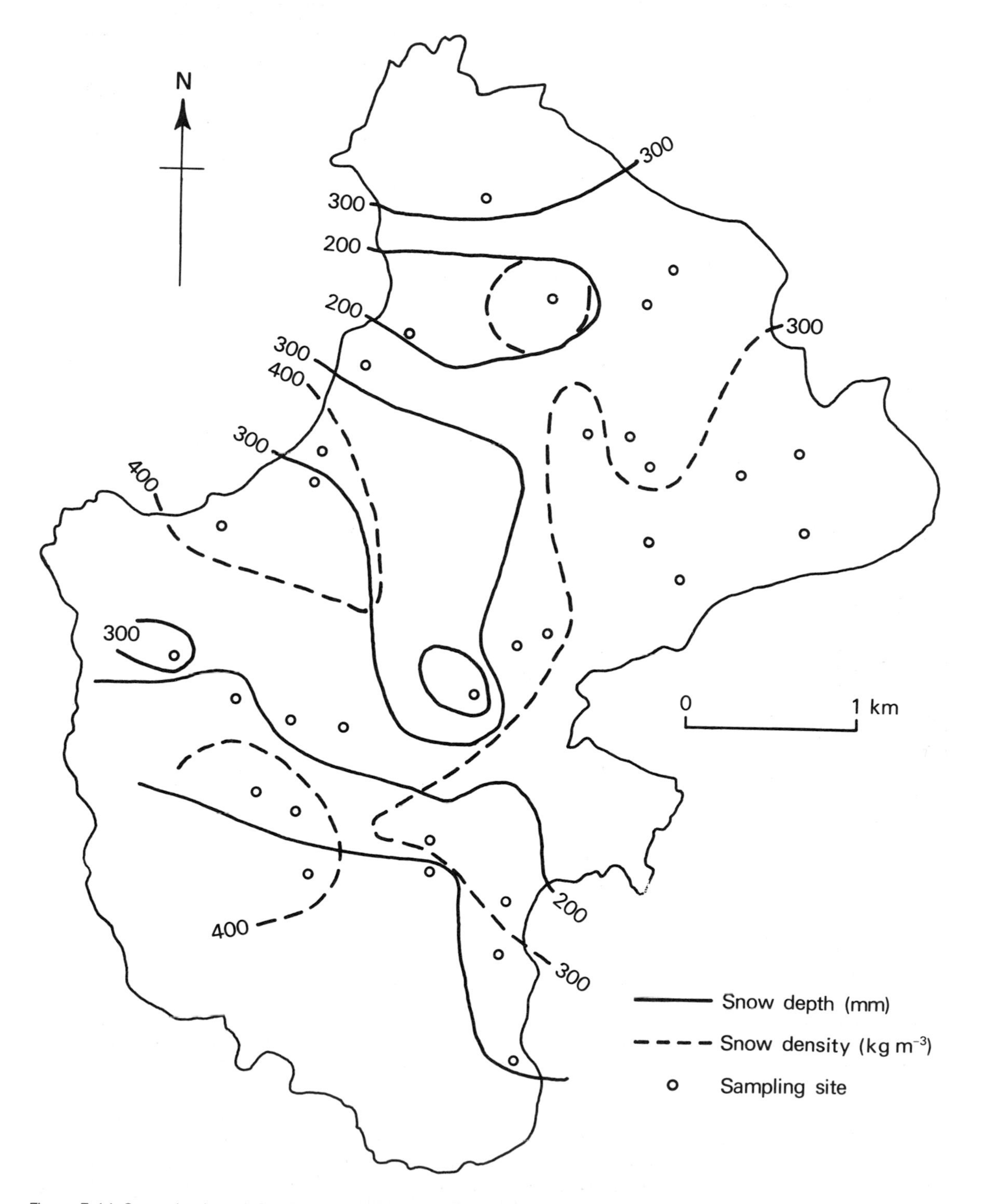

Figure 5.11 Snow depth and density show little correlation in this example from the UK Institute of Hydrology Plynlimon catchments

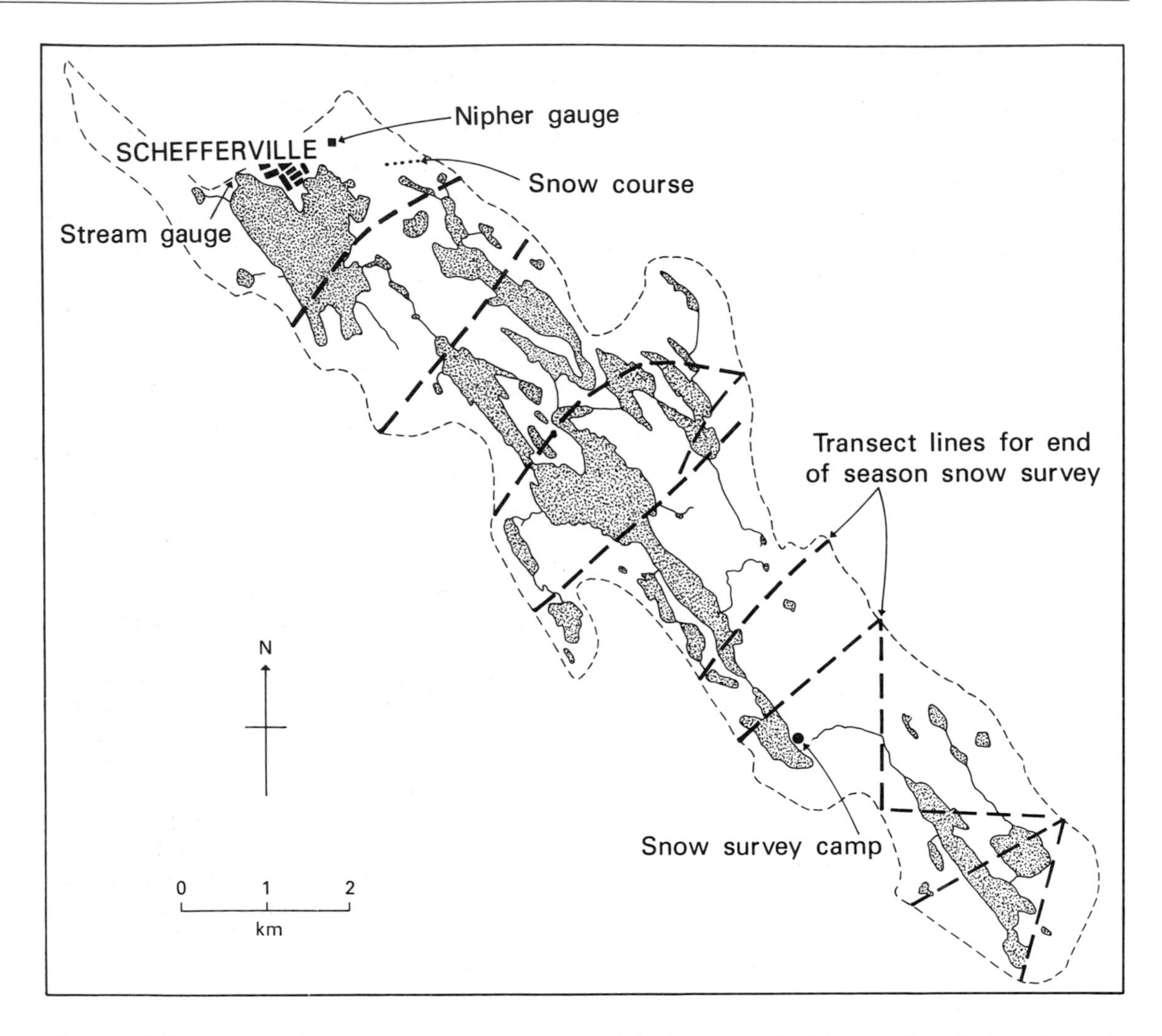

Figure 5.12 Plan for an end-of-season snow survey in the Knob Lake experimental basin, northern Quebec. Transects were laid out to sample the range of cover type and exposure, and snowcover and lake ice were measured at 60/120 m intervals along each transect

water equivalent. Unfortunately, the method never became operational, partly because melting tended to occur before the photogrammetric analysis was completed.

In areas where snow tends to remain on the ground for some time, a wide range of airborne and satellite remote sensing methods are available. **Airborne reconnnaissance** of the extent of snowcover during the melt period used to be commonplace in the western USA before the advent of satellites. Surveys of the changing area of snowcover form an important input to snowmelt flood modelling (section 6.5.3). Unfortunately, lack of penetration of the snowcover by visible and near-IR radiation means that the information they can offer is strictly limited on key aspects like snow depths, profiles, water equivalent and wetness, and often requires considerable ground truth data to calibrate it. Over the last 15 years, however, research using both shorter and longer wavelength radiation has resulted in somewhat greater success. This will be discussed in section 5.2.5.

5.2.3 Design of networks for areal assessment

Most precipitation networks are still based on gauges. Radar and satellite remote sensing are beginning to have a impact, but even these still rely upon conventional, though fewer, gauges for calibration.

Few gauge networks have been designed from scratch for the assessment of precipitation receipts over an area. Many have been dictated by wider meteorological requirements and the distribution of towns and airports. Even those set up for hydrometric purposes have tended to place cost and convenience above accurate areal assessment, perhaps taking daily records only at reservoir sites.

The number of gauges Ideally, the number of measuring sites should be dictated by the desired degree of *accuracy* in assessing the receipts within the area or basin and the *spatial variability* of receipts within that area. Indeed, the conventional statistical formula for the 'standard error of the estimate of the mean', $SE_{\bar{x}}$, i.e. the accuracy, contains just these elements:

$$SE_{\bar{x}} = \frac{\sigma}{\sqrt{n}}$$

where σ is the standard deviation, i.e. the spatial variability, and n the number of sampling sites. If $SE_{\bar{x}}$ and σ are known, then n can be calculated. This has been used as a basis for decisions since R E Horton proposed it in 1923. Unfortunately, σ is the 'population standard deviation', that is, the variability that would be found if we had a complete knowledge of spatial variability or a theoretical means of predicting it. We never have either. Using the formula in this context therefore contains a circular argument: we cannot estimate the variability without taking measurements, but we do not know how accurate those measurements are because we do not know the inherent variability!

There are two or three ways of overcoming this problem, without entirely ignoring it as a 'statistical nicety'. The US Corps of Engineers produced a regionalized version of the formula based on large samples from different hydrometeorological regions of the USA:

$$\log SE_{\bar{x}} = k.\log\left(\frac{a}{n}\right)$$

where a is the average area considered to be represented by each gauge and n/a would be the density of gauges in an area. The value of k varies from one climatic zone to another and effectively subsumes the spatial variability in each region. The UK Meteorological Office produced a similar formula (Bleasdale 1965).

An alternative strategy is to make a staged response, beginning by deploying as many gauges as possible and then rationalizing them to a manageable level. The initial deployment could provide the best estimate of the population standard deviation, which would hopefully indicate that fewer gauges are needed for the required accuracy. In practice, the most common approach has been to use the observation that nearby gauges tend to have similar readings, and so to weed out gauges that are similar and leave only one to represent each group. This is an example of **spatial autocorrelation** and the statistical correlation, r, between gauges is the normal criterion for rationalization, e.g. if $r \geq 0.9$ (Figure 5.13). Hendrick and Comer (1970) used isocorrelation (equal correlation) lines to establish the distance over which each gauge is representative. With increasing cost consciousness over the last decade or so, many water agencies have been using it as a scientific basis for rationalizing their long-standing networks. Clearly, where spatial auto-correlation is significant, the normal $SE_{\bar{x}}$ criterion is inefficient because it suggests that more gauges are needed than are really necessary.

A related approach is to use all the information that is available on the factors that are likely to influence precipitation receipts in the area, and to 'stratify' the monitoring network accordingly. This is based on a fundamental statistical theory, that if there is reason to suppose that values are not simply going to follow a random 'pattern', but have some real pattern to them, then it will be more efficient to divide the sampling up according to the pattern (or the factors behind it) and to sample similar areas separately. Statistical 'analysis of variance' may be used to test the effectiveness of the divisions (see standard statistical textbooks).

The Institute of Hydrology **stratification** plan in Figure 5.14 illustrates a practical application of this approach (Clarke *et al.* 1973). Three factors were deemed to have a major influence, namely, altitude, slope and aspect (orientation). Each was divided up into a number of bands or 'levels', four for altitude, three for slope and the four cardinal points of the compass for aspect. This gave a total of 48 possible combinations, which were possible spatial strata or 'domains'; hence the term **domain theory** for this approach. This was still more than the number of raingauges available, even without considering the fact that some combinations occur many times in the landscape, so two solutions were adopted. First, repeat occurrences of a domain would be alotted the mean value from the monitored example. Secondly, domains that could not be sampled at all were alotted values derived from a statistical estimate of the sensitivity of rainfall to

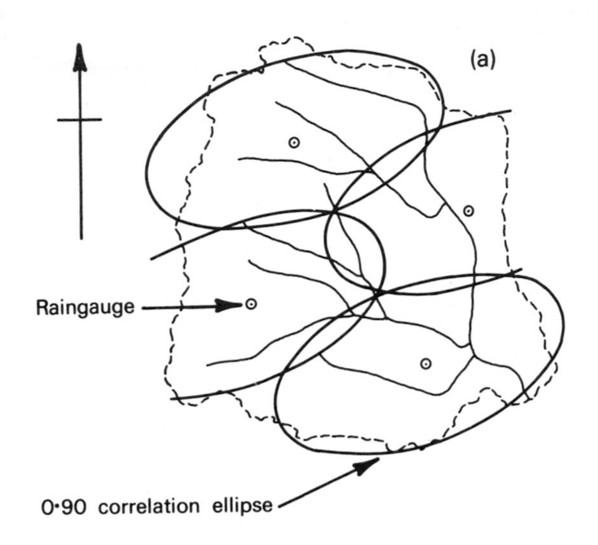

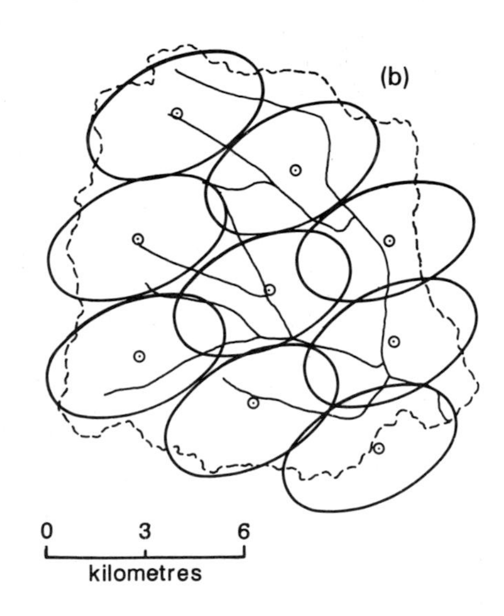

Figure 5.13 Spatial autocorrelation. (a) Each gauge is reasonably representative of the area enclosed by the isolines of 0.9 correlation, allowing surplus gauges to be eliminated. However, (b) summer convective storms are more localized and require more gauges if they are to be covered equally well. Based on Hendrick and Comer (1970)

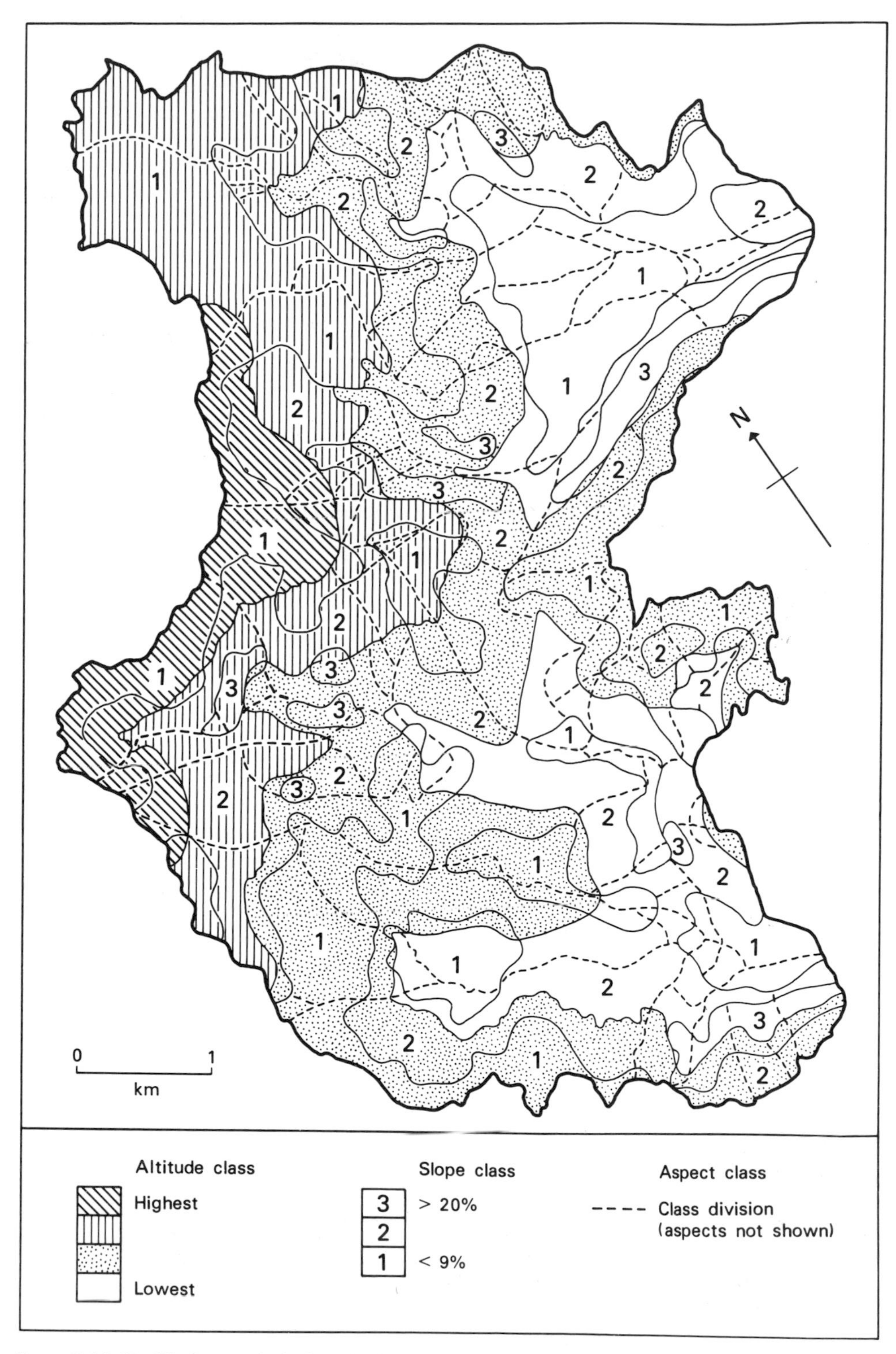

Figure 5.14 Stratified network design: the UK IH Plynlimon plan based on altitude, slope and aspect. According to Clarke *et al.* (1973)

altitude, slope and aspect. Provided every level of every factor has been covered in the domains sampled, it is possible to make a regression estimate without covering every combination of every level.

Where snowcover is being measured, vegetation cover type is also a fundamental factor in stratification. In effect, this is what is being done in setting up a snow course (section 5.2.2).

Thus, in areas with strong and persistent spatial patterns, a much smaller number of gauges is needed than suggested by the simple $SE_{\bar{x}}$ formula. Rodda used this fact to reduce the number of raingauges needed to estimate annual rainfall in the Ystwyth valley in Wales by regressing rainfall against altitude. Sutcliffe (1966) compared the Ystwyth data with the lowland basin of the River Ray in central England. He concluded that the upland basin required fewer gauges than the lowland basin because of the spatial 'persistence' in rainfall patterns. This has important consequences. It has been observed, critically, that the uplands cover 20 per cent of Britain and yet have only 5 per cent of the raingauges (Taylor 1976; Newson, 1981). Moreover, Shaw (1994) indicates that even more gauges should be needed in mountainous areas because of increased variability. This is borne out by the WMO (1994a) recommendations (Table 5.1). However, although there is likely to be greater variability in the mountains it is also likely to be more spatially persistent, with similar patterns recurring from storm to storm, than in lowlands, where localized convective storms are more important and occur more randomly over an area. Therefore, if persistence is taken into account in extrapolating from the existing raingauges within the uplands, the under-representation of the uplands may not be quite as severe as the bald figures suggest. Nevertheless, the situation would certainly be improved by more high-altitude gauges from which to estimate precipitation gradients.

In Britain, the disparity is probably greater with snowfall measurements because of the curious lack of snowgauging at higher altitudes. A survey in Wales suggests that 38 per cent of the principality has a correlation of less than 0.8 with snow records at official meteorological stations, which are mostly sited in the lowlands. Even so, regression relationships based mainly on altitude permitted reasonable extrapolation (Jones and Taylor 1983).

Two of the key contributors to areal persistence are: (1) a dominance of orographically induced precipitation; and (2) a dominance of frontal precipitation. In contrast, a dominance of convective precipitation creates spatial patterns that tend to change more from storm to storm. In general, mid-latitude climates contain a mix of frontal and convective storms that changes markedly with the seasons. In effect, the standard deviation, σ, will vary from season to season.

This means that the designer of any gauge network must be very clear as to the *purpose* of the network as well as the nature of the *climate* they are working in. If only annual totals are of interest, then in climates dominated by cyclonic precipitation fewer gauges are needed to obtain a given accuracy than if summer convective storms are more important climatically or hydrologically, say for flood prediction. Figure 5.13 illustrates the point. In general, more gauges are needed for estimating shorter-term receipts and smaller-scale events.

The International Hydrological Decade paid special attention to improving network design. Rodda's (1969) report showed that for the same density of gauges the percentage errors for areal rainfall estimates are 2.0 to 2.5 times greater in convective storms than in frontal storms. The 1965 WMO/IAHS Symposium on the Design of Hydrological Networks produced useful guidelines for designing gauge densities in relation to required accuracies (Bleasdale 1965; WMO 1994a, first edition 1965) (Table 5.1). Subsequently, Nicholass *et al.* (1981) and O'Connell *et al.* (1977) undertook thorough reviews of data requirements for the UK, indicating preferred instrumentation and gauge densities (Table 5.4).

The location of gauges The siting of gauges can be viewed on two scales. At the microscale, the aerodynamic problems of obtaining a representative catch require a careful compromise between overexposure and shading. The UK Meteorological Office guideline is that gauges should be sited at least twice the height of obstructions, such as buildings or woodland, away from them, in order to avoid shading or excessive turbulence. In contrast, many gauges are sited at airports, where undermeasurement is aggravated by overexposure. Shielding can improve this situation, but rarely provides sufficient compensation.

At the macroscale, the designer needs to consider a rather different range of factors, particularly:

1. Is the data required only for overall averages?
2. Is spatial pattern of interest?
3. Is variability uniform over the area?
4. The balance of cost and benefit in creating and maintaining the network.

There are two potentially important conflicts here,

Table 5.4 Selected examples of monitoring network requirements in the UK

Application	Main type of gauge	Density (km^2 /gauge)	Comment
Precipitation			
water balance – monthly	Daily	64	UKIH (1977) recommendation
soil moisture deficit – daily	Daily, telemetric	16	UKIH (1977) recommendation
flood warning	Auto., telemetric	272	Devon River Authority plan, 1965
river regulation operation	Auto., telemetric	–	River Dee Regulation scheme
reservoir operation – direct supply	Daily, plus 1 auto. per basin	2–20	Institute of Water Engineers recommendation (less dense for larger basins)
Evapotranspiration	Automatic hydrometeorological station	1/8–1/6 of raingauge density	UKIH Welsh research catchments
Riverflow	Auto.	375 (primary gauges)	Water Resources Board target, 1963
Groundwater	Auto.	260 (homogeneous aquifer)	e.g. Triassic sandstone,
		5 (highly inhomogeneous aquifer)	e.g. Chalk

These figures are only approximate and illustrate the range of requirements within a single climatic region.
Detailed specifications can be found in UKIH (1977), Nicholass *et al.* (1981) and IAHS Pub. No. 68 (1965).

which must be resolved according to the purpose of the network. First, there is a potential conflict between the requirements of statistics and those of cartography; statistical theory requires independent, *random* measurements, whereas mapping requires an *even* distribution. Secondly, what is practical and affordable may not be the same as what is scientifically desirable.

A purely **random** selection of sites may often be adequate if only simple mean values are required, whereas a **systematic** distribution, e.g. a fixed number of gauges per square kilometre, ensures similar accuracy for all parts of the area. The **stratified random** approach provides an ideal combination of the two. Stratification may follow a systematic grid or domain theory. In reality, the most common resolution is a combination of areal stratification and a 'pseudo-random' location of the gauge within each cell that is determined by ease of access. A marked difference in variability between one part of the area and another, e.g. between mountains and lowlands, might theoretically require differing scales of stratification, but this is rarely undertaken.

For short-term research projects, **moving gauges** can maximize the number of sites measured by a limited number of gauges; moving may be random or, better, may be determined by intersite correlation. Forest hydrologists first applied this approach in the 1940s, using correlations established between *base gauges* and *roving stations* to predict values at the abandoned roving sites.

Despite all the theoretical research on network design, some problems do not require a 'network'. Eagleson (1967) championed the idea of the **strategic gauge**. One or more gauges located at significant sites in the catchment may be all that is required for flood warning. This is an attractive, cost-effective idea that has been put to practical use by many water agencies. Unfortunately, strategic gauges are the least adaptable for answering other queries, such as estimating total receipts over the whole basin.

5.2.4 Methods of areal extrapolation

For calculating anything other than receipts at a point, the method used to extrapolate or interpolate values is at least as important as the accuracy of direct measurement.

Perhaps the most basic method is using **isohyets**. Isohyets are lines that join points which have the same receipts, and they are usually threaded like contours through the measurement points on the assumption that receipts follow a linear gradation between the gauges, i.e. that the rate of change is constant between neighbouring gauges and is equal to the difference in receipts divided by the distance between them. Computer packages will begin by interpolating values

(a)

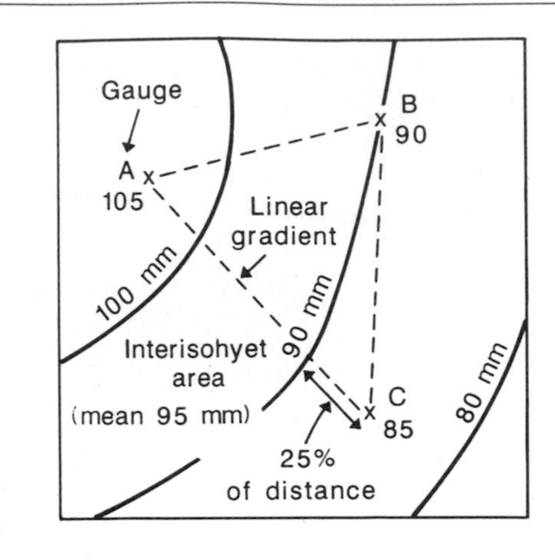

(b)

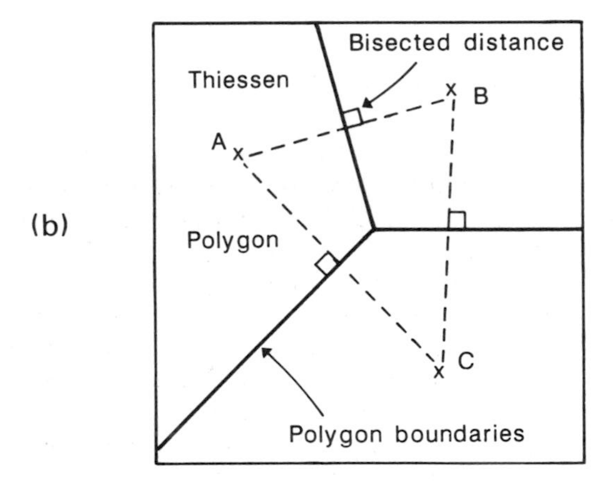

(c)

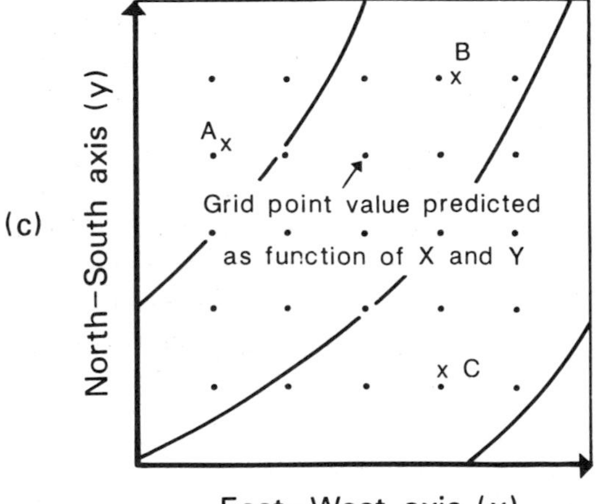

Figure 5.15 Methods for interpolating precipitation receipts and calculating areal totals: (a) the isohyet method; (b) Thiessen polygons; (c) trend surface analysis. See text for explanations

at regular grid points, usually on a linear basis, before constructing the isohyets. Total areal receipts in millimetres are then calculated by multiplying the mean value for each pair of adjacent isohyets by the area enclosed between them, adding them together and finally dividing by the total area (Figure 5.15a).

The isohyet method is fine when there is a large number of gauges. Where the number is more limited, Thiessen's method offers a good solution. This establishes an area around each gauge for which the gauge is considered representative. These **Thiessen polygons** are delimited by bisecting the distance between neighbouring gauges, again assuming a linear gradation, and joining up the bisecting lines (Figure 5.15b). Total receipts in the area can be calculated in a similar way to the isohyetal method, but substituting polygon areas multiplied by actual gauge receipts in each polygon. In areas with marked relief, **height-weighted Thiessen polygons** are better. These are constructed by bisecting the height difference between adjacent gauges, rather than the horizontal distance, and using a topographic map to locate the contour with the appropriate height.

Where gauges have been set up according to **domain theory** (section 5.2.3), areal values are best calculated using the measured or estimated values for each domain and the proportion of the area covered by each domain.

Fitting theoretical surfaces Although polygons and domains work well for calculating total receipts for an area, in reality precipitation receipts vary in a continuous spatial fashion, not in discrete steps. For mapping these receipts and for calculating areal totals from isohyet patterns, some practical basis is required for interpolating and extrapolating the isohyets. Assuming linear gradients is a starting point, but it pays no heed to the processes by which precipitation is created.

More sophisticated solutions may use correlations between precipitation receipts and one or more environmental variables, such as altitude, to establish the gradients before drawing the isohyets. This is commonly used in rainfall atlases. It is superior to simple linear interpolation in areas with high relief, provided there are enough gauges to establish reliable precipitation gradients. However, strong spatial autocorrelation between gauging sites can invalidate the prime assumption of the regression method, i.e. that it is based on independent, random samples. In such a case, the regression equation is likely to give an inflated impression of accuracy. This can be overcome by carefully selecting the sites used to establish the regression equation. The old **Spreen method** developed in the USA in the 1940s used a multiple regression equation which included altitude and relative relief within an 8 km radius, and the exposure and orientation of the site.

Polynomial trend surface analysis uses a multiple regression equation in which the north–south and east–west axes of the map (and possibly derivatives of them) are taken as the environmental predictors (Figure 5.15c). If enough polynomial terms, such as the square or cube of the easterly and northerly coordinates, are included in the equation, then in theory it should be possible to fit a surface to the most complicated of patterns. Unfortunately, the method can easily produce high gradients in areas of the map where measurements are lacking. More fundamentally, the physical processes that produce precipitation are rarely related directly to the map coordinates. Unwin (1969) took the British mountain area of Snowdonia as an example, where 'distance from the sea' can be a spatial factor and one which is in part related to map axes. But in order to obtain a good fit with a cubic polynomial, first he had to remove the dominant pattern imposed by the relief.

An alternative approach is to fit mathematical rather than statistical surfaces. **Fourier analysis** takes regular trigonometrical curves of different amplitude and wavelength and superimposes them until a reasonable fit is achieved to the data values. Shaw and Lynn (1972) proposed **multiquadric surfaces**, in which circular cones are centred on each raingauge and mimic the spatial decline or 'decay' in precipitation around each gauge. A continuous surface is produced which, unlike the statistical regression method, fits every measuring point exactly. Shaw (1994) quotes successful applications for calculating areal receipts over a wide variety of time periods, and as a basis for computer experiments to determine the optimum number and distribution of gauges. Another statistical technique known as **kriging** minimizes estimation error by selecting optimum weightings for each gauge based on the spatial variability (Bastin *et al.* 1984).

From a scientific point of view, all of these methods suffer from the fact that there is no general theory of rainfall distribution. Some potentially helpful observations have been made at the scale of individual rainstorms. For example, Smith and Schreiber (1974) suggested that the central area of updraught within a convective storm cloud varies less between storms of different sizes than does the total area of the storm. They also noted that the typical 'footprint' of a convective rainstorm on the ground follows a 'bivariate normal distribution', because as the raincloud moves the bell-shaped pattern of rainfall intensities within the circular convective cell produces a footprint that is elongated in the direction of movement. However, this is not particularly helpful for extrapolation at storm scale unless we know where the centre of the storm was in relation to the gauge. Similar problems apply to the fitting of any mathematical shape: for example, in process terms, it seems unrealistic to regard each gauge as the centre of a conical distribution.

There seem to be two broad ways forward in the absence of a general theory. One is to make maximum use of what theory is available, for example by selecting predictor variables that are directly related to the likely causal factors. This is best applied to long-term patterns. The other, which offers more for storm-scale patterns, is to use weather radar and satellite technology.

5.2.5 Weather radar, satellites and airborne remote sensing

Modern remote sensing systems, like ground-based radar and satellite imagery, offer an immediate spatial image. This has revolutionized but not completely replaced the use of conventional gauges. Gauges are still needed to calibrate the electromagnetic record. And methods of spatial extrapolation are still needed to extend and refine that calibration over large areas. Moreover, the conventional gauge remains very cost-effective, both as a means of obtaining long-term totals and in its role as a strategic alarm.

Weather radar has been developed over the last 40 years as a refinement of the original active radar navigational system. The most common form of active radar emits pulses of radiation in the microwave wavelength (section 5.1.2) and measures the rate of return of the radiant energy reflected by the falling precipitation, or 'hydrometeors'. Ideally, the power of the backscattered radiation would provide a direct measurement of the mass of the falling precipitation as it approaches the ground. Unfortunately, this presents a number of technical problems, in particular:

1. The strength of the radar echo depends on drop-size and physical phase (whether liquid or frozen).
2. Hills can block the line of sight, so that measurements have to be taken some distance above the ground and are often angled upwards to increase the range of coverage.
3. The amount of radiation returned to the radar antenna falls off with distance and the fall-off rate can be increased by heavy rain along the path.

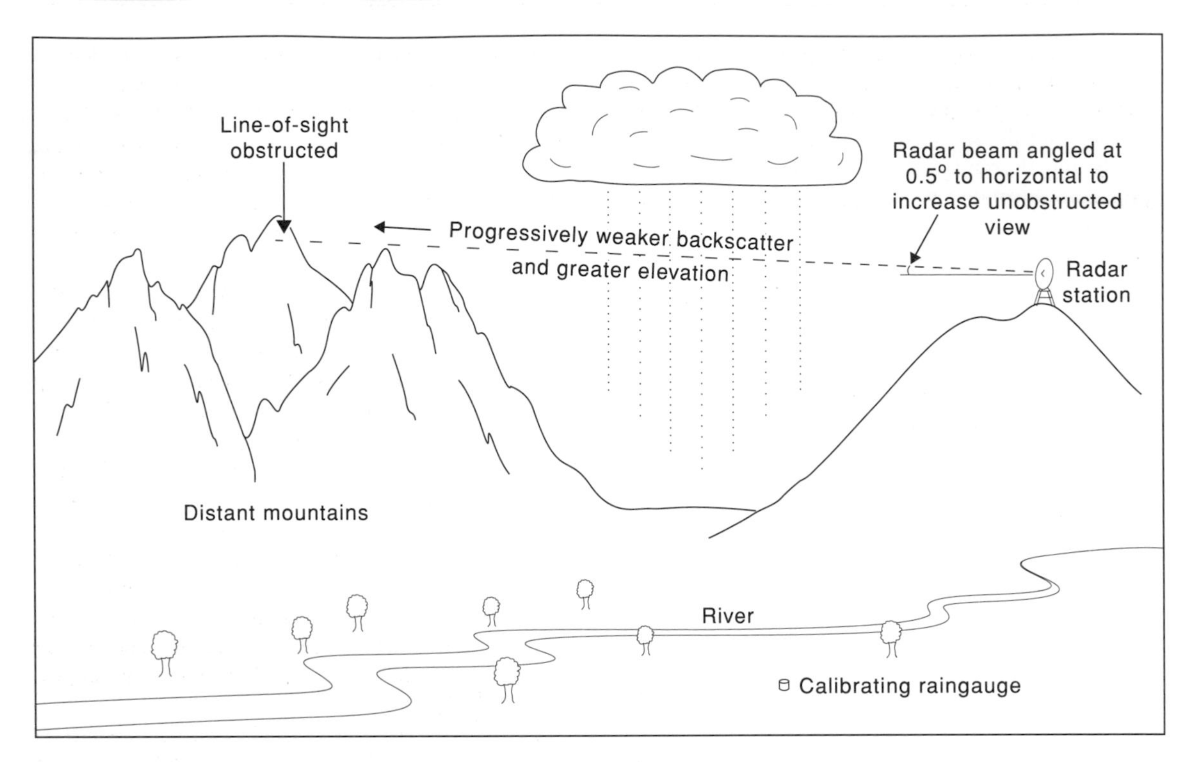

Figure 5.16 Ground-based weather radar

Drop-size distributions vary from storm to storm and, since the reflectivity of a raindrop is proportional to the sixth power of the diameter, the backscatter is very sensitive to these differences. Fortunately, rainfall intensities, I, are also proportional to average drop size, so that the overall reflectivity ranges from a linear function of intensity to a function of I^2; $I^{1.6}$ is commonly used as an average in temperate regions. This uncertainty in the calibration is increased by the presence of snow, especially by melting snow, which has a much higher reflectivity. Hence, snowfall assessment is still less accurate than rainfall.

The problems with line of sight and the progressive loss of signal over a distance mean that each radar unit has a limited range (Figure 5.16). In addition, measuring precipitation some hundred or more metres before it reaches the ground means that significant redistribution and evaporation losses can occur between the radar beam and the ground. Even raindrop growth can occur beneath the beam if the radar passes through the clouds. It is therefore not surprising that radar measurements are only accurate to within about 10 per cent for rainfall and 20 per cent for snowfall.

The UK Meteorological Office overcomes many of these problems by undertaking continuous 'real-time' calibration of the radar data (Collinge and Kirby 1987). Real-time calibration is based on three or four key gauges around the country and separate calibrations are performed for different types of rainfall, such as convective and frontal storms (Battan 1973; Collier *et al.* 1983). The Meteorological Office fits local quadratic polynomial trend surfaces to extrapolate this calibration away from the reference gauges. In areas where radars overlap, there are 'preference maps' for deciding which radar is best. Eventually, the UK network will be integrated with a network covering the whole of Western Europe.

Radar has transformed early warning systems. In the USA, a prime reason for the target network of 100 radars nationwide is to monitor intense convective storms, especially hurricanes. In the UK, at the Clywedog Dam river regulation scheme on the River Severn (section 9.3), early warning of potential flood hazards used to be obtained from a raingauge on a remote automatic weather station that transmitted data to Meteosat. This gave a 2-hour warning period in which water level in the reservoir could be lowered to accommodate the flood wave. Now, radar images from the Meteorological Office give more than half a day's warning.

Satellites are not currently capable of measuring precipitation directly, but useful information may be gleaned from aspects of the cloud cover. Barrett and Martin (1981) discussed a variety of approaches. The USDA's AgRISTARS (Agriculture and Resource Inventory Surveys Through Aerospace Remote Sensing) programme adopted an interactive scheme for estimating rainfall based on combining satellite information with climatic data, local meteorological reports and numerical weather prediction. This is used to calculate the water budget during the growing season. Moses and Barrett (1986) divide the approaches into:

1. *Objective and automatic:* fully computerized, using a single IR threshold to identify rainclouds and then using a weighting to convert the estimated days with rain (raindays) into rainfall amounts, which is empirically derived from local climatological records.
2. *Interactive*, such as the 'Bristol' method, which relies on a manual nephanalysis that is calibrated with local rainfall records.
3. *Manual*, which requires considerable skill to relate cloud type to the amount of rainfall.

The latter requires highly trained personnel. But Barrett's University of Bristol remote sensing laboratory has been particularly successful in developing the intermediate approach, which can be described as a **cloud indexing** method. It requires very little other than a few raingauge records and contemporary satellite images. The method establishes a multiple regression equation relating raingauge records to cloud properties, like percentage cloud cover, cloud altitude, pattern and reflectance/emission as indicated by the tone or 'grey scale value' on the satellite image. Cloud altitude may be estimated from the form of the cloud. In addition, cloud-top temperatures may be estimated from the infrared waveband, and these are a good guide to the status of precipitation-generating processes (section 2.3.1). The equations can be used to estimate precipitation throughout a satellite image or 'scene'. It is a rather indirect method and no real substitute for radar coverage, but it can be of tremendous value where local technological resources are limited, as demonstrated in the East Indies, the Middle East and North Africa. The Earth Satellite Corporation developed another cloud-indexing method, which is based on establishing a regression relationship between cloud temperatures and rainfall.

A detailed interactive system forms the basis for the NOAA Flash Flood Program. In this, the Interactive Flash Flood Analyzer combines nephanalysis with upper air data from the GOES VISSR Atmospheric Sounder (VAS), which gives temperature, humidity and wind profiles at 80 km intervals. Physically based calculations of key parameters, like precipitable water content and strength of uplift currents, are used to produce a regression-based estimate of the probabilities of severe rainstorms, which is disseminated to NWS offices.

Another method, which has been dubbed the **cloud history** approach, can be used where ground radar is available over at least part of the scene. The radar can be used to calibrate the satellite imagery by correlating the area of rainfall as indicated by the radar echo with the spectral image of the overlying cloud. Bellon *et al.* (1980) were able to link the spectral signature of clouds in the visible and infrared bands of a GOES satellite to rainfall measured by radar at Montreal. The radar covered only 10^5 km^2, but the calibration allowed the McGill team to extrapolate rainfall over 2 million km^2 covering most of eastern Canada. The Environmental Research Lab (ERL) used a similar method for convective storms in Florida, which was also adopted by AgRISTARS.

More direct observation of rainfall from satellites is expected from the Tropospherical Rainfall Measuring Mission (TRMM), which should give 12-hourly coverage from 1997, and the Tropical System Energy Budget (BEST), which promises three-dimensional windfield and rainfall rate data by the year 2000.

Visible and near-IR satellite imagery have been used extensively for monitoring the extent of snowcovered area during the melt period, for example using the Landsat MSS (Baumgartner *et al.* 1986), but it lacks the penetration to give much information on the internal structure of the snowpack. In contrast, **passive microwave** sensors can be used to estimate the amount of attenuation of natural microwave emissions from the Earth, and so estimate water equivalents, density layering and the status of snow metamorphosis, which can all be fed directly into process-based techniques for snowmelt forecasting (section 6.5). The scanning multichannel microwave radiometer on Nimbus 7 has been used successfully in Russia and Canada, although its spatial resolution is too crude for smaller basins, with a maximum of only 25 km at the shortest wavelength and 150 km at the longest (Foster *et al.* 1980). As with visible light, microwave radiometers are more use if they are multichannel/multifrequency, because different wavebands have different penetrations and pick up different aspects.

Airborne remote sensing has also been successful in determining key internal properties of snowcover (section 5.2.2). Natural **gamma ray** emissions from the Earth are attenuated by overlying snow and the attenuation is proportional to the water equivalent of the snowpack. Grasty (1981) described efforts in Ontario to develop an operational technique for determining the amount of attenuation employing aerial gamma ray scanners. Grasty's team used a single aircraft overflight rather than 'before-and-after-snowcover' flights, in order to reduce the heavy cost. Kuittinen (1986) linked airborne gamma ray spectrometry with NOAA AVHRR imagery in the red band. He thus reduced aircraft costs by using the satellite imagery to extrapolate water equivalent throughout the melt period.

5.3 Measuring evaporation and evapotranspiration

Evaporative losses are among the least measured aspects of the water budget. This is partly because the processes themselves are complex and difficult to measure (section 3.1) and partly because they are not of primary concern for daily weather forecasting and have therefore received low priority in national meteorological networks. Evaporative losses from the land have become important to meteorologists only with the advent of medium- to long-term forecasting. Otherwise, most instrumental measurements are taken by water agencies or by hydrological and agricultural research bodies.

As a result, extrapolation of instrumental records over wide areas is bedevilled by lack of standardization and lack of centrally maintained reporting networks. Green's (1970) extrapolation over the British Isles based on records from agricultural research stations was a rare undertaking which also illustrated some of the problems.

5.3.1 Instrumental measurement

For most operational work, the hydrologist will be extremely lucky to have any direct measurements within or even in the vicinity of a given basin. Moreover, there are powerful disincentives to establishing a bespoke network, particularly: (1) the high cost of installing many of the instruments; (2) the relatively low level of automation available for many of them; (3) substantial problems with the accuracy of most of them; and perhaps most importantly (4) the relative ease with which modelling can substitute for measurement (section 3.1.1). A further problem is that the high spatial variability in losses from different cover types and slope aspects would ideally require large numbers of instruments to account for it (section 5.2.3). Fortunately, in many humid and perhumid regions, this high spatial variability is partly offset by the fact that the *absolute* amounts of evaporative losses are not so great in relation to precipitation, so that there may be some justification for sampling only a 'typical' site. As a result of all these considerations, 'network design' is not a question normally associated with evaporation measurements.

Direct measurements may be taken by one of four categories of instrument. Evaporation may be measured by (1) tanks or pans, or by (2) atmometers, and evapotranspiration by (3) potential evapotranspirometers or (4) lysimeters.

Tanks and pans are the most common instruments. **Tanks** are buried in the ground, whereas **pans** are set above ground on a small plinth (Figure 5.17). As Table 5.5 reveals, tanks tend to give a more accurate assessment of true evaporation. There are two main sources for this. First, the exposed walls and shallow water of a pan tend to be heated more by sunlight and cause excessive evaporation. Secondly, the smaller surface area of many pans causes higher evaporation rates per square millimetre of surface. Both tanks and pans normally present isolated bodies of water set in a large surrounding area of vegetated soil surface; the air blowing over them will tend to have a lower relative humidity than if it had crossed a lake surface and so it will take up more water vapour. This is known as the '**oasis effect**'. The smaller evaporation pans, such as the old American 300 mm diameter pans or the Japanese pan, are more prone to this effect than the larger tanks. This would not be so bad if the correction factors shown in Table 5.5 were constant, but they vary with climate, sites and synoptic situation. Even so, the WMO's decision to designate the US NWS Class A pan as the *interim* international standard is defensible on the grounds that it is relatively cheap and easy to install and reasonably 'true'.

Atmometers measure evaporation from a small filter paper or black ceramic disc. They use a small glass reservoir graduated in millimetres of water evaporated from the plate or paper surface. Both tend to flood in strong winds and so overestimate evaporation, simply through vibration at the join between the glass and the plate. For this reason, the small Piche atmometer is normally mounted out of the wind in a Stevenson screen and so measures 'potential evaporation in the

(a)

(b)

Figure 5.17 A selection of evaporation tanks and pans: (a) the US NWS Class A pan; (b) the British tank; (c) Russian tank; (d) Chinese design to reduce edge effects by introducing a buffer ring. White finishes reduce solar heating

Figure 5.17 (c) (left) and (d)

shade'. In this setting, it fails to take account of the important effect of insolation on evaporation, but it has been used successfully as a substitute for the mass transfer term in Penman's formula (section 3.1.1). The larger Bellani atmometer gives an unusual measurement for the opposite reason; its black plate is intended to *maximize* the effect of insolation. This has been termed the **latent evaporation**.

Atmometers, tanks and pans all measure potential evaporation. Apart from estimating reservoir losses, however, the processes of evapotranspiration tend to be of most concern to the hydrologist. Potential evapotranspiration is measured by **evapotranspirometers**, which come in two forms, constant water table or constant feed. Near-constant water tables can be maintained by regular addition or removal of water via a tap set in the tank at the appropriate level to allow capillary water to rise to the surface (Figure 5.18). Constant feed may be best achieved by inverted bottle feeding or by artificial sprinkler irrigation. For many hydrological purposes, Penman's restricted, agricultural definition of 'potential' relating to a short green crop (section 3.1.1) is not as valuable as an estimate of the maximum rate of loss possible from the particular local combination of soils and vegetation when water supply is unrestricted. This is essentially Thornthwaite's definition, which emphasizes an extended moist surface covered in vegetation in order to minimize the oasis effect, but does not restrict the vegetation type. Thornthwaite's evapotranspirometer maintains saturation within the root zone over a 4 m^2 area. For non-agricultural land, the aim is to reproduce the actual surface as closely as possible, including undisturbed soil profiles and natural vegetation. Unfortunately, as Figure 5.18 illustrates, this can be a difficult task for forest vegetation.

Actual evapotranspiration (AET) is of more direct value for water balance calculations and this is measured by **lysimeters**. In small free-drainage or **percolation lysimeters** AET is taken to be the difference between measured precipitation and the water that drains from the lysimeter into a collecting bucket, assuming no change in soil moisture levels. The little Popov lysimeter in Figure 5.19c can be removed and weighed to take account of changes in soil moisture, but it is rather small for a good estimate of

Table 5.5 A comparison of the specifications and performance of various evaporation tanks and pans

Instrument	Dimensions	Pan coefficient (correction factor)*
British standard tank	1.83 × 1.83 m, 0.61 m deep	0.93–1.07
US Bureau of Plant Industry tank	1.83 m diameter, 0.61 m deep	0.91–1.04
Russian/Soviet GG1-3000 tank	0.62 m diameter, 0.55 m deep	0.75–1.00
Colorado tank	0.91 × 0.91 m, 0.49 m deep	0.83
US NWS Class A pan	1.22 m diameter, 0.25 m deep	0.60–0.80 (range dry to wet climate/season)

*An average factor needed to correct measurements to 'true evaporation', based on various sources including WMO (1994a).

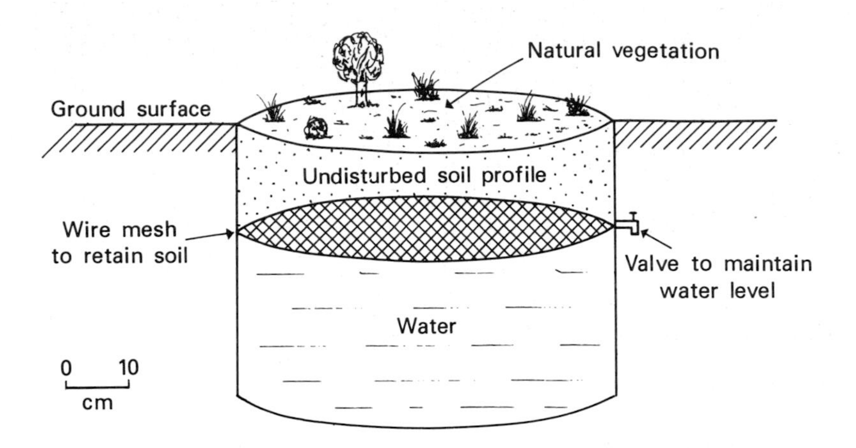

Figure 5.18 An evapotranspirometer for measuring potential evapotranspiration

AET. **Weighing lysimeters** are designed to provide a continuous record of weight, either hydraulically, like the Australian Commonwealth Scientific and Industrial Research Organisation (CSIRO) version, or by balance. One of the largest of this ilk was established at Atmospheric Environment Service headquarters in Toronto. Figure 5.19a attempts to portray this gigantic structure.

Each of these lysimeters contains an element of artificiality, like restricted vegetation type, transplanted soil blocks, or restricted drainage. One of the best designs of **natural lysimeter** was developed by Frank Law (1956) for experiments on water use by forests. Figure 5.19b illustrates his design in use at the UK IH research catchment on Plynlimon. This is a slice of natural forest floor, underlain conveniently by a natural layer of impermeable clay. Rainfall was measured by canopy and surface gauges, soil moisture by neutron probe (see page 164) and seepage by tipping bucket.

More recent developments include direct measurement of vapour fluxes in the air using ultrasonic waves or infrared radiometry, but these are restricted to experimental applications. Alternatively, field estimates of soil moisture can be combined with precipitation, river discharge and groundwater monitoring to calculate evapotranspiration losses.

5.3.2 Areal estimates

Because of the general lack of instrumental networks, basin-wide estimates are normally undertaken either by water balance calculations, in which ET is the one unknown component (section 2.4), or by applying a formula such as Penman's to local meteorological data (section 3.1.1). A detailed guide to water balance calculations is provided by Thornthwaite and Mather (1955) and Dunne and Leopold (1978).

In Britain, the **Meteorological Office Rainfall and Evaporation Calculation System (MORECS)** provides a sophisticated areal estimate of ET, which supplements a Penman approach with simulation of soil water flux and a consideration of local vegetation cover. MORECS includes the results of recent work on the evaporation of intercepted water, which has been found to occur at up to 1.6 times the Penman estimate in summer. To calculate AET, its soil moisture extraction model is based on a two-layer soil. Radiation balances are normally estimated from sunshine hours and an average albedo for the grid square. MORECS provides an online service to water agencies and agricultural interests giving not only PET and AET, but also areal rainfall receipts (P), soil moisture deficit (smd) and 'hydrologically effective rainfall', that is, P–AET–smd, for 40 × 40 km grid squares on a weekly basis.

There is now a substantial and valuable archive of MORECS calculations for the UK back to 1961 that can be used for risk analysis (Chapter 4). The Meteorological Office has hoped for some time to increase spatial resolution to 20 × 20 km grid cells. Only then will it be realistic to make effective use of some of the recent refinements in the calculation of stomatal resistance, the aerodynamic resistance of the local vegetation types and the gross aerodynamic roughness of the topography. But this will require matching improvements in estimating net radiation and some detailed local data collection for topographic and vegetational parameters.

Figure 5.19 Lysimeters for measuring actual evapotranspiration: (a) a very large example of a weighing lysimeter at AES Toronto (surface and entrance to weighing chamber); (b) a section of forest floor isolated to form a lysimeter by the UK Institute of Hydrology, where surplus water drains via the large tipping bucket at lower left, and standard raingauges and neutron probe access tubes are set up to measure throughfall and soil moisture; (c) a small Popov lysimeter in Germany, in which soil moisture is determined by removal and manual weighing – two identical cylinders are weighed to improve the accuracy

Figure 5.19 (c)

Indeed, one of the most critical aspects for the successful application of ET formulae to questions of local water balance is the matching of *spatial resolution* in data collection and estimation with the *spatial variability* in processes. Actual ET rates vary greatly over relatively short distances, depending on vegetation species, soils and topographic position. The use of a single parameter, for example for stomatal resistance, over a 1600 km^2 area is a major simplification. Satellite imagery could be used to get a more detailed coverage of vegetation and other surface controls. It is even possible to infer rates of evapotranspiration directly from imagery, because leaf temperatures can be converted into AET by energy balance models or by empirical relationships. Unfortunately, the infrequent coverage from high-resolution SPOT or Landsat Thematic Mapper imagery continues to be an obstacle.

Despite major advances in monitoring evapotranspiration, it has to be admitted that greater errors are currently introduced into runoff estimates through lack of theory or data covering other parts of the runoff-production process.

5.4 Measuring runoff

The task of measuring riverflow is much simpler than evapotranspiration or precipitation. There are no problems with complex exchange processes or areal extrapolation for a river. Questions of areal extrapolation could arise in estimating overland flow, but this is really only undertaken on a research basis and runoff is normally only measured within well-defined channels. Nevertheless, the hydrologist is frequently hindered by the lack of streamflow stations and the high cost of installing long-term gauging structures. There is the additional problem of extremely high-magnitude events, floods, which can be much more difficult to measure accurately than heavy rain or extreme drought. The case of the Amazon has already been mentioned in section 2.4.

Questions relating to water quality will be postponed to Chapter 8, since a thorough appreciation of the problems requires an understanding of the sources of pollution, the biological and chemical processes involved and the recent legislation on environmental protection, which the recent expansion in water quality monitoring is designed to support (section 8.3).

In order to monitor water quantity, hydrologists measure either the **stage**, which is the water depth, the **velocity** or the **discharge**, the volume of flow. In general, discharge is the ultimate goal of measurement and stage and velocity measurements are only intermediate steps in its determination. Discharge may be measured by building a gauging structure, like a weir, in the river or by direct measurement in an unobstructed channel.

5.4.1 Non-structural gauging methods

Stage is the easiest aspect to measure, but on its own it is the least valuable. Floods were recorded in ancient Egypt in terms of stage. Stage is still commonly recorded by regular visual inspection of graduated **staff gauges** set into the riverbank or some structure in the river. **Crest (stage) gauges** can automatically record peak flow levels by means of some substance like granular cork that floats up inside a small tube and sticks to the side, leaving a debris line at maximum stage. Continuous, fully automated records can be obtained by **water level recorders**, either on chart, punch-tape or electronic record (Figure 5.20).

Stage records are of little hydrological value unless they can be related to a **rating curve**, which allows the discharge to be inferred from the stage (Figure 5.21). Rating curves can be established from either theory or observations, but theory applies only to specially engineered structures, such as weirs or flumes (section 5.4.2). Even then subtle differences commonly occur between the lab-tested blueprint and the actual field installation, so that empirical checks are normally carried out and the structure is recalibrated if necessary. In these cases and in sections of stream that do not contain a gauging structure, rating curves are established by taking a series of measurements of discharge at different levels of flow. It may take some years to obtain sufficient measurements of very high and very low flows, but once a simple curve has been fitted to the data it is assumed that the relationship can be extrapolated beyond the range of actual measurements. In order to ensure the the rating curve is reliable and unchanging when no gauging structure is present, the selected **rated section** should have stable banks and bed and should be on a straight portion of channel, well away from bends, which will cause asymmetric flow patterns.

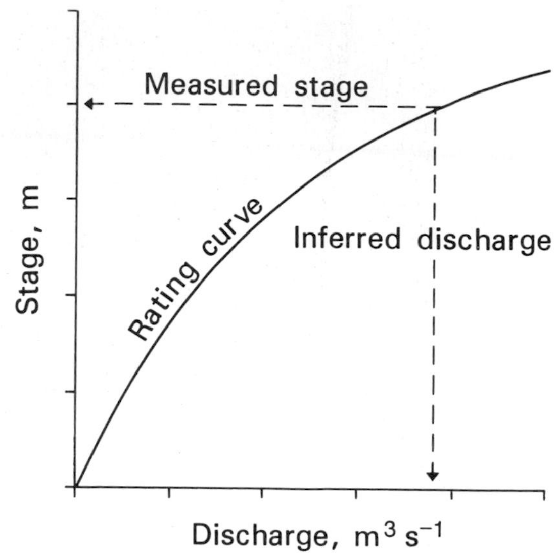

Figure 5.21 A rating curve allows measurements of depth of flow, 'stage' or water level, to be converted to discharge

Figure 5.20 Water level recorders are set on a 'stilling well' linked to the stream, which reduces water turbulence. The photograph shows a horizontal Stevens chart recorder and, in the foreground, a newer digital shaft encoder that is linked to a data logger. Both monitor water level in the well by float and pulley

Discharge can be measured in a variety of ways. The **velocity–area** method is commonly used in rated sections. In this, the average velocity in the channel is multiplied by the surveyed cross-sectional area of flow. Velocities are generally measured by **current meter**, of which the most common type is a small propeller (Figure 5.22). These flow or current meters are immersed in the river by wading or by hanging them from a cableway, cable car, bridge or boat. In each instance, the objective is to minimize any interference with the natural flow and to obtain a representative set of measurements across the river (Figure 5.23). Typically, this involves measuring at a number of points across the channel (20 or more for a large river) and at a number of depths (e.g. 0.2 and 0.8 of the depth). Where the stage is low, a good estimate of

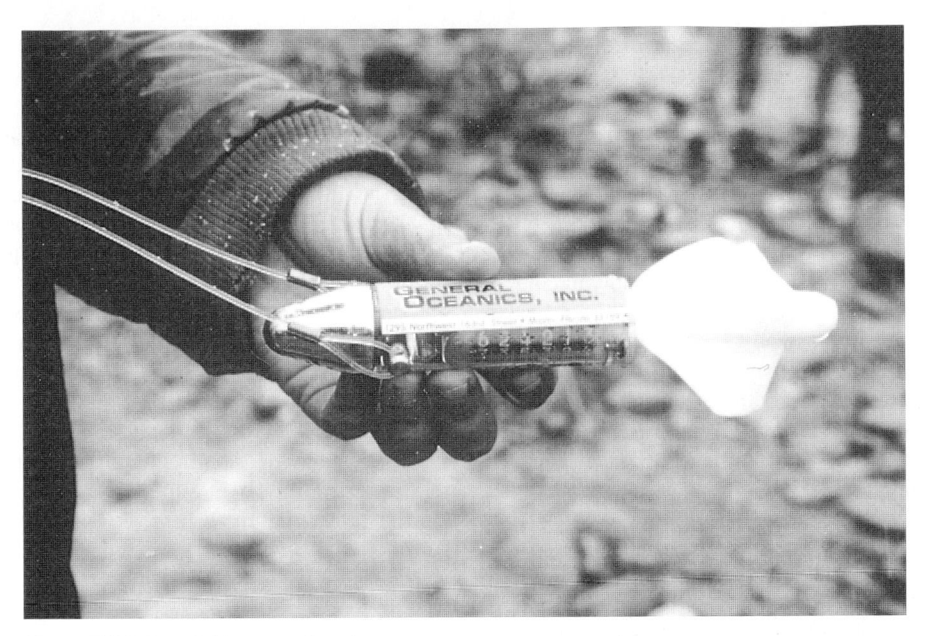

Figure 5.22 A portable current meter

average velocity within the profile can be obtained by a single sample at 0.6 of the depth. Small 'pygmy meters' can also help in low flows. Floats and other meters are described in Herschy (1978).

Dilution gauging offers less mechanical interference and can be undertaken in very low flows, in mountain streams and wherever the bed roughness elements are very high compared with the depth of flow. This operates on the principle that if a known concentration of a substance is introduced into the stream, the concentration measured after it has been thoroughly mixed with the streamwater indicates the amount of water available to dilute it in the stream. This can be achieved using either solutions, dry granules or powder.

The substance should dissolve readily, remain stable and not react easily with other chemicals in the water or be removed by vegetation and sediments, and, most importantly, it should be non-toxic in the concentrations

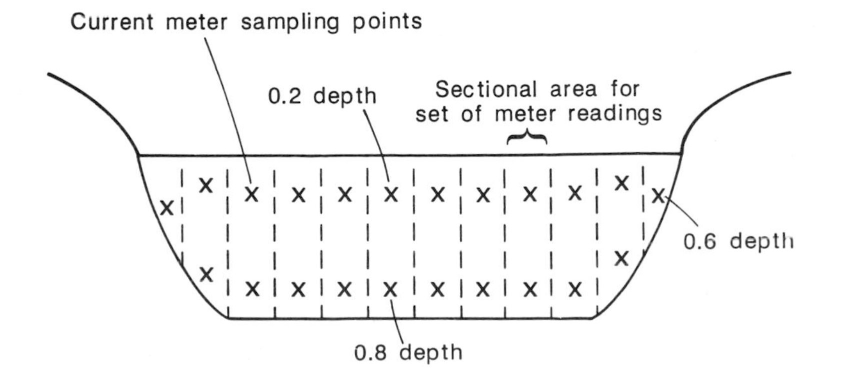

Figure 5.23 Rated cross-sections: a plan for the standard sectional method of discharge measurement – the cross-section is divided into sections, each ideally about 5 per cent of the total width and containing less than 10 per cent of the total discharge, and the current meter is set in the middle of each section (mid-section method) or at either side and velocities averaged (mean section method)

used. In many cases, it is important that it can be detected in very small quantities, and, ideally, it is easier to measure if it has a low **background concentration**, i.e. its natural concentration in the stream is low. If the substance is naturally present in the stream, then pre-injection samples will be needed to establish background levels. Substances not normally found in natural streams are preferable from this viewpoint. Sodium dichromate and lithium chloride were once frequently used because of this. However, they can be environmentally toxic and may be banned in sensitive rivers, and they have largely fallen out of favour on grounds of both cost and environmental hazard. Radio-isotopes have also been used, but environmental concern has largely limited them to either naturally occurring tritium or rapidly decaying bromine-82, with a 35-hour half-life.

The cost and availability of both the materials and the measuring instruments are important considerations. One of the cheapest and safest substances to use is common salt, which can be measured with a simple conductivity meter. However, its background concentration may be quite high and variable, and in any case the conductivity measurement will be affected by the whole range of naturally occurring solutes. It is often helpful to be able to see the substance in the stream, to know when to sample and to confirm that the mixing is complete. Hence, a wide range of **dyes** may be used. The most expensive, fluorescent dyes, can be detected in very small quantities with fluorometers, but again rhodamine-B and fluorescein have a relatively high environmental toxicity. There is much to be said for natural vegetable dyes. They can be measured in the field by portable colorimeters. They are cheap and non-toxic and can be biodegradable.

The substance is injected into the stream either at a constant rate or in a single 'gulp'. **Constant-rate injection** requires more sophisticated equipment, but analysis is simpler. Once the substance has been thoroughly mixed in the stream, just one or two water samples should be enough to establish the amount of dilution. **Gulp injection** needs no special equipment for injection, but requires frequent sampling throughout the whole time that the injected material is passing the sampling point. Figure 5.24 illustrates the procedure. The area beneath the curve indicates the discharge. In each case, samples should be taken at a sufficient distance downstream from the injection point to allow thorough mixing. This critical **mixing length** can be tested beforehand.

One of the greatest disadvantages of both dilution and current meter gauging is that they are very demanding of manpower. Three 'high tech' techniques, developed over the last two decades, revert to the velocity–area approach, but permit unmanned monitoring. **Ultrasonic gauging** uses pulses of high-frequency ultrasound, which are transmitted from both banks at a 45° angle to the flow, one upstream, one downstream. The difference in the time taken for the sound waves to travel in either direction is proportional to the average velocity of flow across the stream. Sampling can be at one or more depths. The technique is affected by suspended sediment and detritus, but is otherwise reliable and environmentally friendly, although the channel is usually smoothed and lined to create a stable rectangular cross-section.

Electromagnetic gauging requires an electric cable to be buried in the streambed. The electric current induces an electromotive force (emf) in the water flowing over it that is proportional to the average velocity in the entire cross-section and this is monitored by bankside probes. It is expensive and requires a mains electricity supply. The **integrating float technique** uses bubbles of compressed air in a sophisticated, two-dimensional update of the old 'Pooh-stick' principle of floating sticks. Bubbles are released at regular intervals from pipes set in the streambed at a number of points across the channel. Computerized photographic monitoring reveals the amount of displacement of the bubbles, and the vertical pattern of displacement is proportional to the stream velocity profile.

5.4.2 Weirs and flumes

Fixed gauging structures are of two types, weirs and flumes. Each offers a stable cross-section with a simple geometry and is designed to give a smooth rating curve relationship, so that regular or continuous monitoring of stage is all that is needed. For long, the great advantage of structures for permanent monitoring has been that they offer a constant cross-section, with a mimimum requirement for recalibration and human labour. This advantage has been somewhat eroded with the possibility of lined rectangular cross-sections used in electromagnetic or ultrasonic gauging, but they still retain the advantage that water level recorders are relatively simple and cheap to operate. The distinction between a lined cross-section and a flume is that the latter is hydraulically designed to give a smooth rating curve, whereas the lined cross-section is merely designed to provide a constant channel shape.

Weirs obstruct the flow to varying degrees. The founding principle for all weirs is that the head of water

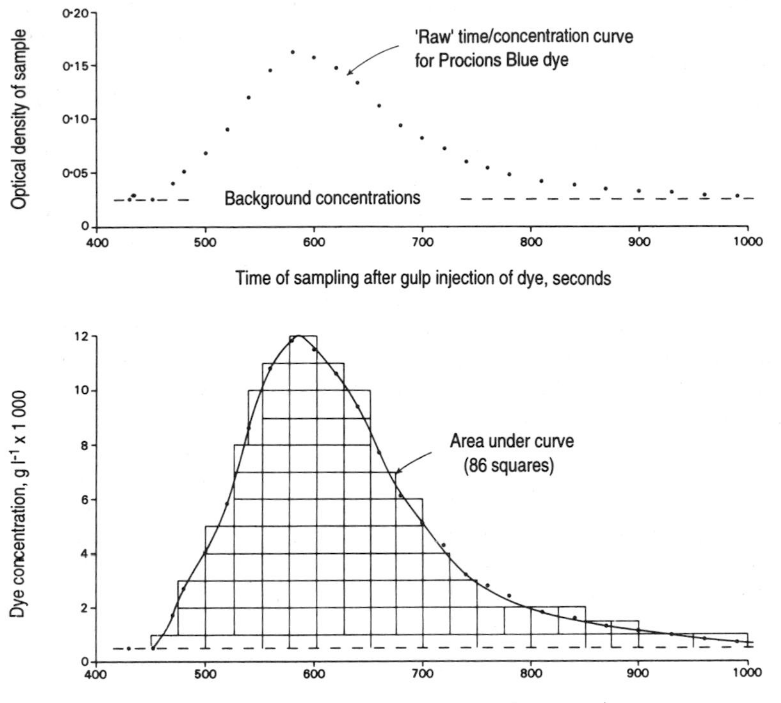

Figure 5.24 An example of the dye dilution method: optical measurements of dye density taken with a colorimeter (above) are converted into dye concentration (below), based on a laboratory calibration curve relating the colorimetric values to dye concentration. Discharge is then calculated from the area under the concentration curve

(or stage) in the weir pool upstream of the obstruction is proportional to the discharge over the crest of the weir. **Sharp-crested weirs** obstruct the flow most of all and are only really suitable for small streams. These consist of a metal plate cut with a rectangular, trapezoidal or V-shaped notch, or a combination of these, set in a concrete dam structure. A rectangular plate is better for high discharges and may be set above a V-notch. The V-notch is designed so that water level recorders can measure low discharges with higher accuracy, by creating a greater rate of change in stage towards the bottom of the V. The included angle of the V can be set according to the expected range of flows and the accuracy required for very low flows: 90° and 45° are common. It is important that water leaves the plate cleanly and with air beneath it, as in Figure 5.25a; the crest has to be set high enough to prevent back-up of water below the weir 'drowning out' the structure and affecting the calibration. Even with the best design, however, it is generally not possible to prevent severe floods from overtopping the structure and flowing over the floodplain. Some structures have walls built across part of the floodplain to try to contain these events, but nature can usually find a way around on the odd occasion and invalidate the measurements. These weirs are also natural sediment traps and must be cleared as required to prevent the deposits affecting the calibration.

Broad-crested weirs are lower structures with a broad lip, usually of concrete or fibreglass, commonly moulded into a rectangular cross-section. These are better suited to medium to large rivers and accumulate less sediment. However, they are more prone to 'drowning out'. The **Crump weir** was invented to allow accurate measurements to continue even when it is drowned out. It has a triangular lip rather than the standard rectangular profile. A compound Crump weir, like that in Figure 5.25b, has two or more separate sections with rectangular cross-stream profiles and triangular downstream profiles, to provide better

Figure 5.25 Weirs: (a) sharp-crested, combined V-notch and rectangular; (b) a compound Crump – the centre section is lower to give better measurement of low flows

measurements in different ranges of discharge. A **flat-V weir** carries this idea further and uses the V-shape principle to improve resolution in very low flows by moulding the broad crest into a shallow V.

Flumes offer a lateral constriction to flow but do not obstruct it. This means that they are self-flushing and can be used in streams that carry high sediment loads. The large trapezoidal flume in Figure 5.26a has a broader top to accommodate larger discharges. Rectangular or U-shaped flumes are adequate for lower

(a)

(b)

Figure 5.26 Flumes: (a) trapezoidal; (b) UK Hydraulics Research Station mountain flume

discharges and small portable versions can be used to measure overland flow. Flumes are designed so that the point of transition from subcritical to critical flow, which is accompanied by an increase in velocity and a lowering in water level, occurs at a fixed location at the upstream end of the throat (the constricted section of the flume). This creates a constant relationship between discharge and water level. The UK IH Mountain Flume was designed for steep, mountain streams. The extra baffles had to be added to stabilize the transition point (Figure 5.26b).

For both weirs and flumes, the general form of the stage–discharge relationship is:

$$Q = KbH^a$$

where H is the measured depth (head) of water, K and a are coefficients reflecting the design of the structure (BSI 1981), and b is the width of flow in the throat or over the crest (variable in V and trapezoidal shapes).

5.4.3 Network design for riverflow

As with precipitation, the design of networks for riverflow measurement frequently needs to reconcile conflicting requirements and resources. Ideally, it should be integrated with the design of precipitation networks, but attempts to achieve this have been rare. Similarly, most networks are 'user specific', serving local needs and established predominantly for water quantity rather than quality (section 8.3.1). National and international networks, like GEMS or the IHP representative basins, tend merely to select from these pre-existing stations. The high cost of establishing gauging stations also means that they are far less numerous than precipitation stations, and a major application of riverflow modelling techniques is in predicting flows in *ungauged* rivers (Chapter 6).

The WMO (1994a) has issued guidelines for minimum densities of streamflow stations according to the spatial variability of flows in different climates and different topographic regions. Table 5.1 reveals two general trends: variabilities are higher in regions with (1) higher relief and (2) higher rainfall. These regions therefore require higher densities of gauging stations, i.e. fewer square kilometres per gauge according to Table 5.1, in order to achieve the same accuracy of areal runoff measurement.

The global distribution of gauging stations follows these guidelines only in terms of broad climatic regions. Gauging is more intensive in humid regions in general. Even so, whereas North America and Europe easily meet minimum requirements, Asia struggles to reach a quarter of the required gauges. The drylands are reasonably covered as a whole, but mountain regions are generally undermonitored in all climates.

Network design tends to be iterative, as experience indicates where measurements are most necessary. A system of primary (permanent) and secondary (roving) gauge sites has been proposed, as sometimes used with raingauges, although realistically only primary stations would merit concrete gauging structures. Ideally, Hall (1986) proposed that network design should balance the need to sample all major rivers (with drainage areas of >1000 km^2, or >500 km^2 in mountain regions) against the need to predict discharges from a variety of different-sized basins in a range of climatic, vegetational and landform environments.

5.5 Measuring water in the ground

Traditionally, long-term hydrometric networks have only been established to monitor groundwater as a water resource in aquifer zones. However, the increasing refinement of physically based models for storm runoff and major advances in monitoring by remote sensing are resulting in more widespread measurement of soil moisture and the hydrological properties of the soil.

5.5.1 Soil moisture

Most direct measurements of soil moisture have until recently been undertaken in experimental basins, but measurements are becoming more widespread. Measurements may be taken either *in situ* or by returning a sealed sample of soil to the laboratory.

Soil moisture content This can be determined in the **gravimetric** method by drying the soil sample at 105°C and subtracting the dry weight from the initial weight. More elaborate measurements may be taken to determine the **soil moisture retention curve** (Figure 3.12), for example using a centrifuge, or pressure membranes and porous plates, in which water is slowly squeezed from the soil sample. These can be used to convert field measurements of soil moisture suction into actual water content. Because of the cost and effort involved in taking such measurements, many scientists have adopted a statistical method of estimation, using multiple regression to establish a relationship between water retention and soil properties like texture and bulk density (e.g. Carsel and Parrish, 1988).

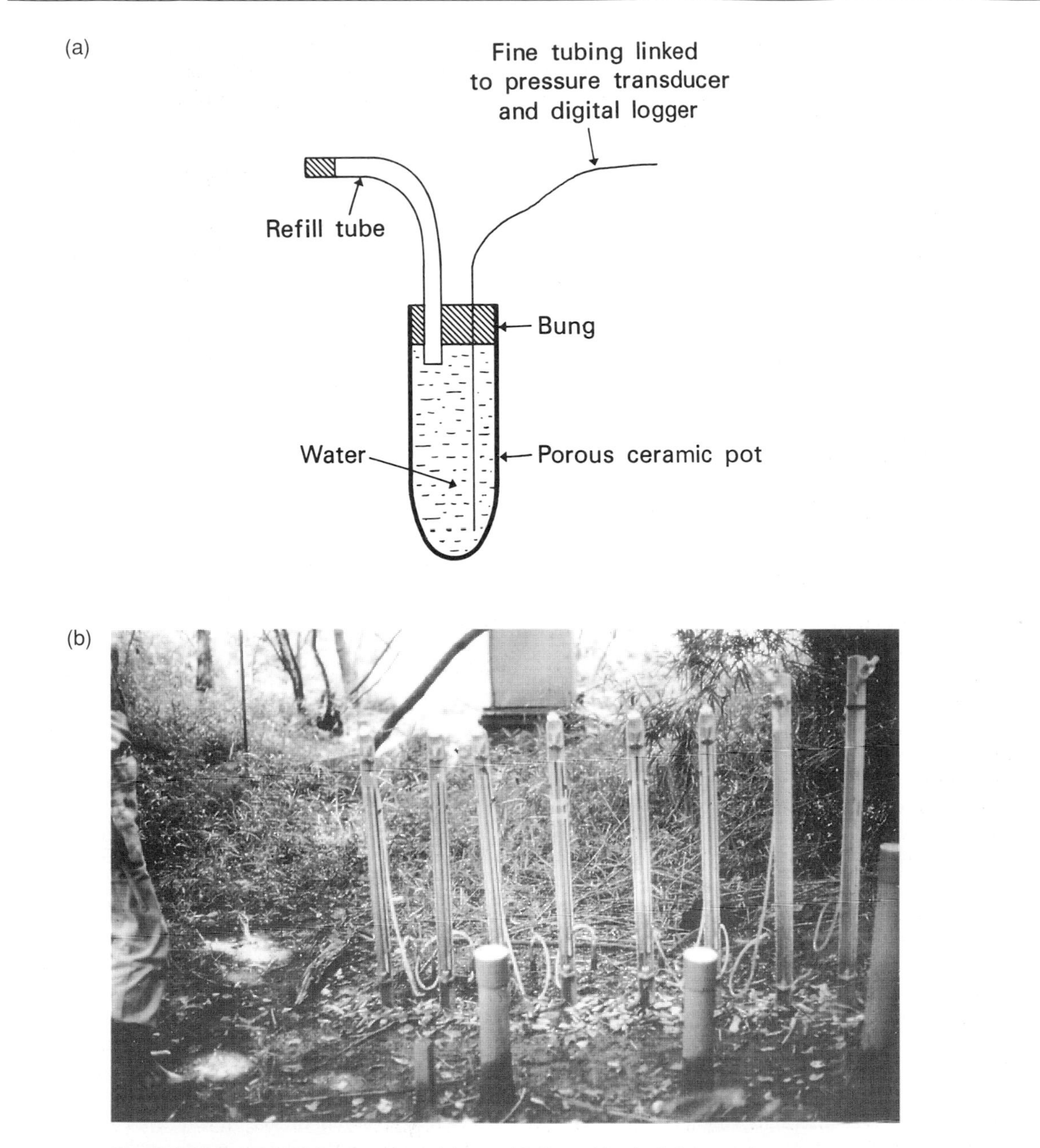

Figure 5.27 Porous pot tensiometers: (a) pot with fine tubing to link to a pressure transducer and data logger; (b) buried pots linked to mercury manometers for visual reading

One of the most popular instruments for *in situ* field measurement is the **tensiometer. Porous pot tensiometers** consist of ceramic pots filled with deionized water; the pressure within the pot equalizes with the tension or pressure in the surrounding soil and is measured via a hydraulic link to a manometer, a pressure measuring device (Figure 5.27). Anderson and Burt (1977a) used a medical scanivalve system to scan 24 porous pot tensiometers at regular intervals and link them to a data logger (section 3.3.4). Porous pots operate best in relatively wet conditions, between p*F* 2.93 and saturation, and under positive pressure, when a vertical bank of tensiometers can be used to monitor the height of the phreatic surface. In drier conditions, there is a danger of air entering the pot and affecting pressure transmission.

Electrical resistance blocks are better in such conditions. These have electrodes embedded within porous blocks made from gypsum (**Bouyoucos blocks**), nylon or fibreglass mesh. The electrical conductivity or capacitance of the blocks is proportional to soil moisture status. Unfortunately, electrical conductivity is also affected by dissolved salts in the soil water, and gypsum blocks tend to disintegrate in very wet conditions. A third approach is to measure water vapour pressure within the soil. In unsaturated conditions, vapour pressure is in equilibrium with liquid water content and can be measured with a **psychrometer**, consisting of 'wet' and 'dry' thermocouple thermometers, but this will not work in saturated soil.

Widespread use is now made of **neutron probes**. These contain a radioactive source that emits fast neutrons. They operate on the assumption that the return of slow neutrons from the soil is proportional to the hydrogen content of the soil, which is taken to be proportional to the water content. Such assumptions are only partially correct: slow neutrons are returned by other atoms, such as aluminium and silicon, and the hydrogen can be in organic matter as well as water. Nevertheless, once the instrument is calibrated for a particular soil, it offers a rapid means of surveying a large number of sites (Figure 5.28). Access tubes are sunk at each site to facilitate repeated measurements (Figure 5.19b). A Russian probe uses the attenuation of gamma rays from a radioactive cobalt source between access tubes 0.5 m apart.

Figure 5.28 The UK Institute of Hydrology neutron probe for measuring soil moisture

Remote sensing may be used to extrapolate point measurements on the ground. Landsat MSS near-IR imagery and the IR bands of the airborne Daedalus scanner can be particularly helpful for inferring soil moisture patterns from surface temperatures and leaf conditions.

For process-based runoff modelling, there is also an increasing need to determine infiltration capacities and percolation rates.

Infiltration capacity and hydraulic conductivity
Again, both these properties can be measured in the field or in the laboratory. Greater control can be exercised in the laboratory, but it is extremely difficult if not impossible to bring back a totally undisturbed soil sample and laboratory measurements often overestimate conductivities because of cracking and disturbance of the pore networks. It is also difficult to remove samples large enough to minimize edge effects and to cover spatial variability adequately.

Figure 5.29 makes a very important point about the minimum size of sample needed to encompass the scale of spatial variation in soil properties. Beven and Germann (1982) called this the minimum **representative elementary volume** of soil. In addition, capacities and conductivities may change quite significantly from season to season, owing to seasonal desiccation or faunal activity. In practice, few measurement programmes are presently undertaken outside research projects, yet for environmental impact assessment infiltration is one of the features most sensitive to changes in land use and husbandry.

Similar techniques can be applied to both these properties, since infiltration is really only a special case of hydraulic conductivity. Both can be measured for (1)

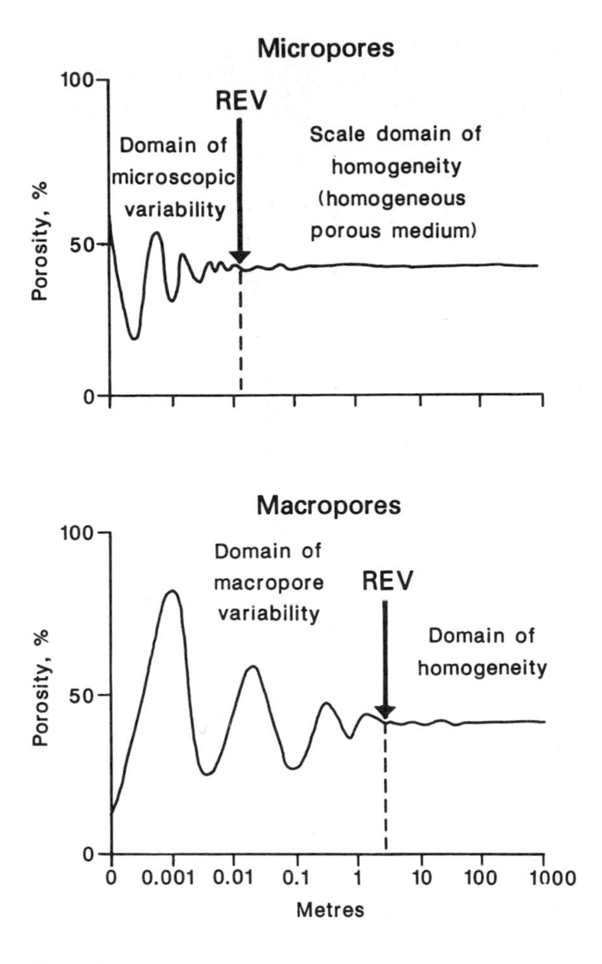

Figure 5.29 The size of sample needed to encompass spatial variations in soil porosity. REV is representative elementary volume. Based on Beven and Germann (1982)

saturated or unsaturated soil, and (2) a steady state flow or a transient flow situation. The most common measurements are for **steady state flow** (i.e. a constant rate of flow) in saturated soils, and these are achieved by first ensuring complete saturation of the soil and then forcing water through it at a constant rate, normally using a **constant head** of water above the sample or sampling site. Most runoff models to date require only saturated, steady state values, and these are normally the most important ones in storm runoff situations.

However, a **falling-head** method is less demanding in terms of the instrumentation needed and requires less water, which can be a major consideration where water has to be carried to a field site. A falling head means that the pressure forcing the water through the soil and therefore the rate of flow are continually changing. Nevertheless, such **transient flow** estimates may be acceptable, especially in slowly permeable soils where it would take a very long time to reach a true steady state. Similarly, a constant rate of sprinkling (**constant flux**) has to be substituted for a constant head when the measurement is intended to apply to an unsaturated situation.

Figure 5.30 illustrates a few of the main types of instrument. **Infiltrometers** can use sprinkling or flooding approaches. The most common form of infiltrometer consists of a ring or tube that is pressed into the ground and filled with water. The **double-ring infiltrometer** gives better measurements of vertical infiltration rates in the inner ring, because it provides a saturated buffer zone that minimizes the lateral spreading which can cause overestimation in a single ring. More replicate measurements may be needed with a ring device than with a sprinkler, because of the small area covered (Knapp 1978), although single rings are perhaps the least troublesome of all devices.

An indirect estimate of infiltration can be achieved by subtracting the volume of direct runoff from total storm rainfall, as recommended by the American Society of Civil Engineers (1970). Theoretically, this should give an average basin-wide estimate. In practice, it relies on the veracity of Horton's theory of runoff (section 3.3.1) and on the difficult art of hydrograph separation (section 6.1.2).

Permeameters are instruments that can be inserted to any depth in the soil. Again, a **double-tube** version is superior to a single tube for measuring vertical hydraulic conductivity, for the same reason.

Infiltrometers and permeameters all add water to the soil. However, where there is a water table within the soil, hydraulic conductivity can be calculated by removing water and measuring the rate of refilling. This is known as the **unlined auger-hole** method. Whether water is added or removed from an unlined auger hole, it is predominantly a measure of *horizontal* hydraulic conductivity, since the walls of the hole form the greatest surface area.

In the laboratory, saturated hydraulic conductivity can be measured by maintaining a constant head of water over a soil sample and measuring the rate of discharge into a collecting cylinder beneath. Hendrickx (1990) offers a good technical review of most of these techniques and a number of others.

Throughflow is rarely measured directly, if at all. Most estimates are based on monitoring the soil water potential, and taking measurements of hydraulic conductivity and soil cross-sectional area, which are then entered into Darcy's Law (section 3.3.4). The soil

Figure 5.30 Infiltrometers: (a) a double-ring infiltrometer (left); (b) a sprinkler, which simulates rainfall over a miniature plot of ground

water potential may be measured by a network of tensiometers, dipwells or piezometers.

Dipwells are observation holes that are either unlined or lined with a perforated tube to prevent collapse. The water level within the dipwell is a reasonable indicator of the surrounding water table level and can be monitored with a water level gauge (section 5.4.1). **Piezometers** are unperforated tubes that are inserted into the layer or horizon of interest and the water level indicates the pressure potential in that layer. The water level in a piezometer will rise above the general phreatic surface if water is confined beneath a horizon of low permeability.

Throughflow pits will give direct measurements of flow in different soil horizons (Figure 5.31a). As with overland flow, such measurements are only samples from a very broad front, so that questions arise as to how representative they are and how to extrapolate them. It is crucial that the pit collects water only from a contributing zone equivalent in width to the upslope face of the pit (Figure 5.31b), so that the volume collected can be multiplied according to the total width of hillslope. Unfortunately, pits can create a local increase in the hydraulic gradient that attracts water from a broader and variable contributing zone (Atkinson 1978). In addition, the amount collected from a specific horizon can be very sensitive to the way the collector has been inserted into the soil, especially its location relative to the divisions between horizons (Anderson and Burt 1977b).

Spatial patterns of throughflow can be so variable that pits cannot possibly provide an adequate sample. This is most obvious where soil pipes or major macropore networks are important (section 3.3.4). In

(a)

(b)

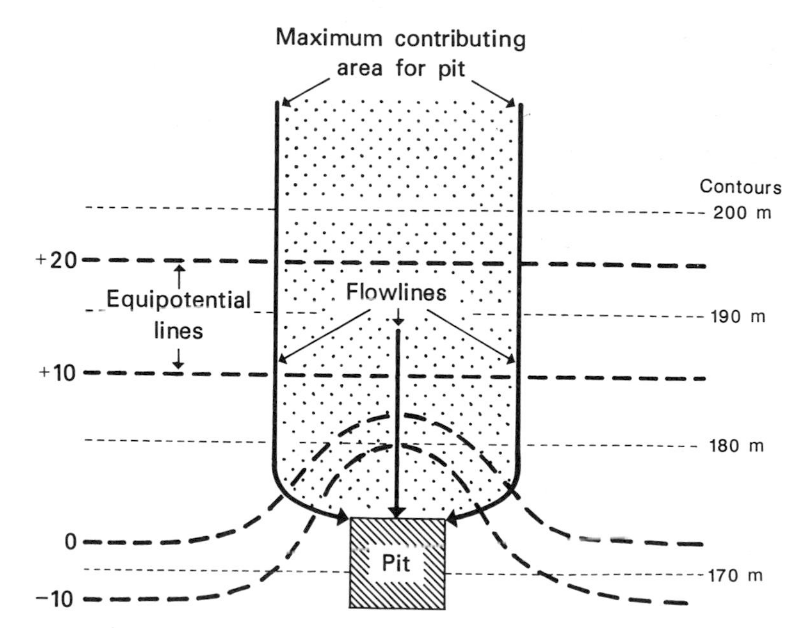

Figure 5.31 Throughflow pits: (a) sampling flows in different soil horizons; (b) the problem of variable contributing areas

such cases, macropore flow tends to dominate (Jones 1997a) and the best method of monitoring is to intercept the subsurface routes and to adapt stream gauging methods, such as flow metering (Gilman and Newson 1980), dye dilution or sharp-crested weirs (Jones and Crane 1984).

In devising a monitoring scheme for soil moisture or throughflow, it is worth remembering that, although overall spatial variability may be high, it tends to have an underlying pattern, which is dictated by topography and soils. This opens up the prospect of stratified sampling schemes based upon 'pedogeomorphic' zones or vegetation.

5.5.2 Groundwater

There are many similarities between the behaviour of water in the soil and in rocks and sediments. Similar basic parameters need to be either measured or calculated, like hydraulic conductivity, transmissivity, storage capacity, pressure gradients, and recharge and seepage rates. But the scale and importance of groundwater resources are altogether greater, and regular monitoring systems are more established.

Standard methods of monitoring and exploration employ observation wells and piezometers. A **pumping test** is used for exploratory work to assess the potential yield of a water supply well. This tests the response of the water level (or the piezometric head) in the well to the rate of water extraction by the pump. Pumping tests enable the hydraulic conductivity and transmissivity to be calculated according to Darcy's Law (section 3.4). As water is pumped out of a well, the water table in the immediate vicinity tends to be lowered in a **cone of depression** (Figure 5.32). The higher the specific capacity of the well (section 3.4), the less the environmental impact of this drawdown will be. The **sustainable well yield** is the discharge that can be sustained without progressive lowering of the water table and can be taken as a conservative measure of the environmentally safe yield. This safe yield tends to fluctuate seasonally. Specific capacity and sustainable well yield can be calculated from routine well logs in existing wells.

During the 1980s a new approach was developed using **geoelectrics**. This is based on correlating hydraulic conductivity with electrical conductivity (Bardossy *et al.* 1986). It is then possible to estimate hydraulic conductivity, transmissivity and specific capacity from a surface-based survey of the electrical resistivity of the aquifer.

The high spatial variabilities in groundwater yields and high costs of drilling test wells mean that maximum use has to be made of geological data collected from a variety of sources. The hydrogeological sections of many national geological surveys specialize in archiving this information and normally

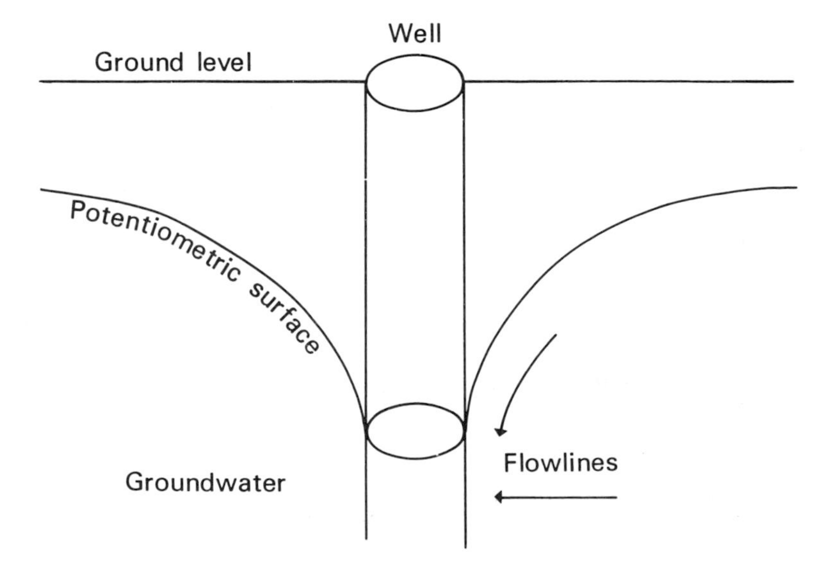

Figure 5.32 Cone of depression around a groundwater well

have legal powers to obtain data from civil engineering firms. The US Geological Survey and many state surveys in America produce regional maps of piezometric head based on collated data, which can form a framework for local test drilling. Fetter (1994) provides a good introduction to methods of field observation and modelling.

Modelling groundwater Modelling has formed an integral part of groundwater assessment from the outset, largely because groundwater resources are not directly observable like riverflow. This is because flow occurs over broad surfaces rather than in discrete channels and because flows are highly interlinked, so that abstraction at one point can affect yields at other points.

Maps of the piezometric head can be used to estimate yields at a specific location, given a measure of hydraulic conductivity, k, using either Darcy's Law or the Dupuit–Forschheimer approximation for flow per unit width, the **specific discharge**, q, in an unconfined aquifer:

$$q = k\frac{(h_1{}^2 - h_2{}^2)}{2x}$$

where the last term is the hydraulic gradient over the distance, x, between two point measurements of height, h_1 and h_2, on the water table, as illustrated in Figure 5.33. The approximation assumes uniform, steady state flow throughout the section of aquifer.

Modern computer modelling methods provide solutions for more complex situations. These predict changes in heads and flows at discrete points or cells set in a regular grid across the aquifer. Each cell or block around a point is considered to be homogeneous, but aquifer properties can differ from cell to cell or point to point. This still requires a considerable amount of field measurement of aquifer properties, estimates of the initial head and accurate specification of boundary conditions in the aquifer. Wang and Anderson (1995) provide a detailed introduction to the mathematical techniques, which are termed either **finite difference** (point grids) or **finite element** (cell networks). Since their development for groundwater prediction, they have been applied to soil water and to surface processes. They now form the basis of the latest models of surface runoff and will be described more fully in the next chapter.

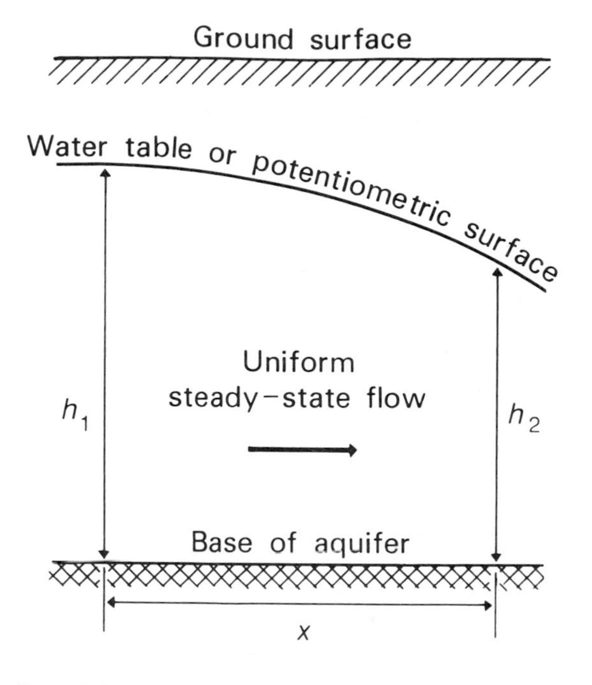

Figure 5.33 The Dupuit–Forschheimer approximation for determining groundwater flows

Conclusions

Rapid advances have been taking place in measurement techniques and in the integration of hydrometric networks, especially through satellite technology. New systems hold the prospect for more global integration, but there is still considerable room for improvements in standard and basic measurements in most parts of the world, and a need for networks carefully attuned to local and national requirements.

Further reading

More details on specific measurement techniques can be found in the following books:

Collinge V K and C Kirby (eds) 1987 *Weather radar and flood forecasting*. Chichester, Wiley: 296pp.

Fetter C W 1994 *Applied hydrogeology*, 3rd edn. New York, Macmillan: 691pp.

Herschy R W (ed.) 1978 *Hydrometry, principles and practices*. Chichester, Wiley: 511pp.

Johnson A I (ed.) 1986 *Hydrologic applications of space technology*. Wallingford, International Association of Hydrological Sciences Pub. No. 160: 488pp.

Kirkby M J (ed.) 1978 *Hillslope hydrology*, Chichester, Wiley: 389pp.

Moss M E (ed.) 1986 *Integrated design of hydrological networks*. Wallingford, International Association of Hydrological Sciences Pub. No. 158: 405pp.

Full descriptions of methods and observational networks are given in:

WMO 1994a *Guide to hydrological practices*, 5th edn. WMO Pub. No. 168, Geneva, WMO: 735pp.

WMO 1994b *Observing the world's environment: weather, climate and water*. Geneva, WMO: 42pp.

Discussion topics

1. To what extent have satellites made standard gauging techniques obsolete?
2. Discuss the criteria that should be used to design an ideal, integrated hydrometric network.
3. Assess the value of radar and microwave techniques.
4. Critically evaluate the methods in use for measuring any one hydrological component.

CHAPTER 6

Modelling runoff processes

Topics covered

6.1 'Black box' or macroscopic models

6.2 Stochastic or synthetic models

6.3 Simulation models

6.4 Flow-routing models

6.5 Modelling snowmelt and icemelt runoff

6.6 Modelling urban runoff

Models of runoff processes have been developed for a wide variety of purposes, from the 'one-off' design of engineering structures and water supply systems to modern *real-time* models used continuously in river regulation schemes. They are also proving valuable for studying the potential impacts of changes in landuse or climate. The variety of uses and the rapid increase both in scientific understanding and in technical support, from data collection systems and computer technology, have produced an enormous range in levels of sophistication.

Model outputs vary from predictions of peak discharges or total volumes of flood flow to the complete specification of the distribution of flow over time, either for individual storm events in **event models** or for continuous sequences of flows in **continuous** or **sequential models**. Peak discharges are required for flood protection and they are normally calculated from a single equation, which often takes little account of the generating processes. In contrast, the calculation of flow **hydrographs**, the time distribution of discharge, requires more knowledge of the generating processes, including the transformation of rainfall into runoff on the hillslopes in a **rainfall–runoff** model and its transport through the channel network by a **flood-routing** model.

The most sophisticated models are **physically based simulation** models, which incorporate a linked system of submodels simulating the transfer processes and storages within the river basin. Other models use past data on the input and output of rainfall and runoff in lieu of detailed scientific knowledge of the processes. Such models range from simple equations generalizing the behaviour of a large number of basins to wholly **stochastic** or probabilistic models, which allow a more complete specification of the magnitude–frequency distributions of flow events and sequences based on data from a specific basin. These stochastic models produce a **synthetic** record of flows based on probabilities derived from past data. They are commonly used to design reservoir systems that will continue to provide water supply through the 'worst' sequence of drought.

In the wider sense, the term 'stochastic' is used to refer to models of systems or subsystems that contain a large random element, and generally sample randomly from a range of probable values. A clear distinction can be made between **deterministic** and **stochastic** models. A deterministic model is one in which a given input of rain *must* produce a fixed output of runoff in a certain physical environment: its output is *determined*. It could be a simple, empirical or 'black box' model derived from input–output analysis of past data. It could be a conceptual model in which processes are represented in a very simplified manner, perhaps including some empirically calculated elements. Or it could be a true physically based model, in which processes are modelled according to fundamental scientific laws. In contrast, the output from a stochastic model is subject to chance.

The deterministic model is not necessarily scientifically superior to such a stochastic model. It is not easy to model the system in a totally deterministic way and a deterministic model may, in fact, ignore natural stochastic variations in the real world. This is partly because it is impossible to measure everything everywhere, partly because not all processes are fully understood, and partly because of the inherent variability of nature. An important part of this indeterminacy is produced by *spatial variability* in properties and processes, for example in rainfall receipts or soil permeability. Process-based models have adopted a range of different levels of *spatial discretization* to account for this (Figure 6.1), as will be discussed in section 6.3. In addition, there remains the more fundamental philosophical question of the ultimate limit to the predictability of nature; the question of at what scale does the system become chaotic? Becker and Serban (1990) recognized these problems when they said that 'hydrological processes always include deterministic and stochastic elements', and they identified a category called **coupled deterministic–stochastic models** (Figure 6.1). In such a model, a stochastic model might be used, for example, to generate rainfall input to a deterministic runoff model, or to generate a representative range of infiltration capacities to be used in the deterministic model.

It is therefore useful to look at the art of the modeller as simplifying reality in three main ways: (1) representing processes in a simplified form; (2) simplifying spatial variability; and (3) breaking down continuous processes into manageable intervals of time (*time discretization*). The simplest peak flow models described in secpion 6.1 effectively ignore all three, process, space and time. The most advanced physically based distributed models described in section 6.3.2 consider a variety of processes within a finely discretized space–time framework.

With the increasing need to ensure both safe supplies and 'safe abstractions', i.e. minimizing impacts on the environment and downstream users, models of the natural system are being linked with models of water resource operational systems. The UK Institute of Hydrology has constructed a model based on the NRA rainfall–runoff model and Thames Water's water resources system model, which covers abstraction, regulation and return water, in order to assist in water management decisions during drought years. The model is fed with rainfall scenarios randomly selected from a century of records to give an assessment of the risks of drought, and can be fine-tuned during the year by long-range forecasts from the Meteorological Office.

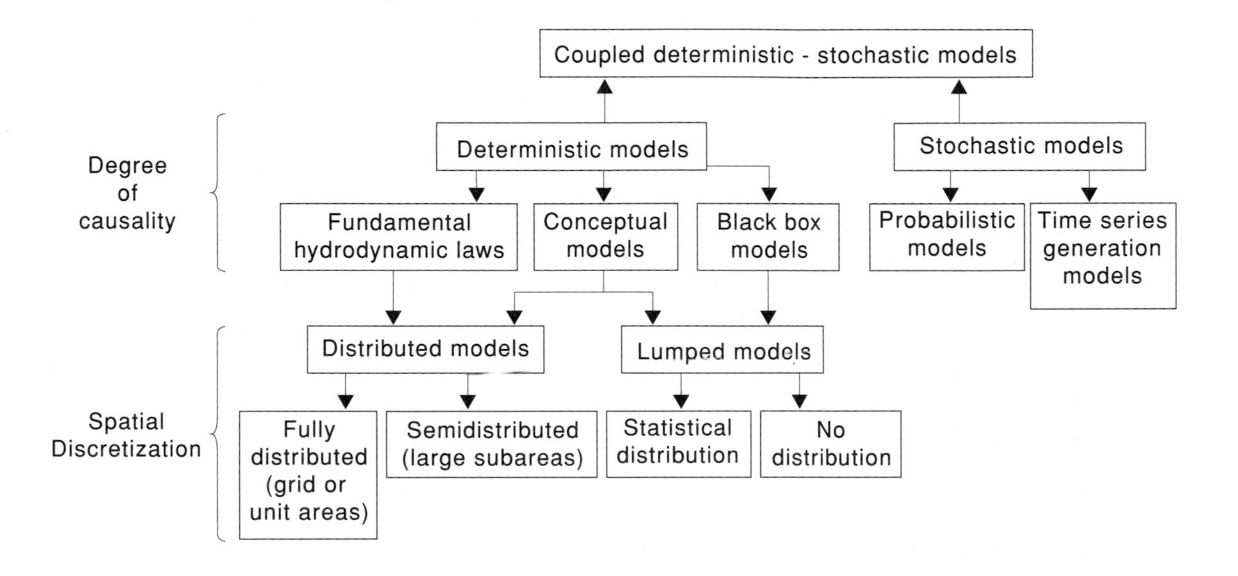

Figure 6.1 Classification of hydrological models. Based on Becker and Serban (1990)

6.1 'Black box' or macroscopic models

The earliest, pre-computer models were of this type. They took no explicit account of processes, and consisted largely of regional generalizations of input–output relationships determined from a representative range of basins. The modern examples of this genre take much greater account of processes by using **parameters** or indices which represent certain factors that affect processes and that have been shown to be statistically related to runoff. These models have been called 'macroscopic' because they take a broad view and treat the entire drainage basin as a single unit. All variables or parameters are assumed to relate to the catchment as a whole and no attempt is made to take account of spatial variability within the catchment by dividing the basin up into sub-basins which may have different characteristics. Technically, such models are therefore called 'lumped' models.

6.1.1 Total, peak or low flows

The first group of models to be discussed is concerned simply to predict either the peak discharge or else the total storm or low-flow period discharge. Each model predicts a single value. More sophisticated models, discussed in section 6.1.2, consider the time distribution and provide complete hydrographs of flow.

Table 6.1 Selected values for runoff coefficients

	Sandy	Soil type Loamy	Clayey
Rural land			
Arable	0.20	0.40	0.50
Pasture	0.15	0.35	0.45
Woodland	0.10	0.30	0.40
Urban areas			
Business district		0.75–0.90	
Industrial zone		0.50–0.90	
Residential suburb		0.25–0.40	
Open land		0.10–0.30	
Urban surfaces			
Roofs		0.75–0.95	
Roads		0.70–0.95	
Parks and lawns		0.10–0.35	

Source: Based on American Society of Civil Engineers (1970) and Dunne and Leopold (1978).

The rational method This is the oldest and simplest approach. It predicts peak runoff rate, Q_{pk}, as a fixed proportion or ratio of the rainfall input. Rainfall input is calculated as the product of spatially uniform rainfall falling with intensity, I, over an area, A, for long enough for storm runoff to reach the basin outlet from the remotest parts of the basin. The time required for this is called the **time of concentration**. Hence:

$$Q_{pk} = CIA$$

where C is the coefficient of proportionality known as the **runoff coefficient**. Values of C have been empirically determined for a wide variety of land uses (Table 6.1). For basins with mixed land use, it is possible to calculate C as an average weighted according to the proportions covered by the different land-use types. Normally, rainfall intensity, I, is selected as one which has a certain return period, and it is assumed runoff events have the same frequency distribution as rainfall events (which may be questionable: see section 4.2.2).

Clearly, the method takes no account of the fact that in the 'natural' landscape a wet basin will respond differently to a dry basin. Consequently, although the formula has been one of the most successful and widely used, its use nowadays is generally restricted to calculations of urban drainage requirements for relatively small, impervious areas of paved surface. It could be used as a quick method of estimating the impact of a land-use change, but the overlaps in runoff coefficients between one land use and another in Table 6.1 tend to restrict its usefulness.

Regression equations Numerous formulae have been developed from the basic rational method, using simple or multiple regression to determine the coefficients and employing the more general power relationship model, which does not assume that peak discharges are linearly related to rainfall inputs. A common form is thus:

$$Q_{pk,t} = CA^b$$

where C and b are the regression coefficients (b = 1 being the linear case) and $Q_{pk,t}$ is the peak discharge with a return period of t years. Following the disastrous Lynmouth floods of 1953 in Devon, the UK Institution of Civil Engineers used scatter plots of catastrophic floods versus basin areas to establish **envelope curves** which enfold the largest recorded floods in the different physiographic provinces. The equations for these curves followed the general form:

$$Q_{max} = CA^{0.5}$$

The **USGS Index Flood** formula for the mean annual flood (which has a return period of 2.33 years) is:

$$Q_{2.33} = CA^{0.7}$$

for which C has been mapped regionally. A formula

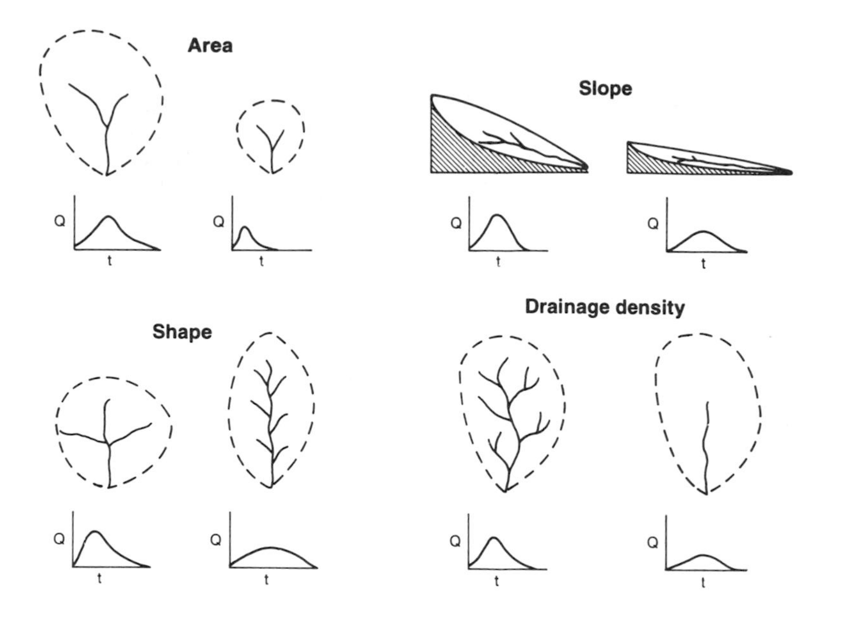

Figure 6.2 The effects of basin morphometry on storm hydrographs

that has been used successfully in the Canadian prairie provinces actually uses the return period, T, as a variable:

$$Q_{RP} = C(82.7A^{0.5}T^{0.444})$$

Other formulae consider additional 'morphometric' properties which affect runoff response as well as basin area.

The geometric properties of a basin, such as size, shape and slope, are clearly important in shaping its runoff response (Figure 6.2). Larger basins tend to catch more rainfall, but the time lag between rainfall and runoff peaks, the **time-to-peak**, and the length of storm runoff will both tend to be longer. For a 250 km^2 basin storm runoff typically lasts a couple of days, whereas in a 25 000 km^2 basin it can go on for 6 days. Steeper basins have a faster runoff and their hydrographs tend to respond more quickly to rainfall. More elongated basins tend to have a more 'dampened' or diffuse response compared with highly peaked storm hydrographs from compact basins. This is because rainwater draining from around the edge of a compact basin will have similar distances to travel from all quarters and so it arrives at the outfall at about the same time, whereas in elongated basins water takes much longer to flow from the top of the basin to the outfall than from the bottom. Some common morphometric parameters are described in Box 6.1.

Box 6.1 Morphometric analysis of drainage basins

Measurements of the 'morphometry' of drainage basins are regularly used in hydrological models.

Basin area (A or AREA) is one of the most important and forms the basis for many other parameters. **AREA** is the prime variable in the UK Flood Studies Report (FSR) formulae (Box 6.2).

Hypsometric integrals or curves represent the distribution of area within the basin according to various altitude bands. The most general form of hypsometric curve plots percentage area against percentage height range within a basin. A simpler form would be a direct graph of the 'altimetric distribution' of measured areas versus contour bands. Hypsometric information was considered in the FSR, but it is basically too cumbersome to use in simple predictive formulae.

Stream lengths clearly affect rates of water collection and transmission. Measurements can be used in a variety of ways, such as **length of the main stream (MSL)**, **total length of stream channel** (ΣL), or as part of drainage density or stream frequency calculations (see below). Horton (1945) and Strahler (1964) defined a number of morphometric 'laws' relating the number of streams, their lengths, slopes and drainage areas in a normal basin to **stream order**, where, according to Strahler's definition, fingertip streams

Box 6.1 (cont.)

are 'first order' and two fingertip streams meet to form a 'second order' stream and so on (Figure 6.3). A **stream segment** is a length of stream between junctions where the stream order changes (Strahler 1964). Unfortunately, Horton's laws arise largely from the methods of ordering the streams, which impose an unreal regularity on patterns by effectively discarding 'adventitious' streams, the lower-order tributaries which do not raise the order. However, they can be used as a guide as to what is 'normal', and major deviations could indicate unusual geological controls.

Nevertheless, the concept of 'ordering' has potential for summarizing and comparing streams. Shreve (1966) introduced **stream magnitude** in order to better represent the continuous rather than steplike increase in natural stream sizes. The magnitude is equivalent to the number of fingertip tributaries feeding a given point on the stream, and a length of stream with a certain magnitude is called a **link**. Shreve has proposed that stream magnitudes may help in: (a) designing efficient and representative sampling schemes; (b) determining the probabilities of locations within the stream network as sources of pollutants monitored downstream; and (c) using it as a basis for structuring streamflow databases. Scheidegger's (1965) consistent order magnitude (Figure 6.3) seems to have no practical advantage.

Drainage density (D_d) is another fundamental property of a basin, which controls the efficiency of drainage:

$$D_d = \frac{\Sigma L}{A}$$

Horton (1945) defined $1/D_d$ as the **constant of channel maintenance**, i.e. the average area of land needed to support a unit length of stream channel, and suggested it could distinguish different types of hydrological regime, with smaller values in more humid regions. Similarly, Horton's **mean distance of overland flow (MDOF)** is equal to

$$\frac{1}{2D_d}$$

Stream frequency is a non-metric measure – the number of 'streams' in a unit area, where a 'stream' could be a Strahler stream segment or, more commonly now, a link. The variable **STMFRQ** in the FSR equations is based on links, by simply counting the number of junctions per square kilometre.

Basin slope is rather difficult to summarize because of the large variation in most basins. A variety of techniques have been proposed, from taking a number of random transects to estimating the slope at the vertices of a grid placed over the map and calculating mean values.

Stream slope is an important control on the velocity of flow, but it also tends to change downstream. The FSR adopted the slope measure **S1085** (in m km^{-1}) invented by Taylor and Schwartz (1952), who found that most streams tend to have relatively constant slopes if you ignore the top 15 per cent and the last 10 per cent of their lengths.

Basin shape is probably the most illusory measure of all. Most definitions compare basin shape with some ideal shape, but some of these ideomorphs are more hydrologically meaningful than others. Gravelius's (1914) compactness ratio compares the perimeter length with the circumference of a circle with the same area as the basin. Miller's (1953) circularity ratio compares the area of the basin with the area of a circle with a circumference equal to the perimeter of the basin: the inverse of Gravelius's index. The elongation ratio compares the length of the basin's long axis with the diameter of a circle with the same area, whereas Horton's form factor compares the area of the basin with that of a square with sides equal to the long axis of the basin. Recognizing that the 'ideal' shape of a basin is neither circular nor square, but pear-shaped, Chorley *et al.* (1957) proposed the more complex lemniscate loop index. Nevertheless, none of these shape indices has found particular favour in hydrological modelling thus far.

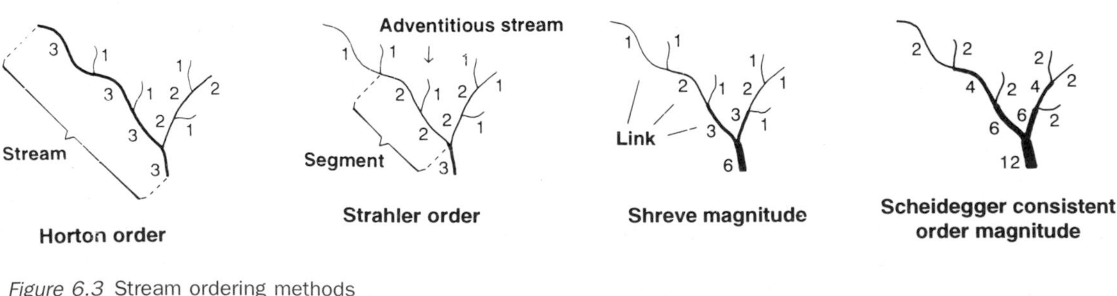

Figure 6.3 Stream ordering methods

The US Corps of Engineers has established a range of power law multiple regression formulae incorporating a variety of morphometric variables. These equations enable flood discharges to be calculated for any rainfall intensity, *I*, or duration, *D*, which may represent storms of a particular return period. The equations take the general form:

$$Q = KI^aD^bA^cS^d, etc.$$

where values for the coefficient *K* have been regionalized and mapped for the USA. Each value of *K* identifies a 'hydrologically homogeneous region' that exhibits similar hydrological responses as a result of similar topography and climate (Figure 6.4).

UK flood studies formulae The Flood Studies Report (FSR) was the result of collaboration between the UK Institute of Hydrology, the Meteorological Office and the Soil Survey of England and Wales (NERC 1975) and, despite updates (e.g. ICE 1980), it remains one of the most valuable and comprehensive national surveys in the world.

The FSR aimed to enable hydrologists to predict mean flood discharges of any return period for any ungauged basin in the UK. This was achieved along similar lines to the US Corps of Engineers, by selecting a representative sample of the longest and most reliable records from basins in the various regions of the country and correlating flood discharges with a selection of topographic and climatic parameters derived for each basin. Multiple regression equations were produced covering mean annual flood, with special versions relating to small basins and a method for estimating extreme events. Most of the local parameters that need to be entered into these equations can easily be measured directly from topographic maps, e.g. basin area (AREA), stream frequency (STMFRQ), mainstream slope (S1085), or the fraction of the basin draining into lakes (LAKE) or covered by urban surfaces (URBAN). Other variables were specially mapped by the FSR team, such as the complex variable RSMD, which is a measure of 'effective rainfall', or else regionalized, as in the case of the runoff coefficient, *C* (Figure 6.4). Details of selected equations are given in Box 6.2.

Two overriding principles were used in selecting variables to construct the FSR equations. First, data sources had to be readily available to the user, either through published maps or through the FSR's own additional maps. Hence, floodplain storage was not included, although it can be important in flood routing, because it cannot be easily calculated from topographic maps. Secondly, they had to be statistically significant, i.e. high levels of statistical correlation between the variables and discharges suggest that they have an important control on riverflow. Newson (1978) outlines the bases for selecting the morphometric variables. Chidley and Drayton (1986) demonstrated how Landsat near-IR imagery may be used to estimate STMFRQ and AREA for applications in Developing Countries.

To enable floods of any return period to be predicted, the FSR established regionalized relationships between mean annual floods and floods of return period up to 500–1000 years. Thus, entering Figure 6.5 according to the return period of interest on the abscissa and finding the intersection between this and the curve for the appropriate region (as numbered in Figure 6.4), the user can read off the proportional adjustments that must be made to the mean annual flood value, $\overline{Q}$. For example, a 100-year flood in region 6 (the Thames basin) would be approximately three times the mean annual flood: $Q/\overline{Q} = 3.0$.

UK low-flow studies formulae A similar approach was adopted for low-flow prediction in a subsequent study by the IH (1980). In this case, separate multiple regression equations were established for predicting the 95 per cent flow-duration discharge with a 10-year return period, *Q*95(10), in each of five UK regions (section 4.1.2). The most important variable is the permeability of the basin and this is expressed by the **baseflow index (BFI)**, which is a weighted function of the proportions of the basin covered by each WRAP soil type (Box 6.2):

$$BFI = (0.6 + 0.23S_1 - 0.03S_2 - 0.12S_3 - 0.17S_4 - 0.20S_5)$$

The FSR low-flow equations show logical differences between upland and lowland regions, and within the lowlands between region 5, where groundwater aquifers are particularly important, and region 4. Low flows in eastern England (region 5) were predicted on the basis of the BFI and the MSL. In all other regions the BFI and the **annual average rainfall (SAAR)** proved the best predictors. There is a marked change in coefficients between region 4 and the uplands. Within the uplands there is a gradual reduction in the negative coefficients as low flows become progressively higher towards the Scottish Highlands (region 1).

A related study by Pirt and Simpson (1982) also developed a regression equation for estimating low flows with a different return period. They used eight catchment variables to predict the difference in discharge between the mean annual 7-day low flow

Figure 6.4 Homogeneous hydrological regions and runoff coefficients in the UK Flood Studies Report equations. From NERC (1975)

Box 6.2 UK Flood Studies Report and related formulae

From a study of over 500 catchments in the British Isles, the FSR produced a simple set of multiple regression equations that can be used to predict flood discharges of any return period in any basin. These are mainly of use in ungauged basins or where records are inadequate.

The key equation predicts the discharge of the mean annual flood, $\overline{Q}$, also referred to as **BESMAF (the best estimate of the mean annual flood)**:

$$\overline{Q} = C[\text{AREA}^{0.94}\text{STMFRQ}^{0.27}\text{SOIL}^{1.23}\text{RSMD}^{1.03}\ \text{S1085}^{0.16}\,(1 + \text{LAKE})^{-0.85}]$$

in which AREA, STMFRQ and S1085 are defined in Box 6.1, and LAKE is the proportion of the catchment draining into any lakes (ignored if less than 1 per cent). Large-scale (1:25 000), maps should be used. RSMD is the maximum daily rainfall occurring once in 5 years minus average soil moisture deficits, as read from the FSR maps. The value for the SOIL parameter has to be taken from the special hydrological response map produced by the Soil Survey. A simple classification scale was developed for this called the **Winter Rainfall Acceptance Potential (WRAP)** by Farquharson *et al.* (1978). WRAP classes run from 1 (high infiltration and low runoff coefficients) to 5 (low infiltration and high runoff). SOIL is a weighted mean based on the proportions of the basin, S_1 to S_5, covered by the various classes of soil:

$$\text{SOIL} = \frac{(0.15S_1 + 0.30S_2 + 0.40S_3 + 0.45S_4 + 0.50S_5)}{S_1 + S_2 + S_3 + S_4 + S_5}$$

The same approach was used by Farquharson *et al.* (1975) to estimate 'maximum floods' for designing flood protection works, based on maximum flows in 80 basins calculated by the unit hydrograph method (section 6.1.2). The **estimated maximum flood (EMF)** is given by:

$$\text{EMF} = 0.835\text{AREA}^{0.878}\text{RSMD}^{0.724}\text{SOIL}^{0.533}\ (1 + \text{URBAN})^{1.308}\text{S1085}^{0.162}$$

The FSR method for calculating flows of any return period is described in the text together with the recent update of WRAP, the HOST system. Further details are available in NERC (1975), Wilson (1990) and Shaw (1994).

(estimated by the method described in section 4.1.2) and weekly low flows with a 20-year return period.

The UK HOST System: A new hydrological classification of soils called HOST ('Hydrology of Soil Types') has been developed to replace WRAP and to improve it, particularly for low flows (Boorman *et al.* 1995). HOST recognises 29 different types, based on the three-dimensional properties of the soil, including the presence of impermeable layers and the thickness of groundwater stores. Classification is based on conceptual models of soil water drainage, which are much closer to a truly physically based model (sections 6.3.1 and 6.3.2). A database is available on a 1 km grid, based on the 1:250 000 soil series maps, to enable hydrologists to estimate parameters. And empirical equations have been developed by relating these soil properties to measured hydrological response in selected basins. These allow parameters like mean annual flow, flow durations and frequencies, the BFI and Standard Percentage Runoff (the runoff coefficient) to be calculated.

Graphical regional correlations A slightly older approach, but one still widely used in the USA, presents the user with a set of tables and graphs that summarize data collected from a range of research plots or catchments. This can be particularly useful for designing small flood-control structures in rural areas and for predicting the effects of altering land use or tillage methods. Its critics would suggest that a more sophisticated method is used where human life may be at stake.

The US Department of Agriculture **Soil Conservation Service (SCS) method** enables total stormflow, or 'direct runoff', to be estimated for a given total rainfall. The rainfall will normally be of a specified return period, derived either from regionalized values given on maps such as those in the US NWS Rainfall Atlas or from magnitude–frequency analysis of local data. Figure 6.6 flowcharts the procedure. The central part of the technique is the selection of a **curve number**, which is really a runoff coefficient. This is selected from a table like the one outlined in Figure 6.6 according to land use/cover type, tillage method, hydrologic soil group and its permeability or degree of drainage (its 'hydrologic condition'). Four soil groups are recognized, ranging from highly permeable soils with low runoff potential like sands to impermeable, high runoff potential soils like clays or very shallow soils on impermeable rocks. The curve number applies

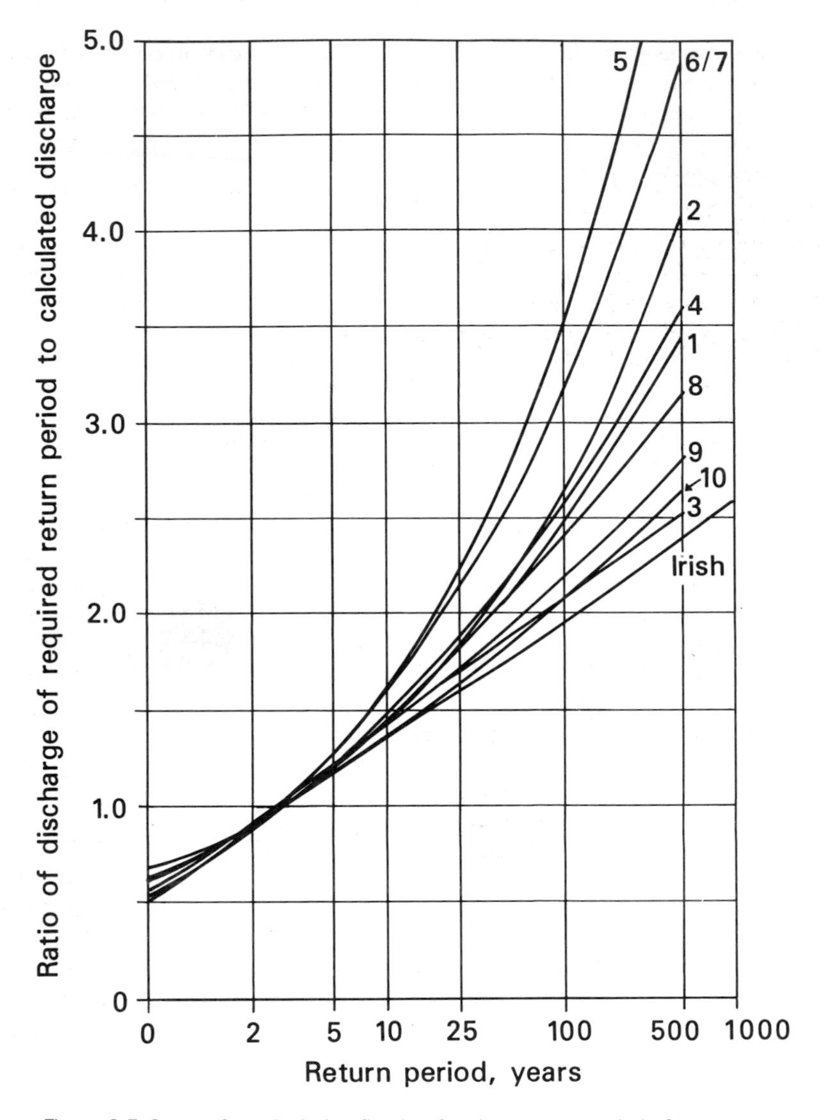

Figure 6.5 Curves for calculating floods of various return periods for regions of the British Isles (in Figure 6.4), based on the ratio of the given flood to the calculated mean annual flood, $Q/\overline{Q}$

to soils with 'average soil moisture', which is the average condition that has preceded annual floods in a number of basins. A supplementary table can be used to adjust the curve number for wetter or drier conditions. A graph like the first one in Figure 6.6 is then used, entering via the appropriate rainfall, following this line up until it intersects the correct curve and reading off the direct runoff. The second graph in Figure 6.6 enables this runoff to be given an approximate time distribution, as described in the next section.

6.1.2 Time-distributed flows

The next level of sophistication introduces the time distribution of runoff. Indeed, the shape of the flood hydrograph is as important as the total or peak discharge for designing reservoirs and spillways for flood control. Sherman (1932) devised an extremely valuable method for determining the average shape of storm hydrographs, which has become known as the **unit hydrograph**. Both the original method and its later variants continue to be widely used and may be found at the heart of many modern and highly sophisticated computer models. In Sherman's original version, the hydrologist uses rainfall and runoff data from a basin to determine an average pattern of runoff response for a rainfall *unit* of one inch. In metric versions the unit is usually one centimetre. The pattern of runoff from larger storms, say two centimetres, is determined by simply adding identical hydrographs for each extra rainfall unit together, allowing an interval between the start of each hydrograph which is proportional to the rainfall intensity, i.e. the second unit hydrograph begins once the first unit of rainfall has fallen.

The method is outlined in Figure 6.7. The analyst begins by selecting a number of representative, single-peaked runoff events from the record as near as possible to one unit of net or effective rainfall, preferably within ± 1/2 unit, and of similar duration. Effective rainfall is what remains after all losses, such as infiltration (section 3.2). The next step is to separate the direct runoff from the baseflow component. There are many different methods of achieving this **hydrograph separation**. Unfortunately, the only truly scientific method would require a thorough knowledge of runoff pathways and processes, which is generally not available. Some commonly used practical solutions are outlined in Figure 6.8 and described in detail in engineering texts such as Wilson (1990). Fortunately, there is broad agreement that for practical purposes the actual method used is not as important as maintaining a consistency in the method used for every storm.

The resultant hydrograph of direct storm runoff will consist of discharge values at regular time intervals; these are the ordinates in the graphs in Figure 6.7. Each ordinate must then be scaled so that it becomes a measure of the relative proportion of total direct runoff occurring at each point in time. This is achieved by first dividing each ordinate and the total by the area of the basin to get the depth of runoff, e.g. centimetres per square centimetre of basin, and then by scaling the results so that they add up to exactly one unit depth of runoff. The final unit hydrograph (UH) is taken as the average of about five such analyses. This can then be used to predict the distribution of runoff for any rainstorm even if the intensity or duration differs from the UH, as shown in Figure 6.9.

The method presumes that the UH empirically embodies all the peculiarities of a particular basin. However, it makes two major types of simplification:

1. It assumes that net rainfall intensity remains uniform throughout all storms and over the whole basin. But rainfall intensities are normally very variable and convective storms can often cover only a fraction of a basin. Storms can produce very different responses depending on where they fall in the basin.
2. It assumes that the basin responds in a linear fashion, e.g. without jumps or exponential relationships. Hence there is no provision for a case in which heavier rainfall or wetter antecedent conditions may cause new drainage pathways to be activated and thus alter the rate of runoff. Examples of such pathways on the surface and in the upper soil have been discussed in section 3.3.

Modern computer simulation models tend to employ a special version of the UH which enables them to take more realistic account of varying rainfall intensities. This is known as the **instantaneous unit hydrograph (IUH)**. In this, the time base is reduced, say, to one minute, i.e. it is the hydrograph resulting from a fixed unit of water being distributed evenly over the basin in one minute. The direct runoff hydrograph can then be constructed from a sequence of hydrographs produced by differing amounts of rain during each minute of the storm. This is one of the main uses of UH theory nowadays, although it still does not necessarily solve problems with nonlinearity.

A third version of the theory was developed by Snyder (1938) to enable UHs to be calculated for basins where no rainfall–runoff data exist. This is known as the **synthetic unit hydrograph (SUH)**. The shape of the SUH is determined by just a handful of points. These points are plotted using equations that have been constructed for a given hydrological region from empirical relationships either with morphometric variables or between different aspects of hydrograph shape. Snyder's method was adopted by the US Corps of Engineers and is illustrated in Figure 6.10. Rodríguez-Iturbe and Valdes (1979) used the Hortonian network structure to develop a **geomorphic instantaneous unit hydrograph**, which has been used to derive flood frequencies (Hebson and Wood 1982) and further developed into a **geomorphoclimatic IUH,** in which the pdfs of IUH properties are related to rainfall characteristics (Rodríguez-Iturbe *et al.* 1982).

Both the FSR and SCS methods allow storm hydrographs to be estimated using an SUH. The SCS approach uses five 'families' of hydrograph. The appropriate shape family is found by entering the second graph in Figure 6.6 according to the rainfall and the curve number, and tables specify the shape parameters for each family of synthetic hydrographs. Engineers can use these hydrographs to route the flows

①

Select storm rainfall
of specified frequency and duration

②

Determine curve number
for average antecedent soil moisture from table

Land-use/ cover type	Tillage treatment	Soil drainage or hydrologic condition	Hydrologic soil group			
			High permeability		→	Low permeability
			A	B	C	D
Pasture/ rangeland	–	Poor	68	79	86	89
		Good	39	61	74	80
Woodland	–	Poor	45	66	77	83
		Good	25	55	70	77
Row crops	Straight rows	Poor	72	81	88	91
		Good	67	78	85	89
	Contoured	Poor	70	79	84	88
		Good	65	75	82	86
	Terraced & contoured	Poor	66	74	80	82
		Good	62	71	78	81

(Other cover/treatments/conditions available) Curve numbers

③

Adjust for appropriate antecedent soil moisture
according to table

Curve number for average moisture condition (from first table)	**Curve number for dry soil**	**Curve number for near-saturated soil**
100	100	100
80	63	91
60	40	78
40	22	60
20	9	37

↓

continues overleaf

Figure 6.6 An outline of the USDA Soil Conservation Service method for calculating peak discharges and storm hydrographs for the design of minor structures. Details are given in USDA SCS (1968) *Engineering Handbook,* Hydrology Supplement, or Viessman and Lewis (1996).

through reservoirs and determine spillway dimensions given allowable velocities or maximum discharges.

The British FSR method uses just three parameters, the time from the start of direct runoff to peak flow, *time-to-peak*, T_{pk}, peak flow, Q_{pk}, and the duration of stormflow, the *time base* or *baselength, TB*. Time-to-peak can be calculated from basin characteristics or, if some data are available, from an empirical relationship with the measured lag time (Figure 6.10).

6.2 Stochastic or synthetic models

A large group of more sophisticated models relies on using local hydrometric data to predict future flows. The theory behind such models is statistical or *stochastic* rather than based upon a knowledge of processes. It centres on the notion that future events will follow a similar pattern to past events. On this

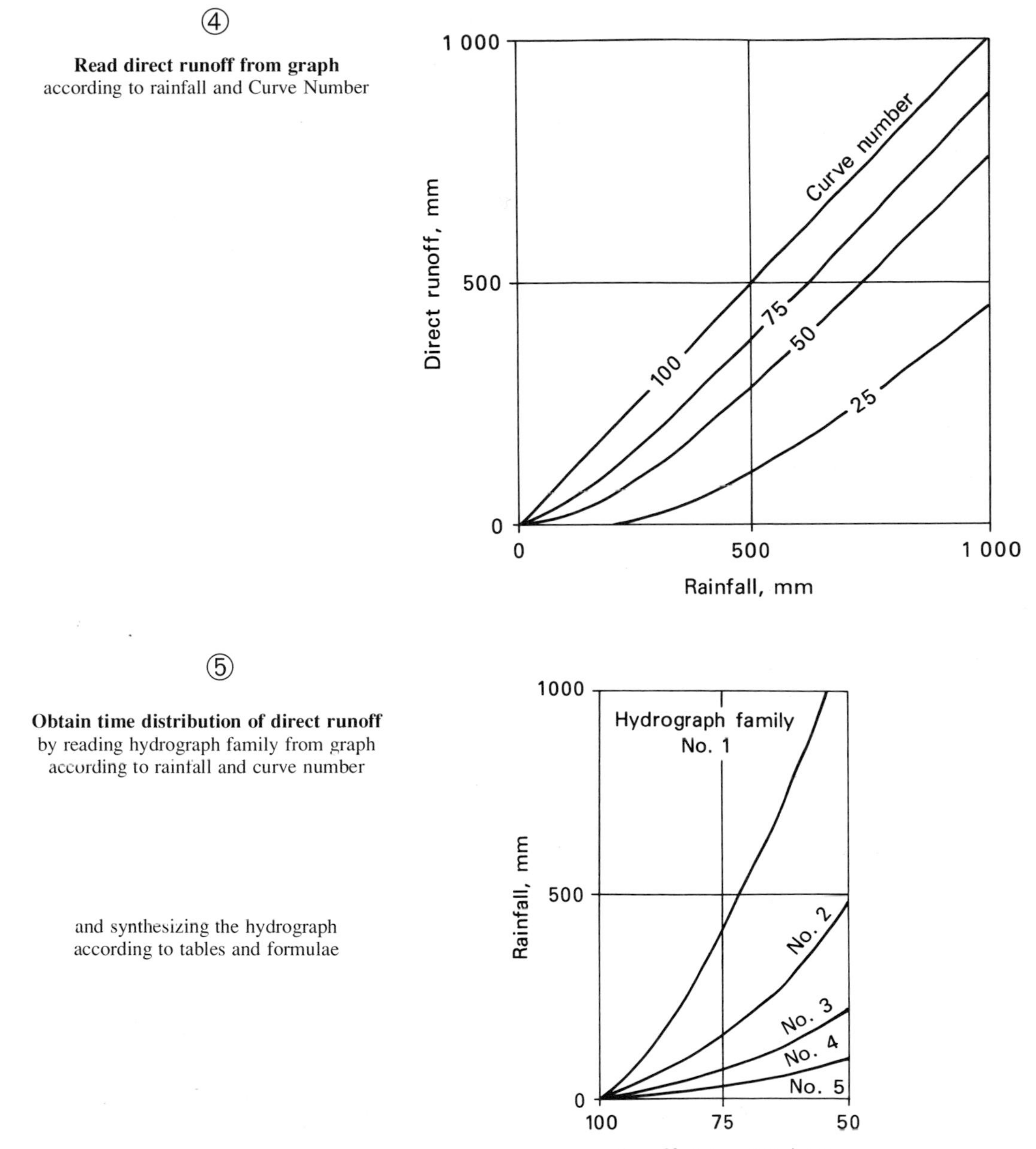

Figure 6.6 (cont.)

basis, if the pattern or the statistical distribution of past events can be established, the 'record' can be extended. It can be extended or *synthesized* backwards historically just as easily as forwards into the future, and this can be done using either the observed distribution or some theoretical distribution that it approximates to. A well-proven theoretical distribution may well be better for exploring extreme events that have not occurred in the historical record (section 4.2). The only data required are rainfall and/or river discharge. Since rainfall records are generally more numerous than discharge measurements, especially in newly exploited areas, some techniques use rainfall alone. Detailed reviews of synthetic methods have been provided by Lawrance and Kottegoda (1977) and Beaumont (1979, 1982).

Although the scientific content of these models may

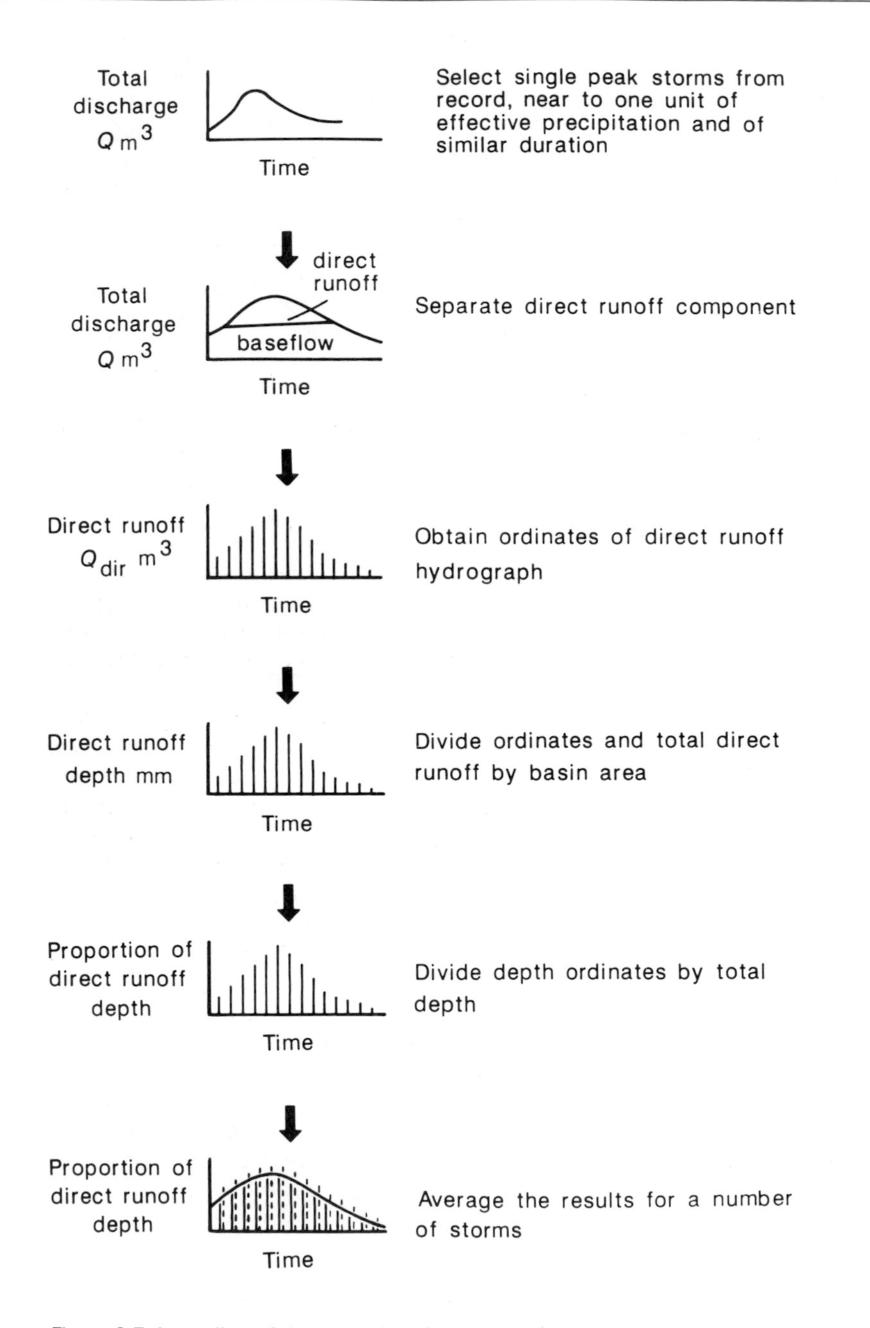

Figure 6.7 An outline of the procedure for calculating unit hydrographs

be limited, widespread practical use continues to be made of them and their value has been amply demonstrated. The whole branch of 'stochastic hydrology' is based on these methods. Indeed, because of their long and widely established practical use, engineers often apply the apparently broad title of **operational hydrology** to the specific field of utilizing historical records to enable them to make *operational* decisions, such as the design capacity of a reservoir. The basic philosophy predates the recent trend towards the scientific modelling of processes, yet it continues to develop and to contribute to effective water

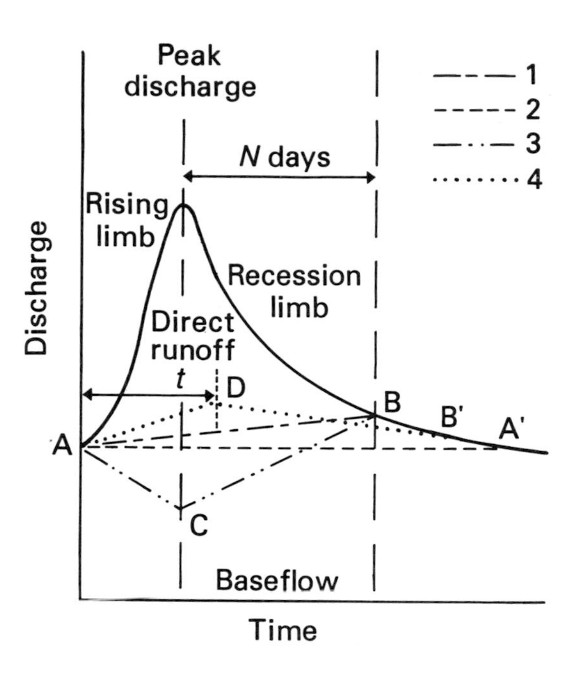

Figure 6.8 A selection of common procedures for hydrograph separation: (1) rising baseflow component ending *N* days after peak discharge (line *AB*); (2) no increase in baseflow (*A* = *A'*); (3) baseflow depletion, rising after peak discharge (line *ACB*); (4) baseflow rise over time, *t*, followed by depletion (line *ADB'*). See texts such as Dingman (1994) and Wilson (1990) for guidance on estimating *N* or *t*

management. Indeed, as the theoretical basis for most hydrometric networks, it continues to justify parsimonious data collection as well as to be self-perpetuating, since more comprehensive data covering other processes are not being collected.

The heart of hydrological synthesis dates back to the attempts last century to design upland reservoirs to provide reliable, disease-free water supplies for the industrial cities of Britain. However, Rippl's technique was based solely upon the historical record (section 4.2.2). In reality, historical sequences do not repeat themselves exactly, as the mass curve technique assumes, and modern operational hydrology is dominated by statistical techniques normally carried out by computer.

Random generation techniques These use the past records either as a source of randomized sequences or else to determine the frequency distribution of events. Thus, a synthetic sequence can be constructed by selecting monthly flows by random numbers from the historical record. The analyst may then look for extended periods of low flows in the synthetic sequence and, because it has been compiled according to the laws of probability, it should be possible to make an assessment of the risks or frequency of occurrence for different lengths of low-flow period, provided the synthetic sequence is long enough to include rarer combinations. The alternative strategy is to begin by determining a probability distribution and then to sample from this distribution in order to select events for a synthetic sequence. The probability distribution may be determined by an empirical fit to the distribution of events in the record or, if the basin is ungauged, it may be transposed from a similar basin or selected according to national or regional guidelines. We have already discussed this type of approach in section 4.2.2, but in the present context the hydrologist is concerned with the probabilities of *sequences of events*, not of individual events. Hence, the distribution of events is merely the starting point for synthesizing sequences of events from which new probabilities can be determined.

Random synthesis is clearly most appropriate where the events themselves are independent of each other. Daily or hourly flows tend to be dependent upon preceding flows; selecting hourly flows by random numbers would produce a very odd storm hydrograph. Randomization techniques are therefore most applicable to annual sequences of flow and possibly monthly sequences, depending on the climate. They are more applicable to short-term rainfall than short-term runoff, because of the nature of the generating processes. Random distribution models such as Poisson or Truncated Negative Binomial are typically applicable to short-term rainfall. Pearson Type III is often used for annual riverflow. In each case, the parameters of the theoretical distribution, such as the mean or other statistical moments, most appropriate for a certain area may be taken from published regional estimates.

Techniques for nonrandom sequences Two approaches form the main basis for generating non-random sequences. One determines the extent to which an event, such as daily flow, depends on previous events and then uses this relationship to create a synthetic sequence. The other decomposes the historical record into a set of components and then recomposes a new sequence using these components.

Markov generation models acknowledge the presence of *persistence* in the flow sequence, so that

Intensity	Duration	Technique	Illustration
UH	UH	(1) The standard UH: multiply total rainfall by the ordinates of the UH.	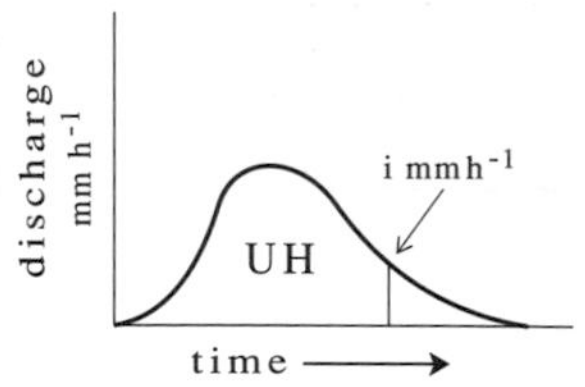 discharge mm h^{-1}; i mm h^{-1}; UH; time →
Different	UH	(2) Multiply (1) by ratio of actual net rainfall intensity to UH intensity, i.e. n times as much rain in a given time causes a runoff rate n times the ordinates of UH.	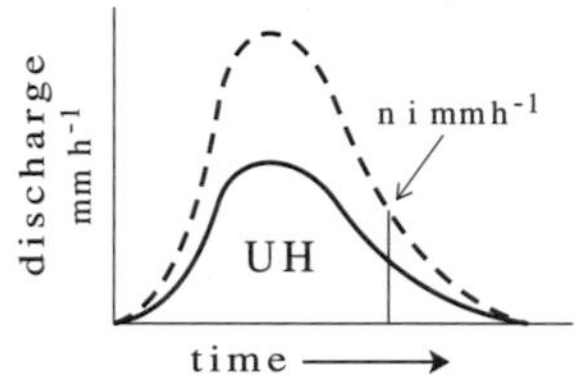 discharge mm h^{-1}; n i mm h^{-1}; UH; time →
UH	Different	(3) Superimpose UHs in sequence, beginning second UH when first unit of net rainfall has fallen.	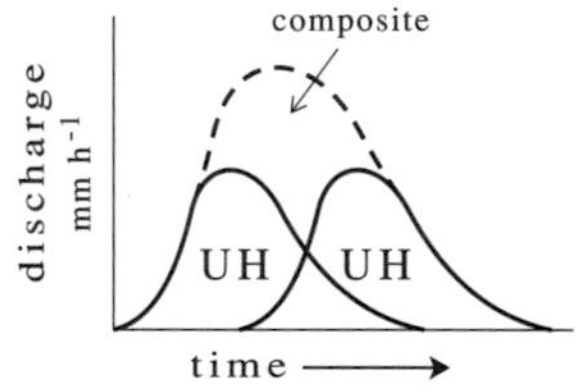 discharge mm h^{-1}; composite; UH; UH; time →

Figure 6.9 Application of the unit hydrograph method to standard and non-standard storms

low flow on one day tends to be followed by low flow on the next, and perhaps the next. The number of steps over which a measurable persistence is deemed to exist is termed the *lag*. A lag-1 Markov model of daily flows considers flow on the previous day, but not the day before that. A lag-2 model would consider the two previous days.

The Markovian argument runs as follows. If we had no better information, the best predictor of the next value in the sequence is the average next value in the records, $\overline{Q}_{j+1}$. However, this ignores the inherent variability of flows, just as a random selection of different values from the records ignores persistence. Markov therefore suggested that the next predicted flow in a daily sequence, Q_{i+1}, is also partly dependent on the last:

$$Q_{i+1} = f(Q_i)$$

Combining the two elements:

$$Q_{i+1} = \overline{Q}_{j+1} + f(Q_i)$$

We can define the function, *f*, better by using the serial correlation or *autocorrelation* in the recorded series, which is the correlation between present and next values, Q_j and Q_{j+1}:

$$Q_{i+1} = \overline{Q}_{j+1} + b_j(Q_i - \overline{Q}_j) + e$$

where b_j is the autoregression coefficient of Q_{j+1} on Q_j, and the brackets enclose the difference between the last predicted value and the average last value in the record. The final *e* is the 'error term', which covers the variation that remains unexplained by the two main terms. This may be estimated on the basis of the variance in the recorded values and the strength of the autocorrelation.

Once all these statistical parameters have been calculated from the record, and some reasonable starting value has been selected for Q_i, like the mean recorded flow, the final equation can be used to create a synthetic sequence. The earlier part of the synthetic sequence will be strongly influenced by the arbitrary starting value and should be discarded. This Markov model was refined by Thomas and Fiering (1962) to make it more applicable to monthly flows. Monthly flows require a 'multiperiod' version, because each month tends to have a different mean flow, $\overline{Q}_j$.

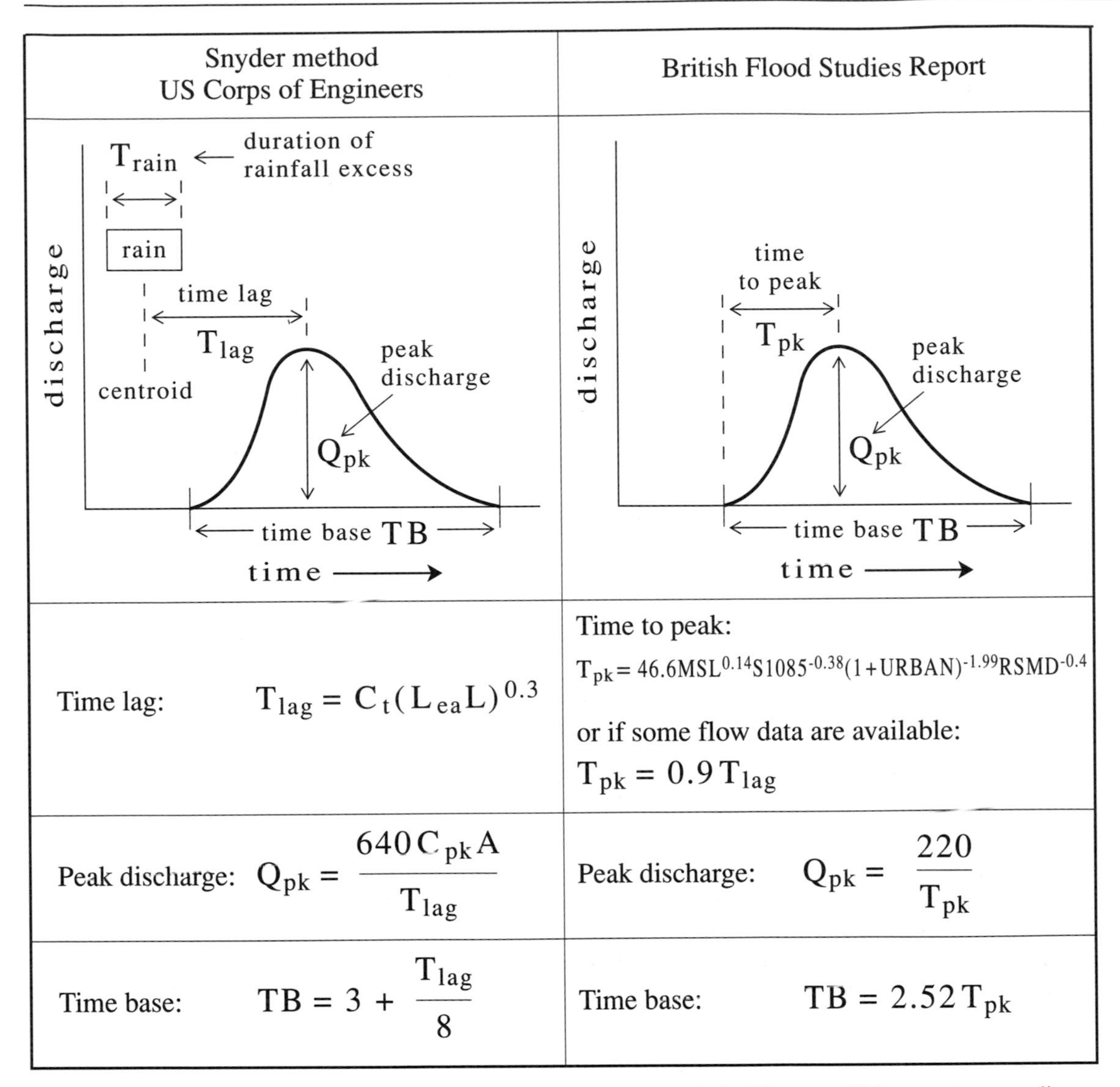

These equations are intended to be purely illustrative of the approach. Many coefficients vary according to hydrological region and basin size. Details of applicability and alternative versions may be found in texts such as Shaw (1994), Wilson (1990), Viessman and Lewis (1996) or Linsley *et al.* (1982)

Figure 6.10 Examples of calculations for synthetic unit hydrographs

One of the most important criticisms of the methods of synthesis described up to this point is that they take no account of climatic change or variability. They are therefore of little value to current questions of designing water management schemes in an environment affected by global warming. Equally, any estimates based on shorter records can be strongly influenced by both random and cyclical oscillations, as demonstrated by Hurst *et al.* (1965) or the Colorado Compact (section 4.3). Similarly, a lag-1 or 'first-order' Markov model is clearly a 'short-memory' model and is not capable of modelling the degree of persistence found in the Hurst effect. One solution to this was proposed by Mandelbrot and van Ness (1968) using a

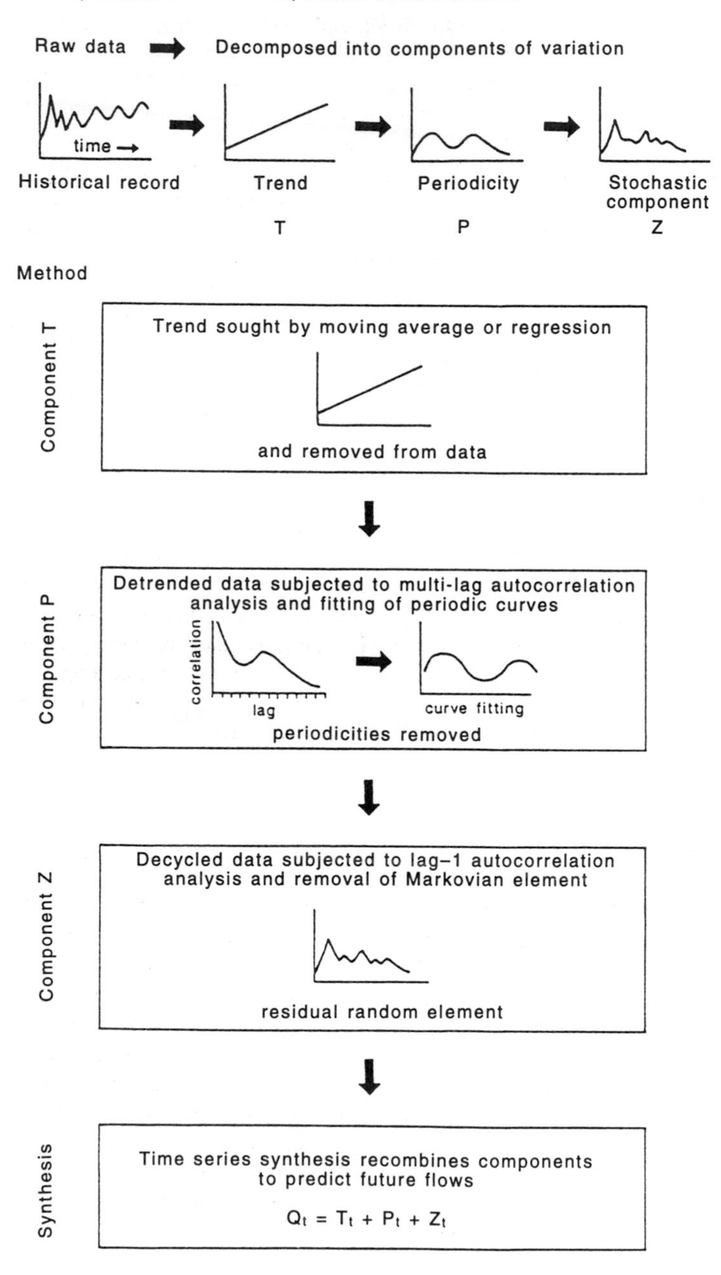

Figure 6.11 The components of time series analysis and synthesis

complicated autoregressive model called Fractional Gaussian Noise, which has a memory covering the whole record.

Time series synthesis also addresses some of these criticisms. First, time series analysis decomposes the recorded series into three components: the trend, and periodic and stochastic components (Figure 6.11). The *trend* picks out any overall shift in average values between the beginning and the end of the period of record. Technically, it is evidence of 'non-stationarity' in the sequence. This is normally due to climate, but a similar effect could be created by land-use change or river works. The *periodic* component picks out oscillatory shifts, which could be seasonal or, for annual data, cyclical, covering a number of years. The *stochastic* component may be divided into persistent or Markovian elements and independent random residuals. Analysis begins by removing the trend (detrending) with a smoothing technique, such as n-year moving averages or regression against time, followed by removal of any periodicity. The search for periodicities involves calculating autocorrelations for a series of lags and plotting the correlation coefficient against the lag in a 'correlogram'. This will indicate the frequencies of any oscillations, say every ten years, whence various techniques may be used to estimate the amplitude of the waves in order to subtract them from the detrended data set (Shaw 1994). The stochastic variations remaining at each time interval, e_t, are then transformed into normal deviates, Z_t (by subtracting each e_t from $\bar{e}$ and dividing by σ_e), and autocorrelation is performed between successive Z values to search for any remaining persistence.

The synthetic series can then be constructed by recombining the three components:

$$Q_t = T_t + P_t + Z_t$$

for each time step, t, in which the stochastic component, Z_t, is the sum of the *autoregressive* element, rZ_{t-1}, and the random element, e_t.

The synthesized flow series may be used in a number of ways. The increased length of record may be used to obtain more reliable flow statistics and return periods to help in planning water projects (section 4.2.3). Alternatively, the time series may be extrapolated into the future.

One of the most important developments in stochastic hydrology in recent years has been the creation of a new family of models known as **autoregressive (integrated) moving average (ARIMA** or **ARMA)** models. These extend the Markov approach by combining the autocorrelation with a smoothed running mean of simulated random oscillations. ARMA models combine weighted values of previous flows, Q_i, with a simulated random element, Z_i, to provide an extended 'record' from which flow probabilities can be calculated for designing water resource systems. The random element is also converted into a weighted version of previously generated random values; this is the *moving average* element. Thus, the discharge at each time, t, is given by:

$$Q_t = \sum_{i=t-1}^{i=t-n} \alpha_i Q_i + \sum_{i=t}^{i=t-(n-1)} \beta_i Z_i$$

where α and β are the autoregression coefficients or weightings for flows and random numbers in each of the time steps, i, e.g. days, months or years. The autoregressive element begins with discharge in the previous period (where i is time $t-1$) back to $t-n$ (where n is the total number of previous periods to be considered). One previous year may be quite adequate for annual river flows. The random element begins with the value generated by a random number generator program for the current time, t, and weights this by a 'running' function of the values generated for previous time intervals. More details are available in Box and Jenkins (1970), O'Connell (1974) or Shaw (1994). ARIMA models provide a better means of modelling the Hurst effect, i.e. long-term persistence, than the original Markov model and require fewer components than traditional time series models. Their current popularity reflects the recognition now given to climatic persistence. The ease with which these models can be used to generate a wide range of possible sequences with similar properties to the historical record, but each one a unique alternative 'record', is extremely valuable in many planning problems. They can also be adapted for *real-time* forecasting, if immediate past measurements are entered into the equation and then regularly updated.

6.3 Simulation models

Simulation models were born with the age of the computer and have transformed deterministic modelling. Model packages for PCs and even laptops now allow the hydrologist to make forecasts or to check the effects on riverflows of actions, such as opening reservoir valves, not only in the office but on site and in the field in order to make 'real-time' decisions. These models attempt to simulate actual real-world processes to reasonable levels of accuracy and are generally capable of simulating not just storm events but *continuous hydrographs* of flow.

Almost at the outset of these developments, Dawdy and O'Donnell (1965) made the oft-quoted, idealistic definition of such a model:

> The ideal model would specify completely the properties of and the processes that occur in all the relevant components of a basin. The specification would be given in terms of physical parameters and would involve all the behavioural relationships within the basin.

With hindsight, it is now generally accepted that such a model is impractical, certainly for the foreseeable future if not for ever. It would make too many demands in terms of data collection, computer time or memory, the degree of scientific specification of processes and their three-dimensional environment, and at present even upon certain aspects of scientific theory. But more than that: if even the nuclear physicist has to resort to stochastic mathematics to predict the behaviour of atomic particles, it seems unlikely that the hydrologist can realistically hope to specify totally the behaviour in all drainage processes within a basin. There is bound to be a degree of indeterminacy, if not because of the simple impracticality of total observation and complete modelling, then because of an underlying 'chaotic' streak in nature. A decade after that idealistic definition, mathematicians and scientists began developing 'chaos theory' with particular reference to moving and turbulent fluids (Gleick 1988), and there is no reason to suppose that runoff is any exception to this theory. Chaos theorists can now even speak of 'the fantasy of deterministic predictability'.

Given that there are limits to predictablity in both theory and practice, there is still much that is laudable in Dawdy and O'Donnell's definition as an ultimate goal. The art of the serious modeller should ideally be to balance as high a degree of scientific verisimilitude as possible against maximum efficiency in data collection and computational algorithms. The fact that so many models fall short of this ideal, or do not even aim for it, and yet are perfectly adequate for their specific job, should not be allowed to detract from the evident truth that the more exactly the model represents the workings of nature, the more versatile, 'transportable' and, ultimately, the more reliable that model should be. Considerable attention has been focused on *transportability* or *transferability* in recent years. This is the ease and accuracy with which a model developed in one basin can be used in another. A highly transferable model would be one which works well in other locations with minimal adjustments. In general, these adjustments should be limited to changes in the value of parameters, rather than involving changes in the structure of the model. They should also be changes that can be rapidly achieved without a protracted period of recalibration against historical records. The importance of versatility is only just being realized, as models developed originally for the traditional hydrological aim of predicting flows are increasingly required for use in water quality and fisheries applications, where they can be combined with models simulating chemical or biological processes.

In analysing the wide range of sophistication within simulation models, two basic aspects are paramount: (1) the degree to which real-world processes are represented; and (2) the degree of spatial resolution used in the simulation. We may view these for simplicity as the questions of (1) causality and (2) discretization within the model. We can recognize an ascending scale of causality or process representation passing from simple assumptions of linear responses within the system, through nonlinear response models to parametric and physically based component models.

Similarly, in terms of spatial discretization or resolution we can identify an ascending scale of sophistication beginning with **lumped** models treating the complete basin as a homogeneous whole, through **semi-distributed** models, which attempt to calculate flow contributions from separate areas or sub-basins that are treated as homogeneous within themselves, to fully **distributed** models, in which the whole basin is divided up into elementary unit areas like a grid net and flows are passed from one grid point (node) to another as water drains through the basin (Figure 6.12).

Early lumped simulation models were produced in the late 1950s. Nash (1957) represented the basin as a series of 'reservoirs' or **tanks**, such as interception storage, surface detention and soil moisture storage, which overflow from one to the other in a cascading system. Outflows from each reservoir were taken to be linearly proportional to the size of the reservoir, i.e. **cascading linear reservoirs**, and the outflow hydrographs were modelled by instantaneous unit hydrographs based on flow measurements or morphometric correlations. Dooge (1959) added the effect of flow routing or translation between the reservoirs in a Nash-cascade model, modelling the links as 'channels' with a linear response. These models were soon followed by more general versions with nonlinear storages and 'channels' with nonlinear responses, in which, for example, outflow may be seen as a power function of storage and vice versa.

Distributed modelling began with semi-distributed versions, which essentially link together a set of lumped models representing each sub-basin in the catchment (Figure 6.12). These were first developed in the early 1960s by Eagleson at the Massachusetts Institute of Technology as models with **multiple linear elements**, i.e. reservoirs and 'channels' representing different areas of the drainage basin, each with a fixed storage constant relating storage and outflow volumes and a fixed translation constant controlling the time distribution of the outflow (Figure 6.13). These were soon extended by the introduction of **multiple**

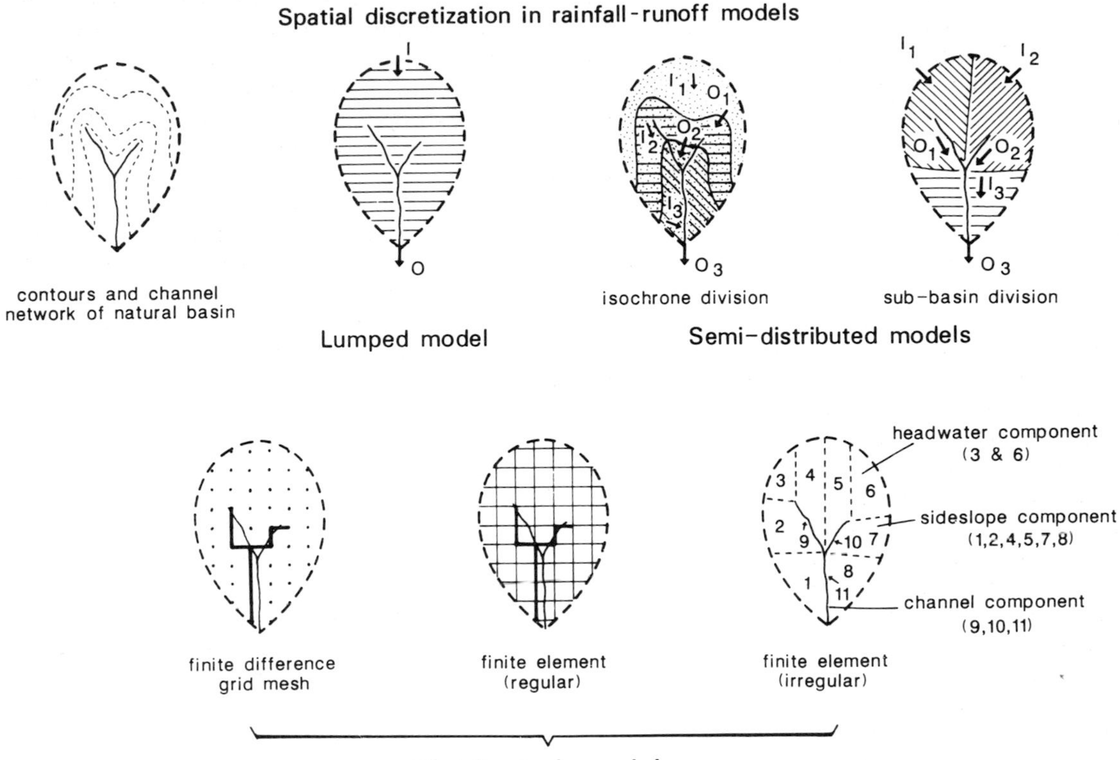

Figure 6.12 Scales of spatial discretization and process identification. I is input and O is output

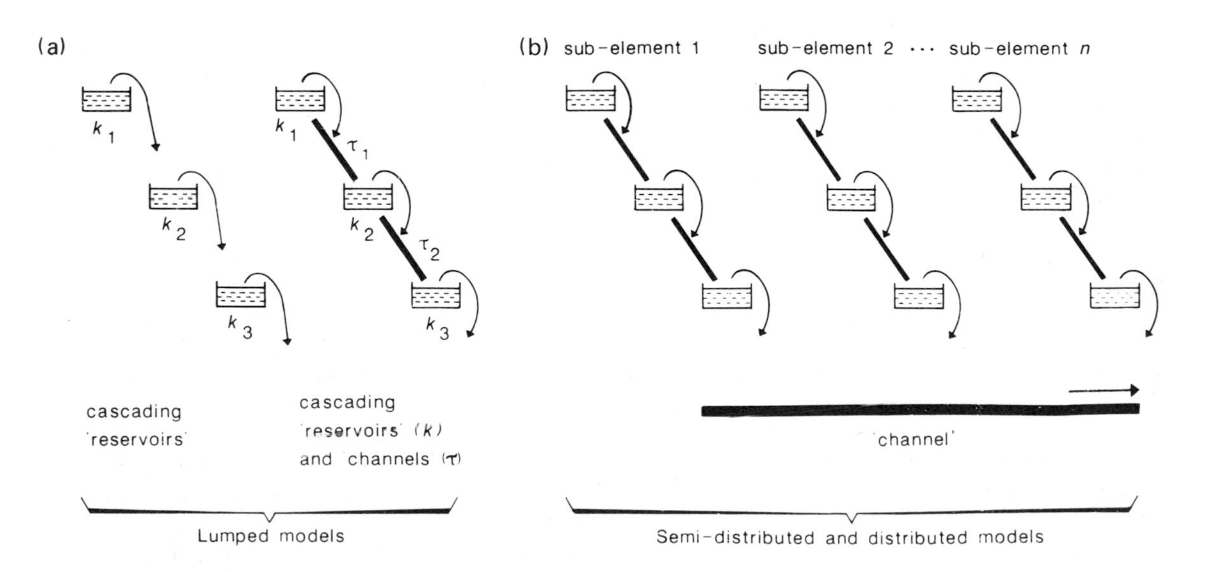

Figure 6.13 Cascading reservoirs and cascading reservoirs and channels in lumped and distributed models. 'Reservoirs' are storages, such as soil moisture, and 'channels' are transfer processes

nonlinear storages and Laurenson (1964) used isochrones, delineating the average time for flows to reach the basin outfall from around the basin, to define zones that could be modelled as separate homogeneous contributing areas. The storage coefficients and exponents were estimated from flow records. The zoned TANK model developed in Japan by Sugawara (Becker and Serban 1990) defines zones as subareas which have different water storage behaviour.

A common form of multiple nonlinear storage model nowadays calculates the streamflow hydrograph at a central point in each tributary basin, based on unit hydrographs for each central collection point, and then 'routes' the water into the main channel using some translation formula to represent the effect of the linking channel on the shape of the hydrograph. The British FSR (NERC 1975) offers the option of calculating UHs if data are available, or else using regression formulae relating flows in the sub-basins to catchment characteristics, followed in each case by applying hydraulic routing formulae that relate discharge to channel geometry.

6.3.1 Parametric models

Parametric models are the embodiment of a conceptual model, where the author has the notion or concept that certain processes or catchment characteristics are important for runoff generation. A **parameter** is a numerical value representing that process or attribute within the model. Infiltration capacity or rates of seepage through the subsoil might be examples of parameters. Parameters tend to reflect the fundamental laws of water movement in a simplified manner and are often statistically or empirically derived rather than being directly measured.

It is useful to make the distinction, based on James (1973), between parameters that are: (1) readily measurable basin characteristics; and (2) those that are known to vary but are not specifically related to any readily measurable features. Parameters in class 1 are easily determined for a new ungauged basin, but those in class 2 can only be estimated by some degree of trial and error. Trial and error usually involves using a mathematical algorithm to 'optimize' the value of the parameter by iteratively altering it until predicted flows 'fit' observations. Blackie and Eeles (1985) discuss various **optimization** techniques. The accuracy or 'goodness of fit' of the output can be measured by an **objective function** (since the 'objective' is to get the best fit): the sum of squares of the deviations between the observed and the predicted values is a common measure. Alternatively, the fit could be progressively improved by interactive computing, in which the user skilfully 'tweaks' the value of the parameter to get the closest fit between the output and the observed flows. A relatively new optimization method called **neural network** analysis attempts to automate the thought processes of the human brain in deriving optimal values. Nevertheless, optimization contains a number of potential pitfalls. Different optimization methods may identify different sets of parameters or give them different values. Skill is required in selecting an initial value for the parameter before optimization can begin, especially in subjective methods. More crucially, optimization normally presupposes some flow data: if that does not exist then it might be transposed from a similar basin. James himself managed to overcome this problem by establishing a correlation between a class 2 parameter and a class 1 parameter in a version of the Stanford Watershed Model (see below) and transferring this to his study basin. In the HYSIM model developed by Manley (1975), most of the 17 parameters were class 1 or relatable to class 1, except the groundwater parameters, which are less likely to be available but may be estimated from data on adjacent basins. Perhaps most critically, where there are a number of parameters in the model that need to be estimated by optimization, the selected values will not be independent of each other. In automatic optimization procedures, changing one of the optimized parameters will cause another to change and the statistical 'black box' procedure will take no account of physical logic. The construction is therefore inherently unstable and more may depend upon the skilful selection of initial values than upon physical reality. Clearly, therefore, the guiding principle must be to maximize the proportion of class 1 parameters when the model is being designed.

The Stanford Watershed Model This was the first great success in combining all the main hydrological processes within a computer model, and was developed at Stanford University, California, by Crawford and Linsley (1966). It essentially founded a new type of model which made a real attempt to model actual processes in the correct relationship to each other, and it spawned a whole family of direct descendants that have adapted the model to differing requirements, with differing time resolutions or differing emphases for different hydrological regions. In essence, the models are of the tank-cascade type and use lumped parameters. In the 1960s these simplifications enabled

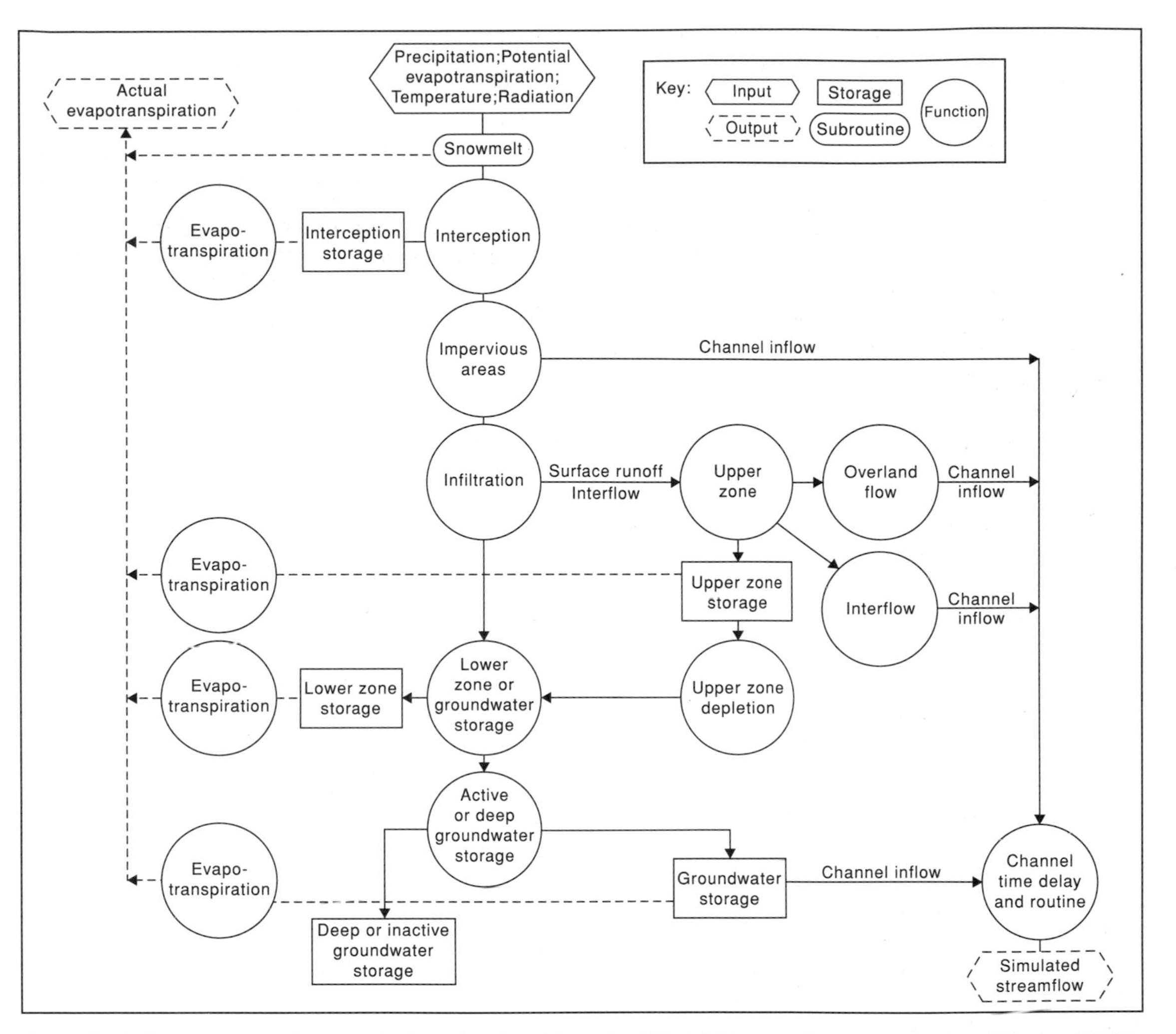

Figure 6.14 The conceptual framework of the Stanford Watershed Model IV, according to Crawford and Linsley (1966)

flows to be modelled on a continuous (or sequential) basis, even though the hydrological system was modelled in a reasonable amount of detail. Despite considerable advances in modelling and computer capabilities since, version IV of the Stanford Watershed Model (SWM) and many of its relatives are still used throughout the world today. It also deserves consideration for the excellent insight it provides into the modellers' art.

Figure 6.14 shows the basic structure of the model. SWM IV uses this to calculate hourly streamflows. It contains 34 parameters, most of which are measurable and represent water input or catchment characteristics. For some parameters where specific measurements are not available, regional values are available within the USA. Four of the parameters are derived by calibrating the model against flow measurements, namely, infiltration capacity, the storage capacities of the upper and lower soil stores, and an index of subsurface flow rates. These are clearly parameters for which data are normally very difficult or impossible to obtain. Calibration is normally carried out using a **split sample**, that is, the period of flow records is divided in two; one part is used in calibration and the other in testing the calibrated model in order to check the accuracy of the predictions. The original SWM IV required 3–6 years of data for calibration. The more sophisticated version developed by Crawford's commercial consultancy firm, Hydrocomp International, requires only 2 years or even less. The Hydrocomp Simulation Program improved on the reservoir and channel-routing methods and added water quality modelling. In order to achieve this, it introduced a high temporal resolution both for the input data and for tracking the progress of flows within the system.

SWM IV is based on solving the fundamental **continuity equation**:

$$P = E + R + \Delta S$$

(where P = precipitation, E = evaporation, R = run-off and S = storage)

for each discrete time interval. Realistic simulation of evapotranspiration processes is achieved by allowing evaporation from interception, the upper and lower soil zones, surface water and from groundwater via phreatophyte vegetation or groundwater outcrops. Overland flow is simulated using average lengths of slopes (MDOF), gradients and surface roughness values. Infiltration and subsurface drainage, however, were modelled by empirical rules and optimized or calibrated parameters. Despite considerable improvement in the scientific understanding of these latter processes during the last 25 years, data are still very difficult to obtain and spatial variation is among the highest. Channel routing is based on the simple mass balance:

$$Q_{\text{out},t_2} = \overline{Q}_{\text{in}} - kS(\overline{Q}_{\text{in}} - Q_{\text{out},t_1})$$

where Q_{out,t_2} is the outflow at the end of the time interval and Q_{out,t_1} at the beginning, $\overline{Q}_{\text{in}}$ the average inflow rate during the time interval, and k is the coefficient for the storage, S.

It is worthwhile looking for a moment at just two examples of empirical rules employed in SWM IV, which illustrate the art of simplification. These are rules used in estimating evaporation and contributing areas. In both examples, simplifications are introduced to deal with: (1) the problem of spatial variability; and (2) complex physical processes which would require an impractical amount of detailed measurements to fully specify.

It is relatively easy to calculate PET rates (section 3.1.1), but difficult to estimate AET over a variable land surface. SWM IV therefore begins by allowing PET to be satisfied first from water held in interception storage, then from surface and near-surface water (in the 'upper soil zone'). The modellers then defined a variable called **evapotranspiration opportunity** as the maximum amount of water held available in the lower soil zone during a specified time interval at any location in the basin. The maximum available water is empirically determined by the ratio of current soil moisture content in the lower soil to its long-term

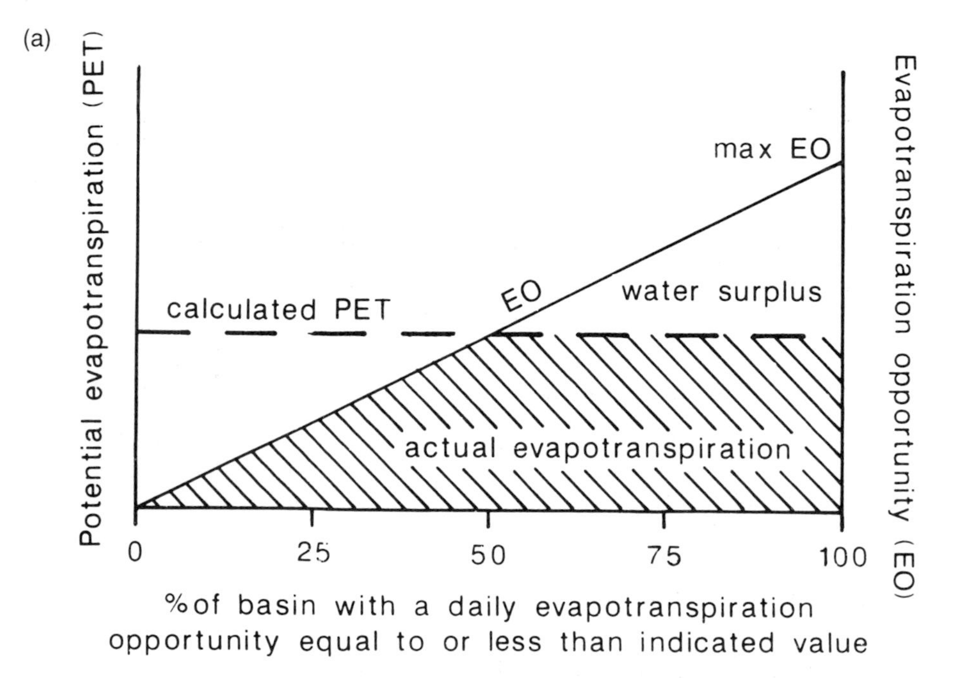

Figure 6.15 Examples of pragmatic solutions in the Stanford Watershed Model IV to the problems of estimating spatially variable amounts of (a) evapotranspiration losses and (b) runoff contributions from different sources. See text for full explanation. After Crawford and Linsley (1966) but updated in terms of terminology and the concept of dynamic contributing areas

median value. This allowed them to model the effects of limited soil moisture on evaporation rates by making the assumption that soil moisture status within the basin is distributed in a linear fashion, i.e. that the percentage of the basin with different levels of daily evapotranspiration opportunity is linearly distributed between 0 and 100 per cent. Figure 6.15a makes this clear. The actual amount of evapotranspiration loss is then given by the area on the graph bounded by the combination of the ET opportunity line and the PET line.

The solution for contributing areas follows a similar line of reasoning. The proportion of the total basin area which has infiltration capacities equal to or less than a certain value is again linearly distributed between zero and some maximum value (Figure 6.15b). In this case, two maxima are identified, which represent the separate infiltration capacities of the upper and lower soil zones at a given soil moisture status. These maximum values are empirically related to the ratios between current soil moisture storage and the median values for upper and lower soil zones. Once the model has calculated the amount of water available for infiltration in a given time interval by subtracting ET losses from rainfall (the net 'moisture supply'), the schema in Figure 6.15b enables the model to distinguish three separate drainage processes and their contributing areas. These are defined by the intersection of the net moisture supply line and the two infiltration capacity lines. The total amount of infiltration is defined by the shaded area in the graph. The amount remaining on the surface is represented by the area above the upper soil infiltration line and below the net moisture supply line. Once detention storage has been filled up, the remaining water creates overland flow. The amount retained in the upper soil zone and potentially contributing throughflow lies between the two infiltration lines and below the net moisture supply line. The amounts of lateral seepage through the soil and the deeper groundwater flow are determined by the amount in storage (calculated from the input–output continuity equation

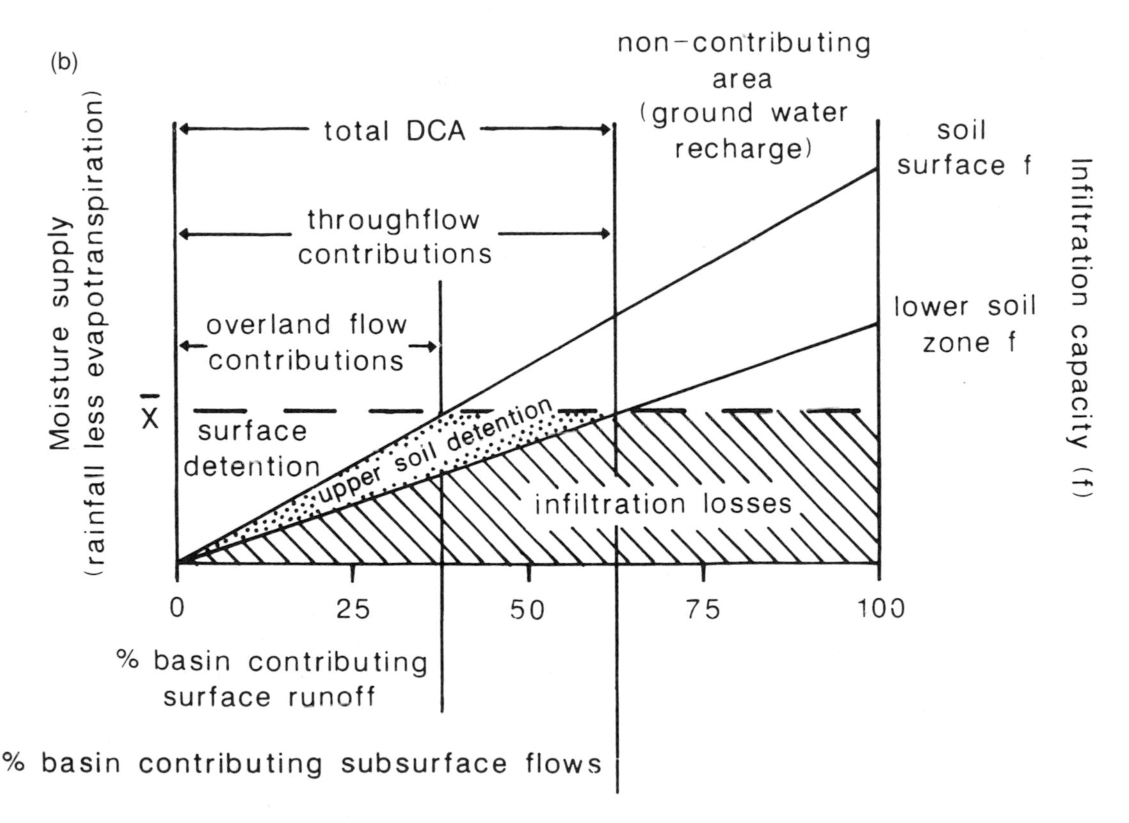

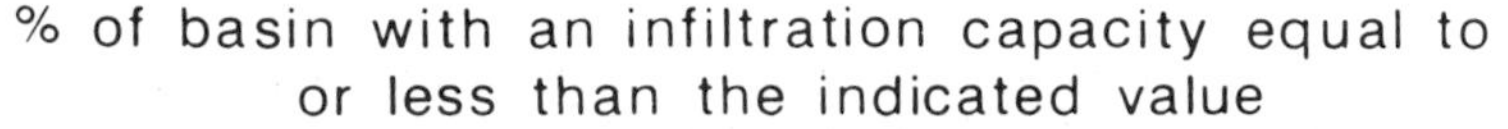

Figure 6.15 (cont.)

for each store) and the assumption of a simple exponential decline in outflow from the store once it has been filled by a storm, i.e. by assuming a simple recession curve. Full introductions to SWM are given by Crawford and Linsley (1966) and Viessman and Lewis (1996).

Many assumptions are made in this modelling. Some of them could be more realistic and some have been improved in subsequent developments of the model. The important point to remember is that modellers will always have to make similar decisions at least for the foreseeable future. The art lies in finding a workable balance between simplification and the accuracy of predictions, which is part and parcel of striking the correct balance between the cost and effort involved in devising and running the model and the value of the results.

The main reason for developing these parametric models has been to predict riverflows as a continuous series. However, they may also be used as a way of deriving flow probabilities for use in risk assessment and system design. A model that accurately simulates the processes that govern the transformation of rainfall into runoff should be capable of simulating runoff events that are more extreme than those contained in the historical records of either runoff or rainfall. They therefore have the potential to provide a better assessment than a Gumbel-style analysis of either historical records, with all the attendant questions of length of record and the most appropriate pdf (section 4.2.2), or synthetic series, which are only an extension of past records. The Hydrocomp Simulation Program is frequently used for this purpose.

6.3.2 Physically based component models

In models that are labelled 'physically based', 'physics based' or 'component' models all the parameters have real physical meaning and can be derived from assessable catchment characteristics. These models should therefore ideally meet the requirements of Dawdy and O'Donnell (1965) and eschew mathematically optimized parameters. Klemes (1985) has extolled the virtues of such models as the most geographically and climatically transferable, and ideal for the study of the effects of climate change or for 'what if' assessments of the effects of land-use change or planning decisions.

In practice, however, the transition from the archetypal parametric model to the archetypal physically based model is gradual. On the one hand, this is because over the years designers of parametric models have followed the guiding principle of reducing the number of parameters that have to be assessed by iterative approximation within their programs. On the other, it is because there remain stochastic elements and indeterminacies even within the best constructed physically based models that require some approximating assumptions.

Typically, the modern physically based model is also fully distributed and employs either a **finite difference** or a **finite element** technique for calculating flows and storages at points set in a fine mesh across the drainage basin. The finite difference method computes these values for 'nodal points' set either at the intersections of a grid mesh or at the centre of each block or cell within the grid (Figure 6.16). The finite element method not only calculates values at nodes but also interpolates values in between the nodes. It can also use a variety of mesh configurations, not just the standard rectangular grid of the finite difference method, and is thus able to present a wider range of solutions. The added sophistication of the finite element approach can be particularly useful in modelling the transport of pollutants or in dealing with 'moving boundary' problems such as the rise and fall of the water table. Because rainfall intensities are continually changing during a storm, physically based models need to operate on short time intervals and take account of 'nonsteady state' flow patterns; again, the finite difference or element approach is ideal for this. A good introduction to the mathematical techniques is provided by Wang and Anderson (1995).

The most sophisticated models take a three-dimensional view of water exchange, with meshes superimposed vertically (Table 6.2 and Figure 6.17). These techniques have opened the way for major advances in modelling by linking them with elevation data held on similar grids, as digital terrain or elevation models (DTM/DEM) derived from maps, or with other data derived from raster-based satellite imagery, which may indicate vegetation cover, soil moisture patterns and lines of subsurface drainage.

Two models will be taken as illustrations: the Système Hydrologique Européen (SHE) and the UK Institute of Hydrology Distributed Model (IHDM).

The Système Hydrologique Européen This was developed during the 1980s in a multinational programme stimulated by the European Community. It involved the UK Institute of Hydrology, which undertook the modelling of the interception,

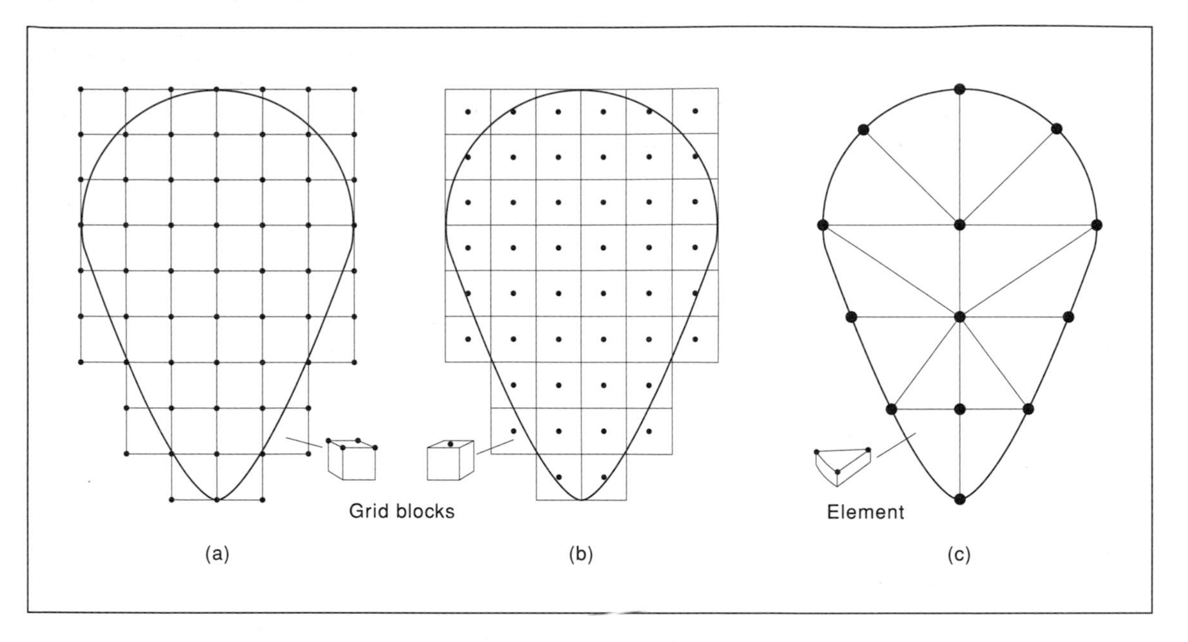

Figure 6.16 Finite difference and finite element representations of a drainage basin: (a) finite difference nodes at grid points; (b) finite difference nodes at cell centres; (c) finite element method, in which values are interpolated within the cells. Grid meshes are not necessarily rectangular, especially in finite element representations

evaporation and snowmelt component subprograms, the Danish Hydraulic Institute, which covered hillslope seepage processes, and the French Société Grenoblois d'Étude et d'Applications Hydrauliques, which produced the surface flow subprograms.

SHE uses a finite difference approach to model flows. This approximates the continuous exchange of water throughout the basin, which in reality is continuous in both time and space (the type of process to which a mathematician would apply differential equations), by breaking it down into discrete but small units of space and time. The model uses a square grid of cells laid over the basin and calculates flows and storages in each cell at successive intervals of time. The cells are repeated in a set of overlays for different layers within the system, such as the soil surface and the subsoil (Figure 6.17), and the program passes flows from one cell to another vertically and horizontally as appropriate. Each cell has its own characteristics, which are held within the model, specifying its elevation, vegetation, surface roughness and soil properties, including information on hydraulic conductivities and layers with restricted permeability. The river runs along the boundaries of the grid cells which cover the hillslopes.

The model is very flexible in terms of data requirements and use. Interception is determined using the Rutter model (section 3.1.2), but evapotranspiration can be calculated in a variety of ways, depending on the data available. The Penman–Monteith approach will provide the best results if sufficient data are available. Flow in the unsaturated zone is modelled by the Richards equation (section 3.3.4) and lateral flow is assumed to be negligible. In contrast, flow in the saturated zone is assumed to be only lateral in order to minimize the amount of computing needed. In use, it is possible to concentrate modelling on certain components and omit others for a specific application or to operate components independently.

The model has been proven to be quite transferable, with successful applications, for example, in Europe and New Zealand. Its proponents regard it as one of the most reliable tools for predicting the impacts of land-use or climate change. However, as with any model it is not without its detractions. These centre on the large amount of data required and, ironically, also on the simplifications made. There are a wide variety of different parameters and functions to be quantified in each cell, such as the aerodynamic resistance of the vegetation for the Penman–Monteith formula. In practice, however, it is possible to reduce data requirements by making some simplifying assumptions, such as taking measurements at 'type sites' and extrapolating from these to similar cells.

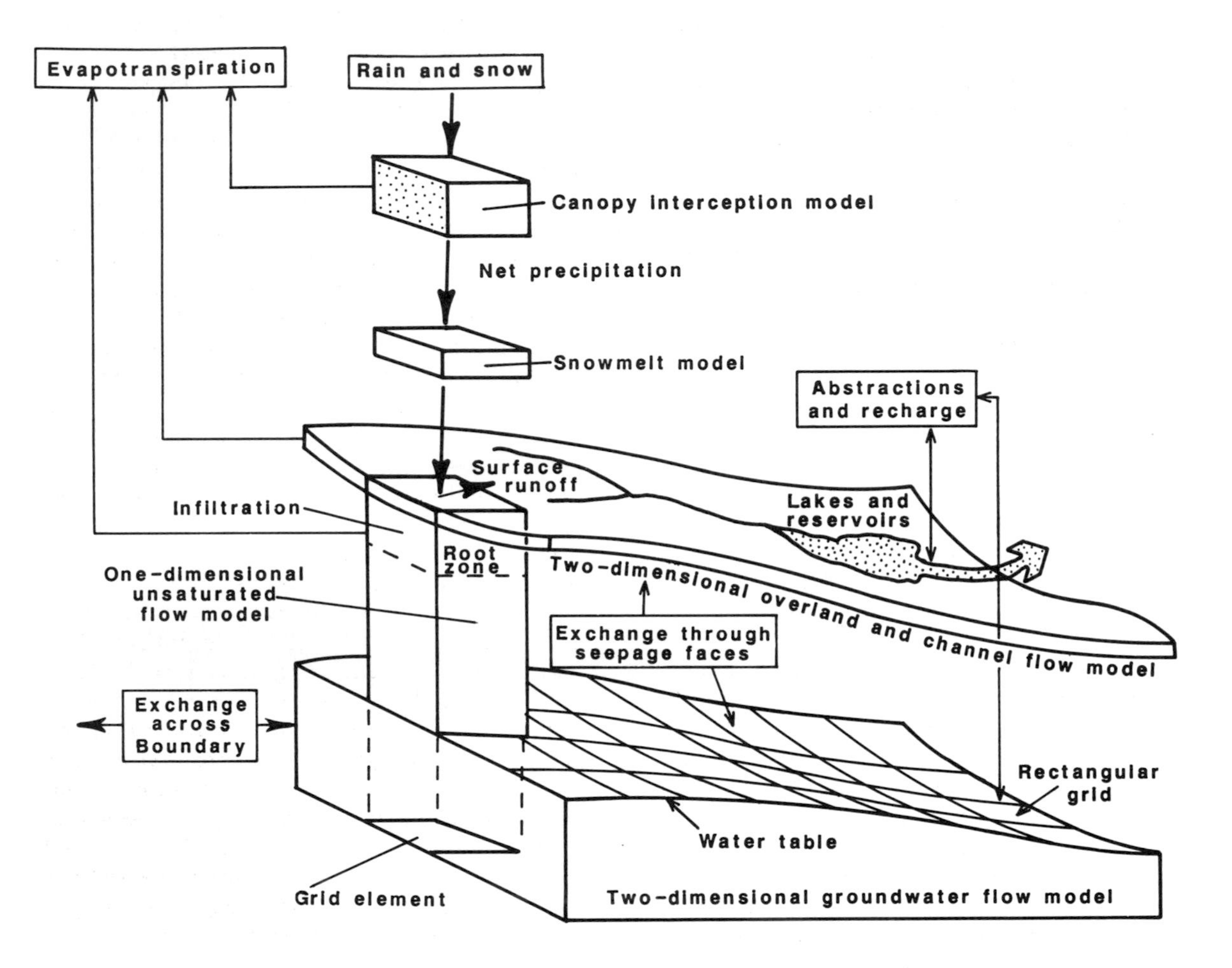

Figure 6.17 The structure of the Système Hydrologique Européen (SHE) model, according to Abbott *et al.* (1986)

Table 6.2 Spatial discretization and dimensionality in different types of model

Model type	Spatial discretization in basin			Dimensional detail		
	Distributed	Semi-distributed	Lumped	One dimension	Two dimensions	Three dimensions
Physically based	✓	(✓)		✓	✓	✓
Parametric/conceptual	(✓)	✓	✓	✓	✓	
Black box			✓	✓		

Source: Based on Becker and Serban (1990).
Ticks indicate normal levels of discretization and dimensional detail. Brackets indicate possible under certain conditions.

Even so, operating the model requires quite considerable training and skill, particularly in laying out the grid cells and in selecting initial values. The model may not represent topographic convergence too well because of its fixed pattern of uniform grid squares, especially if the grid cells are set a little too large. Again, if the selected grid size is too coarse, the model will not trace the expansion and contraction in the dynamic contributing areas very well. On the other hand, if it is too fine a mesh, then the operator pays for it with a much increased computational time. Abbott *et al.* (1986) provide a good introduction to the original model.

Recent work by Wicks and Bathurst (1996) has added a sediment transport capability and it has been renamed SHETRAN-UK. Bathurst's group are so confident of the physical basis of the model that they are attempting to raise calibration techniques to the next logical level by replacing the standard split sample approach (section 6.3.1) with **blind testing**. Blind testing involves using measured basin parameters alone to predict runoff, without calibrating against streamflow records. This is the ultimate aim of physically based modelling and accords with the ideals of Dawdy and O'Donnell (1965) and the practical principles of James (1973) outlined earlier. This is asking a lot of any model, but after nearly 30 years of developments it seems that physically based modelling is on the verge of this major breakthrough. The initial work with blind testing shows reasonable success in predicting monthly discharges and long-term total volumes, but the results are not yet acceptable for peak discharges or continuous flow simulation. This is nevertheless a major step towards achieving the long-term objective.

Bathurst's group is also addressing the perennial problem of obtaining grid-scale parameters as efficiently as possible, i.e. of getting reliable values for land characteristics and hydrometeorological inputs at grid points without having to measure at every point. They have successfully scaled up from sample plot scale measurements to a regular 50 m grid interval in the Cobres basin in Portugal, but have understandably experienced some difficulty in scaling up to a 2 km grid. It is also easier to extrapolate parameters affecting surface runoff to a larger scale, especially Hortonian overland flow, than those affecting subsurface flows. This seems to be due to the variety of subsurface processes, not all of which are represented in the model, the high spatial variability in subsurface properties and the difficulties of obtaining measurements. Consequently, the extrapolation of parameters generating saturation overland flow is still unsatisfactory.

The Institute of Hydrology Distributed Model IHDM version 4 is a finite element model which offers a significant improvement on SHE in terms of its representation of the three-dimensional nature of the basin. The improvement involves using a flexible grid mesh which is stretched or compressed to fit the topography. This mesh of finite elements is defined by the subprogram TOPO, based on the contour map of the basin. While SHE is adequate where widespread contributions from groundwater or Hortonian overland flow are important, IHDM-4 is better able to represent partial area contributions from saturation overland flow, since these depend more on areas of topographic convergence (Beven *et al.* 1987). SHE might approximate this with a very fine grid, but the high resolution would be paid for in increased computing. Beven *et al.* (1987) suggested that SHE may be more appropriate for lowland catchments and IHDM for the uplands.

At the heart of the model lies the concept of **hillslope planes**. These are strips of land that run the whole length of the hillslope and issue into the channel head or bankside (Figure 6.18). The boundaries of the planes run down the maximum slope, perpendicular to the contours, and it is assumed that there is no lateral exchange of water between adjacent hillslope planes. This is a reasonable general assumption, although it is possible for subsurface flow directions in some basins not to conform with the surface topography, because of differing 'topography' in subsurface layers or soil piping (Jones 1986; Jones *et al.* in press). The width of each plane depends of how undulating the topography is and the degree of downslope convergence or divergence (section 3.3.4).

The concept of hillslope planes makes for a very efficient handling of spatially varying land surface properties, limiting the number of elements needed in the computation by dividing the basin up on the basis of the main factors that control runoff generation rather than according to a spatially invariable mesh. Each plane has its own values for slope, soil depth and hydraulic conductivity in different horizons. The subsurface elements within each hillslope plane are also variable, so that greater resolution can be achieved at the bottoms of the hillslope near the stream, which are more important for runoff generation.

At the same time, spatial variations in input from rainfall or snowmelt or in evapotranspiration losses are accommodated by the concept of 'input zones'. **Input zones** may cross hillslope planes and each hillslope plane may have up to 12 different input zones (Figure 6.18). The routine PLANE calculates overland flow and throughflow (using the Richards equation) for each

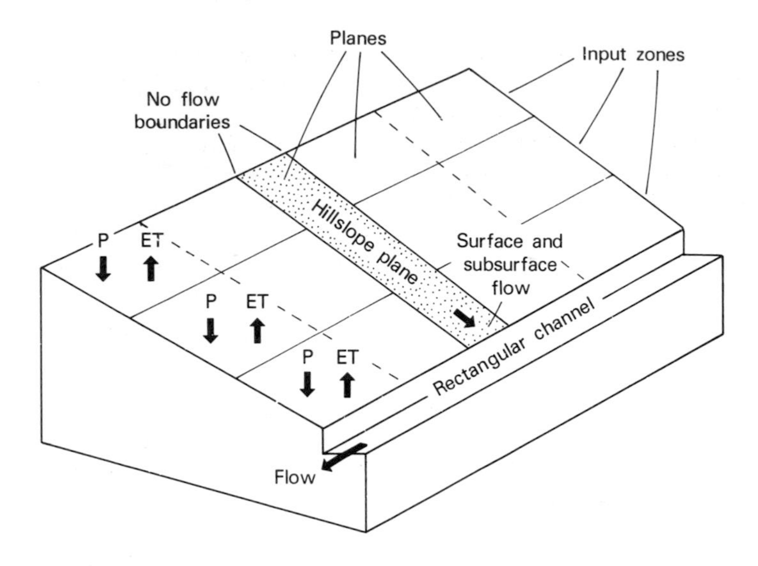

Figure 6.18 The spatial structure of the UK Institute of Hydrology Distributed Model. Based on Beven *et al.* (1987)

separate plane. The water is discharged from these planes into channel sections which are treated as units of constant slope and rectangular cross-section, and a kinematic wave-routing equation (section 6.4.1) is then applied to the flow.

The complete calculating system consists of a set of linked models. The MET model calculates the atmospheric input into the different input zones by adjusting measurements taken at a meteorological station according to differences in altitude and aspect. GLI then calculates the effective surface inputs according to slope angle, vegetation cover, and albedo, using a Penman–Monteith equation for evapotranspiration and a Rutter model for interception losses. IHDM finally takes over with the computation of the surface cascade of flows. Almost every process is covered within the models, including snowmelt, interception, throughfall and stemflow, and is simulated by physically based routines, except for soil moisture content, which uses a simpler conceptual model. The finite element procedure calculates flows for each cell at hourly intervals.

Like SHE, IHDM makes quite prodigious demands on data, including, for example, both vertical and horizontal hydraulic conductivities for each soil horizon in each hillslope plane. However, it is possible to obtain acceptable simulations in certain basins with far less data. It depends largely on the degree of spatial variability within a particular basin.

6.4 Flow-routing models

Once drainage water reaches a river channel the problems of modelling are simplified quite dramatically. The water is now essentially flowing through one routeway in a channel which has maximum 'permeability', and the rate of flow is now governed by the balance between the momentum generated by gravitational forces and the frictional resistance to flow within that channel. The physics of this flow are probably better understood than for any other aspect of hydrology and are defined by the Navier–Stokes equations for unsteady, turbulent fluid flow and the mass continuity equation (Cebeci and Bradshaw 1977).

In practice, however, these equations are far too complex for normal requirements and demand data that are not available. Hydrologists have therefore simplified the computation, either in conceptual hydrological models or in more mechanistic hydrodynamic models. It is easier to follow the logic if we take the more detailed models first, for a change.

6.4.1 Hydrodynamic routing models

These channel flow models are based on the Saint-Venant equations, which model the real physical processes but simplify the formulation by disregarding lateral and vertical flows in the channel and only consider flow in the downstream direction, x, over time, t. The equations consist of: (1) the **continuity equation**, which expresses the necessary balance between input, storage and output in a section of stream, i.e. the changes in the cross-sectional area of storage and in discharge downstream are the product of the amount of water laterally entering the section:

$$\frac{\partial Q}{\partial x} + \frac{\partial A}{\partial t} - q = 0$$

with Q as output discharge and $\partial Q/\partial x$ representing the rate of change in discharge downstream, A the cross-sectional area of flow, and q the amount of lateral inflow within the section; and (2) the **momentum equation**, which relates the rate of change in momentum to the applied forces:

$$\frac{\partial Q}{\partial t} = \frac{\partial \left(Q^2/A\right)}{\partial x} + gA\left(\frac{\partial h}{\partial x}\right) + gA(s_f - s_b) = 0$$

where g is the acceleration due to gravity, h the depth of water, s_b the bed slope and s_f the 'friction slope' calculated from the Darcy–Weisbach friction formula:

$$s_f = \frac{Q^2 f}{8ghA^2}$$

Here f is an empirical coefficient which expresses the retarding effect of channel roughness and eddies in the flow.

Bathurst (1988) and Becker and Serban (1990) list the many simplifying assumptions made by these equations, especially that the channel remains fixed and that there are no sudden downstream or cross-channel changes. Models using the full Saint-Venant equations are termed **complete hydraulic** or **dynamic wave models.** Even so, some terms in the momentum equation pose problems for solution and further simplification is often made. Fortunately, the first two terms tend to be very much less important than the bed slope. The **kinematic wave model** therefore considers only the last term and models the downstream transmission of water as a simple, unchanging wave. The **diffusion wave model** considers only the last two terms and therefore allows for changes in the height of the wave as it moves down the channel.

6.4.2 Hydrological routing models

These are based on the first equation, the **continuity equation**, in the Saint-Venant formulation, with some empirical method being used to cover aspects in the momentum equation. From the continuity equation,

$$\frac{dS}{dt} = Q_{in} - Q_{out}$$

where S is the volume stored in a given reach of the channel, and Q_{in} and Q_{out} are the input and output discharges. Over a certain time interval or **routing period**, t, this becomes:

$$\frac{Q_{in,1} + Q_{in,2}}{2}t - \frac{Q_{out,1} + Q_{out,2}}{2}t = S_2 - S_1$$

where subscripts 1 and 2 refer to the beginning and end of the time interval. Knowing the inflows, the initial storage, S_1, and the outflow, $Q_{out,1}$, it is then possible to calculate the outflow and storage at the end of the time interval. The US Corps of Engineers pioneered this approach for routing flows through reservoirs. In rivers, the situation is complicated by the downstream procession of flood waves, which can normally be disregarded in reservoirs. Hence, storage in rivers is not totally a function of discharge, but can be divided into two parts: the part that is defined by the input and output discharges (known as **prism storage**) and the extra storage contained in any waves within the reach (called **wedge storage**), which would be accounted for by the momentum equation. Various empirical methods have been adopted to take account of this.

One of the earliest and most well-known methods is the **Muskingum model**, in which:

$$S = k[xQ_{in} + (1 - x)Q_{out}]$$

with k being an empirical **storage constant** for each reach derived from measurements of input and output hydrographs, and x another proportionality constant for the reach with an average value of 0.25 suggested by Carter and Godfrey (1960). This approximation can then be used to calculate flows that are passed downstream from reach to reach in each time interval. Details of the method are given in Wilson (1990).

In general, both hydrological and hydrodynamic models assume that the system is linear. Once overbank floods occur, this assumption is immediately invalidated, because the flows then behave differently inside and outside the channel. Multilinear models have been successfully introduced to cover this situation. Becker and Kundzewicz (1987) describe a **threshold model** which introduces a parallel linear model for overbank flow when it occurs. Such models are a significant advance on older hydrological models and provide a cost-effective alternative to the fuller hydrodynamic models.

6.5 Modelling snowmelt and icemelt runoff

Snowmelt and icemelt are important sources of streamflow in many areas of the world and can be seasonally significant in many more (section 2.1.2). As a result, simple models for predicting snowmelt discharge were developed at an early stage relying on air temperature as a surrogate for a full energy balance approach. They are still widely used, especially for warning of snowmelt floods, even though the trend towards a more process-oriented evaluation of the energy budget began in the 1950s and despite a considerable amount of scientific exploration over the last two decades into the role of spatial variations in snow properties and in seepage through and under the snowpack (section 3.3.5).

6.5.1 Temperature-based models

The **degree–day method** requires local measurements of daily snowmelt to be correlated with mean daily air temperatures, $\overline{T}_a$. A regression equation of the form:

$$M = a + k\overline{T}_a$$

is established using measurements of the melt rate, M, which relate either to total basin discharge or to the melt rate estimated for sample areas of the snowpack by snow sampler or meltwater lysimeter. As with all such simple black box models, the model is not very transferable and, although some transfer of the empirical coefficient, k, may be possible within the same hydrological region, ideally local data are needed. Dunne and Leopold (1978) quote a k value of 40 mm day^{-1} °C^{-1} as commonly used in Europe and North America, but work by Fitzgibbon and Dunne at the McGill Subarctic Research Station in northern Quebec indicated much lower local values. These could be as low as 6 mm day^{-1} °C^{-1} in 'closed lichen woodland', that is, spruce forest with a canopy cover of 25 per cent. The value of k increases as the canopy cover decreases and so provides less shade (Figure 6.19). Melt rates are thus very sensitive to vegetation cover as well as to the major geographical factors that affect the radiation balance, latitude, aspect and season, and the albedo of the snow surface. In basins with a variety of cover types, a more realistic approach would be to establish degree–day formulae for each cover type. The sensitivity of accumulation rates to vegetation cover means that such basins can experience a high spatial variability in meltwater yields. Open areas without tree

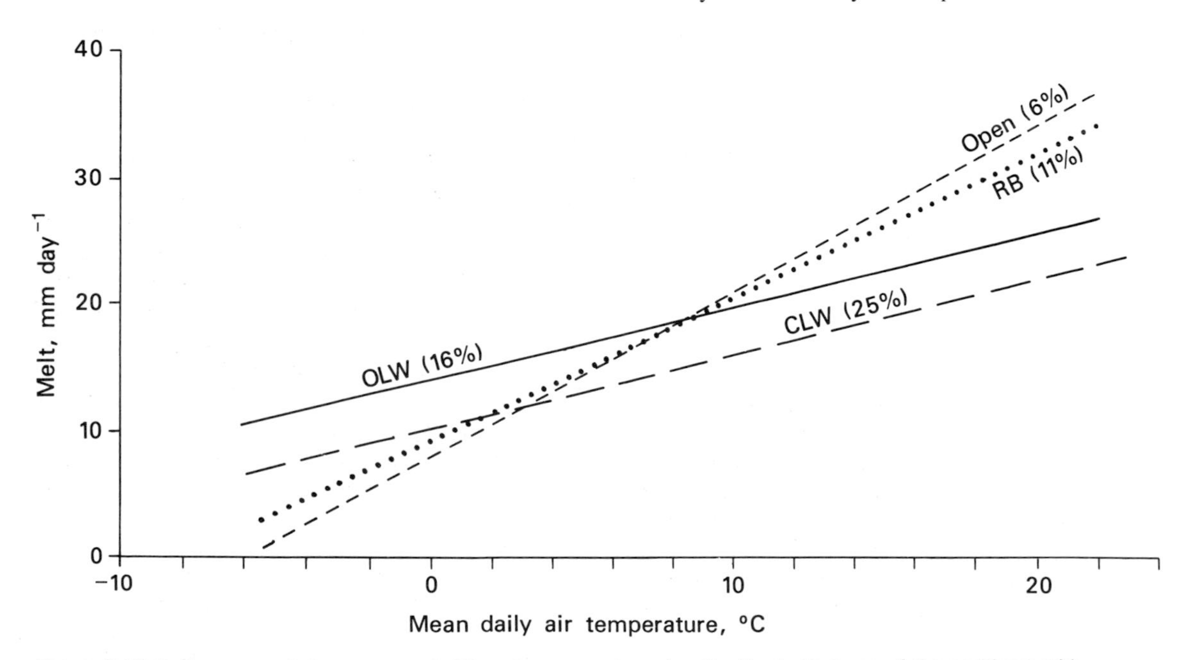

Figure 6.19 Daily snowmelt from areas of different canopy cover density, illustrating one of the problems with attempting to predict melt rates from temperature alone. The curves relate to closed lichen woodland (25 per cent canopy cover), open lichen woodland (16 per cent), regenerating burnt woodland (11 per cent) and open areas (6 per cent) at the McGill Subarctic Research Station, northern Quebec. Based on Dunne and Leopold (1978) from Fitzgibbon's data

cover will have the fastest melt rates but these will be maintained for a much shorter period because of a thinner snowpack.

Correlations with temperature can also be affected by the amount of advected atmospheric heat and by how representative the temperature measuring site is of conditions throughout the basin. The work of Fitzgibbon and Dunne also revealed the disturbing fact that the equations cannot necessarily be relied upon from one year to another, even in the same basin. They quote errors of the order of 35 per cent in the second season. This underlines the fact that daily air temperature is only a partial and largely indirect cause of snowmelt, which among other things takes no account of the structure and liquid water content of the snowpack.

Meltwater runoff, as opposed to melt rates, is also subject to all the other factors that affect rainwater runoff: basin slopes, soil properties and drainage networks. Scientifically, it would be more realistic to model snowmelt and runoff as two distinct processes.

6.5.2 Energy budget methods

The local energy budget is the most fundamental factor in snowmelt. A detailed investigation by the US Army Corps of Engineers in the 1950s (1956, 1960) produced a definitive account of the physics of snowmelt, from which a widely used set of simplified equations was derived.

The melt rate from an isothermal snowpack depends on the amount of surplus energy per time unit, Q_m, and the fixed amount of energy needed per cubic millimetre for the phase change. However, extensive and expensive instrumentation is needed to specify all the terms in the budget accurately.

US Corps of Engineers equations These consider in a simplified empirical way the six processes by which heat may be gained or lost: (1) absorbed solar radiation; (2) longwave radiation losses; (3) convective and conductive heat exchange with the air; (4) condensation or sublimation of water on to the snowpack from the air; (5) conduction of heat between the ground and the snowpack; and (6) the heat content of new precipitation. Although they were initially developed in the mountains of the western USA, the equations have been successfully modified and applied in the eastern and northern lowlands of North America. Guidelines set down by the Corps of Engineers allow coefficients to be modified for basin aspect and forest cover. Nevertheless, in adapting the approach to a new environment it is always advisable to check the calibration against some local or transferable data.

Separate equations are applied to heat exchange on rainy days and fine days. During periods of rain, the equation assumes a complete cloud cover and this allows some processes to be disregarded, such as net shortwave solar radiation, $Q_s(1 - \alpha)$, and variations in relative humidity:

$$M = T_a(L + b_1 K\bar{u} + b_2 P) + S(1 - F) + G$$

where L is an index of net longwave radiation, K is a constant for the basin which assesses the effect of forest on the mean windspeed, $\bar{u}$, P is daily rainfall depth, S a factor which expresses the effect of the orientation of slopes within the basin upon shortwave receipts at the snow surface, F the fraction of the basin covered by forest, and G the amount of melt induced by geothermal heating at the base of the snowpack. For open areas $K = 1$, i.e. there is no reduction in windspeed, and K is reduced as forest cover increases. Similarly, the weighting factor S is set at 1.0 for a horizontal basin or one with equal north- and south-facing slopes. In some areas, geothermal heating can also be ignored, but it can be important where snowcover lies trapping the heat for most of the year.

Under clear skies:

$$M = S(1 - F)\{b_3 Q_s(1 - \alpha)\} + b_1 K\bar{u}(b_4 T_a + b_5 T_{dew}) + b_6 F T_s + G$$

where the net solar radiation is now included in the first term. The longwave balance is no longer positive without the radiation returned from a cloud base and is ignored. The windspeed term now includes a weighting which is a simplified representation of the sensible and latent heat fluxes, which are determined by temperature and humidity profiles above the snowpack. A truly process-based evaluation would require detailed measurements of micrometeorological profiles, which are not normally available (section 3.1.1).

The equations are flexible and can be modified or used in conjunction with others to accept differing levels of information. Solar radiation, net longwave or net all-wave radiation may be either measured or estimated. For cases in which micrometeorological measurements of windspeed are available at 2 m, along with similar temperatures and humidities, a range of equations is offered for different degrees of forest cover, in addition to the general version given above. These show that the equations can be simplified as forest cover increases. Under dense forest covering more than 80 per cent of the catchment, only sensible and latent heat fluxes are important and radiation and windspeed can be ignored. The original publications or Dunne and Leopold (1978) give details.

UK Meteorological Office Flood Studies Report equation The Meteorological Office FSR team chose a similar approach (NERC 1975), but with an important shift in emphasis.

Despite the received wisdom that the local radiation balance is a dominant factor in snowmelt, in the British climate advected heat coming largely from oceanic sources can be very important. In this context, air temperature is not simply a surrogate but important in its own right. Moreover, in relatively thin snowpacks, which rarely happen to have temperatures more than a few degrees below freezing point, the small amounts of heat added by rainfall appear to have a disproportionate effect on melting. Heavy rain has preceded some of the most severe snowmelt floods in Britain (Archer 1981), such as those in England in 1947 or the flood on the Yorkshire Ouse that cut off the city of York in December 1981 and resulted in a major flood barrier around the city being completed in 1985. In contrast, the great snow-up of January 1982, when even coastal towns in West Wales were cut off for about a week, melted slowly under clear skies and produced no significant flooding. Snowmelt floods have troubled the River Severn in its lowland reaches since then with far less snow accumulated in its Welsh source areas, but these have been accompanied by rainfall. Of course, it is not simply a question of the heat provided by the rain. Perhaps more importantly, the rain in Britain tends to be accompanied by winds coming off the warm ocean and carrying advected heat in with a moist, mild, cloudy airstream.

The Flood Studies Report equation is:

$$M = T_a(1.32 + 0.394Ku) - 0.30Ku(T_a - T_{dew}) + 0.0126PT_a$$

The equation looks like an amalgam of the rainy and clear-skies formulae in the Corps of Engineers approach, but one which omits the shortwave and longwave radiation balances, geothermal heat and forest effects, except in so far as they may be incorporated into the friction factor, K. We are left with variables that represent sensible and latent heat transfer and any heat derived from the rain, which is assumed to have the same temperature as the air. Archer (1983) used a parametric model which incorporated the degree–day approach of the UK Flood Studies team for predicting snowmelt floods in northeast England.

6.5.3 Simulation models of snowmelt runoff

A variety of models exist nowadays that undertake physically based simulations either of the snowmelt alone or of the complete runoff process. Snowmelt may be modelled as an 'add-on' component to a general runoff model, or it may be a component in a model specifically designed for snowmelt runoff. In either case, it tends to be modelled as a self-contained element, although models designed specifically for snowmelt runoff may take more account of spatial variations in the snowpack (section 3.3.5). SHE contains a layered process-based energy budget submodel which feeds the meltwater into its distributed finite difference flow model (section 6.3.2).

Both approaches involve linking together a series of submodels covering different elements, such as: (1) the meteorological input, including snow accumulation, interception and heat budget; (2) the snowmelt processes, including snowpack ripening, water accumulation and movement; (3) subsequent overland flow and throughflow; and (4) channel flow.

As with rainfall–runoff models, there is now a tendency for increased attention to process detail to be accompanied by greater concern for spatial variability, and for both to be embodied in distributed models. Nevertheless, a fully distributed approach is still rare, as it was at the time of the WMO (1986b) review of snowmelt runoff models. Semi-distributed models are more common, largely because of their simplified structure and lower data requirements. These normally divide the catchment into altitudinal zones.

Bloschl *et al.* (1991) presented a fully distributed model based on a 25 m mesh DTM in an Austrian Tirolian basin. Ranzi and Rosso (1991) developed a fully distributed model in the Italian Alps and demonstrated the great importance of spatial variability in snowpack melt. A major factor in this is the two-way relationship with shortwave flux. Slope aspect affects insolation receipts. But the net shortwave balance is also controlled by the albedo, which is affected by the wavelength and angle of incidence of the radiation, and by metamorphism of the surface snow crystals, which is partly a by-product of radiation receipts.

In many basins, an additional source of spatial variability is the redistribution of snow by the prevailing wind in accord with topography and vegetation cover (section 3.3.5). Even differences in sublimation rates on slopes of various aspects can add to the variability in Tianshan (Yang *et al.* 1991). The optimum grid spacing for a finite difference mesh will depend on the ruggedness of the basin and on the scales of spatial variability in the main hydrological processes (which may or may not be closely linked to surface topography).

One of the most crucial, yet difficult to model aspects, is the spatial depletion of the snowcovered area. Ferguson and Morris (1987) suggested that an extra areal depletion submodel is needed. Such a submodel would clearly have large data requirements, including a detailed topographic base, such as a DTM, as well as a map of vegetation cover type, and details of the spatial patterns of snow depth, density and layering. In practice, the longstanding general solution to this problem has been to monitor rather than model spatial depletion patterns. On the St John's River in New Brunswick satellite imagery has been used to monitor the transitory snowline and this information is combined with frequent ground measurements of water equivalent and snow depth at representative sites and fed into a snowmelt runoff model for flood warning purposes. Prior to the advent of satellites, aircraft were often used to determine the proportion of the catchment contributing snowmelt, particularly in the western USA, but this is costly and has been largely superseded.

Many of the fascinating microscale processes of snowmelt outlined in section 3.3.5 have been incorporated into physically based models. Unfortunately, the quality of data normally available for setting the initial conditions within such models probably does not merit the detailed modelling at present. The situation is better in the limited number of basins where highly detailed monitoring takes place. It is also continually changing as monitoring technology improves with, for example, the monitoring of snow density from the air or of snowfall and rainfall by radar during the melt period.

Akan (1984) described a model that simulates water movement within the snowpack in both liquid and vapour phases in the unsaturated zone and seepage through the saturated basal zone. Marsh and Woo (1984) modelled the migration of the wetting front through the snowpack, including Wankiewicz's (1979) fingering process.

One of the key fields for future progress lies in the search for successful 'parameterization' of detail at the microscale, i.e. in determining just how far models need to go in simulating the real physical processes and at what point these processes can be simply summarized by a parameter. Ranzi and Rosso (1991) suggest that some parameterization of topography, perhaps statistical descriptors of slopes and aspects, could be used to improve simple degree–day models.

6.5.4 Modelling glacier runoff

Glacier runoff is perhaps the most complex of all runoff processes to model. It usually includes elements of snowmelt and normal runoff processes as well as the element of icemelt. The conceptual glacier runoff model tested in the upper Indus basin by Ferguson (1985) distinguished between off-glacier and on-glacier snowmelt. In the long term, year to year differences in glacier mass balance can produce greater fluctuations in storage volumes than in 'normal' basins and current melt volumes can be influenced by historical accumulation rates. In contrast, the occasional sudden release of meltwater entrapped within or around the glacier can be problematic for short-term predictions.

Simple models may use empirical correlations between melt rate or runoff and air temperatures. Kang (1991) produced a better statistical model by using multiple regression and establishing different relationships for the glacier and ice-free areas. This showed that daily discharge from the glacierized area could be modelled from temperature and vapour pressure, whereas that from the ice-free area was correlated with precipitation.

The ideal physically based simulation models would give separate consideration to: (1) meltwater production based on energy budget modelling; and (2) routing the meltwater drainage over, through and beneath the glacier. However, the second of these presents major problems because of the hidden, varied and continually changing nature of englacial cavities. The englacial drainage network is three-dimensional, with both anastomosing and dendritic links, and evolves by a combination of thermal erosion (Shreve 1972) and the opening and closing of cracks due to glacier movement. It is therefore considerably more difficult to monitor and model than macropore networks in the soil, yet it is capable of transmitting far higher discharges and of generating sudden releases of water which are unrelated to the short-term heat balance.

In addition to the regular cycles of drainage of meltwater generated by the daily and seasonal heat budgets (Figure 6.20), the englacial drainage network can cause glacier outburst floods or **jökulhlaups**, by abruptly draining ice-dammed lakes alongside, above or within the glaciers through new cracks in the ice. These can happen in most major regions of valley glaciers from the High Arctic to the Alps. Hewitt (1982) estimates that 30 glaciers in the Karakoram Himalaya may generate jökulhlaups sporadically, chiefly during periods of glacial advance. They tend to occur irregularly, although the Gornersee in Switzerland

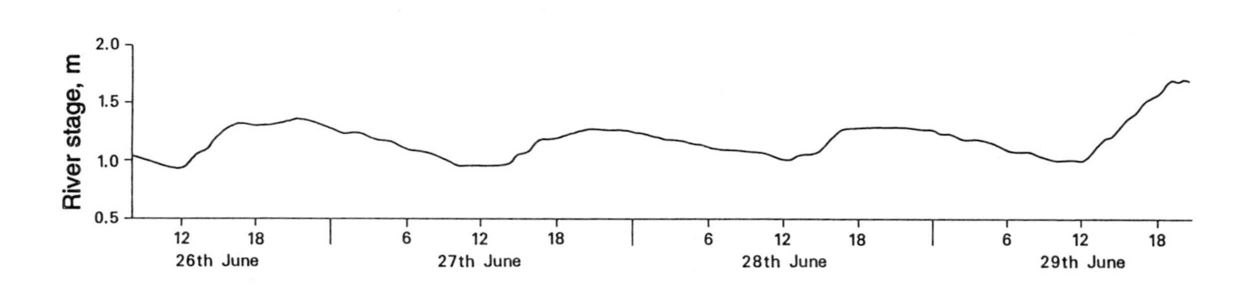

Figure 6.20 Diurnal fluctuations in summer discharge in a Himalayan river controlled by glacier melt. Measured at the highest gauging station in Tibet, *c.* 4500 m, in 1993

currently follows an annual cycle of drainage. The Lowell and Tweedsmuir Glaciers in British Columbia periodically advance at an accelerated rate in **glacier surges** and temporarily dam the Alsek River. Surging itself can have a variety of causes, from endogenetic factors related to glacier mass balance to exogenetic factors like earthquakes, and it can vary in pattern from quasi-regular to highly irregular and from frequent to very rare. To achieve a successful model of either surges or jökulhlaups requires a considerable amount of local data on the ice mass and its bed topography. Radio echo sounding has helped in collecting the necessary data.

Even so, despite the general recognition of the importance of englacial water and internal drainage that has emerged over the last two decades (Nye 1976; Hodge 1979), the scientific theory is still far from complete and is holding back the development of physically based models. Inability to verify theory by direct observation is as much of a hindrance as the great variety of physical circumstances encountered.

A valuable analysis of the problems was provided by the report of the IAHS Working Group on Prediction of Runoff from Glacierized Areas (Young 1985). In this report, Fountain and Tangborn (1985) analyse seven representative glacier runoff models. Only one of these, designed for the Vernagtferner Glacier in the Austrian Alps by Baker *et al.* (1982), attempts a direct assessment of the heat balance. Most use some version of the degree–day approach for melt calculations, although a number do divide the basin up into altitude/melt zones. Power (1985) describes the FLOCAST model developed by BC Hydro for the Columbia River, which includes 18 elevation zones for snow and ice melt. Melt rates are calculated from a simple degree–day formula, but sophistications include different melt rates for glacier ice and for on- and off-glacier snow, with on-glacier snow melting at half the rate because of the 'cold content' of the underlying glacier.

Internal drainage tends to be 'modelled' at a very crude level at present. Among the more elaborate treatments of internal drainage, the Vernagtferner model uses three parallel linear reservoirs for snow, ice and firn zones on the glacier. Lundquist (1982) described a model that used a set of two linear reservoirs representing the accumulation and ablation zones respectively connected in parallel and in series to provide a very simple routing for drainage. Other models may treat the glacier as a very thick snowpack or use lagged regression relationships between calculated melt and measured runoff as a simple 'blackbox' model for the throughput process.

6.6 Modelling urban runoff

As mankind develops ever more artificial environments, so planning needs transferable models for urban runoff (section 7.4).

The majority of predictive formulae have been developed to help in drainage design. Both the rational method and the unit hydrograph can be used and are probably more appropriately used within the city environment (sections 6.1.1 and 6.1.2). The **rational method** is commonly used to design small storm drainage systems covering up to 250 ha. The USGS pioneered its use, with the runoff coefficient based on the percentage of impervious cover (Carter 1961). Leopold (1968) compiled a graph for the USGS synthesizing the dual effects of paving and drainage (Figure 6.21a). The **SCS method** (section 6.1.1) can also be used with a **composite runoff curve number** based on the percentage of impervious area and the curve number for the pervious area. The UK Flood Studies Report found the crude term 'percentage covered by urban development' (URBAN in Box 6.2), without specifying the degree of permeability, was sufficiently specific for predicting *river* discharges from urban areas, within its broader multiple regression formulae (section 6.1.1).

For designing *drains*, however, the American Society of Civil Engineers (1970) elaborated upon the single index approach by differentiating between different types of urban cover and provided standard values for the runoff coefficient for each type (Table 6.1), so that the capacity of every drain segment can be designed according to the local cover type.

The USGS also extended the rational method in a number of ways that were subsequently taken up by the UKFSR. Anderson (1970) produced a formula for mean annual flood using a coefficient, K, based on the percentage of impermeable area, and lag time, T_{lag}, based on the **basin ratio**, the ratio between the length of main stream and the square root of its average slope (MSL and S1085 as used in UKFSR, section 6.1.1):

$$\bar{Q} = KA^{a}T_{lag}^{-b}$$

where A is basin area, and a and b are coefficients which he determined empirically. Anderson proposed that this mean annual flood could be converted into a flood of any return period by taking a regional analysis of return periods for rural catchments (as in UKFSR), but adapting it for urban areas according to the percentage of impervious area (Figure 6.19b).

Unit hydrograph techniques are commonly applied to storm sewer systems covering more than 250 ha (section 6.1.2). The peak discharge, Q_{pk}, of the unit hydrograph can be calculated by an empirical formula similar to Anderson's mean annual flood, but using the time to peak, T_{pk}, rather than lag time (Figure 6.10). T_{pk} can be estimated from the length, slope and roughness of the main stream (Viessman and Lewis 1996).

The UKFSR synthetic unit hydrograph determined T_{pk} by:

$$T_{pk} = 46.6\text{MSL}^{0.14}\text{S1085}^{-0.38}(1 + \text{URBAN})^{-1.99}\,\text{RSMD}^{-0.4}$$

(variables as defined in section 6.1.1). Subsequent work at the Institute of Hydrology (Packman 1979) showed that the effect of urbanization on both time-to-peak and mean annual flood can be related simply to urban area, for example:

$$\frac{T_{pk(urban)}}{T_{pk(rural)}} = (1 + \text{URBAN})^{-1.95}$$

Since the 1970s, however, the limitations of general black box models have been increasingly recognized. Wide deviations can be found in urban response and clearly not all paved areas are equally impermeable or storm sewer systems equally efficient. Dutch slab pavements set in sand have been shown to have infiltration capacities of 25 mm h^{-1} when dry and new, but this reduces considerably when they are wet and the joints have silted up. Some asphalt can even have a degree of permeability. Moreover, the geographical pattern of urban development and characteristics of the sewer network can have a major influence on runoff response. Instantaneous unit hydrograph models can be very sensitive to assumptions made about the distribution and rate of losses from the sewers.

A number of **simulation models** have been developed or adapted for urban areas. Ostensibly, these are capable of modelling the spatial variations in inputs and losses, and can be linked to routing programs (section 6.4) which use information on sewer layout, although even in the most developed countries such information can be sadly lacking. The US Environmental Protection Agency's **Storm Water Management Model (SWMM)** has been widely used. SWMM is a semi-distributed model, using lumped parameters for each subcatchment. It uses Horton's infiltration equation and standard hydraulic formulae, like the continuity and Manning equations, to calculate overland flow and route it through the sewers. This requires a considerable amount of data on the geometry and cover type for each urban subcatchment (Table 6.3).

The Canadian Inland Waters Directorate used SWMM to address the problem of defining 'zero urban impact' as a planning criterion for new building. Where urbanization has begun, lack of pre-urban data can make it very difficult to establish whether there has been an impact. Marsalek (1977) therefore used SWMM to simulate both the pre-urban and post-expansion responses by first calibrating the model on

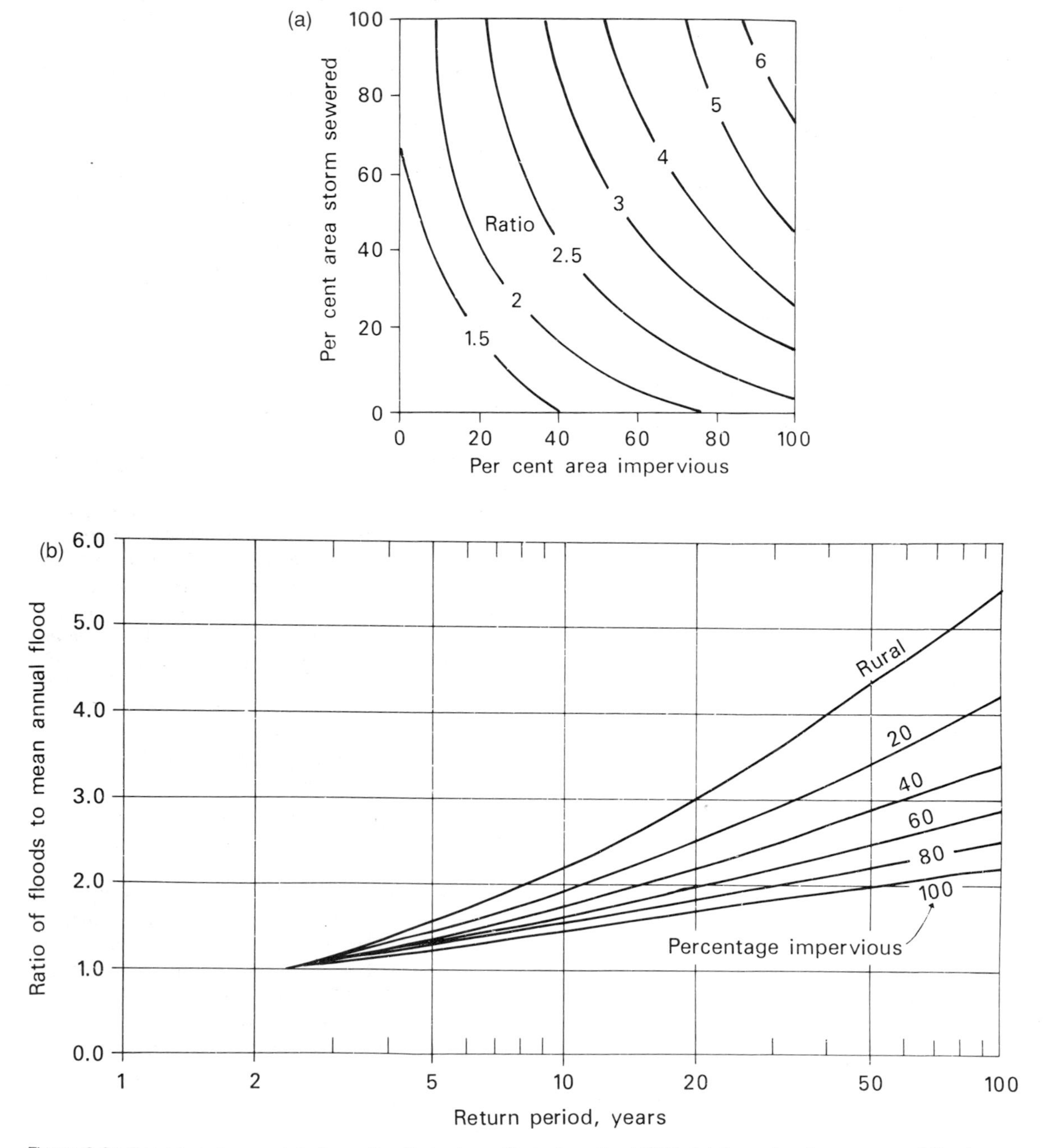

Figure 6.21 Two black box models for estimating urban effects from the USGS: (a) the ratio of mean annual flood after urbanization to that before development (Leopold 1968); (b) the effect of different percentages of impervious area on flood size (Anderson 1970)

the current urban situation and then changing the parameters that would be affected. The model was then used to calculate the volumes of detention storage that must be provided by artificial ponds in order to maintain peak discharges at or below pre-development levels (Figure 6.22).

This application highlighted some of the problems of legislating for environmental impact. Should the zero

Table 6.3 Data required by the EPA Storm Water Management Model

Type of data	Examples
Basin characteristics	Area, slope, detention storage depth, percentage impervious area, infiltration capacity, Manning's coefficient for overland flow
Gutter characteristics	Slope, roughness coefficients, shape, network layout, maximum allowable depths
Pipe characteristics	Slope, shape, Manning's coefficient, network layout
Inlet characteristics	Elevations, locations
Storage facilities	Capacity, location
Rainfall characteristics	Full history of rainfall (hyetograph)
Other data	Frequency of street cleaning, treatment devices, pattern of land-use types

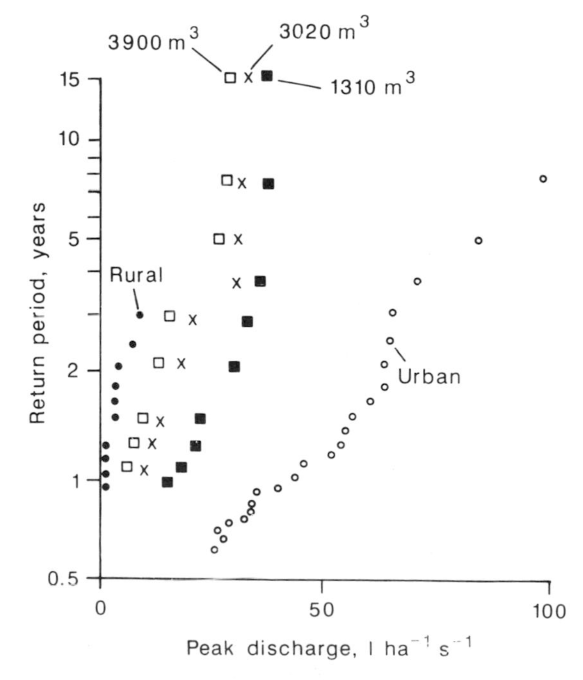

Figure 6.22 Example of SWMM used to determine the size of detention storage needed to negate the effect of urbanization in the Malvern catchment, Hamilton, Ontario, by Marsalek (1977). Three different sized reservoirs are assessed

impact apply to storms of all return periods or just to common events, like the mean annual flood or less? The former criterion is probably impossible to meet, witness the 1993 Mississippi floods (section 7.6), so legislation must be soundly based upon scientific advice as to both practicality and enforceability.

The UK **Road Research Laboratory model (RRL)** and the Illinois State Water Survey's modification of it, the **Illinois Urban Drainage Area Simulator (ILLUDAS)**, are semi-distributed models which use isochrone divisions rather than sub-basins (section 6.3). Whereas the original RRL model limited its concern to impervious areas that are linked to the central drainage system, ILLUDAS included the intervening permeable areas and therefore also the impermeable areas that are only indirectly linked via permeable areas during extreme events. Hence, the RRL model proved more unreliable in events with return periods of 20 years and over, when both saturation and Hortonian overland flow may issue from grass and soil surfaces within the urban area.

Both SWMM and ILLUDAS/RRL have proved endurable and generally reliable. SWMM has the best routing scheme and this has assured it prime position for many years, especially for large urban areas, where routing is an important element. However, as is usually the case with the more elaborate models, it requires considerably more computing power and data.

Conclusions

Considerable progress has been made over the past three-quarters of a century towards the scientific ideal of predicting streamflow based upon the understanding of runoff processes. This has brought physically based modelling firmly within the realms of application to practical problems such as predicting pollutant movement or the effects of land-use change. Much of this has been achieved in combination with detailed field investigations of the processes. It has been made possible by major advances in computer technology and techniques of data collection. Perhaps the most pressing questions being addressed by these modellers at the present time are: (1) questions of appropriate scales in measurement and calculation; and (2) judging the most important processes to model well. Beven and Binley (1992) have advocated keeping an open mind in modelling so that initial judgements as to the best set of parameters to incorporate within the model might be re-evaluated in the light of experience with the model, and they propose that this could be done objectively by mathematically assessing the likelihood of different sets of parameters giving the desired accuracy of prediction. This proposal can be linked with the broader question of 'equifinality' in both the real world and in models; it is often observed that different models or different sets of parameters can produce similar results. This is particularly true when the values of parameters have been established by mathematical optimization. Thus, different models or sets of parameters might equally well fit the real world within an agreed tolerance level. Indeed, Hata and Anderson (1990) point out that a model containing a large number of parameters is generally able to fit observed hydrographs, but not necessarily via the right processes.

The extent to which different combinations of processes in the real world can produce similar streamflow reponse is perhaps more of a moot point, but this probability certainly exists. Thus, the 'physical' modeller's art is one of continued convergence on reality.

This is not to belittle other forms of modelling, which have also undergone major advancement and which have generally proved their worth in addressing practical requirements to a much larger extent over past years. The simplicity of black box models can be extremely useful when data are lacking and a rapid answer is required. Because of the problems of quantifying spatial variability and subsurface properties within a catchment, a black box model is not necessarily an inferior predictive tool. Within the usual practical constraints for producing an Environmental Impact Statement for land-use change, for example, models such as the UK Flood Studies equations may provide a more realistic approach than a scientifically superior physically based model. Similarly, where long records exist of rainfall and/or runoff without data on catchment characteristics, stochastic models can provide the most cost-effective answers as well as covering some aspects which the standard physically based runoff model does not address, such as enabling probabilities to be attached to flow forecasts. Blackie and Eeles (1985) noted that stochastic models are more frequently used than lumped conceptual models in a short-term operational role where rapid decisions have to be made. Nevertheless, exceptions are not difficult to find, such as the Dee conceptual model (section 9.3) and the Dutch Rhine management system (section 10.2.2), in which conceptual models provide the basis for operational decisions. In fact, Blackie and Eeles (1985) quote tests that showed no significant difference between the results from stochastic models and a simple conceptual model. The choice of approach tends to be governed by considerations of data availablity and collection, and by cost–benefit decisions rather than purely scientific considerations. For some problems, a combination of approaches will be needed.

Nevertheless, in the realm of pollution control a physical process approach seems to be the most viable method (section 8.3.2). Fawcett *et al.* (1995) argue strongly for *internal validation* of distributed models to ensure that parameters remain within physically realistic bounds and that processes are correctly represented, that is, validating models on *process representation* rather than the traditional *prediction of catchment outflows*. Even so, validating subsurface processes remains particularly difficult: Wheater *et al.* (1993) suggest that different soil structures might require different models.

Further reading

Good discussions of the main issues can be found in:

Anderson M G and T P Burt (eds) 1985 *Hydrological forecasting*. Chichester, Wiley: 604pp.

Becker A and P Serban 1990 *Hydrological models for water-resources system design and operation*. Geneva, WMO Operational Hydrology Report No 34: 80pp.

Dawdy D R and T O'Donnell 1965 Mathematical models of catchment behavior. *Proceedings of American Society of Civil Engineers*, HY4, 91: 123-37.

Shaw E M 1994 *Hydrology in practice*, 3rd edn. London, Chapman and Hall: 569pp.

Wilson E M 1990 *Engineering hydrology*, 4th edn. London, Macmillan: 348pp.

For a comprehensive review of morphometric indices, see:

Gardiner V and C C Park 1978 Drainage basin morphometry – review and assessment. *Progress in Physical Geography*, 2(1): 1–35.

Discussion topics

1. What are the arguments for and against using physically based simulation models?
2. Why do we need to understand hydrological processes to produce a transportable model?
3. Outline the value and the limitations of using black box or stochastic models. Discuss examples.
4. Discuss the particular problems in modelling snowmelt, icemelt or urban runoff.
5. Consider the problems in legislating for zero hydrological impact.

CHAPTER 7

Inadvertent impacts on hydrological processes 1: water quantity

Topics covered

7.1 Vegetation change

7.2 Desertification

7.3 Irrigation and reservoirs

7.4 The effects of urbanization

7.5 Overexploitation of groundwater

7.6 Land drainage and channel modification

The role of humanity in altering the hydrological cycle may be minuscule compared with the range of natural climatic change over geological time. It has nevertheless proved critical in some areas of the world and it continues to pose significant threats to local and even regional water resources. There is also evidence that broader-scale changes have occurred and that we are now seeing a global acceleration of these effects.

Even before civilization began deliberately harnessing water, prehistoric Man probably had a significant indirect effect upon hydrological processes simply by altering the vegetation. Much of the tropical savanna grassland may have been created by hunters setting fire to woodland to clear the land for hunting from the Mesolithic period onwards. The vegetation cover of at least 20 per cent of the land area of the globe, amounting to 30 million km^2, has been drastically changed over the millennia by grazing and agriculture. In mid-latitudes alone, about 20 million km^2 of forest and forest-steppe have probably been cleared.

The process is accelerating today. Over the last 300 years, the world's forests have been devastated by European colonization and the development of world trade on the one hand, and by expanding native populations, agriculture and the search for fuelwood on the other. With the humid tropical forests now being cleared at the rate of 6 per cent per decade and only 0.6 per cent being reafforested, there is the very real fear that all 20 million km^2 of these forests could be lost in 200 years time (Mannion 1991).

The overwhelming effect of the destruction of natural vegetation, from deforestation to desertification, must have been to slow down the hydrological cycle by reducing evaporative losses. Conversely, over the last 100 years urbanization, industrialization and the expansion of irrigated agriculture have all tended to accelerate the cycle. Even if these conflicting trends were to cancel each other out in global terms, and perhaps we are now approaching this situation, the changes have wrought major shifts at local and regional scales, with not a few significant disasters along the way.

7.1 Vegetation change

The removal of wildland vegetation over the ages is likely to have affected the hydrological cycle through biological, thermal and physical effects. As trees and shrubs have been replaced with grassland and agricultural crops, the rates of interception and evapotranspiration have been reduced. *Potential evaporation* is likely to have been reduced further by the increase in surface albedo, and the resultant reduction in net radiation balances, as woodlands are cleared. Similarly, cultivation and improved land drainage will tend to reduce soil moisture levels and so further reduce *actual evaporation* losses. Kayane (1996) estimates that vegetation clearance in the humid tropics reduces actual evapotranspiration by over 400 mm a^{-1} and about half this in humid temperate climates.

Trees are also a significant water storage medium. The tropical rainforests not only evaporate 60 tonnes ha^{-1} of water daily, they also store 1350 tonnes of water for every hectare of forest.

All these effects are likely to have increased the rate of runoff, at least in the short term. In the longer term, however, the effect on runoff is less easily determined, because there could be a negative feedback on rainfall. If the forest clearance covers a broad enough area, then the lower evapotranspiration and net heat balances might eventually reduce precipitation downwind by reducing atmospheric moisture levels and convective cloud formation.

7.1.1 Experimental evidence of vegetation change

A considerable body of evidence on the effects of vegetation change has been collected from over 100 experimental catchments worldwide. Deforestation and afforestation have actually been the favourite topics for catchment experiments; so much so that critics once referred to a presumed hydrologists' syndrome of 'calibrate, cut and publish' – to measure the water balance for a number of years, to clearcut the trees, to monitor the subsequent hydrological changes for a few years and then to publish the results.

In practice, hydrologists have traditionally chosen either to monitor sequential changes within an individual basin or else to monitor two similar basins concurrently, in one of which the vegetation has been altered. Each of these approaches has its drawbacks as well as its advantages. The latter approach, which is known as the **paired catchment** strategy, can pose difficulties in finding two basins that are identical in every respect other than the vegetation cover. However, it does overcome the main criticism of the 'before-and-after' strategy in **sequential experiments**, in which some of the observed changes could be due to a parallel climatic change. In one case, the researcher needs to prove physical identity, in the other stationarity of climate.

Unfortunately, as Bosch and Hewlett (1982) discovered from their review of nearly three-quarters of a century of such experiments, it is extremely difficult to extrapolate from the results and to use them as a basis for quantitatively predicting the effects in another basin. The main reason for this, and perhaps the biggest scandal in hydrological science, is that so many of these experiments failed to investigate the hydrological processes involved.

Most have been content with a 'black box' view of the system, measuring precipitation input and runoff output and very little in between. Black box formulae have been used quite widely in the USA, based on Hibbert's (1967) collation of the results of these experiments. Yet the problem is really a classic case for developing a simulation model (section 6.3). Only a simulation model, like that described by Lockwood and Sellers (1983), based upon the key internal processes that are affected by the land-use change, has the potential to become a universal, transferable predictive tool. Ironically, after years of wasted effort, hydrological science is now approaching this ideal not so much from whole catchment experiments as from detailed investigations of the component processes. Fully distributed, physically based finite element simulation models like IHDM (section 6.3.2) now offer the best way forward.

The best of the catchment experiments do, nevertheless, permit a number of broad generalizations. Reviews by Hibbert (1967), Dunne and Leopold (1978), Bosch and Hewlett (1982) and Trimble *et al.* (1987) all show a general relationship between changes in the forested area and annual streamflow.

Removing the forest clearly increases overall runoff, although the proportional increase differs from one experiment to another (Figure 7.1). The paired catchment experiment run by the UK Institute of Hydrology in Wales showed that the grassland Wye basin lost only 18 per cent of its mean annual precipitation of 2415 mm through evapotranspiration, whereas the Severn basin, which is 80 per cent afforested by coniferous plantations, lost 30 per cent of 2388 mm a^{-1} (Clarke

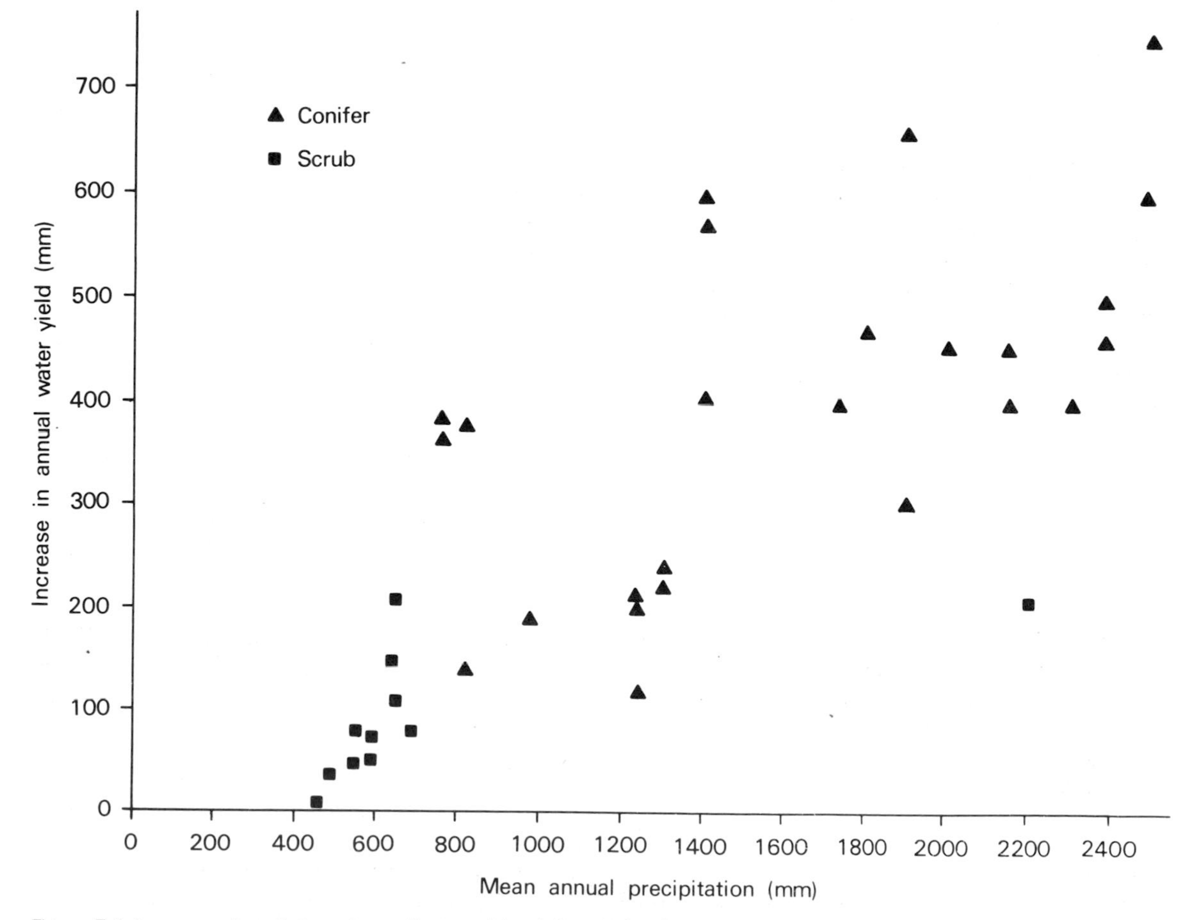

Figure 7.1 Increases in catchment runoff caused by deforestation in a range of rainfall regimes. Collated by Bosch and Hewlett (1982)

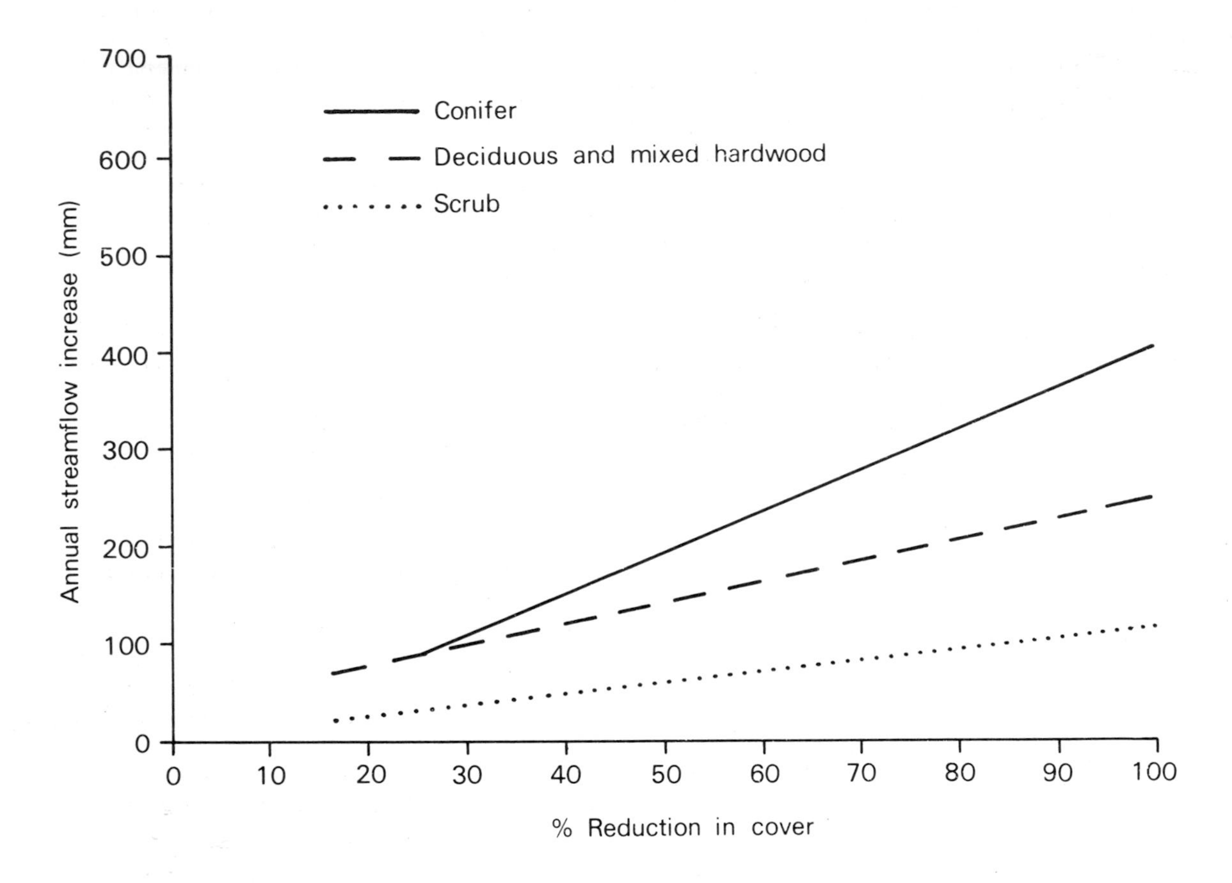

Figure 7.2 The effect of degree of forest reduction and tree type on water yields. Collated by Bosch and Hewlett (1982)

and Newson 1979; Blackie and Newson 1986). The Forestry Commission estimates a 2 per cent reduction in annual runoff for every 10 per cent of the catchment afforested. In the foothills of the Sierra Nevada in California, conversion of oak woodland to annual grass increased annual runoff by 114 mm and reduced evapotranspiration losses by 24 per cent from 500 to 380 mm a^{-1} (Lewis 1968). Increases of up to 200 per cent have been recorded in Alberta (Swanson *et al.* 1987).

It is possible to make a rough estimate of the effects according to the tree type involved (Figure 7.2). Felling of pine and eucalyptus forests shows an average increase in yield of 40 mm per 10 per cent reduction in forest cover, whereas deciduous hardwood shows an average rate of 25 mm and scrub just 10 mm per 10 per cent change in cover. The differences are largely due to interception rates and depth of rooting (section 3.1.2). Maximum contrasts occur in high rainfall areas, but more rapid regeneration of the vegetation in those environments also makes the effects shorter lived.

Removing the forest also tends to reduce concentration times, creating higher peak discharges and so increasing the risk of flooding. At the USDA experimental catchment at Coweeta in the Appalachian Mountains of North Carolina, clearfelling has increased runoff by 11 per cent and raised peak flows by 7 per cent (Hewlett and Helvey 1970). At Jamieson Creek, British Columbia, tree harvesting caused a 13 per cent increase in winter peak discharges (Henderson and Golding 1987). In Texas, Wood *et al.* (1989) report significant increases in sediment movement associated with higher peak discharges. In general, higher peak discharges are due to more overland flow from grass-covered basins, where infiltration and surface roughness tend to be lower than under forest (section 3.2).

Numerous reports from European colonists in the late nineteenth and early twentieth centuries confirm these effects, especially in America, Australia and New Zealand where clearance of the wildland vegetation was often followed by severe flash flooding and soil erosion. In New Zealand, the Soil Conservation and Rivers Control Council was set up in the 1940s to contain this joint problem. In some cases, increased flooding and soil erosion were followed by an overall reduction in water resources. In Victoria, deforestation in the 1870s was soon followed by increases in the number and discharge of springs, as a result of reduced evapotranspiration. Eventually, however, the uppermost springs began to dry up as the lower springs drained the water table. This caused problems for agriculture by aggravating the impact of extreme natural droughts, as in 1902. The increased throughflow also caused salts to be leached and deposited as **saltings** on the lower slopes below the spring outlets, destroying the protective tussock grass (Marker 1976). More recently, lumbering has so reduced evapotranspiration losses in parts of Western and South Australia that saline groundwater has risen into the soil cover, destroying agricultural land and increasing the salinity and sediment content of the runoff; runoff is increased, but water quality has been reduced to such a degree that it has become a hazard (Peck 1978).

Fortunately, replacing the forest can reverse many of these trends (Trimble *et al.* 1987). Successful rehabilitation of damaged catchments in non-saline areas of Australia has frequently involved planting with deep-rooted trees, like *Eucalyptus viridis* or *Pinus radiata*, and shrubs with high evapotranspiration rates in order to restore the water balance. This is supplemented by undersowing ryegrass and clover to stabilize the soil. Restoring soil structure with lime, organic matter or other conditioners can be equally important, not only through reducing erodibility but also because it increases infiltration and internal drainage and reduces the amount of overland flow.

The experimental evidence also shows that nature often begins the rehabilitation process itself, so that the increase in water yield decreases exponentially over the years and may even reverse eventually. The decline and reversal are mainly caused by regeneration of the vegetation cover, which increases losses and retards runoff. Data collated by Hibbert (1967) suggest that the rapid increase in yield has a 'half-life' of 2 to 7 years if natural regeneration is allowed. In certain circumstances, the reversal can come early on. Hibbert (1971) actually reported a case where yields were lower than under forest during the early stages of a seeded grass cover, because of excessive losses during the period of rapid growth in the fescue grass. After this initial period, the yield settled down to a net increase of 125 mm a^{-1}.

The most notable cases of 'abnormal' behaviour occur where snowfall is a major component of annual precipitation. Soviet observations often indicated an overall reduction in runoff after clearfelling. This can be explained by reduced snowcover retention and accumulation after the forest has been removed (section 3.3.5). The aptly named Fool Creek in Colorado, where 75 per cent of annual precipitation is snowfall, is a good example. Here, clearfelling of lodgepole pine and spruce–fir forest resulted in a 40 per cent reduction in yield, giving 100 mm a^{-1} less. Far less effect was found at Wagon Wheel Gap, Colorado, where just 50 per cent of annual precipitation is snow and felling resulted in a small increase, although after 4 years this amounted to only 13 mm a^{-1}.

In spite of the high degree of unpredictability, the general conclusions drawn from these catchment experiments have been put to practical use to increase water resources, particularly in the United States. Not surprisingly these have had mixed results, with some unfortunate and unplanned side-effects. Some cities successfully clearfell parts of their water-supply catchments on a regular rotational basis, in order to increase water yield but limit sediment production. In arid areas of the USA, particular concern has been directed towards **phreatophyte** vegetation, which survives on phreatic water and so depletes groundwater resources and baseflow. Phreatophytes cover 6.5 million ha of valley floor in the western USA, transpiring over 30 billion m^3 a^{-1}. In some basins they reduce yield by 20 per cent. Many projects have shown increased yield after clearing woody phreatophytes. Such programmes, however, can destroy wildlife habitats that offer crucial refuges in the drylands. Perhaps the worst cases of such destruction have been the removal of the phreatophytic piñon–juniper woodlands and cottonwood trees from riparian sites in Arizona. Dunne and Leopold (1978) described the destruction of the bankside cottonwoods as a 'major environmental degradation' on purely aesthetic grounds. More worrying still, neither clearance has been demonstrated to give a sustained increase in yields.

7.2 Desertification

The most extreme case of an atmosphere–vegetation feedback occurs in desertification. Desert presently covers 37 per cent of the land area of the globe, some 4.5 billion ha. But, according to the UNEP, this is 7 per cent more than would be sustained on climatological grounds alone: the difference is man-made. This is increasing by over 6 million ha per annum (UNEP 1977). By the millennium, mankind will have added an area twice the size of Canada to the world's deserts. Deforestation, overgrazing, overcultivation and poorly designed irrigation schemes have all contributed to desertification (Middleton 1991).

A key factor in desertification is destabilization of the topsoil. One-third of India's arable land is currently threatened with total loss of the topsoil. Even in the USA, 40 million ha of agricultural land has been damaged beyond practical repair. Desertification affects rich and poor countries alike. By 2000, one-third of all the land that has been used by agriculture will have either been totally lost or become drought-ridden.

With desertification, the reduction in atmospheric moisture may be accompanied by a more stable atmospheric boundary layer that inhibits the development of convective rainstorms. The American climatologist, Reid Bryson, has postulated that devegetation along the margins of the great natural subtropical deserts causes an artificial extension of the tropical inversion layer, which is formed by air descending within the subtropical high-pressure zones and spreading towards the equator. According to Bryson and Murray's (1977) hypothesis, large amounts of soil particles get entrained as the winds blow over the bare ground. Lifted aloft by the wind, this dust increases the atmospheric albedo, so cooling the air aloft, which causes the air to subside, heat up adiabatically and form an inversion, which acts as an effective 'lid' to convection and the vertical development of clouds. Figure 7.3 shows some of the complexity in the responses and feedbacks between vegetation, soils, atmosphere and surface hydrology.

By speeding up runoff, reducing soil moisture storage and inhibiting rainfall development, the processes combine to create a one-way drift towards a permanently drier environment. The Sahel seems to be a good example of this trend, although many question the extent of human input into the process (Nicholson 1989).

Bryson and Murray (1977) also give a graphic description of the case of the Rajputana Desert, on the borders of India and Pakistan. Here, atmospheric moisture levels remain high, but the extreme dustiness of the air effectively stifles convective raincloud development. Atmospheric moisture levels in the Rajputana are comparable with those of humid tropical forests, but it is the dustiest desert in the world. This was the land of the great Indus civilization between 2500 and 2000 BC, but the causes of its demise are still uncertain. Bryson speculates that a decline in the irrigation systems (for reasons unknown) may have initiated the cumulative process of desertification, which eventually caused the ancient cities to become buried in the drifting sand. Even so, the River Indus is still a major water resource today, and it is fed not by local rainfall but by the rains and meltwaters of the Himalayas; at least for the ancient city of Mohenjo-Daro, the critical factor may have been a shift in the course of the river away from the city and its supporting flood-irrigated land rather than a climatic change.

7.3 Irrigation and reservoirs

Artificially created water surfaces and elevated soil moisture levels stimulate evaporation by lowering the albedo, increasing net radiation receipts, and by allowing evaporation to take place at or near potential rates. The 'oasis' effect can add to these losses (section 5.3.1). Currently, 75–90 per cent of the 2000 km^3 of water used annually for irrigation is evaporated. Soviet hydrologists calculated that the increased evaporation from the area irrigated by the aborted Arctic river reversal scheme (section 9.4) would stimulate rainfall in the Asian Republics. But a country has to be large to get a significant return: Australia is too small (UNESCO 1978).

Fels and Keller (1973) estimated the overall effect of evaporation from reservoirs at about 10 per cent of that caused by irrigation, and the proportion is probably little changed now. Though small overall, it can be important in certain areas. Extremely large amounts used to be lost in Australia before more effective control techniques were introduced during the 1960s (section 10.3). Plans for the enlargement of Lake Chad capitalized upon the amelioration of the climate over a large area that would result from increased evaporation (section 9.4).

One of the great environmental failings of irrigation throughout history has been salinization of the soil and drainage return waters (section 1.2.3). There has been no more tragic case this century than the salinization of

Box 7.1 Case study: The Aral Sea crisis

The plan

The Rakitin and Litvak Plan aimed to increase the irrigated area to 10 million ha by 1990, and eventually to 23.5 million ha (equivalent to the total irrigated area in the CIS in the early 1990s). Most of this would be used to develop cotton production as a stratgic resource, but it also aimed to raise employment and food production. The plan actually envisaged the shrinking of the Sea and its possible demise by 1980, although this could be delayed by the river reversal scheme.

The results

In 1961 it was the fourth largest inland sea in the world, containing 1064 km^3 of water and covering 66 000 km^2. It was fed by 56 km^3 of annual riverflow from the mountains of the Hindu Kush, the Pamirs and the Tien Shan via the Amu Darya and Syr Darya. By 1975 only 7–11 km^3 was reaching the Aral. Inflows from the Syr ceased altogether from 1974 to 1986 and the Amu failed to reach the Sea for half the years during the 1980s (Figure 7.4). The Sea briefly divided in two in 1987; in 1989 this division was re-established, apparently as the future norm. The Sea had lost 69 per cent of its water by 1990 and barely covered 36 000 km^2. Yet it continues to lose almost 60 km^3 a year in evaporation and to receive virtually no inflows. By the mid-1990s the Lesser Aral may disappear and by 2005 the remainder is likely to be split into three lakes.

Over this period, the population in the Amu and Syr basins almost trebled to 30 million and the irrigated area and cotton production approximately doubled (Glazovsky 1990). The cotton uses 10 000–15 000 m^3 of water per hectare, but the rice crop needed to feed the burgeoning population uses 25 000–55 000 m^3 ha^{-1}. Over 10 000 m^3 of water is needed to produce 1 tonne of rice. At the same time, applications of mineral fertilizers have increased up to six-fold and large amounts have been washed into the rivers and groundwater, along with high concentrations of salts and pesticides. The permanent cotton monoculture encourages weeds and pests, which have been tackled by spraying chemicals at up to 50 kg ha^{-1}, 25 times the CIS norm. These chemicals, like the defoliant Agent Orange, are subsequently ingested by cattle and humans through food or drink.

The net result of this profligate use of resources and disregard for the environment has been desertification, salinization, declining agricultural fertility, decimation of fisheries and wildlife, dangerous levels of water pollution, death, disease, deformity and loss of livelihood (Saiko and Zonn 1994). Eighty per cent of the population do not receive a glass of clean water a day. For the people of Karakalpak around the old Amu delta the risk of contracting paratyphoid is 23 times higher than in the rest of the CIS. Gastric typhoid increased 20–fold in just 5 years in the Kzyl-Orda region east of the Sea. Dysentry is rife and infant mortality is as high as in Paraguay.

The people drink water and inhale air polluted with pesticides, herbicides and salt dust. Not surprisingly, three-quarters of the population suffer from illness. Banned chemicals like hexachloran are present in near-fatal concentrations in the lower reaches of the Amu and Syr. The former coast and bed of the Aral Sea have become a new desert, the Aral Kum, where dust storms have become up to three times more frequent over the past 25 years. During the 1980s the incidence of cardiovascular disease increased 1.6 times. Cancer of the oesophagus increased ten-fold. Tuberculosis doubled. The incidence of deformed babies has increased dramatically and general mortality has risen fifteen-fold.

A million hectares in the lower floodplains of the Amu and Syr have dried out, decimating the wetland reed beds and reducing production from 3–4 tonnes of reeds per hectare to 7–130 kg ha^{-1}. Productivity has also been falling in the irrigated area as a result of waterlogging and salinization. The gross yield of cotton is not increasing in Kazakhstan and Kirgizia despite expansion of the irrigated area. Production per hectare has fallen up to 40 per cent in Turkmenistan. Near the former Sea, atmospheric fallout is causing secondary salinization: the best estimates suggest that 40 000–150 000 tonnes of salt is blown from the former seabed on to the land every year (Glazovsky 1990), though some suggest it runs into millions of tonnes (Zarjevski 1987).

Wildlife has suffered similarly. Only 38 of the former 178 species survive. The fish have suffered most as the Sea's salinity has increased from 10 per cent to 24 per cent and both riverflow and seawater have accumulated a 'cocktail' of poisons. Bream, sturgeon, chub and pike-perch have all disappeared. Even on land, rootcrops contain 12 times the permitted level of DDT.

Box 7.2 Possible solutions to the Aral Sea crisis

Any rehabilitation is bound to be expensive. The Commission set up to 'save' the Aral calculated at least 35 billion roubles for engineering works alone. Total rehabilitation costs could reach 60 billion or more, some eight times the cost of the Chernobyl nuclear disaster.

The Commission proposed to divert the Syr to fill a small reservoir ringed by damwalls on the old seabed near Aralsk, accompanied by a similar project around the mouth of the Amu. According to the plan, these would restore fisheries, yielding 200 kg ha^{-1} of fish, stimulate the tourist industry and commercial exploitation of musquash for fur, and allow horticulture to be re-established. An alternative plan recognizes the futility of re-establishing the Sea and proposes that salt-tolerant tree species be planted as a 'green barrier' to stabilize the seabed.

In reality, the problems now facing the Aral Sea basin demonstrate the need for an integrated plan to cover the whole environment. Glazovsky (1990) discussed a variety of proposals, the best of which might be combined in a multi-pronged approach:

1. *Withdraw land from irrigation* 15 per cent of this land is in an 'extremely unsatisfactory state'. Ceasing irrigation on this land could save 15–20 km^3 of water a year and would be enough to be a valuable stimulus for ecological rehabilitation. This could be achieved by:
   (a) *Reducing rice production* – taking a million hectares out of production and importing the rice would save 3 km^3 of water and cost only 1.7 kopecks for each cubic metre saved.
   (b) *Reducing cotton production* – simply reducing the high (36 per cent) rejection rate for cloth could reduce requirements and save 10–15 km^3 in irrigation.
2. *Improve irrigation systems* These are mainly primitive and only 55–67 per cent efficient, whereas modern automated sprinkler technology could make them at least 88 per cent efficient and save 40–70 km^3 a^{-1}.
3. *Improve management of drainage waters* 46 km^3 of return waters currently drains into rivers, lakes and groundwater, adding to salination. This could be diverted directly to the Aral Sea or barren closed depressions, or else desalinated and reused. A scheme was begun in 1989 involving a 1500-km-long right bank collector canal to drain return waters from the mid-Amu irrigated region to the Sea, but it was poorly designed and most of the water is lost through evaporation or infiltration into the desert sand.
4. *Re-establish soil fertility naturally* This may be achieved by crop rotation and replacing high water demand crops by vegetables and grapes.
5. *Preserve the Aral Sea* 35 km^3 of extra water is urgently needed simply to stop further reduction in sea level. Schemes must determine the ecological requirements, including containment of salt blowout, and the technical limitations; several dozen schemes have been proposed.
6. *Inter-basin transfers* Proposals include:
   (a) *Transfer from the Caspian Sea to the Aral* – but Caspian water is saltier.
   (b) *Transfer from the Volga* – but this would reduce the input to the Caspian, and necessitate dams that would affect migratory fish.
   (c) *Transfer from Siberian rivers* – this seems more feasible, but fears have often been expressed that this would be seen as a panacea and slow down vitally necessary reforms in agriculture and water management (section 9.4). Even so, perhaps no more than 3–12 km^3 of this would finally reach the Aral.
7. *Rainmaking* The USSR State Committee for Hydrometeorology calculated that 25 km^3 could be created in Central Asia, but at a cost exceeding 100 million roubles. This could, however, be at the expense of snowfall on the glaciers, so that gains in rainfall would be partly offset by reduced meltwater (section 10.3.1). A related scheme, proposed in 1989 by Stepanov, envisages stimulating rainfall by increasing atmospheric moisture. These somewhat questionable calculations suggested that large reservoirs on the Ob and Yenisei storing 70 km^3 would increase the temperature of water entering the Kara Sea in the Arctic, so stimulating evaporation, much of which would follow meteorological trajectories into Central Asia.
8. *Intensifying glacier melting* Several cubic kilometres might be gained by methods outlined in section 10.4.2. But this could be at the expense of runoff later and cause ecological hazards.
9. *River regulation* Proposed by Lvovich, this is seen by many as the most promising. It might allow up to 30 km^3 to be retained and

Box 7.2 (cont.)

released in drier years. It would help irrigated agriculture, but not have much effect on the Aral itself, since only the temporal distribution rather than the total discharge is affected.

10. *Exploiting groundwater* Groundwater resources are several times greater than surface waters. Up to 10 km^3 of extra groundwater could be used without causing a reduction in riverflow, and even brackish water might be used. Ultimately, however, the danger of adding to salinization is ever-present and this is likely to be a short-term and ecologically unsound solution.

Most of these proposals involve rehabilitation work on a grand scale. Many are highly original, perhaps to the point of being questionable. Glazovsky said most 'suffer from giantism'. Unquestionably, however, if the problems are to be solved, plans *must* be laid on a large scale. The big questions now are whether the money is going to be available in the post-Soviet era, whether adequate amounts will be allotted to proper feasibility and EIA studies, and whether the now independent states can work together.

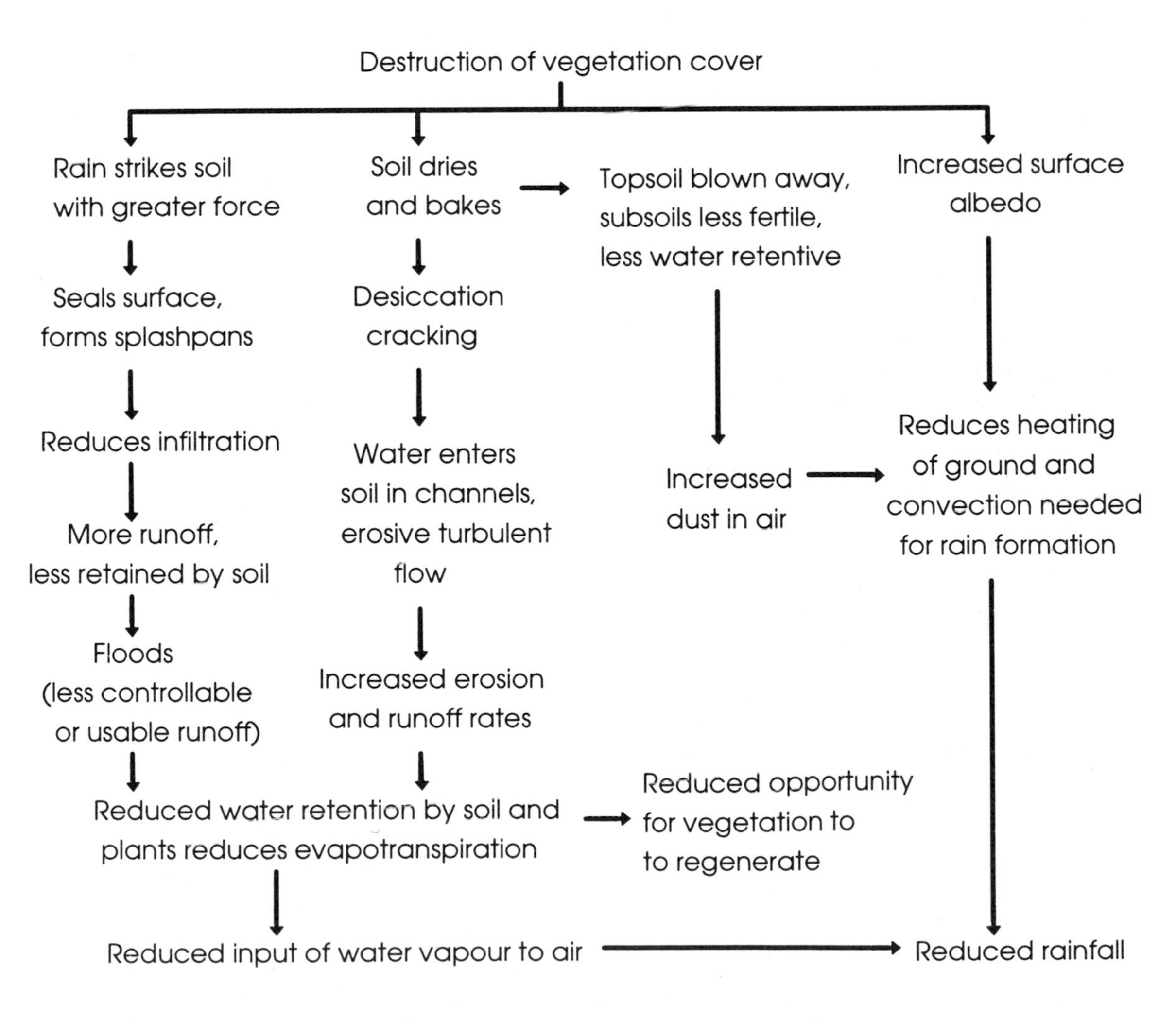

Figure 7.3 Feedbacks on the desertification process

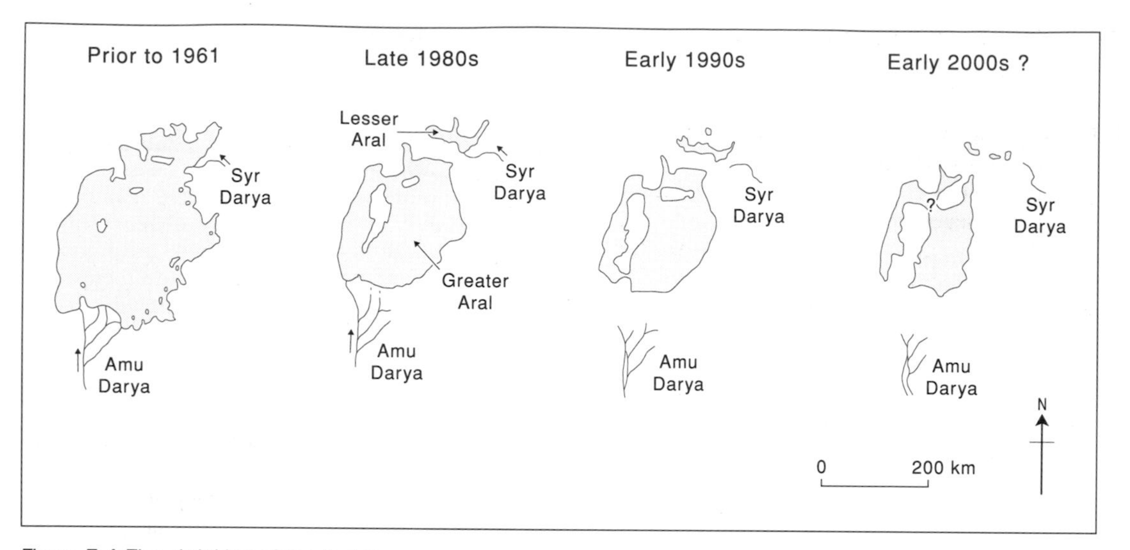

Figure 7.4 The shrinking of the Aral Sea

the irrigated lands of the Aral Sea basin combined with the drying up of the Sea and the rivers that used to feed it (Figure 7.4). The disaster is the result of a flawed 1960s Soviet plan involving inefficient irrigation practices, a damaging cotton monoculture and a supporting scheme to provide extra water from the Arctic rivers that was never implemented (Zarjevski 1987). In some respects, it is the most significant legacy of that grand river reversal scheme (section 9.4).

Because it is a large **internal drainage system**, i.e. a river system not connected to the world oceans, reducing the discharge entering the salt lake or 'sea' increases its salinity and reduces its size. The plan also failed to recognize the fragility of the dryland environment which depended on **allogenic water**, water derived from a different climatic region, in this case the mountains of Central Asia (Box 7.1).

7.4 The effects of urbanization

The spread of urbanization has been a major source of pressure on water resources in the twentieth century (section 1.2.1). It has also had marked effects upon hydrological processes. The potential effects on runoff were recognized in Ontario in the 1970s with the introduction of legislation to ensure that all new urban developments must be designed to have 'zero hydrological impact'.

Some of these effects are due to the modification of local channel networks and land-surface properties. Others arise from the high level of water consumption in urban areas and the need to collect and divert water, often over great distances, and then to return it to the river environment at just a few selected points. The provision of urban water supply typically removes water from rural environments, often in distant hills, and ejects it into major lowland rivers. The source and the destination of this water are often in different drainage basins. Occasionally, source, consumption and release may each be in a different basin. This can affect river regimes to such an extent that they become yet another human artefact (section 7.4.2).

In many respects, cities behave hydrologically like a forest plantation suffering from 'giantism'; their drains speed runoff, the alterations to surface roughness and the heat budget increase evaporation and rainfall, but unlike forests their 'roots' extend far outside the city cover to get their water. Again, the more intense modification of aerodynamic and thermal properties in urban areas creates rainfall that is more specifically associated with them. Urban effects on rainfall are a proven fact (Landsberg 1981), whereas the contention that forests increase rainfall has been the subject of heated scientific debate (Kittredge 1948; Geiger 1957).

The effects on land-surface cover also extend far beyond the urban area, because, as the German economist von Thünen recognized in his theory of agricultural location (Hall 1966), the city support system requires and encourages certain patterns of agricultural activity around. These have their own hydrological implications (sections 7.1 and 7.6).

7.4.1 Effects on precipitation and evaporation

Effects on precipitation and evaporation arise from three main sources: (1) the thermal, radiational and aerodynamic properties of city surfaces; (2) thermal, gaseous and particulate air pollution; and (3) urban drainage. These combine to cause the **urban heat island**, whereby cities tend to become warmer than the surrounding countryside, particularly in late afternoon and at night (Oke 1992).

This is created by the release of heat from artificial sources and the effects of the geometry of the built environment, which traps solar radiation and conserves heat, like the **urban canyon effect** and the **sky-view factor**, and increased **aerodynamic roughness**, which reduces windspeeds (Figure 7.5). The aerodynamic roughness may increase vertical updraughts over the city, akin to the orographic effect, which can trigger conditional instability (section 2.3.2; Atkinson 1979). The differences are strengthened at night by divergence between city and rural surfaces in their pattern of radiational heat losses. The surface air in the countryside cools more and tends to develop a temperature inversion, while in the city artificial heating and the delayed release of entrapped heat allows convective activity to continue. Urban drainage may supplement these effects by creating drier surfaces and therefore less consumption of latent heat.

Precipitation Landsberg (1981) quotes an average increase in precipitation of 5–10 per cent in and around urban areas. Rainfall tends to increase more than snowfall in the warmer city. However, patterns are very variable and average figures are not particularly meaningful, partly because only a limited number of reliable studies have been undertaken within a restricted range of climates. In each instance, proving the urban effect can be difficult. Most investigations are based on comparing otherwise similar rural and urban sites, but there can be problems proving the rural site is truly unaffected or that the urban measurements truly represent the magnitude of the effects. There are also few opportunities for studying the dynamic aspects of urban growth, although interesting correlations have been detected between rainfall and urban expansion in Manchester and Bombay.

The spatial patterns of urban-induced precipitation are complex and they can shift in different synoptic situations. Only a very small number of investigations have been able to afford to establish a spatial network of rainfall sites within the urban area and its surroundings.

The METROMEX experiment and the case of La Porte, Indiana The Metropolitan Meteorological Experiment (METROMEX) was undertaken around St Louis, Missouri, in the 1970s as the ultimate test of the hypothesis of urban-induced rainfall. Evidence from raingauge studies in Europe and North America had been accumulating slowly for decades, without any serious attempt to study either the processes involved or the wider spatial patterns. Most suggested a difference of a few per cent.

Suddenly and quite unexpectedly, data collected by an amateur observer in the small town of La Porte, Indiana, which seemed to show a disproportionately large effect, were published in the 1960s. Rainfall appeared to have increased dramatically since 1933, so that by the mid-1960s La Porte was recording 1270 mm a^{-1} compared with only 914 mm a^{-1} in the surrounding countryside (Figure 7.6). Increased cloud over the town visible on TIROS imagery seemed to offer some corroboration, but a detailed investigation was called for.

The National Center for Atmospheric Research (NCAR) from Boulder, Colorado, flew aircraft through the storm clouds and collected freezing nuclei emanating from the smoke plumes of the Gary steel mills about 60 km upwind on the eastern edge of Chicago. Detailed statistical analyses showed that between 1955 and 1965 La Porte had 31 per cent more precipitation, 38 per cent more thunderstorms and a remarkable 246 per cent more hail-days than the surrounding countryside (Changnon 1968). One-fifth of the thunderstorms occurred only in La Porte, often between midnight and 6 a.m., with no similar storms recorded within 100 km.

The evidence from La Porte clearly indicated that the effects owed little to the town itself. The increasing rainstorm activity was correlated with the expansion of the Gary industrial complex and air pollution in the Chicago region. Moreover, by the time of the investigations the effect seemed to be declining rapidly. This was subsequently found to be due to a shift in the location of peak precipitation out to the southwest of the town. It seems that the development of the precipitation peak over the town was largely coincidental. It was really a delayed product of particulate, and perhaps thermal, pollution from Chicago–Gary, probably further enhanced by the geographic location in which cool air from Lake Michigan meets warm, moist air from the Mississippi

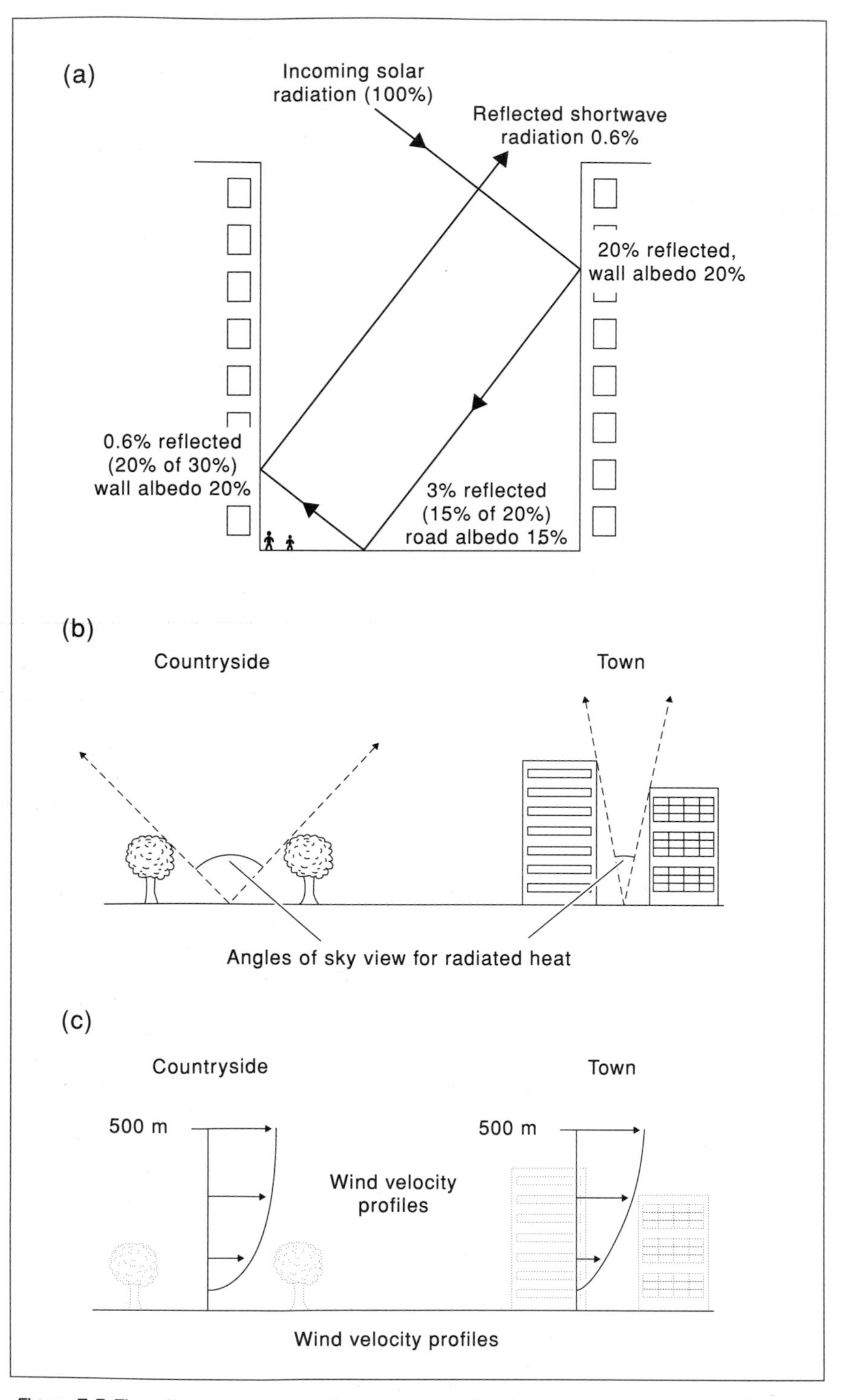

Figure 7.5 The effects of urban surfaces on microclimate: (a) the urban canyon effect traps more solar radiation than suggested by the albedos of individual surfaces; (b) the lack of 'sky-view' traps heat radiated from urban surfaces; (c) rougher surfaces reduce windspeeds

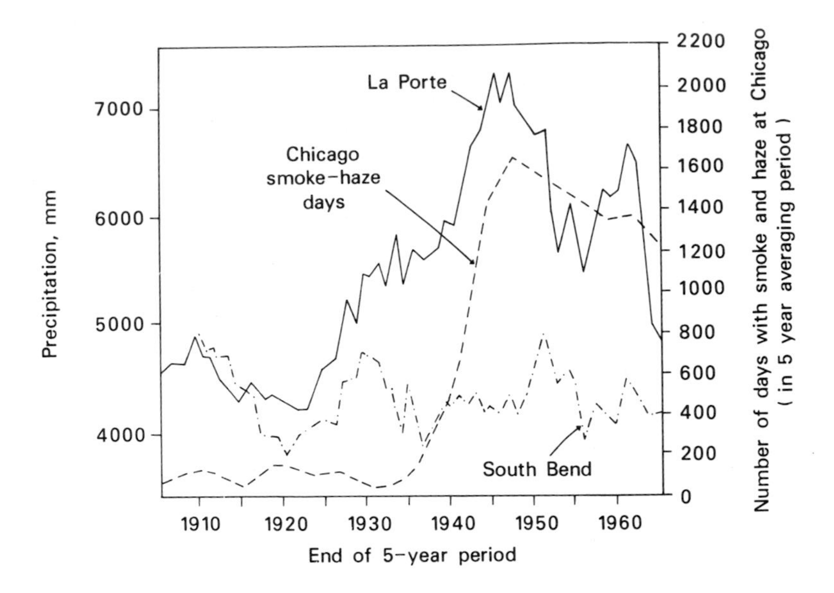

Figure 7.6 Increasing rainfall at La Porte, Indiana, associated with air pollution from Chicago–Gary (as five-year totals)

basin. Slight changes in the contrasts between these airstreams, in the location of the polar front between them and in the trajectories of the frontal depressions can alter the average location of the precipitation peak.

The METROMEX project was therefore designed to provide a full spatio-temporal coverage around a large city in a less sensitive geographical position. It ran from 1970 to 1975 and covered a 5200 km^2 area. It involved 225 raingauges, three weather radars, a lidar (to monitor atmospheric particulate pollution), four aircraft for sampling nuclei and numerous balloon stations for measuring temperature, humidity and wind profiles (Huff 1977).

The observations showed a 10–30 per cent increase in precipitation within and immediately downwind of St Louis, with strongest effects during the summer, May–August, and when the surrounding natural precipitation was moderate to heavy. Storms of 25 mm or more showed a 70 per cent enhancement over the city. The urban effect was greater than the natural orographic effect of the Ozark Plateau (Figure 7.7).

The urban effect was therefore shown to be most marked when it was augmenting already well-advanced natural processes, as in artificial rainmaking (section 10.3.1). The greatest differences occurred during the afternoon, as the build-up of the urban heat island supplemented natural convection, and in cold fronts, where the city may have enhanced vertical updraughts. However, the radar did detect some entirely new rain cells developing over the city area, especially during the afternoon, and there was a significant increase in the number of very localized, short but heavy storms of 5 minutes to 2 hours in duration. Rainfall was also greater on weekdays than at weekends.

Some of the extra rainfall could be linked to aircraft observations that showed there were sufficient extra giant condensation nuclei in the air to initiate the warm-cloud rainfall process. Aircraft observations downwind of other American cities have revealed enhanced concentrations of freezing nuclei, which have led to thunder, hail and brief snowstorms (Schaefer 1969). But there may be some evidence that rainfall can also be suppressed by unusually high levels of atmospheric pollution. In the Detroit–Windsor conurbation, precipitation is actually less than in the rural environs during the main pollution period in autumn and winter, but reverts to the 'normal' urban pattern in summer when the urban surface effects become dominant. This may be because of too many competing nuclei (section 10.3.1).

Evidence from European cities Observations in London confirm an increase in the severity of convective storms over the city, especially in thunderstorms (Atkinson 1979). Weekly cycles have been found in Rochdale and London with most rain on Thursdays and least on Sundays, following cycles in air pollution (Nicholson 1969).

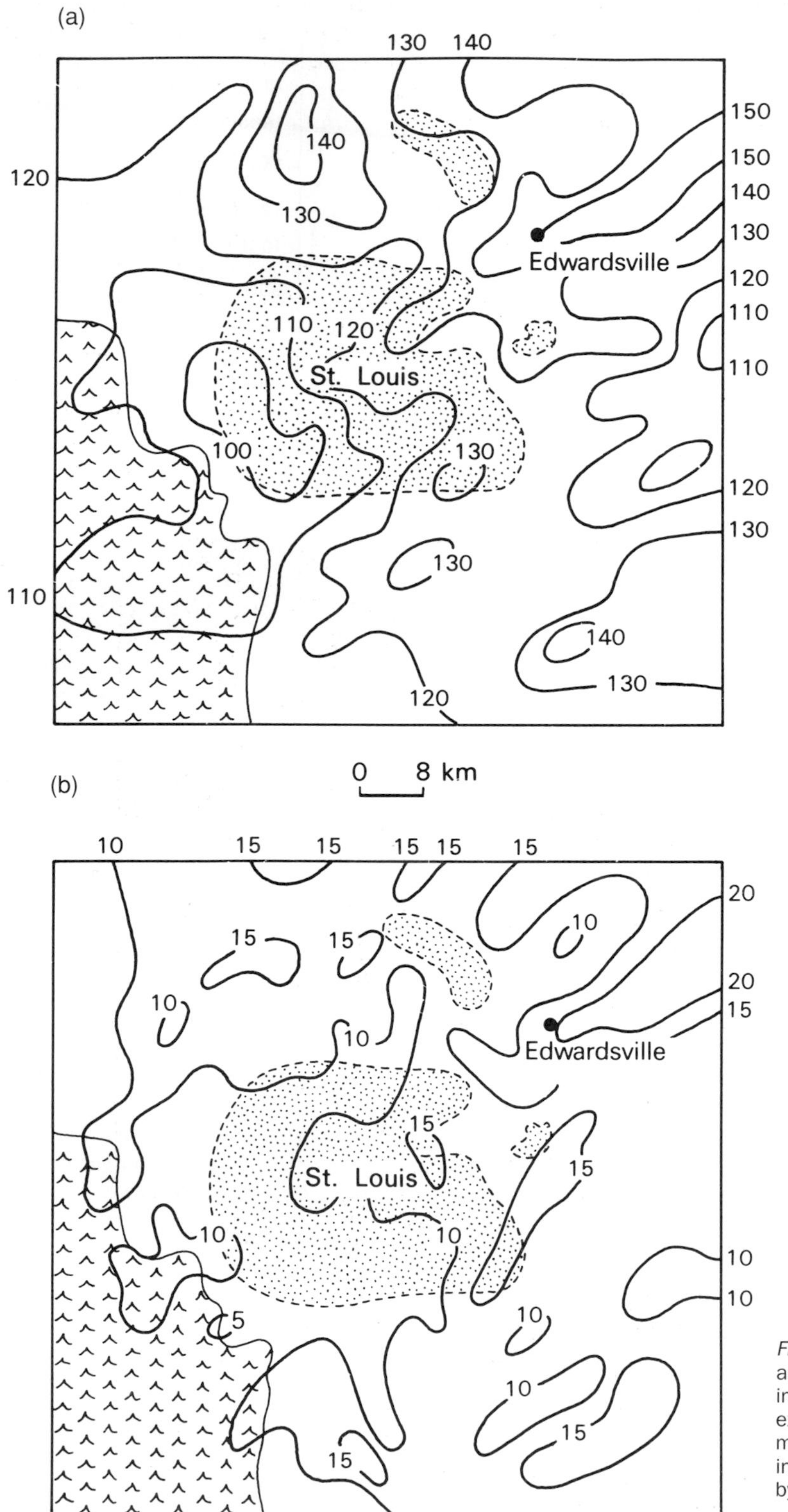

Figure 7.7 Increased rainfall around St Louis, Missouri, indicated by the METROMEX experiment. (a) increase in mm a^{-1}; (b) percentage increase. Ozark Plateau marked by hill symbols. After Huff (1977)

Locally generated rain cells are also in evidence. Munich has 11 per cent more raindays than its environs. Other cities average 6–7 per cent. Nuremberg experiences 14 per cent more thunderstorms than its surrounds. However, work in Holland has identified what amounts to **urban rainshadows**. Figure 7.8 shows a marked rainshadow downwind of Rotterdam balancing the increased precipitation over the city. This is presumably for the same reason that mountains have rainshadows, that the interference is redistributing rather than creating rainfall.

One consequence of localized urban convective storms is an increase in the spatial variability of rainfall, with obvious implications for accurate rainfall measurement (section 5.2). This variability is compounded by interception from tall buildings. The Buildings Research Centre in Britain found that interception at 100 m is twice that at 10 m. This can produce very localized peaks of runoff for storm sewers to evacuate around tall buildings. In Glasgow, which is probably the worst city in Britain for driving rain, discharges of *c.* 0.5 l m^{-2} can occur annually from the walls of high-rise buildings.

Evaporation The effects of urbanization on evaporation rates are less clearly established and may be more variable than on precipitation. On the one hand, city surfaces tend to be better drained and windspeeds at least 20 per cent lower. On the other, the heat island provides surplus energy and creates a greater vapour-holding capacity. The net result seems to be that **potential evapotranspiration** is higher by 5–20 per cent in cities, but **actual evapotranspiration** is somewhat lower. Consequently, **relative humidities**

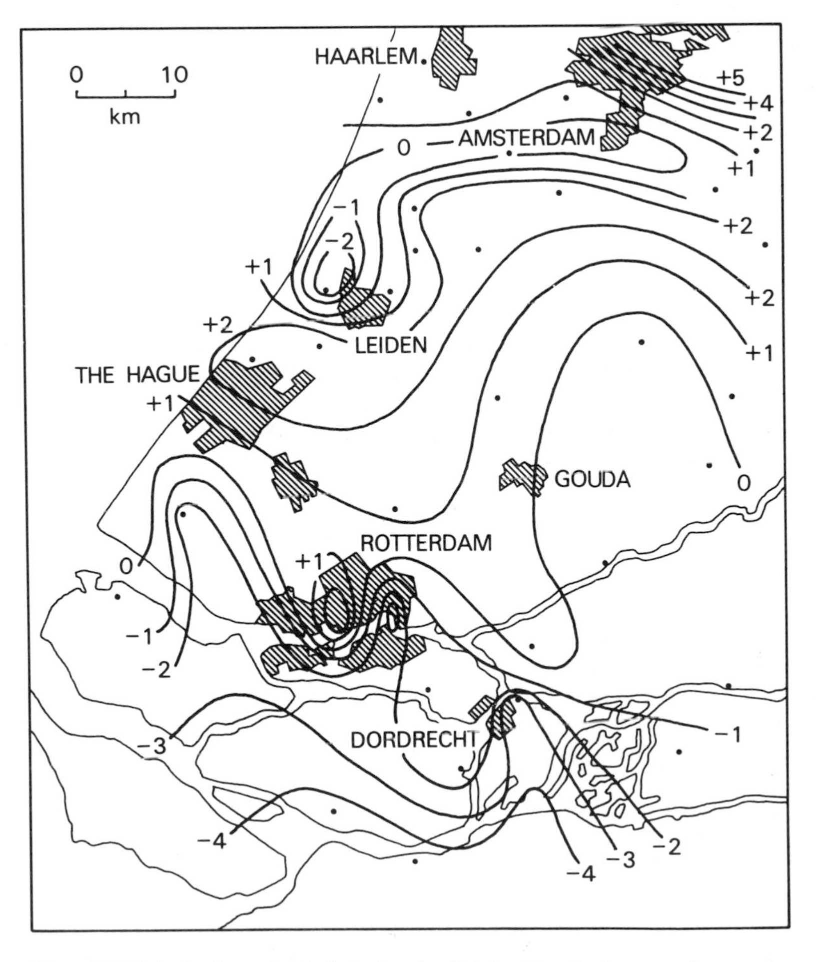

Figure 7.8 Rainshadows downwind of major Dutch cities. Isolines are in standard deviations. After Yperlaan (1977)

tend to be 8–10 per cent lower throughout the day during the season of peak evaporation in summer. In winter, the difference can be minimal or even reversed, perhaps as a result of artificial heating and some water vapour from combustion. **Absolute humidities** show a clearer tendency to be higher than in the country during the winter for similar reasons, but in summer they are also higher at night, when the heat island allows more evaporation. Condensation nuclei from air pollution may also cause more frequent winter fogs.

The overall effects on actual evapotranspiration are not as marked as once supposed, partly because of artificial water surfaces like canals and lakes, and partly because of the area covered by parks and gardens, especially when lawn sprinklers provide 'urban irrigation'. Even small wet areas have a micro-oasis effect.

7.4.2 Urban runoff

The greatest modification in hydrological processes occurs in runoff as a result of paved or low-infiltration surfaces and artificial drainage. The result is swifter velocities and flood peaks perhaps two to eight times higher (Anderson 1970) and 10 times earlier (Kuprianov 1977). Total runoff volumes may be increased 2–2.5 times. Lvovich estimated that every 1 per cent increase in urban cover is amplified to become a 2–4 per cent increase in runoff.

Observations of suburbanization and new town developments in Britain have confirmed this sensitivity. By the time housing had covered 12 per cent of a drainage basin near Exeter, peak discharges had increased two- to three-fold, with higher percentages of rainfall entering the stream as storm runoff and higher total volumes of quickflow (Gregory 1974). At Harlow New Town, Essex, Hollis (1977, 1979) found that not all storms were equally affected. For smaller floodflows there was an increase in maximum monthly discharge, accompanied by a decrease in lag time, time of rise and the duration of the storm hydrograph. Some summer stormflows were 12 times larger. But there was no increase in very large floods with return periods of 20 years or more. Greater increases in the magnitude and frequency of stormflows with shorter return periods were also found at Skelmersdale New Town, Lancashire (Knight 1979). This may well be because rural basins tend to be similarly impermeable in more extreme events because they are wet, so the rural–urban contrast is less. Extensive saturation overland flow can produce a very similar hydrograph to an urban drainage system, especially in basins with a clay soil as at Harlow. This could also explain the seasonal contrasts in response, with greater effects on summer storm volumes at both Harlow and Exeter, because winter storms would have fallen on wetter catchments prior to development. Even so, winter runoff coefficients at Exeter were higher than before development. Seasonal contrasts can be more marked where winter snowfall is important (see below).

Effects on low flows have received less attention, yet they appear to be significant in a number of cities. Around Tokyo, baseflow has been markedly reduced in the smaller rivers by overabstraction of groundwater. Similar effects may be found in Bangkok, southeast England and throughout Denmark (section 7.5).

An important distinction must be made between the effects on the runoff derived from the urban area and the net effects on riverflow downstream. The latter will include the effects of the release or non-release of wastewater fed from local or distant sources, which can have a major effect on river regimes. In the Dutch province of Zeeland, where 10.5 per cent of the area is urbanized and 13 per cent of runoff is derived from urban areas, total river discharge has decreased by 10 per cent and groundwater levels have been falling for decades because most of the wastewater is output directly to the sea. In contrast, Figure 7.9 shows that the daily regime of the River Tame downstream of the main Birmingham sewage works is largely determined not by rainfall but by the daily round of human activity. At Minsk in Byelorussia, mean monthly minimum discharges appear to be double the regional average below the city's effluent outfall in summer, when water consumption is highest, but only 25 per cent higher in the winter. This compares with a 50 per cent increase in average discharge. The extra water is derived from an aquifer source which is not hydraulically connected to the river.

Very few studies have tried to integrate the urban effects on both precipitation and runoff. Lvovich and his colleagues attempted this using the existing hydrometric networks for Moscow and Kursk (Lvovich and Chernishov 1977). Interestingly, they found virtually no overall increase in precipitation. In Moscow, 10 per cent increases in the eastern and western suburbs were balanced by reductions elsewhere, whereas at Kursk there was an overall negative effect. However, both cities displayed markedly more surface runoff. In Kursk, summer rainstorms which were incapable of producing runoff in the countryside could produce runoff within the city area. In Moscow, there was a four-fold increase in surface runoff in the central area

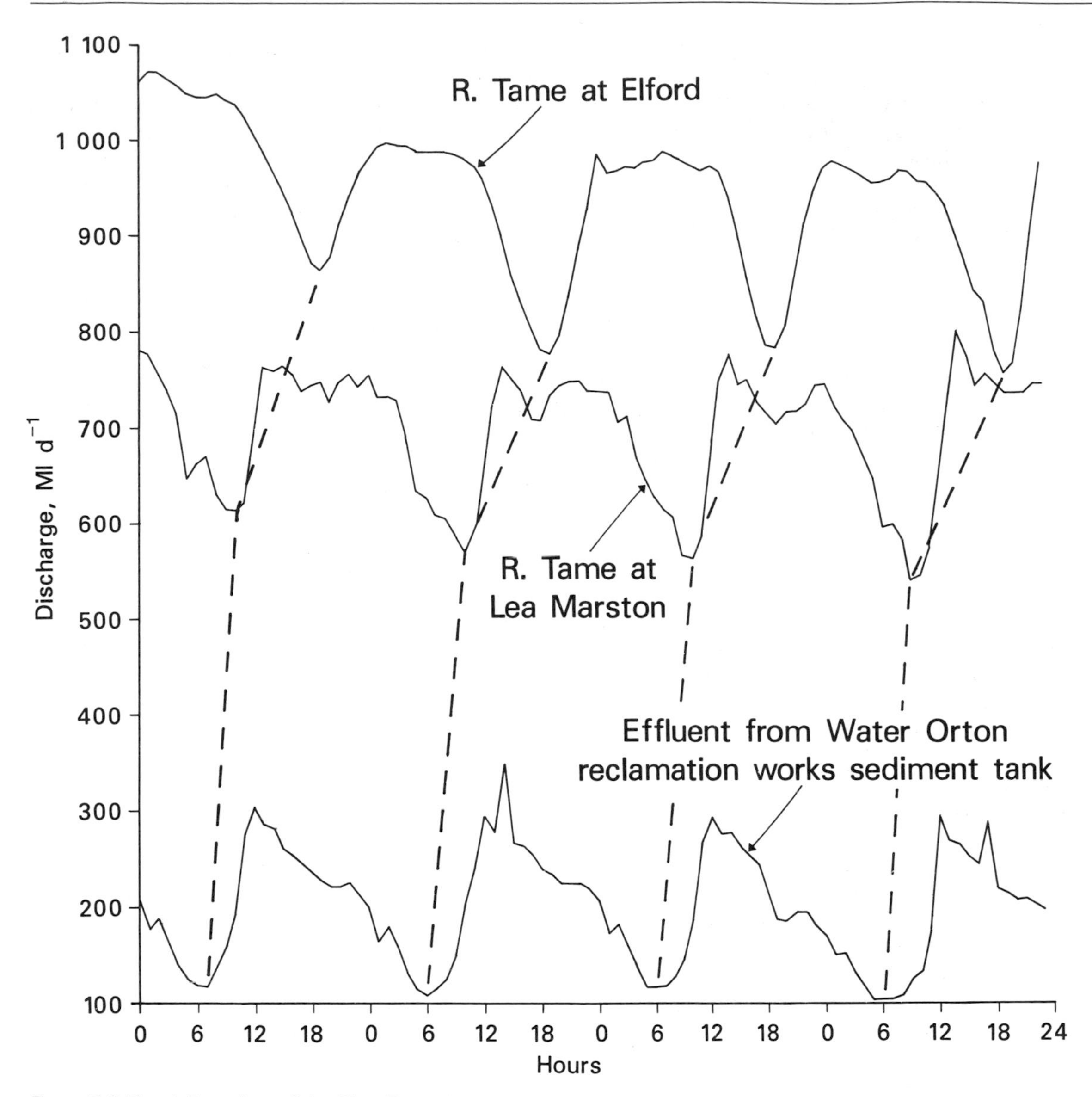

Figure 7.9 The daily regime of the River Tame downstream of Birmingham contains so much effluent it follows the human diurnal cycle: a river truly tamed! See Figure 8.19 for locations

and a two-fold increase in the metropolitan area as a whole. In other words, the area of increased runoff was not correlated with the area of increased precipitation; it was more closely related to the amount of impermeable cover and perhaps the density of drains. The Minsk study suggested that 10 per cent of the increase in average river discharge was due to increased urban precipitation (Kuprianov 1977).

Even so, the Moscow study showed that despite the high increase in surface runoff within the urban area, river discharge downstream increased by only 50 per cent. This was due to overexploitation of groundwater in an aquifer that is hydraulically connected to the river.

Snowmelt in cities Snowcover is generally reduced in duration and amount, despite a 5 per cent increase in snowfall. This is because more snow melts on contact with the warmer city surfaces, snowpack temperatures are higher and because of clearing operations. In Lund,

Sweden, nearly 40 per cent fewer snowcovered roofs have been recorded in the centre of town compared with the outskirts.

This should cause fewer snowmelt floods in urban areas, but clearance operations can alter the trend. Where snow is only a minor problem, salting or albedo changes due to dust or partial clearing will assist the trend. But where snowfall is a major feature, both the duration and the volume of snowmelt runoff can be increased. In Russia, snow is transported to snow dumps, which have commonly been in river channels. In Moscow individual dumps may contain 100 000 m^3 of water. Because they are located in river channels, the runoff coefficient is close to unity. Even outside the river channels, paved surfaces, sewers and compacted ground result in lower infiltration losses than in the countryside. Set against low losses, melting from compacted snow dumps tends to be slower; in one Russian example it took a month to clear compared with one week in the country.

In Canada, observations made by Taylor and Roth (1979) during the development of a suburb of Peterborough, Ontario, showed that the most marked of all urban effects occurred during snowmelt floods. As the built-up area grew from zero to 25 per cent of the basin over 5 years, snowmelt response was tripled. The built-up area had increased the contributing area for snowmelt quickflow from 5 per cent of the basin to 30 per cent. But it had little effect on contributing areas during summer rainstorms because these were largely restricted to the riparian area, which had not been developed.

7.5 Overexploitation of groundwater

An ever-increasing number of aquifers are currently being exploited at abstraction rates which exceed the rate of recharge. Much of the Arabic Region is suffering from this **groundwater mining**, utilizing resources that are not being replenished in the present-day climate (section 2.1.3). Falling water tables are contributing to the expansion of desertification, which begins with individual zones of impoverished vegetation above the cone of depression around each well (section 5.5.2) which subsequently expand and coalesce as the cones join. In Tokyo and Bangkok, falling water tables have contributed to land subsidence. Where the exploited aquifers are hydraulically linked to the rivers, groundwater abstraction can also be self-defeating, because it leads to a parallel reduction in surface water resources, as in Moscow (section 7.4.2).

Many cities in the southwestern USA are currently afflicted by dramatic falls in phreatic levels. In Dallas–Fort Worth the water table has fallen 120 m since the 1960s. Till the Central Arizona Project brought water from the Colorado in the early 1990s, Tucson was the largest US city dependent on groundwater. Local rainfall recharges only 35 per cent of the water currently abstracted, and the overdraft has caused groundwater levels to fall by 50 m.

There is major concern for the whole of the Ogallala aquifer, which extends from Nebraska to Texas, where levels are falling by 1 m a^{-1} in many areas, with only 1 mm a^{-1} in recharge, largely due to overexploitation for irrigation in the Great Plains (Figure 7.10). This is a classic case of mining an aquifer that may have taken 25 000 years to fill. Concerns have been expressed that the groundwater will be totally depleted by the end of the century and that irrigated agriculture will be forced into a rapid decline. Similar overdrafts are occurring in other major areas of irrigated agriculture. In parts of northern China levels are falling at rates similar to the worst areas of the Great Plains. In parts of south India levels have fallen up to 30 m in a decade.

Even supposedly well-watered England appears to be suffering from groundwater overdraft (Box 7.3).

7.6 Land drainage and channel modification

Runoff rates have been affected significantly over most of the civilized world by deliberate and inadvertent alterations to river channel geometry, drainage networks and the drainage properties of the land. Channels have mainly been straightened, cleared, dammed or deepened for navigation, water supply or as a flood control measure. Even deliberate flood control engineering can have unexpected adverse effects on river flows (Box 7.4).

7.6.1 Land drainage

Agriculture and forestry have affected hydrological processes through artificial drainage carried out to aerate the soil and improve crop growth. Agricultural land drainage has expanded markedly since the Agricultural Revolution of the eighteenth century. Early techniques used open ditches, ditches covered by peat sods, brushwood land drains or simply 'landing' the fields with a ridge-and-furrow pattern. These techniques

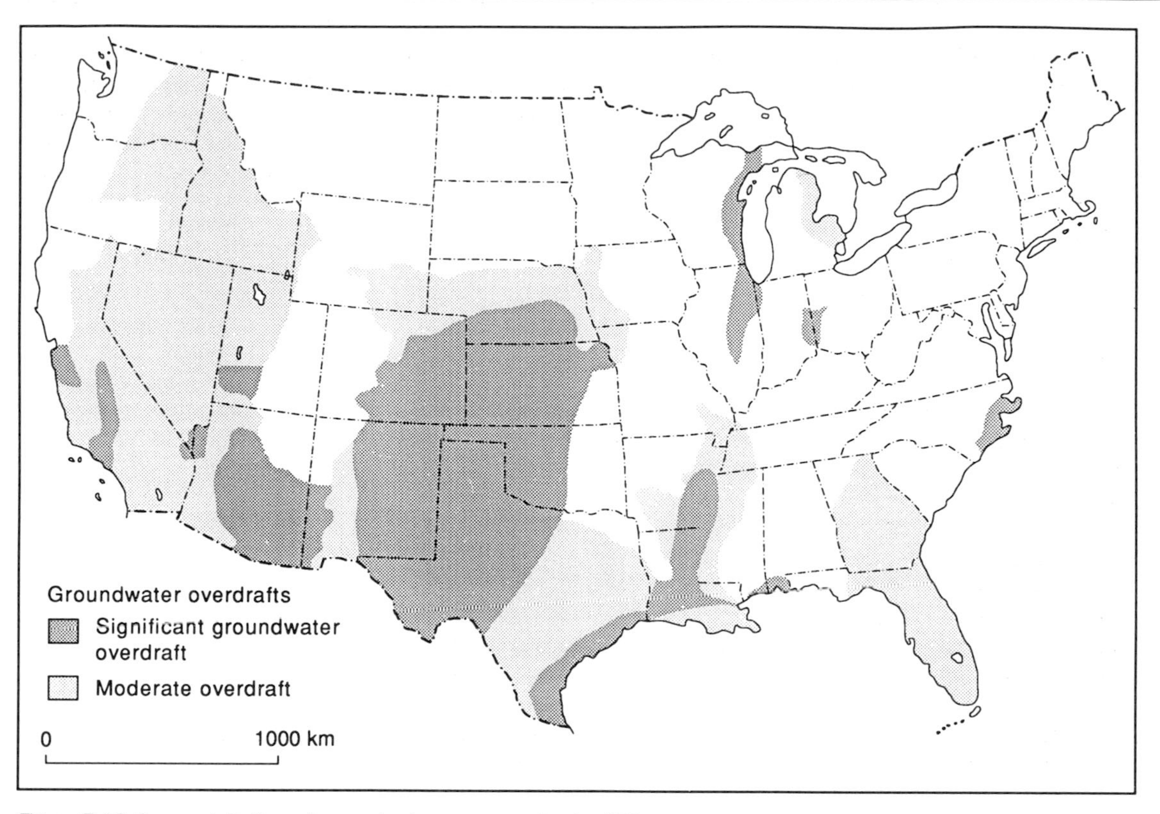

Figure 7.10 Overexploitation of groundwater resources in the USA, especially the Ogallala aquifer beneath the American Great Plains. From *Environmental Science* by Daniel D. Chiras. Copyright 1994 by Benjamin Cummings Publishing Company. Reprinted by permission

continued into the twentieth century, but have gradually been replaced by tile drains and modern perforated plastic tubing. Less obvious and permanent land drainage is created by subsoiling or moling, in which a bullet-like extension to the plough (a 'mole') is drawn through the subsoil, preferrably just before the wet season. Modern agriculture often combines irrigation and drainage in order to maintain optimum soil moisture levels for plant growth and, where appropriate, to reduce the danger of salinization.

Much of this drainage has been undertaken as an essential part of reclaiming **wetlands**, bogs and mires, for agricultural use, a process that proliferated in Europe and North America after the Second World War (Box 7.5). Many British wetlands were lost in the 'dig for victory' campaign during the war, but more have been lost since through continued subsidies to agriculture and particularly under the European Community's Common Agricultural Policy (CAP). About half of UK food production now depends on effective land drainage on more than 11 million ha of farmland. A further million hectares rely on pumping to prevent permanent waterlogging (Hockin *et al.* 1988).

Figure 7.12 charts the expansion of agricultural **underdrainage** in Britain. There is a widespread view that this underdrainage, which acts essentially as an extension of the natural drainage network, increases runoff and peak discharges. Expansion of underdrainage in the basin of the Yorkshire Ouse was cited as a possible contributing factor in the near catastrophic York floods when the river rose 5.05 m above normal in December 1981 (section 6.5.2). Underdrainage certainly increases annual stream discharge and there are many instances of increased flood hazard (Penning-Rowsell *et al.* 1986). However, Green (1979) concluded from his experimental catchment at Bury Brook, Cambridgeshire, that it can actually reduce flash flooding, because pre-storm soil moisture levels are lowered; in fact, peaks-above-threshold actually increased, but he considered this was due to changing rainfall patterns. The British evidence suggests that peak flows can be reduced in permeable mineral soils, because of pre-storm aeration, but open drains (or 'gripping') in peat soils increase peaks, because aeration penetrates the peat more slowly and the drains intercept more overland flow (Newson and Robinson

Box 7.3 Case study: Groundwater overdraft and drought in southeast England

During the 1976 drought, the River Thames ran dry at the Teddington gauging site for the first time in its history. Even though the drought was estimated to have a return period of at least 220 years, overexploitation of both surface and groundwater resources seems to have been a contributory factor. Demand is increasing at 1 per cent per annum within the Thames basin and threatens to leave only a 1 per cent surplus in local resources by 2021 (NRA 1992). The long-term increase in demand has caused water levels in the chalk aquifer to fall, with the consequent loss of artesian pressure in London and severe reductions in riverflow draining from the surrounding chalk hills (section 3.4); the fountains of Trafalgar Square were originally artesian. The Thames Groundwater Scheme, introduced in the 1970s to alleviate low flows by pumping water out of the chalk and limestone aquifers, is becoming less effective and overuse could be detrimental. Numerous plans for inter-basin transfers have been proposed in recent decades to support demand (Kirby 1984; NRA 1994). These have included artificial recharge of the chalk aquifer.

The problems extend to much of southern and eastern England, where groundwater is a major source of public water supply and where rising demand will wipe out the current small regional water surpluses by 2021 (Figure 7.11).

In the early 1990s, problems were exacerbated by the worst multi-year drought since 1899–1902. The area received only 80 per cent of average rainfall between summer 1988 and summer 1992, a net shortfall of over 350 mm. By December 1989, the Institute of Hydrology had found the discharge in many rivers in the southeast already the lowest on record. Winter 1991/92 was especially dry with only 50 per cent of normal precipitation in December. By spring 1992 groundwater levels were the lowest in living memory, lakes had disappeared, some boreholes were running dry and rivers reduced to a trickle or completely dry, like the River Darent in Kent or the Kennett in Cambridgeshire. The NRA designated 40 rivers as in a dangerous state and banned abstractions. Hosepipe bans affected 4.5 million consumers. Fears were widely expressed by conservation bodies that the climatic events were only highlighting the potential ecological damage that overexploitation of groundwater might do to rivers over a wide area.

Heavier rains during summer 1992 and the following winter recharge period helped some rivers and halted the crisis in water quality that lack of dilution, particularly of sewage effluents, was causing. However, heavy rain is not so helpful to groundwater, because a higher percentage enters surface runoff; instead, water companies and farmers in Kent were praying for a persistent drizzle. In many areas the water table failed to approach more normal levels until 1994.

The drought event was part of a broader pattern affecting much of Western Europe. In 1990 over half of the French *départements* had to impose water restrictions and many French rivers reached exceptionally low levels in 1992. In contrast, the drought hardly touched Scotland and Wales as depressions skirted northwards around the anticyclones affecting France and southeast England. Scotland received twice the normal rainfall in 1988–90 (Hamer 1990), which emphasized the NW–SE gradient in resources and illustrates the logic of NW–SE interbasin transfers in Britain.

1983; Robinson 1985). The UK Agricultural Development and Advisory Service (ADAS) Drainage Research Unit and the Institute of Hydrology have carried out fuller investigations since and developed models of drainflow (Robinson 1990).

Forestry ditches can have a similar effect. There is direct evidence that pre-planting drainage schemes increase both total runoff and flood frequency (Robinson 1986). In Britain, Binns (1986) reported that Forestry Commission ditches collect large amounts of throughflow. Howe *et al.* (1967) suggested that the expansion of forest ditches in the upper Severn basin in Wales was responsible for increasing flooding during the 1950s and 1960s in and around Newtown (Figure 9.6), although changes in rainfall patterns may also have contributed (Higgs 1987). At Llanbrynmair in

Figure 7.11 Groundwater use in England and Wales, showing the proportion of groundwater currently exploited, its relative importance in regional Public Water Supply (PWS), and the effects of projected increases in demand by 2021. Sources: NRA (1994) and *Digest of environmental protection and water statistics* No. 16 (1994), HMSO

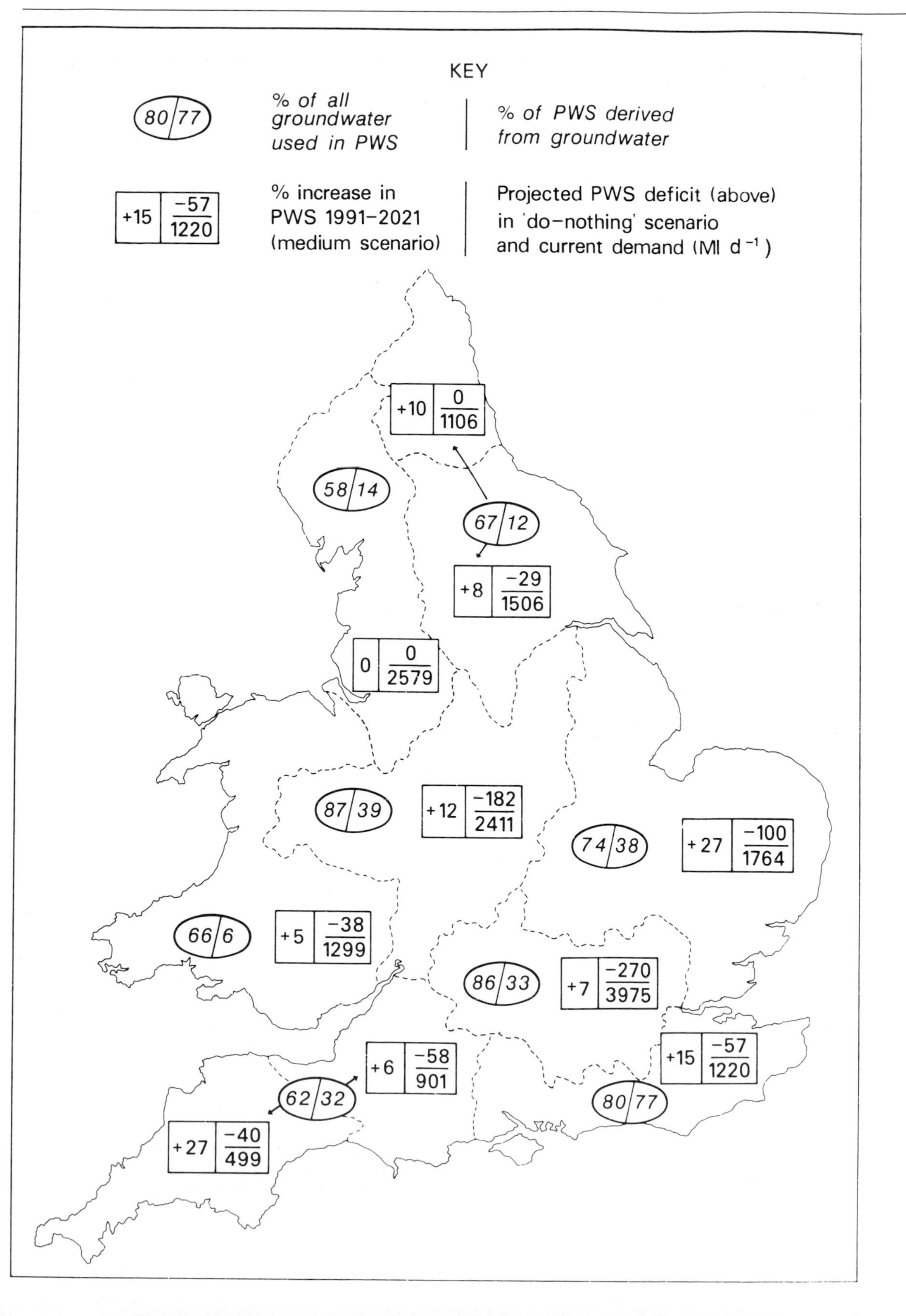
KEY
80/77
% of all groundwater used in PWS
% of PWS derived from groundwater
+15 −57/1220
% increase in PWS 1991–2021 (medium scenario)
Projected PWS deficit (above) in 'do–nothing' scenario and current demand (Ml d⁻¹)
+10 0/1106
58/14
67/12
+8 −29/1506
0 0/2579
87/39
+12 −182/2411
74/38
+27 −100/1764
66/6
+5 −38/1299
86/33
+7 −270/3975
+6 −58/901
62/32
+15 −57/1220
80/77
+27 −40/499

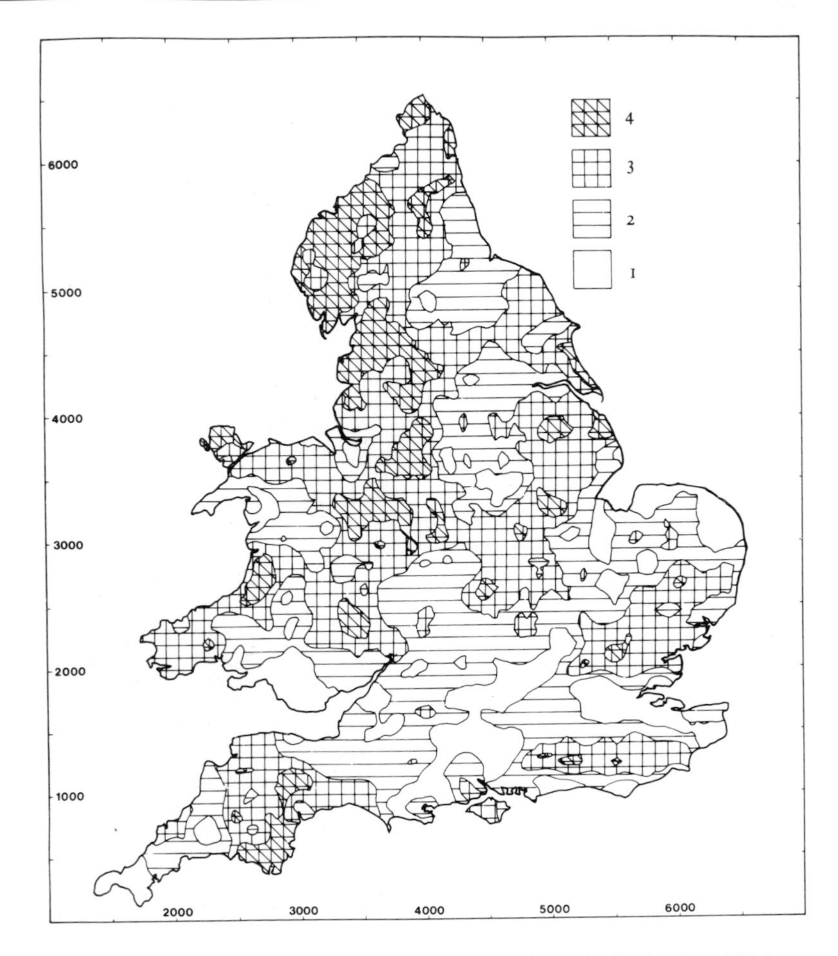

Figure 7.12 The expansion of agricultural underdrainage in England and Wales. (a) Percentage of agricultural land underdrained prior to 1939. 1: under 25%, 2: 25–50%, 3: 50–75%, 4: over 75%. From Robinson (1986b)

mid-Wales, forest ditching increased the drainage density from 3 to 200 km km^{-2} (Francis 1988). Large quantities of sediment can also be entrained in these forest ditches (section 8.2.3).

7.6.2 Channel modification

Most channel modification is undertaken for navigation or flood control. **Navigation** often requires channel deepening and dredging, straightening, clearing obstructions and perhaps regulating locks. The eighteenth century development of transport canals also required river water to be diverted to feed these canals. In effect, canals are a form of inadvertent interbasin transfer, and it is not surprising that the NRA (1992) floated the idea that the British canal network might be used as a quick and cheap means of transferring much-needed water from the north and west to southeast England. **Flood protection** can also require channel deepening, whether by dredging or by artificially raising the banks, and the removal of shoals, bars and organic obstructions, like trees dead or alive, from the channel. But an additional aim of flood control has been to speed the discharge of floodwaters by reducing channel friction and locally even increasing channel slopes.

After decades, indeed centuries, of ignoring the ecological and hydrological side-effects of channel modification, hydrologists are realizing that engineering works, like **channel dredging**, **steepening** or **straightening**, **floodwalls**, **earth levées** (raised banks), or bank stabilization using **riprap** (loose boulders or concrete blocks) and **gabions** (rock-filled wire cages) provide only limited protection for the length of bank covered. This protection is often at the expense of

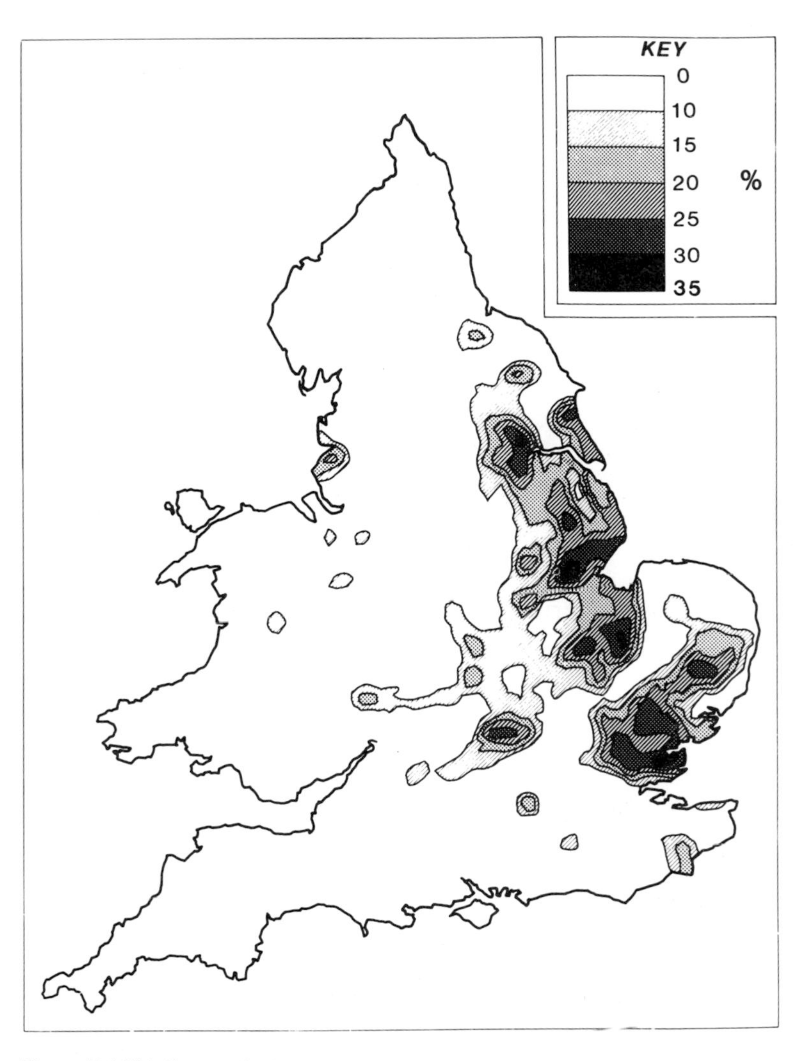

Figure 7.12(b) New underdrainage during the 1980s. The latest expansion has been in the lowlands, sometimes linked with spray irrigation to maintain optimal soil moisture for crops. After Robinson (1990)

unprotected areas downstream, which consequently receive more floodwater and may suffer more bank erosion (section 10.2.1). Moreover, the more of the streambank that is protected upstream, the more flood protection has to be increased downstream. Extending protection upstream creates ever higher flood stages downstream by removing natural overbank storage and increasing velocities.

Active reappraisal and some reversal of established philosophy took place following the 1993 Mississippi floods and the 1995 Rhine floods (Box 10.2). The great floods of July 1993 in the Mississippi basin, the worst since records began in 1895, *may* have been critically enhanced by flood protection schemes. They certainly led to calls for a re-evaluation of the policy of containing floods by building levées and floodwalls along the riverbanks and of speeding up drainage by straightening bends, dredging channels and removing islands and backwater areas (Figure 7.13).

Some 3200 km of earth levées have been constructed in the upper basin. Since the 1930s over 200 flood control reservoirs have been built to hold floodwaters back and reduce pressure on the levées. The US Corps of Engineers also built thousands of stone wing-dikes in the channel to maintain a 2.75 m deep channel for navigation. The wing-dikes narrow the channel and increase flow velocities (Figure 7.13).

By the end of straightening in 1942, the Mississippi

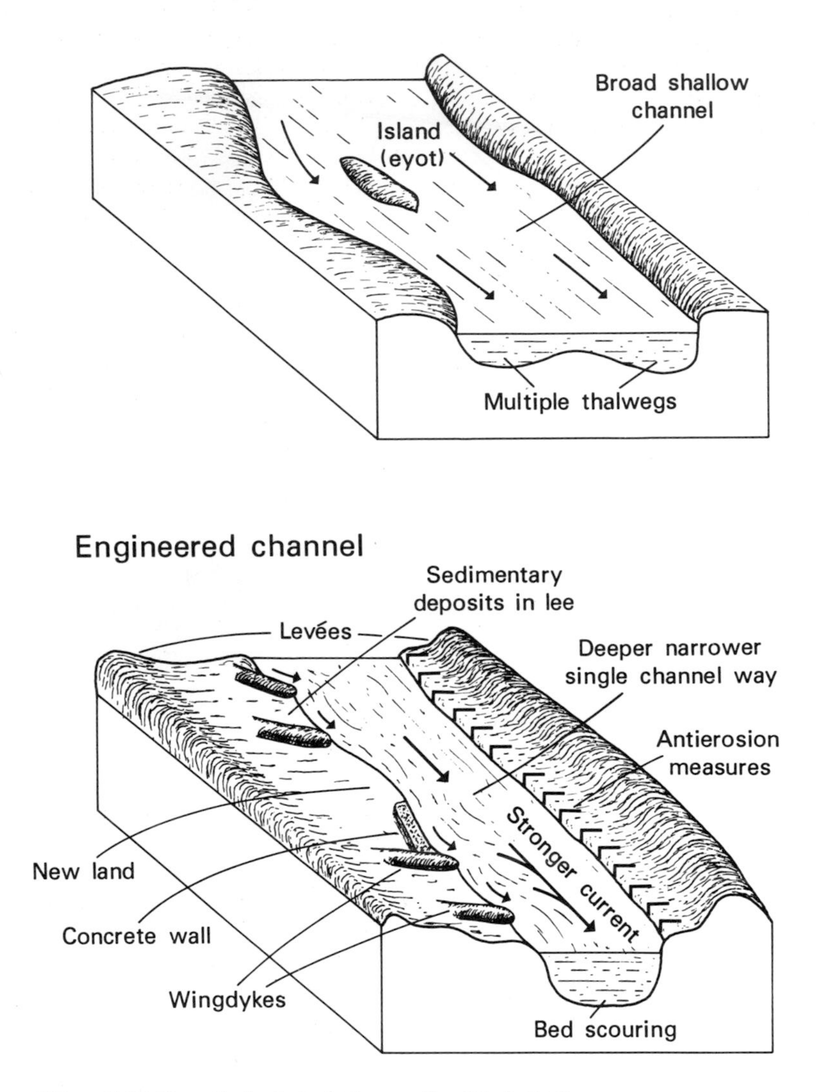

Figure 7.13 The effect of wing-dykes on the Mississippi

was 240 km shorter. The whole scheme cost over $10 billion to construct and requires $180 million a year to maintain. Yet it has probably also caused a fore-shortening of concentration times within the river system, which may have led to peak discharges arriving at about the same time in the main rivers and so created exaggerated peak flows in the Mississippi and Missouri Rivers. Reisner (1993) argues that the policy created a two-fold problem: (1) channelization and wetland drainage exacerbated flooding, to control which the Corps had to build dams; and (2) wetland drainage helped intensive agriculture on the floodplain, which increased soil erosion and required the Corps to dredge the rivers more frequently.

Moreover, the earth levées are more prone to failure when they get saturated during prolonged flood events, and when they eventually burst the water causes more damage because of the artificially elevated hydraulic head. Equally critically, the levées have given people a false sense of security and encouraged occupation of the floodplain – 80 per cent of private earthen levées and 70 per cent of all levées failed in the flood (Figure 7.14). Most federal levées held, but these directed the water to less well protected towns and fields.

Proving or disproving the extent to which human interference was responsible is, nevertheless, extremely difficult. There is no computer model that covers the whole of this enormous drainage system, which embraces half the co-terminous USA. To construct one would be the largest programming task yet undertaken in hydrology and would have to include the erosional and depositional processes that are continually modifying the channels. The Corps has defended its policy, but a limited hydraulic model developed at the Civil Engineering Department of the University of Illinois to cover a 64 km section of the Mississippi near St Louis did suggest that levées *can* raise flood levels in certain circumstances. It indicated that, if the Columbia levée had not failed on the opposite bank, water levels would have overtopped the St Louis floodwalls by nearly 0.5 m. Because of the difficulties of computer modelling, much of the modern engineering work has been based on concrete models, in which the important processes of sediment movement and channel migration have been either ignored or only poorly simulated.

Evidence from historical records is equally limited, because of the lack of reliable measurements. The 1844 and 1903 floods were probably larger but river stages were lower than in 1993. However, neither of these floods was properly measured. Prior to the introduction of the Price current meter in the 1940s, velocities were overestimated by up to 33 per cent, and before modern sonar equipment the shifting channel cross-sections were difficult to monitor. The Corps maintains that there has been no upward trend in flood levels over the last 50 years. However, much of the engineering work had been completed before this era of reliable measurements.

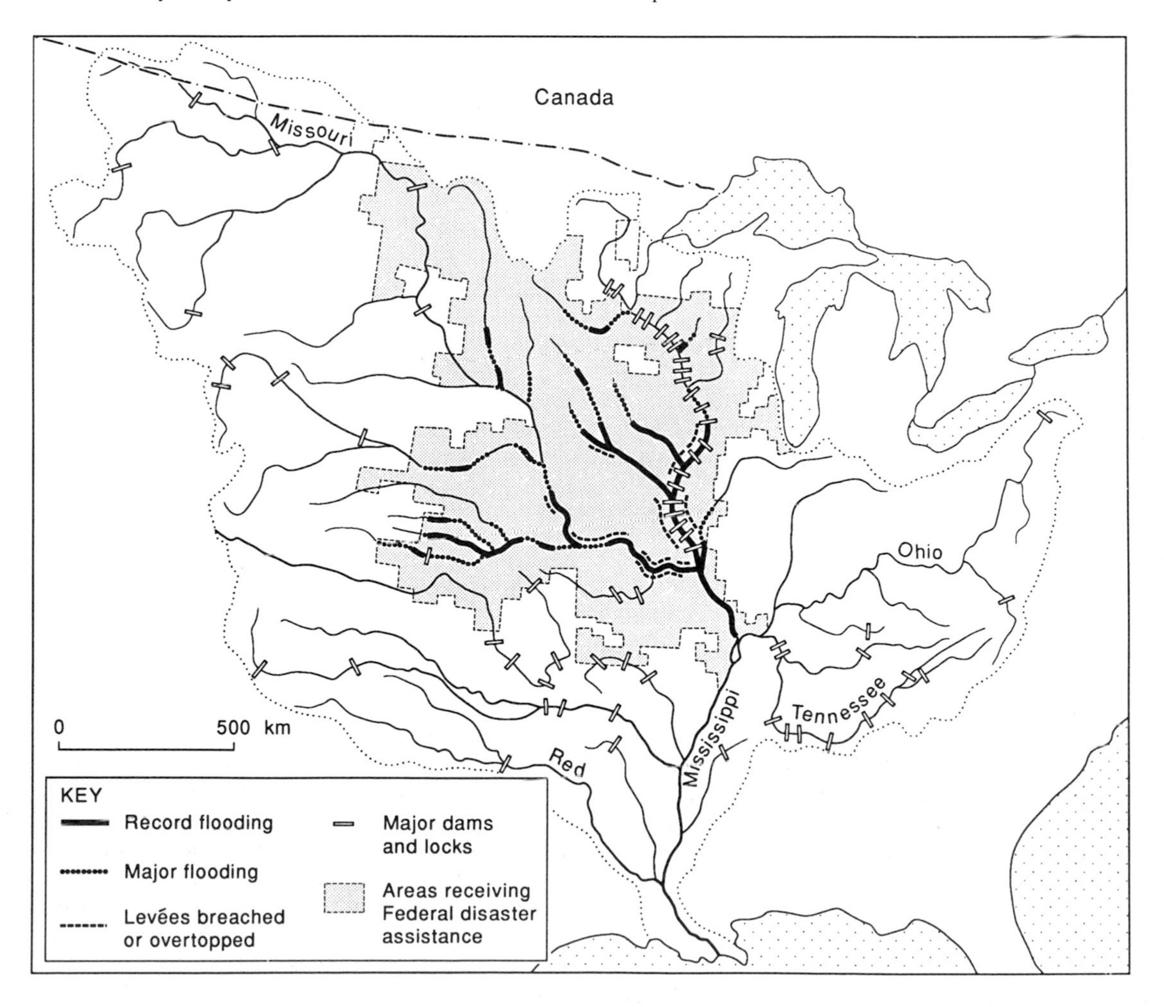

Figure 7.14 The 1993 Mississippi basin floods

Box 7.4 Case study The 1993 Mississippi floods

The flood

NOAA (1994) called it 'the most costly and devastating flood to ravage the United States in modern history' with estimated losses of $15–$20 billion (Figure 7.14). Springtime floods are a regular occurrence within the basin, but the 1993 floods followed eight months of unusually high rainfall, with up to twice the normal amounts, plus exceptionally heavy winter snows in the Rockies. Some areas received more than 1200 mm of rain over the period. Cedar Rapids, Iowa, received nearly 900 mm, a year's rainfall, in just four months. By late June the Mississippi was overbank in Minnesota, at that stage a 30-year record. Then on 8 July and 11 July, a total of over 160 mm fell on saturated land in parts of Iowa. Upstream at Des Moines the water purification plant was flooded by the Raccoon River and the city's water supply became contaminated (Bryson 1994) – 250 000 people had no water supply for more than two weeks and the water was not fit to drink for over a month. Downstream at Hannibal, Missouri, the peak discharge on the Mississippi exceeded the new 9.5 m levée built to contain the 500-year flood level, itself an exceptionally high specification, by 0.6 m.

More worryingly, just upstream of St Louis the flood crests of the Mississippi, Missouri and Illinois Rivers all met simultaneously for the first time since records began. The Missouri gouged out a new channel link to the Mississippi at Portage des Sioux. Large areas of St Charles County were flooded and 8000 people made homeless. Luckily, St Louis itself was saved by a 16 m high floodwall, but a second wave of flooding followed renewed heavy rain on 20 and 21 July, and created a new record peak of 14.3 m at St Louis, 5.2 m above average flood level.

Many mid-western rivers remained in flood stage till September, with levels topped up by sporadic and localized heavy rain through August in what for many upstream states was the wettest summer ever (WMO 1994). The soils of the entire basin upstream of the Ohio River confluence at Cairo, Illinois, remained saturated into September and models at the Midwestern Regional Climate Center indicated this was likely to continue till March 1994.

Fortunately for the states downstream of Cairo, and especially for New Orleans, which stands below sea level, the Ohio River was unusually low. This offered crucial compensation, combined with less local rainfall and a wider, deeper river channel along the lower Mississippi.

Overall, the floods resulted in 50 deaths, property damage estimated at over $7 billion, with around 37 000 people evacuated from their homes, and over 30 million ha of farmland flooded, with around $3 billion in crop damage (Adler 1993). Congress voted $3 billion for initial disaster relief with a similar sum to follow, but the final total cost was estimated to exceed $11 billion.

The floodplains are highly attractive to farmers because they are more fertile, and to poorer households, because the land is cheaper to buy. For the farmers, fields on the floodplains normally yield 20–40 per cent more per hectare than the higher ground. But the poor, who often cannot afford flood insurance, were the most affected. Ironically, the net cost to US agriculture as a whole was not that great, because losses in the floods, and in the complementary drought in the southeast, were balanced by higher prices and some bumper crops on higher ground. Nevertheless, the net 30 per cent fall in the maize harvest and the 19 per cent fall in soyabean production left America only just able to satisfy domestic and export requirements.

The weather situation

Antecedent conditions had been extremely wet, with excessive winter snowpacks in the Rockies and persistent and excessive rainfall throughout spring and early summer. The final trigger for the floods was an unusually strong quasi-stationary jetstream running northeast across the upper basin with a strong blocking anticyclone (a 'Bermuda high') centred over the eastern USA for more than five weeks. This caused a persistent southerly airstream on the western flank of the anticyclone, carrying warm moist air from the Gulf of Mexico up the Mississippi basin. As it travelled northwards, this very moist air became increasingly unstable, spawning thunderstorms. It also fed a stationary weather front fixed by a persistent low-pressure system associated with the jetstream, creating widespread heavy frontal rains over the upper Midwest.

The US National Meteorological Center has produced a close simulation of this synoptic situation by setting the contemporary ENSO-related sea surface temperature (SST) anomalies in the tropical Pacific in its numerical climate model. The US Climate Analysis Center concluded that the event resulted from the superposition of the ENSO trigger on top of the wet antecedent conditions.

Box 7.5 Possible solutions to flood problems on the Mississippi

Improving warning systems

The warning systems worked reasonably well. The weather radar and the network of rainfall and river stage gauge DCPs linked by satellite to offices of the US Corps of Engineers provided data for local *real-time flood-routing models*, which offered valuable short-term warning. Fortunately, the new Chicago and Kansas City radars had already been installed and this certainly saved lives. In addition, the Corps had been aware of the danger from *long-range weather forecasts*, and the gates of the 29 regulation dams on the Mississippi had been opened to release as much water as possible since the spring.

However, NOAA's (1994) review of NWS warning procedures concluded that longer warning lead times and better hydrological/hydraulic modelling are needed. The programme of improvements now officially embodied in the *NOAA 1995–2005 Strategic Plan* will be invaluable. The new generation of doppler weather radar (WSR-88D) forms the foundation of the strategy, along with new interactive mosaicking and processing facilities for radar data at River Forecast Centers. To this will be added very substantial improvements in basin-wide runoff modelling, including new models of the complex hydraulic changes during a flood, which present a major problem in the mobile channels of the basin. Meteorological forecasts will be integrated into the hydrological models to give longer lead times. NOAA also proposes incorporating confidence limits into hydrological forecasts and running the models for a range of meteorological forecasts to help decision making.

Methods of flood reduction

Clearly, the heavy rain falling on a saturated basin was an extreme natural event. However, it seems equally evident that the flood containment policy and channel modification programme aggravated the problems. The loss of natural overbank storage shortens concentration times and increases flood peaks; bargees plying the river claim they now need twice the engine power they needed 20 years ago because of increased flow velocities (Mairson 1994).

Flood levels may also have been made worse by water levels being kept high in reservoirs on tributary rivers (Adler 1993). Many reservoir dams have a dual role to check the floodwave and to guarantee water supply and/or leisure activities like sailing. But there is a tendency for water levels to be kept as high as possible for these other uses: a classic conflict of interests.

The policies of building levées, draining wetlands, clearing the channels and building towns on floodplains are now all being questioned (Tickell 1993). Under the Corps' supervision, 16 000 km of levées and 500 dams have been built in the USA this century at a cost of $25 billion. In addition, half of the nation's wetlands, 475 000 km^2, have been drained over the last 200 years, yet the wetlands, the backwaters and the floodplain itself provide valuable storage areas that delay runoff. The national Wetlands Reserve Program was established to reverse this trend and had aimed to pay farmers to restore 2.5 million ha by 1995, yet by 1994 barely 125 000 ha had been rehabilitated. After the floods, there were calls for Congress to spend more money on the scheme and speed up funding for 500 000 ha of restoration by farmers already prepared to join the scheme.

The US National Wildlife Federation has also been campaigning to stop the subsidized National Flood Insurance Program, because low premiums encourage building in flood-prone areas. In fact, only 13 per cent of home-owners who live in flood-prone areas use the scheme. The false security given by the dikes is probably more to blame, and so, ironically, in some eyes is the Federal Emergency Management Agency (FEMA), which guarantees compensation. One of the more ambitious plans is to move most of the town of Grafton, Illinois, which has suffered six major floods in the last 20 years, off its floodplain site near the Illinois–Mississippi confluence at a cost of $25 million, with perhaps 40 per cent coming in FEMA aid (Mairson 1994).

Environmentalists are advocating moving the dikes back from the river to allow controlled flooding of riparian meadows. In addition, building sluices into the dikes would allow even more extensive but controlled flooding over croplands. Other control structures might even be removed altogether.

Action has already begun to rehabilitate the natural flood storage areas along the Missouri River in Nebraska. Although no amount of restoration could have totally prevented this particular event, it would have reduced peak discharges and it might prevent problems from many lesser storms in future. It should also help re-establish the wildlife, and especially help some of the birds for whom the basin is one of the most important routes for north–south migration in the world.

Impounding and regulation dams indirectly modify the downstream sections of the river channel by altering the total discharge or the frequency distribution of discharges. Where water is removed from the river system for public water supply or irrigation, the hydraulic geometry of downstream channels tends to adjust to lower flows, with a contracted cross-sectional area (Petts 1980a and b, 1984). Consequently, when extreme events occur which the reservoir is unable to contain, there is increased danger of overbank flooding. Evidence to date suggests that significant contraction in channel capacity may be observed within a few years of dam construction.

Regulation dams seem to have less effect on channel capacity, but their effect on the median discharge may be important for wildlife and for sediment movement. Figure 7.15 shows how a hydropower regulation scheme has increased the frequency of medium discharges and reduced low flows, but has little effect on extreme events which exceed the storage capacities of the reservoirs. It also shows the effects on sediment movement.

Part of the feasibility study for the creation of the largest man-made lake in Europe at Craig Goch in Wales (Figure 9.4) looked at the effects of the proposed interbasin transfer releases on channel stability in the

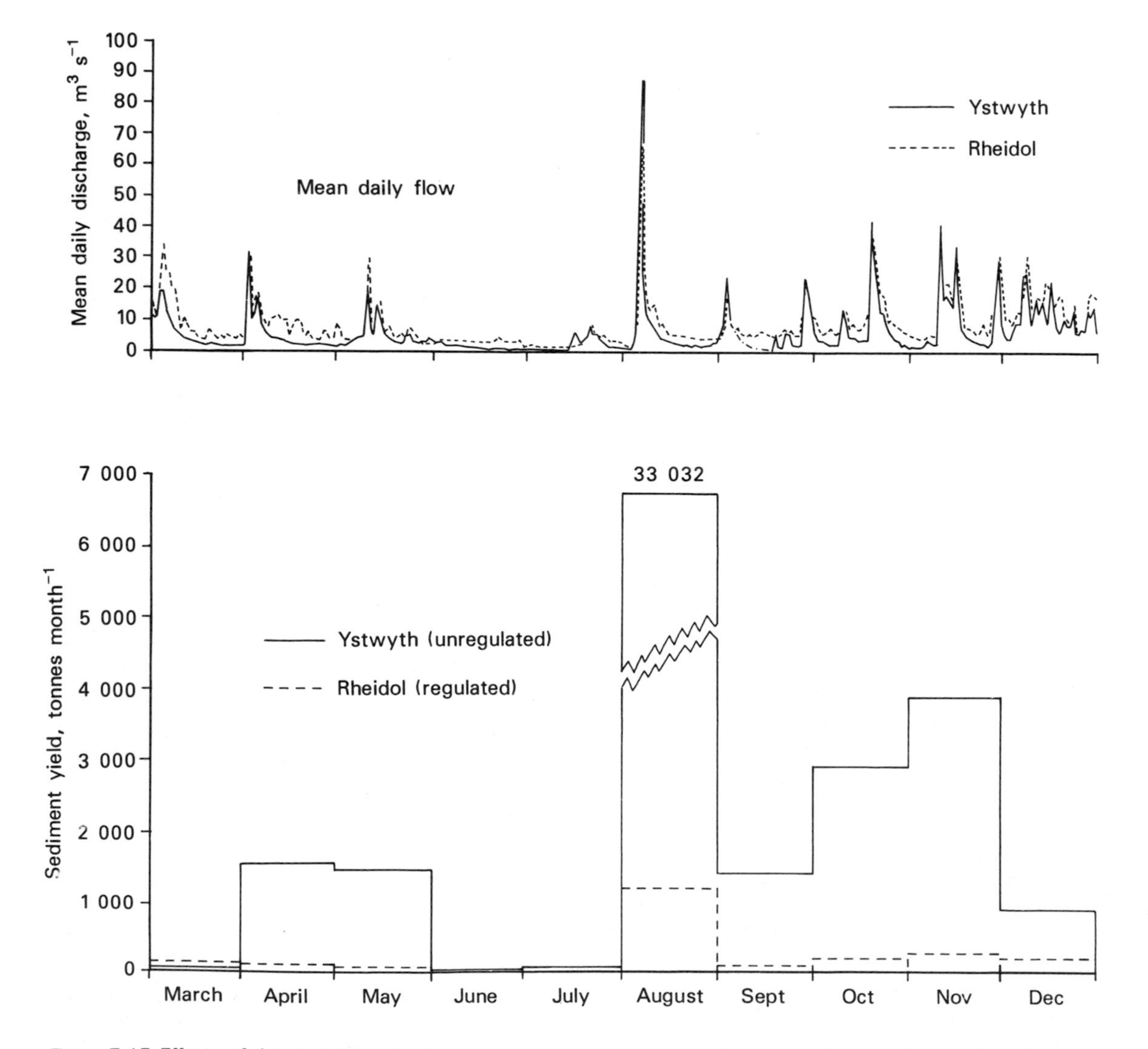

Figure 7.15 Effects of river regulation for hydropower development on discharges and bedload in the River Rheidol, Wales, compared with the adjoining unregulated River Ystwyth

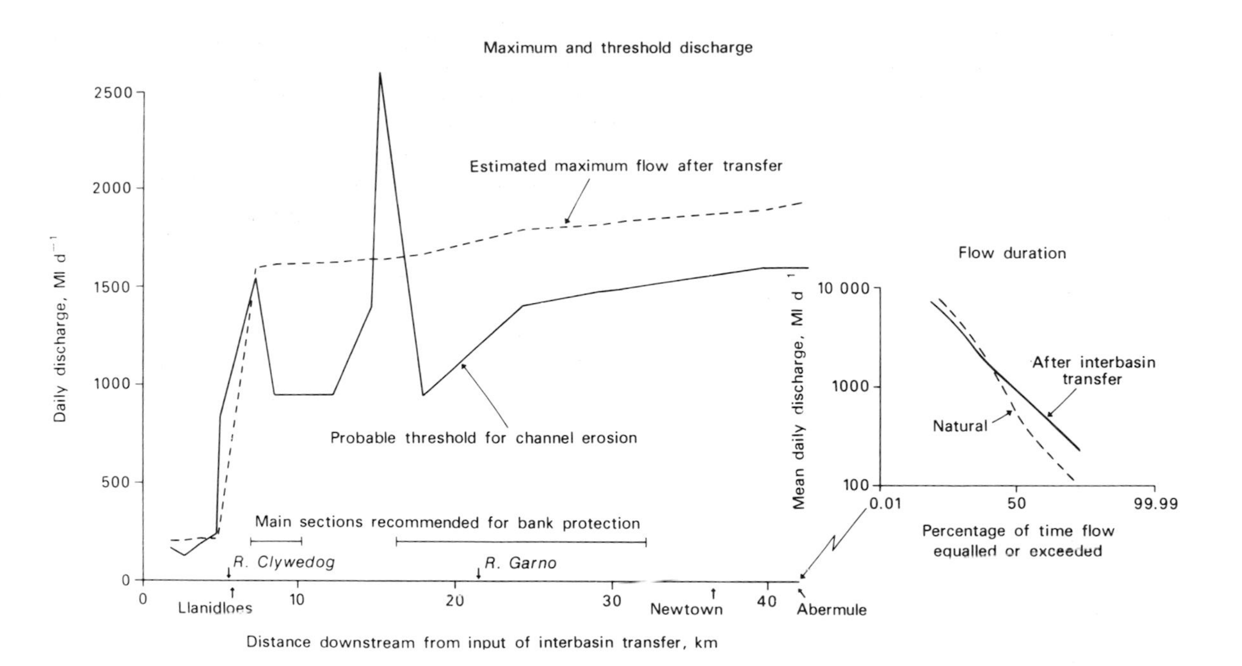

Figure 7.16 The effect of interbasin transfer on erosive flows. The River Severn system downstream of proposed release of regulating flows from an expanded Craig Goch reservoir (Figure 9.4), showing estimated maximum flows, erosion thresholds and changes in the flow duration curve. Data from Hey (1986)

receiving rivers (Hey 1986). The study compared field determinations of the threshold discharges for bank erosion and bedload transport with the effect of regulation on the frequency of flows (Figure 7.16). There are numerous difficulties in estimating these impacts, including complications like the possible development of **armouring**, a surface layer of coarser sediment that protects erodible material beneath. However, the results suggest that the receiving channels are likely to show a widespread increase in bedload movement at the maximum level of regulated flows and certain sections would need extra protection against erosion.

Conclusions

It is abundantly clear that, after millennia of small-scale and gradual hydrological changes wrought by human interference, impacts have taken a giant leap to an altogether different level since the mid-twentieth century. More than ever before, we need to manage our abilities to interfere with great care and insight.

Further reading

A broad introduction to processes and effects, with their policy implications, is offered in:

Newson M D 1992 *Land, water and development*. London, Routledge: 351pp.

Review articles, summaries of literature and research papers covering a wide range of human interference can be found in:

Arnell N 1989 *Human influences on hydrological behaviour: an international literature survey*. Paris, UNESCO.

Hollis G E (ed.) 1979 *Man's influence on the hydrological cycle in the United Kingdom*. Norwich, GeoBooks: 278pp.

International Association of Hydrological Sciences 1977 *Effects of urbanization and industrialization on the hydrological regime and on water quality*. International Association of Hydrological Sciences, Pub. No. 123: 12–19.

Particularly useful summaries of forest effects are to be found in:

Bosch J M and J D Hewlett 1982 A review of catchment experiments to determine the effect of vegetation changes on water yield and evapotranspiration. *Journal of Hydrology*, 55: 3–23.

Swanson R H, P Y Bernier and P D Woodward (eds) 1987 *Forest hydrology and watershed management*. Wallingford, International Association of Hydrological Sciences Pub. No. 167.

The effects of dams are covered in:

Petts G E 1984 *Impounded rivers: perspectives for ecological management*. Chichester, Wiley: 326pp.

UNEP/UNESCO 1990 *The impact of large water projects on the environment*. Paris, UNESCO: 570pp.

Discussion topics

1. Review the effects of vegetation change on hydrological processes.
2. Assess the importance of urban effects.
3. Analyse the arguments for and against large dams.
4. Assess the benefits and disadvantages of established river engineering practices.

CHAPTER 8

Inadvertent impacts on hydrological processes 2: water quality

Topics covered

Deteriorating water quality is a threat to water resources and wildlife in many parts of the world. Much of the problem has been generated by industrialization and urbanization, from surface and atmospheric pollution. But in recent years agriculture has become a major source. Efforts to control agricultural sources require basin-wide vigilance. Despite recent legislation, the peak in nitrate pollution may lie some years ahead because of the quantities still stored in groundwater.

The earliest human impact upon water quality came from clearing the land for agriculture, which increased runoff velocities and sediment yields. This was soon followed by irrigation and salinized return waters. Nevertheless, throughout most of the last millennium the worst effects came mainly from urban areas. Prior to the Industrial Revolution in Europe and North America during the eighteenth century, these were largely due to sewage and to chemical activities like tanning. During the Industrial Revolution, the new chemical by-products, the increased use of fossil fuel and the rapid concentration of humanity in industrial cities all contributed to a rapid decline in both surface and groundwater quality (Figure 1.8). Most pollution inputs were **point** sources, like sewers or factory outfalls. By the late nineteenth century, cholera and typhoid contracted from local water supplies were forcing many industrial cities to seek clean water supplies from pristine mountain sources.

Since the Second World War, there has been a rapid rise in **nonpoint** sources of pollution, i.e. pollution entering the drainage system over a wide area. Nonpoint sources are inherently more difficult to deal with because they are more spread out and varied. More than half the pollutants entering US rivers come from nonpoint sources (Figure 8.1). Mechanized agriculture, artificial fertilizers and the overall intensification of farming have played a major role. Ironically, so has legislation designed to combat the effects of fossil fuels on air quality.

Agriculture is now the dominant source of nonpoint pollution in over 50 per cent of American rivers (Figure 8.1). Agriculture has long been a source of water pollution. History is littered with evidence of accelerated soil erosion and flooding caused by poor husbandry, which must have significantly raised sediment loads in the rivers (Chapter 7). But both the range and degree of agricultural pollution has increased dramatically over the last half-century.

It is important to remember that virtually no water anywhere in the natural world is totally pure chemically. There is therefore no absolute by which to define pollution, except perhaps where the pollutant substance itself is man-made. **Pollution** is normally defined in relation to *average concentrations*, as found in the supposedly natural or undisturbed environment, or of *acceptable levels*. What is acceptable is normally defined in relation to its known effects upon human health. But as medical science advances, so acceptable levels change, generally downwards. Likewise, average concentrations vary from environment to environment and from one period to another, and the average figures depend very much on the quality of the records. These measurements should define **background concentrations** or **baseline levels**. Pollution will normally be defined as concentrations that exceed these levels.

Pollution may be found in all realms of the hydrosphere, from the atmosphere to groundwater, and surface waters can be polluted from both above and below as well as from the surface itself.

8.1 Acid rain and acidification of surface waters

The main atmospheric source of pollution causing concern is acid rain, although local cases of alkaline rain have been reported. The term is broadly used to cover not only liquid rainfall but also **acid snow** and the deposition of dry **acid aerosols**. Indeed, pollutants may reach the surface by a wide variety of processes. They may be removed from within the clouds by **rainout** or **snowout**. They may be collected from the air beneath the clouds by **washout**, or they may be deposited directly on to surfaces by gravitational **sedimentation** or wind-induced **impaction**. Impaction is particularly effective if either the aerosol or the depositional surface is wet. Small cloud or fog droplets can be especially acidic if the condensation nucleus was acidic, e.g. a sulphate particle, or dissolved acids have not been diluted by excessive condensation. This fog may impact upon trees to cause acid **fog-drip**. **Dew** can also collect acid aerosols.

Snow is particularly good at scavenging pollutants from the air because of free ionic bonds at the ends of the crystal lattices that attract aerosol matter. It is also a good collector when it is wet and aerosols stick to it. This can happen in the air or on the ground. The **black snow** events in Scotland illustrate this aerial scavenging (Davies *et al.* 1984). In 1986 on Plynlimon mountain in Wales dirty snow yielded runoff with a pH of 3.2, almost as acid as vinegar. The snow was white when it fell, but it lay for a month under easterly winds carrying aerosols from the industrial Midlands.

The hydrology of snowmelt aggravates the problem (section 3.3.5). Pollutants that might normally pass through the system in harmless concentrations over a period of time are accumulated and released in a concentrated **acid flush**. Concentrations are further increased by the property of ice crystals to expel alien molecules so that they concentrate on the surfaces of crystals or between metamorphosed crystals. The pollutants are therefore located right where the first

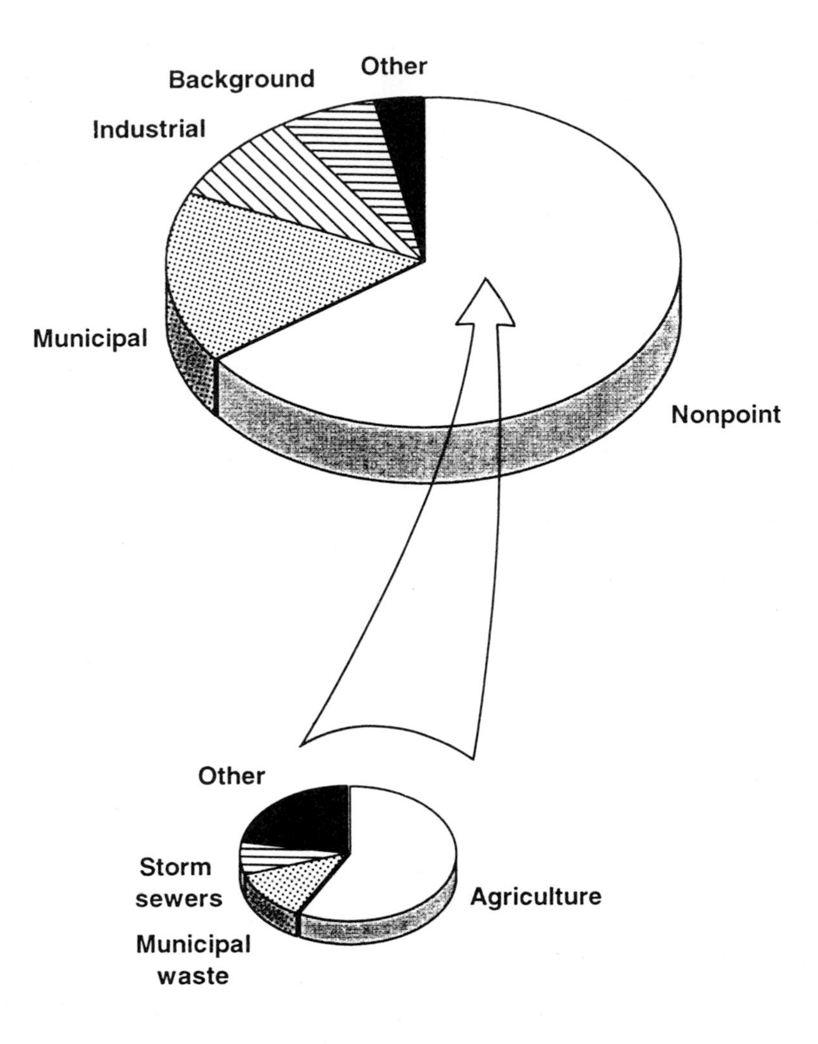

Figure 8.1 Sources of pollutants entering US rivers. Based on Chiras (1994) from EPA data

drainage routes are likely to develop in the melting snowpack. They may also contribute directly to the development of these drainage pathways by helping to concentrate solar energy at the crystal surfaces, especially if the pollutants are dark. By the time the first meltwater trickles through the snowpack, the pollutants are either free or readily released and they are picked up by the meltwater. This causes higher concentrations of pollutants at the beginning of snowmelt runoff.

Catchment hydrology plays a key role in determining the extent to which this acid deposition is damaging the environment (section 8.1.2).

8.1.1 Trends in acid deposition and control

Studies of peat bogs and lake deposits reveal a general rise in acid deposition in Western nations following the Industrial Revolution. Much evidence comes from studying the abundance and diversity of diatoms, algae that are very sensitive to acidity, deposited in lakes (e.g. Batterbee 1990).

Since the 1950s, deposition has intensified and expanded, partly because of the expansion of industrial activity and fossil fuel power stations, but also ironically as a result of clean air legislation. The British Clean Air Act of 1956 set out to stop the city 'pea-soupers' or 'smogs' like the one that killed 4000 people in two weeks in London early in 1953. The Act took two lines of action, restricting total emissions and reducing downwind concentrations. Unfortunately, it contributed to subsequent acid rain problems in a number of ways. It focused primarily upon particulate emissions from coal burning in cities. This did little to contain sulphur dioxide (SO_2) emissions. By following the principle of dispersal, it also spread the pollution problem to the countryside. More recently, another source of acid emissions, road transport, has expanded and shows no signs of being brought under control, despite vehicle emissions legislation.

Similar problems have emerged from US legislation. The Clean Air Acts of 1970, 1977 and 1990 again used the *concentration* principle in establishing **national ambient air quality standards (NAAQS)** but added the policy of **prevention of significant deterioration (PSD)** to prevent industry moving to a cleaner area and raising levels there up to the NAAQS (Elsom 1992). Good though these ideas are, they have not stopped the problem of atmospheric transport of acid pollutants into non-industrial areas. Nor have the 1980 US Acid Precipitation Act and the US–Canada binational control programme. Ontario and Quebec continue to receive 50 per cent of their acid load from the USA.

Transboundary atmospheric transport remains a major problem at all scales, from city limits to national frontiers. Norway receives 80 per cent of its acid deposition from beyond its frontiers. In Sweden, a 50 per cent reduction in sulphur emissions between 1980 and 1987 reduced acidic deposition by only 8 per cent. Even in Britain, it could take an 80 per cent reduction in domestic emissions to effect an overall reduction of 50 per cent, because of receipts from elsewhere in Europe.

Figure 8.2 illustrates the global range of acid rain. It shows the well-established problem areas in the industrial regions of eastern North America and Europe, now extending into Arctic Russia, and Japan–Korea. During the 1980s many tropical and southern hemisphere regions began to show the first effects of acid rain. Clearly, more needs to be done to reduce *total* acidic emissions directly (Lükewille 1994). The '30 per cent' club, in which 21 European nations undertook to reduce emissions of SO_2 between 1980 and 1993, and the EU Large Combustion Plant Directive were a start. Led by Sweden, Europe is pressing ahead with ever more stringent controls. But as the US administration recognized in formulating the 1990 Act, few national economies could stand a sudden, drastic change in permitted emissions.

8.1.2 Catchment hydrology and acidification

Concern over acid rain began in the late 1960s with damage to freshwater fisheries in Scandinavia. By 1980, 4000 Swedish lakes were pronounced devoid of fish and a further 14 000 seriously affected. Sweden now spends millions of dollars each year to lime the lakes. In the early 1980s, damage to forests was also blamed on soil acidification in Germany and Central Europe. Coniferous forests are also a contributory factor as well as a victim.

Surface water acidity is controlled by the interaction between chemical inputs and a wide range of catchment characteristics. Chemical inputs are controlled not only by emissions, but also by the *location* of the catchment in relation to those sources, to the trajectories of airflow, to factors controlling precipitation and to the sea. Mountains that induce orographic rainfall tend to have larger total receipts of acids, a higher **acid loading**, even though it may be more diluted. Hence, the mountains of central Wales receive rainfall that is typically half the concentration of that received in the East Midlands, yet because they also receive more than 2.5 times the annual rainfall their acid loading is greater.

Locations upwind of pollution sources might normally be expected to be less troubled than those downwind. However, occasional reversals of wind may be sufficient to counter this trend, especially over shorter distances. A roughly 3:1 westerly-to-easterly wind balance in central Wales is not sufficient to shield it from acid pollution originating in industrial areas to the east. Moreover, maritime regions like this with predominantly onshore winds face the added problem of sea salt deposition. Although salt (NaCl) is alkaline, chemical reactions in the soil can produce free chloride ions, Cl^-, which add to the acidity, and the sodium ions can displace bases like calcium from the soil, which impares its ability to neutralize acidity (Langan 1987).

The chemical **buffering capacity**, more explicitly known as the **acid-neutralizing capacity (ANC)**, of the surface materials is of paramount importance. Rocks and soils that are high in bases such as calcium and magnesium have a high buffering capacity and tend to have little trouble from acid rain. Agricultural liming is also an effective antidote, and some of the recent increase in acidity in British streams has been blamed on the ending of liming subsidies during the 1970s.

Catchments underlain by acidic igneous rocks or ancient metamorphic rocks tend to have lower buffering capacity. These ancient rocks make the Canadian Shield, the Appalachian Mountains of America and the mountains of Scandinavia and western Britain

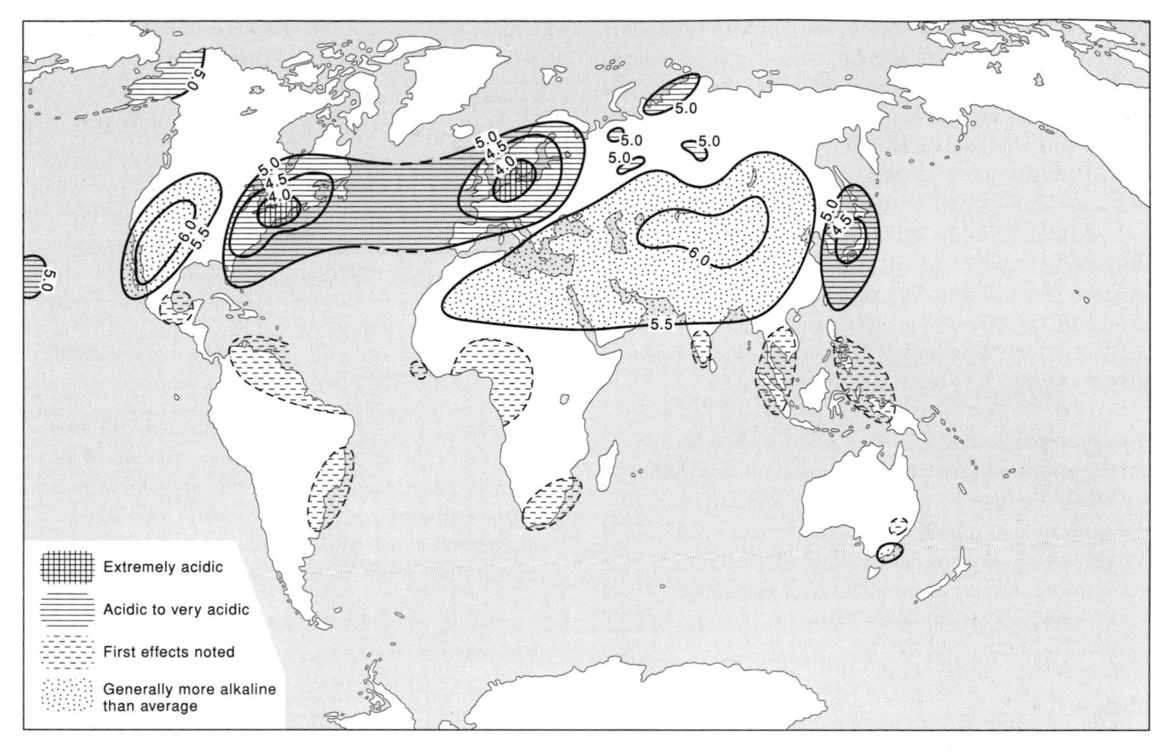

Figure 8.2 Global distribution of acid rain

particularly vulnerable; all are areas that presently receive high acid loads.

It has recently become clear that the *effective* buffering ability of a catchment is not always the same as that suggested by soil analyses. There are two reasons for this. The first is largely a matter of **residence time**, the length of time the water is in contact with the soil. If water drains through the soil too rapidly, chemical exchanges will be only partially complete, chemical equilibrium will not be achieved and the acidity will be only partially buffered. The chemistry of partial equilibrium processes is complex and much remains to be understood. This complexity is compounded by the **hydrological pathways** within the soil and within the catchment.

Hydrological pathways control the *speed* and the *location* of water flows. Hortonian overland flow will tend to offer shorter residence time and less surface area of contact with soil particles than throughflow, and chemical exchanges will be more restricted. Where overland flow runs over organic soil surfaces, the water is likely to have less opportunity of contact with neutralizing bases than if it were to flow across mineral soil horizons. It is also likely to pick up more weak organic acids released by the decomposing organic material. In some catchments this may be the sole source of acidity. About one-third of acid waters in temperate climates are acidic because of natural organic acids (Hemond 1994).

The same principle applies to pathways within the soil (Gee and Stoner 1989). Macropores concentrate flows in a small fraction of the pore space, generally within the surface horizons, where velocities can exceed 1 m h^{-1} (Nyberg *et al.* 1993). One reason why water supplied to Welsh streams by soil pipes tends to be more acidic is that the pipes drain the water through the upper peaty horizons and reduce its contact with weathering minerals (Jones 1997a). Soils with fingered flow may have buffering capacities 'a few dozen per cent' lower than under uniform flow (Winand Staring Centre 1994).

Water also follows preferred drainage routes at the macroscale (section 3.3.4). The chemical properties of the contributing areas, the percolines, swales and channelways, are therefore more important than those of the non-contributing area. And the buffering capacity of these flow zones is more rapidly reduced. This is particularly important when considering rehabilitation work (section 8.1.3). The properties of the non-contributing area are more important to baseflow, but baseflow is generally less acidic because of longer residence times and greater contact with mineral surfaces. There is now considerable evidence that stormflow is the driving force in freshwater acidification and that the **acid episodes** it creates are the source of the greatest damage to freshwater ecology (Edwards *et al.* 1990).

Vegetation and land management are also important factors in chemical responses. Research suggests that afforestation can aggravate acidification in a number of ways. The most marked effects occur in forests 15 to 25 years old, but some effects begin at planting time:

1. *Scavenging* Trees act like scrubbing brushes, trapping acid aerosols as dry or occult deposition. These are later washed off by the rain and cause a pulse of acid runoff. This is most effective at full canopy closure around 15 years after planting. Scavenging may also increase salt entrapment, especially from mountain fog in maritime regions.
2. *Drainage* Where ditches are dug to drain wet soils for planting, these reduce both residence times and concentration times, creating more rapid and acidic stormwater pulses.
3. *Oxidation* Ditching and turning the sod to create mounds of deeper soil for the saplings increase oxidation. In peaty soils this tends to produce acids, including sulphate ions, SO_4–, which are washed out in subsequent storms.
4. *Soil water pathways* Effects here are more ambivalent. Ploughing and root growth could increase infiltration and even break up ironpans and eliminate perched water tables in stagnopodzols, but most commercial conifers are shallow rooting. Moreover, it is questionable whether the pore space created by roots is of much use once there is a closed canopy, because canopy interception concentrates drip and throughfall between the trees, away from the stem and roots, and tends to favour overland flow (section 3.1.2).
5. *Litter leachates* Conifers produce acidic litter, which contributes weak organic acids. These acidify runoff and they assist the removal or **leaching** of bases from the soil, especially by the binding of metals to organic leaf leachates, a process known as **chelation**.
6. *Harvesting* Removing trees interrupts the natural return of base cations taken up by the trees and could cause a long-term reduction in the soil's ANC, if they are not replenished from other sources.
7. *Leaf leaching* Under acid rain stress, even non-harvested deciduous trees may contribute to a reduced ANC, as the plant tries to replace nutrients leached from its living leaves. The nutrients washed from the leaves tend to be washed away into the streams, so the former recycling of nutrients via leaf decay now becomes a one-way removal process.
8. *Acid drip* Trees on the streambank can contribute acid drip directly into the stream.
9. *Increasing evapotranspiration* Increased ET may lower phreatic levels and increase oxidation. It also increases chemical concentrations in soil water and alters soil–solution interactions. This can make aluminium more mobile compared with base cations (see point 10 below).

Other effects of afforestation can aggravate the *impact* of acidification:

10. *Aluminium mobilization* Aluminium is increasingly mobilized below pH3, so all processes that reduce the base status of the soil and increase the acidity of throughflow contribute to aluminium concentrations in streamwater. Aluminium levels in the water are more important for fish life than acidity (see below). Inorganic monomeric aluminium, leached by the strong acids from air pollution, is more problematic biologically than aluminium in organic complexes, leached by organic acids (Neal *et al.* 1990).
11. *Increased bedload* Some damage to fisheries may be caused by increases in sediment loads, especially more mobile bedload, resulting from ploughing, ditching and impoverished soil structure. Mobile bed sediments can destroy the hatching environment.
12. *Lower water temperatures* In Britain, shading by trees close to watercourses lowers water temperatures and may hinder fish reproduction in springtime. In contrast, summertime shading may protect fish in Canada.

The role of afforestation has been hotly debated in Britain. Nisbet (1990) for the Forestry Commission showed that conifer afforestation had no significant impact on records of average streamwater acidity. However, detailed sampling of stormflows gives a different picture. It is now clear that the ecological damage caused by acid episodes can be very persistent (Ormerod and Jenkins 1994). Biota killed off during these events can take a long time to reappear and predator populations can be impoverished through the food chain. Fish kills during acid episodes have been widely reported (Muniz 1991). Much of this rapid response is due to aluminium poisoning, which causes inflammation and hypersecretion of mucus in the gills, and asphyxiation (Rosseland and Staurnes 1994). But less visibly, bacteria, algae, higher plants (macrophytes) and invertebrates may also be killed or show sublethal toxic symptoms from a wide variety of causes. And fish populations are weakened by poorer food supplies. Observations at the Institute of Hydrology, Plynlimon, have found fish to be 16 times more abundant in the moorland catchment, with considerably greater individual body weight and more abundant invertebrate life than in the forested catchment.

Foresters rightly point out that the most important factor in acidification is atmospheric scavenging and inputs, and that similar trends are underway in non-forested areas. But forests can clearly add to the problem and even small additional effects may be significant, because most commercial forestry occurs on soils that are too poor for agriculture and have a low ANC.

8.1.3 Catchment rehabilitation

While efforts to reduce emissions continue, there is an urgent need to halt or even reverse acidification. The human degenerative disorder, Alzheimer's disease, may in part be related to aluminium levels in drinking water, although natural mixing of stormwater in reservoirs, merging water supplies from different sources and correcting the pH during water treatment tend to reduce aluminium levels in all but a few unfortunate areas. Ironically, aluminium sulphate used as a coagulant in the treatment process can be a more important source of aluminium and British water companies, among others, have been moving away from using it.

Concern centres more on the environment. A quarter century of research has produced many guidelines, but the problems are manifold, both in diversity and complexity. The 1992 Dahlem Workshop on freshwater acidification revealed the numerous problems encountered when linking hydrology with biological, chemical and pedological knowledge (Steinberg and Wright 1994). Information often relates to different temporal and spatial scales. Much laboratory research is not applicable because it relates to water conditions not encountered in field situations. Differing patterns of input, different pollutants, different catchment characteristics and different biological tolerances for different species all combine to produce a bewildering array of results.

There are two ways forward: to apply the results of catchment experiments to similar field situations or to use models to guide rehabilitation.

Catchment experiments can look at only a limited range of factors for a relatively short period of time, often too short to get a proper view of environmental processes. Nevertheless, useful guidelines have emerged from such experiments in North America, like the Integrated Lake–Watershed Acidification Study (Goldstein *et al.* 1984), and Europe. Nyberg *et al.* (1993) describe a 'covered catchment project' at Lake Gårdsjön, Sweden, simulating the effects of a total cessation of deposition, as part of ENCORE, the European Network of Catchments Organized for Research on Ecosystems. The Gårdsjön evidence suggests that the same rapid throughput hydrology that creates transit times of under a day and low buffering also leads to rapid rehabilitation, with SO_4– levels falling off quickly when deposition is halted.

Table 8.1 summarizes the main set of British experiments with afforestation, selective liming, ploughing, fertilizing and reseeding moorland. Despite some experimental failures, they offer a number of guidelines.

Liming the land can produce long-term improvements at economical costs as the lime is fixed in the soil. However, nitrogen fertilizers applied during tree planting or to improve moorland grazing may increase aluminium mobilization until the lime penetrates the lower soil horizons. The Swedish procedure of *liming surface waters* provides a more immediate improvement, but it is costly and needs constant renewal, because much is washed away and coarse granules can develop a coating which inhibits solution.

Both the costs and efficacy of liming can be improved by studying the sources of runoff in a basin, especially stormflow. Restricting liming to *source areas* and/or *stormflow contributing areas* makes maximum use of the lime by directing it where it is most effective. The Welsh Acid Waters Project successfully used the *a/s* index to identify source areas, although more direct identification of DCAs might have improved this still further.

Forestry practices can also be improved. Bankside clearance may not be the most effective (Figure 8.3), but planting broadleaved trees near watercourses, stopping ditches before they reach the streams, ditching

Table 8.1 Results of the Llyn Brianne acid waters mitigation experiments

Site	Treatment	Purpose	Result
Moorland	*Standard ploughing* reseeding grass, plus lime and fertilizer in 25% of basin	Return to standard treatment before end of liming subsidy	*Little effect* acid episodes still toxic with Al = 0.35 mg l^{-1}. Al mobilized in subsoil by N fertilizer till Ca penetrated
Moorland	*Contour ploughing* 16% of basin	Increase residence times to reduce acid flushes	*Slightly adverse effect* increase in sulphate and aluminium release following soil aeration
Moorland	*Liming entire basin* 8.8 t ha^{-1} fine ground limestone	Liming soil is more long-lasting than liming waters	*Effective* water pH raised from 5.2 to 6.4. Al lowered from 0.5 to 0.1 mg l^{-1}. Trout population doubled
Moorland	*Source areas only limed* 15-20 t ha^{-1} in 50% of source areas	Concentrate heavy liming of soils in runoff source areas to give long-lasting and economic coverage. Avoids unnecessary lime damage to wider moorland ecosystems	*Effective, economic, eco-friendly* water pH raised from 5.0 to 6.9. Ca raised from <1 to 6–10 mg l^{-1}. Al unchanged at 0.15 mg l^{-1} but may be less toxic at higher pH. Trout increase 1.5-fold
Coniferous forest	*Bankside clearance* and liming riparian zone	Reduce direct acid input to stream and buffer bankside runoff	*Little permanent effect* initially pH raised from 4.6 to 6.0 but episodes unaffected. Felling releases N, which increases Al mobilization
Coniferous forest	*Pelletized limestone* spread by helicopter	Avoid lime damage to leaves	*Failed* relied on frost to break up pellets, but little frost under forest canopy. Some increase in soil Ca but not fed through to stream
Coniferous forest	*Source areas only limed* 25 t ha^{-1} fine ground limestone	Concentrate heavy liming of soils in runoff source areas to give long-lasting and economic coverage	*Limited effect* water pH raised to 7.0 in wet spells but reverted in dry. Perhaps poor identification of sources?

and ploughing more across than downslope and ploughing into the mineral horizons can all help increase the buffering and reduce sediment yields. The Forestry Commission (1988) guidelines incorporate many of these findings. In addition, it makes hydrological sense to designate the DCA as a 'no-go' area for new afforestation (Jones *et al.* 1991).

Modelling can theoretically cope with situations not already studied in the field. However, it is limited by lack of understanding, particularly of some chemical and biological aspects, and by the complexity of interactions. The BIRKENES *event model* simplifies the hydrology into two reservoirs, surface and subsurface with their respective chemistries, but Neal *et al.* (1988) had difficulty reproducing stormwater quality in the Plynlimon catchments. More success may be achieved modelling long-term acidification. Whitehead and Neal (1987) chose the simulation model MAGIC (Model of Acidification of Groundwater In Catchments), developed for the Shenandoah National Park, Virginia. This simplifies the hydrology further by concentrating on *baseflow* quality and models chemical mass balance (input–output) and equilibrium exchanges. Using MAGIC, Whitehead and Neal (1987) projected trends in acidification for moorland and forested catchments in the UK (Figure 8.4). The results suggested that forested catchments will fare worse in all scenarios, and that marine salts exacerbate acidification trends.

Figure 8.3 Bankside clearance to reduce direct inputs of acid rain in the Llyn Brianne acid waters experiment

Modelling the biotic impact is even more complex. Nevertheless, some success has been achieved in combining very disparate *statistical* models of biological response with *process-based* models of flows and chemical exchange (Steinberg and Wright 1994).

The concept of **critical loads** is also being used and refined to aid in planning. Each ecosystem is seen as being able to absorb a certain acid loading without critical damage (Bull 1991; Batterbee *et al.* 1993). Many mountain areas with poor soils passed this critical level years ago, and it no longer makes economic sense to try to rehabilitate them; indeed more harm could be done to semi-natural acid ecosystems (Reynolds and Ormerod 1993). It is better to focus remedial activity on the penumbral zone, the area of advancing degradation between the 'lost' and the safely buffered zone. In this area, **acid episodes** tend to be more important than long-term **chronic acidification**.

Unfortunately, there are many practical problems in defining what is critical. The temporal and spatial distribution of a given load (Hauhs 1994) and the species involved and their sensitivity in different environments can affect its impact, as well as the soils and hydrology. Equally, the critical parameters of acid episodes have still largely to be determined (Ormerod and Jenkins 1994).

A key issue remains to determine the *degree of rehabilitation* that is desirable or practicable. Most attempts to date have been very limited and anthropocentric, concerned with restoring fish stocks. But there is the wider aim of achieving complete restoration of the aquatic ecosystem. The whole community structure may have been changed by acidification. While the lower levels of life, like bacteria, may have found a bolthole in which to survive and recolonize, this is less likely for the higher animals, like fish. A critical job in reconstructing the ecosystem is to provide the top predators that keep the system in balance. One problem with occasional liming, along Swedish lines, is that it may do more biological harm than good by not providing a sustained change in water quality and so interfering with the productivity and structure of the ecological community. Re-establishing natural biodiversity will require different measures in different environments. In some cases, physical reintroduction of species may be necessary, in others it might be sufficient to leave it to nature. In some lakes, acidity disrupts the nitrogen cycle, whereas in others it disrupts the phosphorus cycle.

It is also important to identify and distinguish the natural causes of acidification. It is neither desirable nor perhaps ultimately sustainable to attempt to reverse natural trends.

While emission controls are taking effect, there is a clear need for careful moderation of management practices in forestry, agriculture and hydrology.

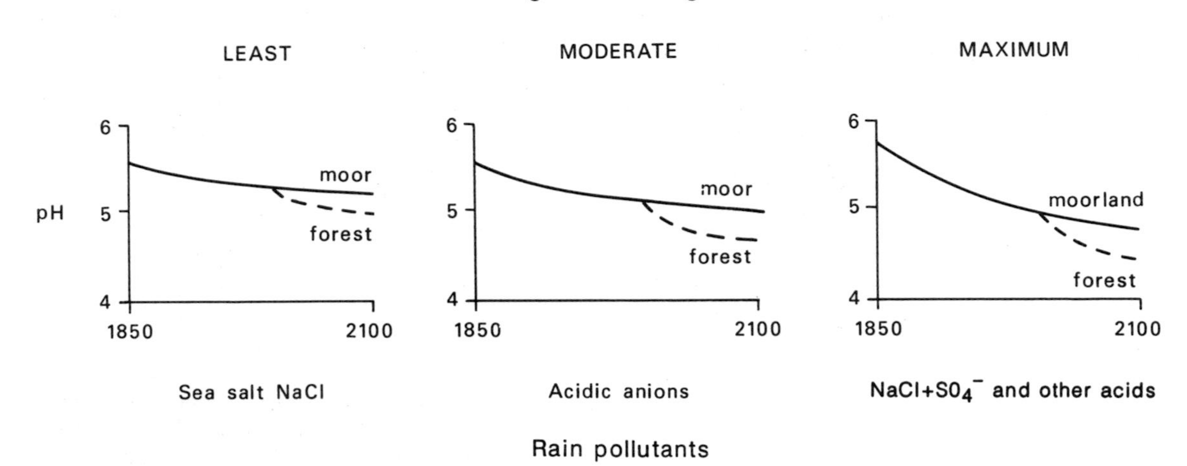

Figure 8.4 Projected trends in acidification in moorland and afforested catchments predicted by the MAGIC model. Simplified from Whitehead and Neal (1987)

Ultimately, it may only be feasible to turn the clock back to a limited degree and even then only in certain areas where the greatest ecological benefit or the best human cost–benefit ratio, e.g. in terms of fisheries, can be achieved (Barth 1987). Even so, the notion of turning the clock back at all remains largely only a scientific prediction: the hard field evidence for sustained success is still very limited.

8.2 Major surface sources of pollution

Despite recent concern over acid rain, surface sources continue to dominate water pollution. Organic pollution tends to be relatively short-lived, but it can produce devastating effects and it continues to be the major concern for drinking water quality in the tropics. Inorganic pollution can be equally devastating in the short term, but it also has the ability to lie around in the environment for long periods of time and be reworked during periods of high runoff.

This section will mainly describe the sources. Solutions will be discussed in section 8.3.

8.2.1 Organic pollution

Organic pollution originates from a mixture of point and nonpoint sources. Traditionally, human sewage has been the principal source, but agricultural manure, slurry and silage liquor have become major sources in the 1980s and 1990s as progress in treating urban sewage has been counterbalanced by intensification of agricultural production. Larger herds, overwintering in confined spaces, with slurry tanks holding the waste, and the trend in Europe towards 'zero grazing' with silage (fermented hay) replacing grazing, have all increased the danger of leakage into streams. Local production of slurry often exceeds the capacity for spreading on fields, yet it is currently uneconomic to transport more than 12 km.

Reported incidents of 'accidental' leakage of farm waste into watercourses in England and Wales rose from zero in 1970 to 2034 in 1992 (NRA 1993). Organic leaks accounted for 73 per cent of all pollution from agriculture. To counter this growing problem the NRA established fast-response units and has followed a strong policy of taking offenders to magistrates courts, where fines of up to £20 000 can be exacted. An estimated 100 000 fish were killed in a single incident in 1994 on the River Camel in Cornwall when farm slurry polluted the river and a nearby fish farm. A similar number were killed in the River Perry, Shropshire, after a tank of pig excrement burst in 1986 (Pearce 1986).

Organic pollution can adversely affect colour, smell and turbidity (from suspended matter), and introduce pathogenic bacteria and viruses into watercourses. But the most ubiquitous problem for the environment is probably loss of oxygen. Oxygen is the most important dissolved gas for wildlife within surface waters and levels are normally maintained by turbulence in flowing water. Where dissolved oxygen levels are low, it is usually the fault of organic pollution. Bacteria use up oxygen as they digest rotting organic material. The biochemical or **biological oxygen demand (BOD)** is defined as the amount of oxygen needed for bacterial decomposition to reach a stable stage. The loss of dissolved oxygen is normally measured by incubating a water sample in the dark for 5 days and is regarded as a key indicator of water quality (Table 8.2).

Crude sewage has a BOD of 600 mg l^{-1} O_2/5 days, compared with under 5 mg l^{-1} for unpolluted water. Figure 8.5 illustrates a typical pattern of BOD decline and restoration downstream from a sewage outfall. Slurry from cowsheds has a BOD ten times higher than sewage. Worst of all, silage liquor from fermented fodder grass can have a BOD 3000 times higher than sewage and is therefore lethal to freshwater life.

The UK National Water Council (1981) classification of rivers is based upon three parameters that are closely linked with organic pollution: BOD, dissolved oxygen and ammonia (Table 8.2). **Dissolved oxygen (DO)** is a measure of the actual status of the water at the moment of sampling. High levels of **ammonia** (NH_3 or the ammonium cation NH_4) are also a clear indicator of organic pollution, particularly sewage. Morc than 0.2 mg NH_3 l^{-1} is toxic for salmon. Even lower levels can reduce growth in trout.

Excessive levels of nitrogen and phosphorus produced by bacterial decomposition of organic waste can lead to **eutrophication**. One of the first signs of pollution is the appearance of velvety clumps of blue-green algae on the bed at N concentrations around 2 mg l^{-1}. High nutrient levels initially cause abundant growth in all the plants, but the algae tend to take over by reducing the supply of light and nutrients to other

Table 8.2 The UK National Water Council (1981) classification of rivers

Class of river	DO % saturation	BOD mg l^{-1}	Ammonia mg l^{-1}	Quality and use
1A	<80	<3	<0.4	High amenity, potable extraction, game fisheries
1B	>60	<5	<0.9	Less high quality, but similar uses
2	>40	<9		Potable after treatment, moderate amenity, coarse fishing, no physical signs of humic colour or foaming at weirs
3	>10	<17		Not likely to be anaerobic, but fish absent or sporadic, low-grade industrial abstraction, considerable potential if cleaned up
4	Likely to be anaerobic			Grossly polluted, likely to cause a nuisance

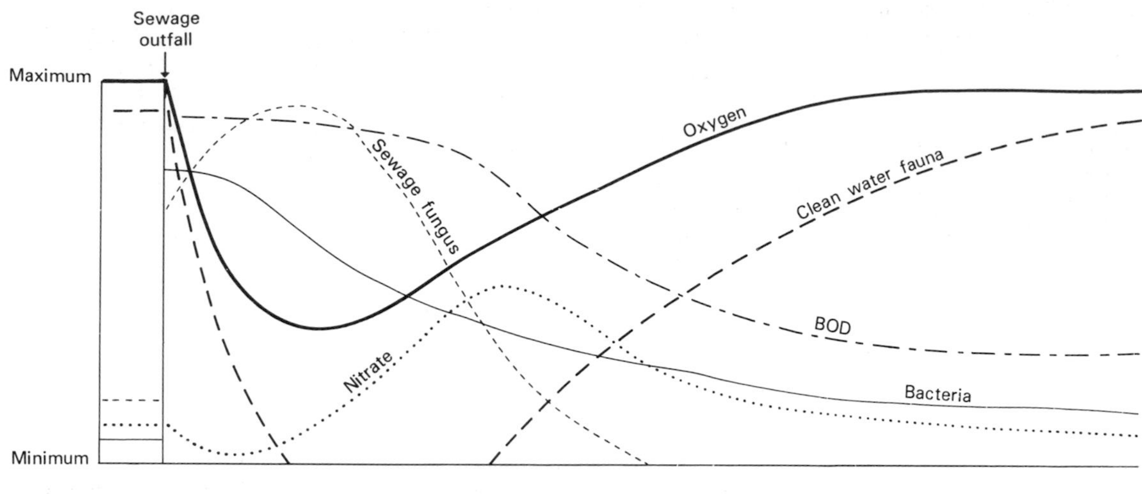

Figure 8.5 The increase and natural decline in BOD following input of organic pollution

plants. This may be aided by reactions in the chain of predators: in healthy waters, water fleas eat algae, bream eat the fleas and pike eat the bream, but when pike begin to die as excessive algal growth consumes oxygen this natural check on algae is disturbed. Fishing out the bream and stocking with pike may help arrest eutrophication in these early stages.

At around 6 mg N l^{-1}, long, waving filamentous algae float on the water, inhibiting photosynthesis in other plants, so they begin to decay. Chlorococcal algae are particularly effective at this around 20 mg N l^{-1}. Bacteria feeding on the rotting plant material lower the oxygen levels further, killing the less hardy fish like trout and leaving only coarse fish like bream or roach. Eventually all higher life is killed by lack of oxygen, in black, toxic, *anaerobic* conditions. Some may also be poisoned by **toxins** from algae like *Prymnesium parvum*, or the blue-green algae, Cyanophyta.

Instances of eutrophication have become more common. It is aggravated by higher water temperatures and slower velocities. Low flows reduce turbulent incorporation of oxygen as well as the dilution of pollution inputs, so overabstraction of water and warmer, drier summers add to the problem. In England during the warm summer of 1993, the Royal Society for Nature Conservation reported eutrophication in 102 water bodies within protected Sites of Special Scientific Interest. Toxic algal blooms have also threatened drinking water supplies, e.g. Rutland Water Reservoir in late summer 1989.

Despite increasing provision of sewage treatment plants in Developed Countries (section 1.2.1), even in Western Europe many rivers have shown a significant rise in pollution from untreated sewage in recent decades. Almost universally, the cause is over-rapid expansion of the urban population, overloading both treatment facilities and natural recovery processes. Rapid rural depopulation in northern Spain after the 1950s led to most of the cities of the Ebro basin producing river pollution levels equivalent to the average expected from much larger cities (Table 8.3). **Detergents** are a significant component of this urban pollution and the high levels of phosphorus produced by their breakdown can encourage algal blooms and eutrophication.

Figure 8.6 shows the rise in **faecal coliform bacteria** from human sewage in the River Seine. The WHO-recommended safe limit for faecal coliforms in drinking water is zero. Yet faecal coliform concentrations in excess of 100 000 per 100 ml are now found in 4 per cent of rivers in Developed Countries and 8 per cent in Latin America, where this is a major cause of infant mortality (World Resources Institute 1991).

Faecal contamination from human and animal sources can carry not only bacteria like the ubiquitous *Escherichia coli* or *E coli*, but also the virus *Rotavirus*, the protozoan *Cryptosporidium* and the bacterium *Salmonella*, all prime causes of diarrhoea and gastroenteritis, the latter coming especially from poultry. Parasitic worms can also be transmitted via ingestion of their eggs, like the beef tapeworm *Taenia saginata*. *Cryptosporidium* contaminated the Farmoor Reservoir,

Table 8.3 Effective pollution from major cities in the Ebro basin, northern Spain

Town	River	BOD (kg m^{-3})	Average discharge (m^{3}s^{-1})	Population (1000s)	Equivalent population* (1000s)	Rank order of pollution centre**
Vitoria	Zadorra	2.70	3.0	120	450	3
Pamplona	Arga	2.50	5.0	130	700	2
Balaguer	Segre	1.35	2.0	12	150	10
Lerida	Segre	1.30	3.5	80	250	4
Tudela	Ebro	0.32	50.0	22	150	7
Zaragoza	Ebro	0.88	45.0	500	2 200.2	1

Source: Based on Confederación Hidrografíca del Ebro (1976)
* Calculated equivalent in terms of pollution.
** Rank position within Ebro basin.

Oxfordshire, in 1989 through runoff entering streams from fields that had been spread with slurry, infecting over 400 people. Another outbreak in the hot summer of 1995 infected over 300 people in Devon. Unfortunately, the protozoan is resistant to conventional disinfection.

Many of these problems can be eliminated by proper storage and disposal; for example, *Salmonella* is largely eliminated if slurry is stored for a month before spreading. But they can also be very persistent once they enter the water system. In Britain, the Badenoch Report (DoE/DoH 1990) proposed tighter control of catchments as the best solution, because chlorine treatment cannot eliminate all microbiological problems and standard quality tests tend to miss them.

8.2.2 Nitrate pollution

Acidification and organic pollution are largely problems for surface waters, as they tend to be neutralized or filtered, respectively, by passage through rocks and aquifers. In contrast, nitrates affect both surface and subsurface waters and are a prime source of nonpoint

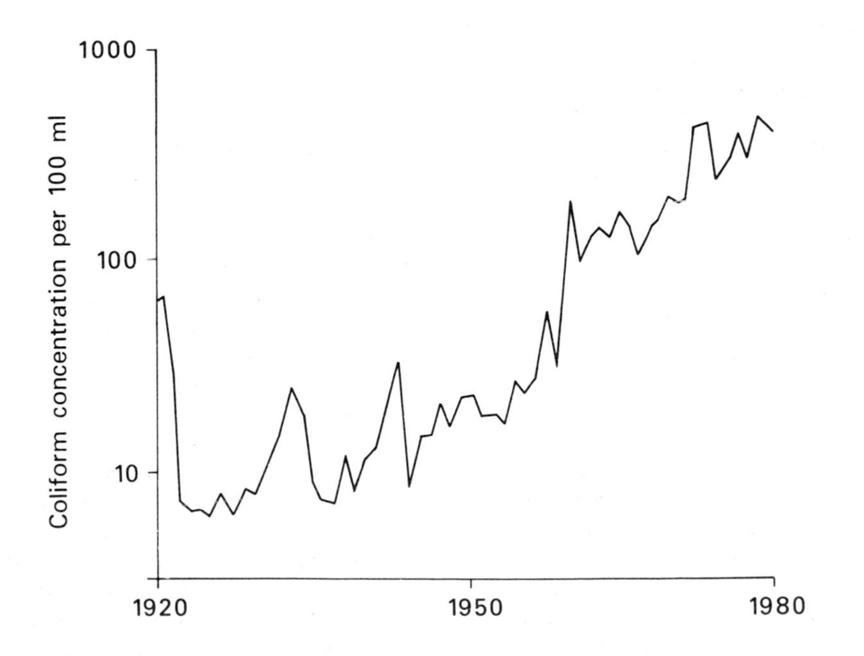

Figure 8.6 Trends in the concentration of faecal coliform bacteria in the River Seine. After Meybeck *et al.* (1990)

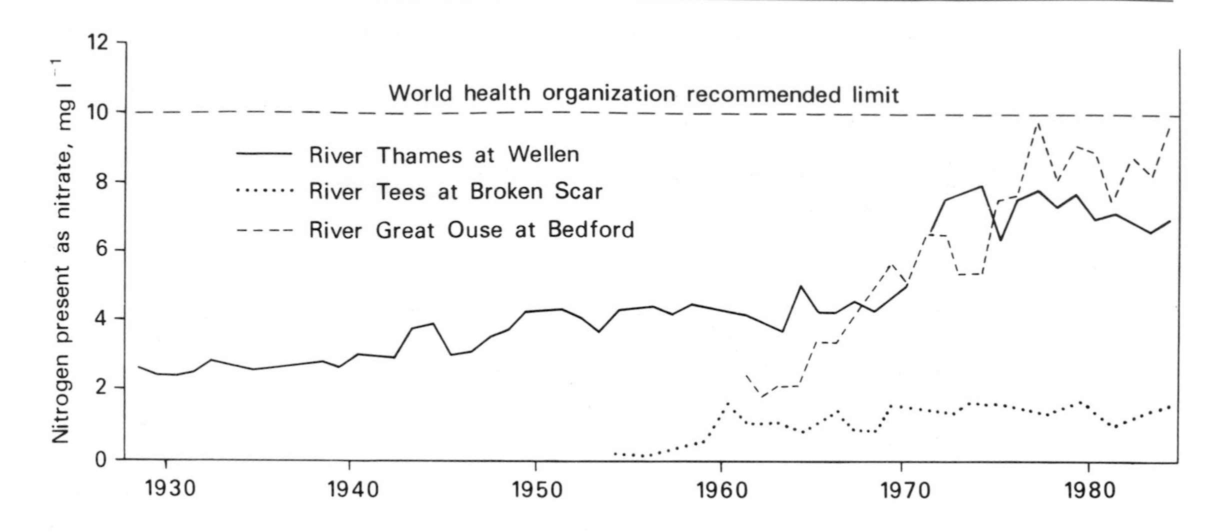

Figure 8.7 Trends in nitrate concentrations in British rivers. After Roberts and Marsh (1987)

pollution of aquifers beneath agricultural land. The main source of nitrate pollution is inorganic fertilizers (DoE 1986), in which nitrate compounds have outstripped phosphate or potash fertilizers and increased three-fold in Britain since 1963 (section 1.2.4). Urea from animal excreta has also been an increasing source.

Rising levels of nitrates in rivers and groundwater (Figure 8.7) have led to fears for human health and concern over eutrophication in watercourses. The 'blue baby syndrome' may be a result, although cases are very rare, and nitrosamines, created in the human gut in the presence of high levels of nitrogen, have been identified as carcinogens. Such fears led to the 1980 EC Directive on the Quality of Water for Human Consumption, which adopted the earlier WHO limit of 50 mg l^{-1} as the maximum permissible concentration of nitrate in drinking water. The WHO further reduced this to 45 mg l^{-1} (or 10 mg l^{-1} of nitrogen present as nitrate – NO_3-N) in 1984 (Gray, 1994). Following the philosophy that prevention is better than cure, the 1991 EC Directive Concerning the Protection of Waters Against Pollution Caused by Nitrates from Agricultural Sources has provided for the protection of aquifers used for drinking water (section 8.3.3).

Figure 8.8 illustrates the main processes and pathways within the nitrogen cycle. Nitrogen enters the soil *naturally* mainly from animal waste and by **fixation** from the atmosphere by bacteria living freely or in the root nodules of plants, especially legumes. Mammals excrete urea, but other land animals excrete uric acid, which is insoluble. Organic nitrogen from these sources is broken down by bacteria into ammonia, carbon dioxide and water and by oxidization of the ammonia into nitrates (NO_3) or nitrites (NO_2) in the process of **mineralization**. These are then available as plant nutrients, to be taken up by the process of **assimilation**, or to be leached away by throughflow. Inorganic fertilizers are available for uptake or leaching without mineralization, although they may be recycled through this route by subsequent plant decay. A certain amount of inorganic nitrogen is also lost to the atmosphere through **denitrification**, whereby denitrifying bacteria convert nitrates and nitrites into nitrogen gas, or else directly as nitrogen oxides by **volatilization**. Atmospheric pollutants, NO_x and ammonia, are now contributing yet more N to the soil system.

The timing and balance of these processes is crucial to the amount of water pollution that occurs. Biogeochemical processes tend to operate more rapidly under higher temperatures. This includes both assimilation and mineralization. Thus in summer during the growing season plant uptake tends to absorb nitrate fertilizers and to balance the increased activity of mineralizing bacteria. However, applications of nitrogen fertilizer to bare soil in autumn in preparation for winter crops carry more danger of causing excessive nitrate leaching. At that time, the artificial fertilizer is supplemented by high rates of mineralization in a soil which is still warm but crop-free, so assimilation is minimal. This is commonly aggravated by higher rainfall in autumn and early winter, which increases **leaching** and may add **overland wash**. Rainfall can also stimulate mineralization by topping up soil moisture levels. But whereas the chemical reactions that produce inorganic N peak at

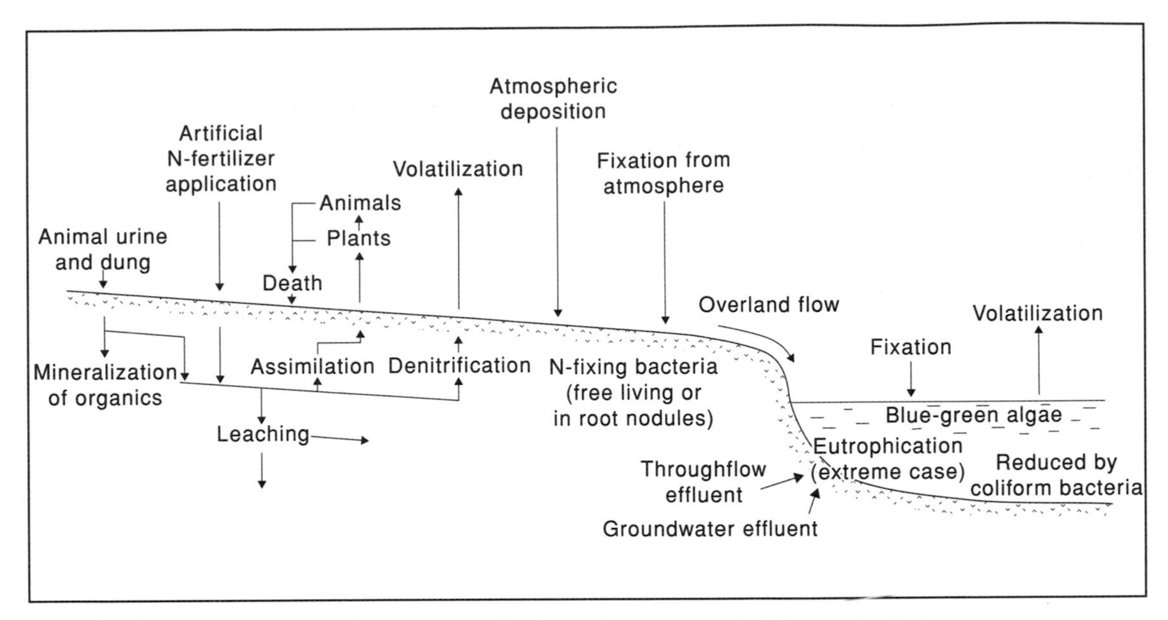

Figure 8.8 Main processes and pathways in the nitrogen cycle on land and within watercourses

moisture contents of about one-third, the leaching of inorganic N-fertilizer is greatest in soils that experience higher soil moisture deficits in summer. This is partly because desiccation will tend to develop more macropores, partly because of lower plant uptake in drier conditions and partly because oxidation leaves the fertilizer available for solution. The results of Scholefield *et al.* (1993) suggest that leachate approximately doubles between soil moisture deficits of 100 and 400 mm at the higher rates of fertilizer application.

Important biochemical processes also operate within the watercourses (Figure 8.8). Fertilizer leachates contribute to eutrophication (section 8.2.1), but blue-green algae also fix atmospheric nitrogen, and coliform bacteria reduce nitrates to nitrites, which are more hazardous for human health.

Nitrate yields in Britain are greatest in the arable southeast of England (Figure 8.9). The East Anglian Rivers Welland and Stour exceeded environmental standards for inorganic N in the late 1980s, with levels in the Stour rising at 1.5 mg l^{-1} a^{-1}. The primary factor is levels of N-fertilizer applications, but this is aggravated by summer temperatures and drought frequency. Most of East Anglia requires irrigation in 9 years out of 10 for optimum crop yields. Summer drought and soils that are prone to cracking, like fen peat soils and clayey pelosols, increase rates of infiltration and throughflow. At the same time, a worsening low flow problem due to the naturally low water surplus combined with rising levels of abstraction, plus low gradients in ditches and streams, increase N concentrations, stagnation and eutrophication.

The nitrates also infiltrate directly to groundwater. Indeed, most summer flows are sustained by groundwater in this region. But because the region contains many important aquifers, it has become the focus for groundwater protection zones (section 8.3.3). Because rates of migration into and through aquifers tend to be very slow, many water companies are only now experiencing problems created by N-fertilizer applied many years, even decades, ago. Current N concentrations in springs average the input over a time period up to the maximum retention time of the aquifer. The peak levels of pollution may still be within the aquifer system, yet to appear. Equally, preventive measures taken now can take years to show results.

This is an excellent illustration of what Rang and Schouten (1988) termed **hydro-inertia**, whereby some hydrological systems respond only slowly to certain environmental changes. They calculated the response of two aquifers in south Limburg, where applications of N-fertilizer had risen as high as 540 kg ha^{-1} a^{-1}. Table 8.4 shows the differences in the response of spring-fed streams above aquifers with short and long retention times. Even if strong measures were taken to prevent nitrate leaching into groundwater, concentrations in the springs will continue to rise. It would take 50 years for

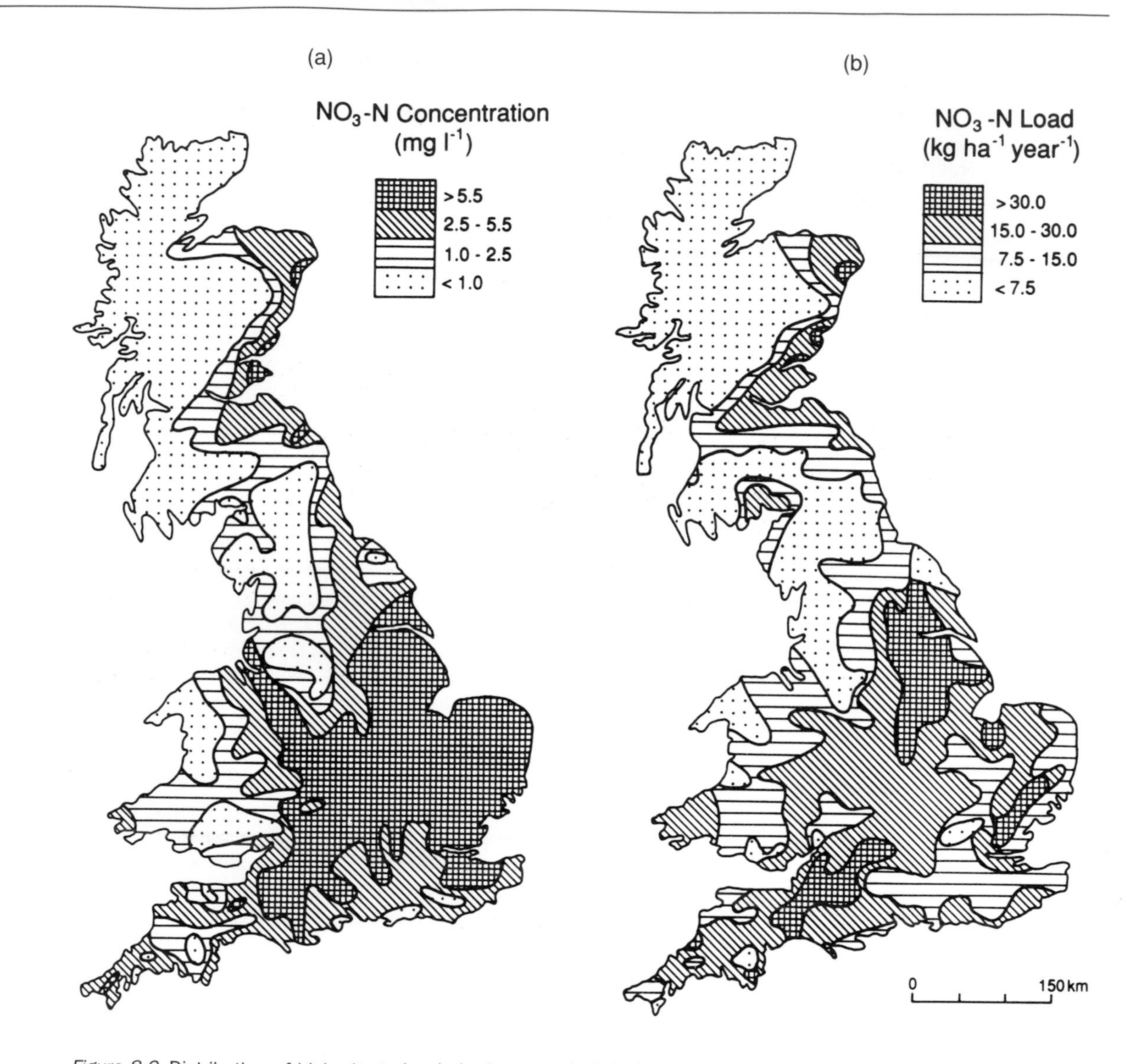

Figure 8.9 Distribution of high nitrate levels in riverwater in Britain. According to Betton *et al.* (1991)

the concentrations in streams fed by the Margrater unit to start to fall (Figure 8.10).

8.2.3 Suspended solids and sediment yields

Any undissolved, inanimate material within a river may be called 'sediment'. The water supply engineer regards all suspended sediment or solids as pollutants, affecting light transmission, clogging intakes and needing to be filtered from the water supply (Table 8.9). Suspended solids may arise from organic matter as well as inorganic sediment *sensu stricto*. Rivers with source areas in soils with a naturally high organic content, e.g. mountain peats, tend to turn brown in stormflow because of suspended organic matter (section 8.3.1). But suspended sediment is normally part of a much wider fluvial system that is variously sensitive to human management of the basin as a whole, and in which background or baseline conditions vary enormously between different geographical regions.

Maps of global sediment yield in rivers (Figure 8.11) indicate that the greatest sediment transport occurs in rivers emanating from regions of high water surplus

Table 8.4 The effect of aquifer retention time on output of nitrate, sulphate and chloride in spring-fed streams in S Limburg

Hydro-geological unit	Aquifer	Maximum retention time (years)	NO_3 (mg l^{-1})	SO_4 (mg l^{-1})	Cl (mg l^{-1})
Margrater plateau	Limestone	80–100	31	37	17
Central plateau	Gravel/sand	20–30	73	86	39

Source: After Rang and Schouten (1988).

and passing through regions with highly erodible soils and sediments. More than half of all sediment transported to the oceans is carried by the 60 largest rivers. The Hwang Ho is a pre-eminent example, in which most of the discharge originates in the Central Asian plateau and most of the sediment derives from the lowland loess deposits. High gradients also contribute to the 'stream power', the kinetic energy that causes the erosion, as in the Andes and Himalayas. But human cultivation and exploitation of the primeval vegetation often plays a critical role (sections 9.2.1 and 11.4.3). In environmental terms, only sediment from human or human-stimulated sources can be regarded as pollution.

Such human influence is by no means limited to especially fragile or Third World environments. Table 8.5 lists a range of examples that have occurred in Britain over the last 25 years. For instance, forest drainage ditches can increase flood peaks by increasing the drainage density, e.g. from 3 to 200 km km^{-2}

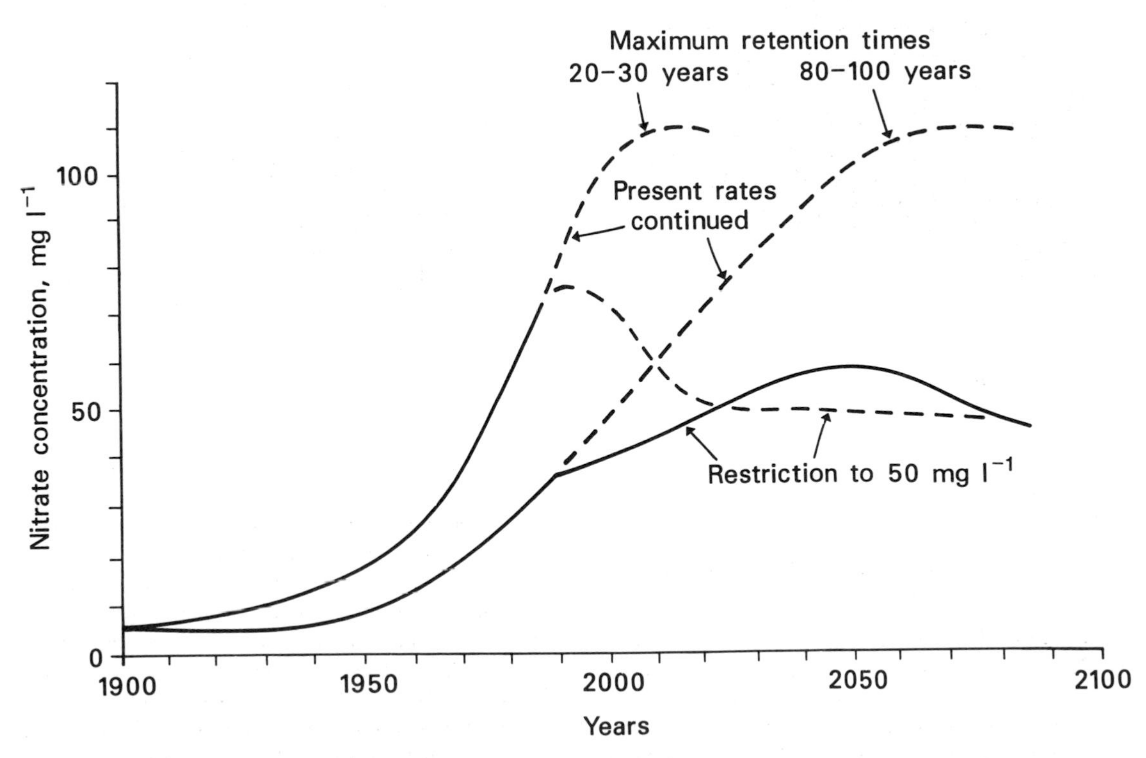

Figure 8.10 Predicted response of water quality emerging from short- and long-retention aquifers after legal restriction of nitrate inputs. After Rang and Schouten (1988)

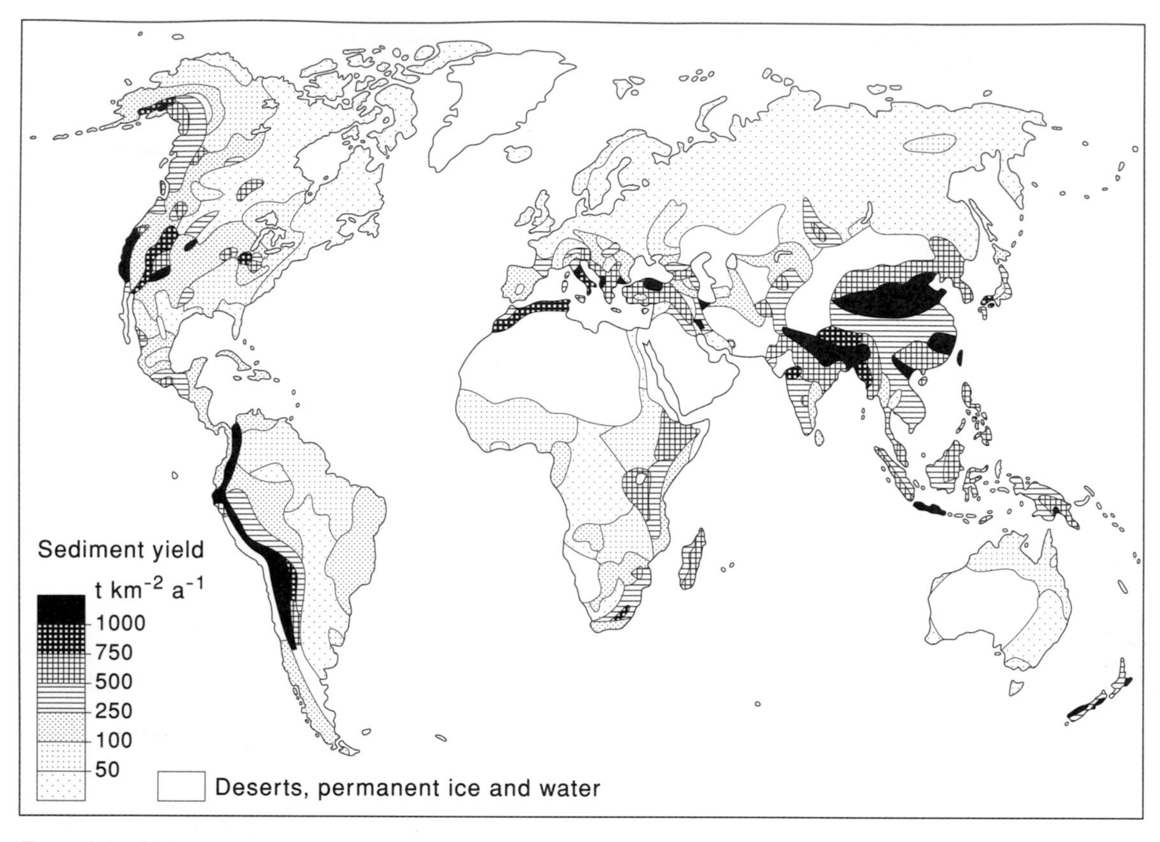

Figure 8.11 Global sediment yield in rivers. After Walling and Webb (1983)

(Francis 1988). Large quantities of sediment can also be entrained in these ditches (Burt *et al.* 1984). The Institute of Hydrology recorded four times higher concentrations of suspended solids in its forested catchment on Plynlimon because of ditching.

Although sediment is among the most widespread 'foreign material' in rivers, it is extremely poorly monitored and understood. **Suspended sediment**, which is mostly less than 0.5 mm in diameter, is only part of the total transport load in a river: coarser sediment travelling as **bedload** on or near the channel bed is far more difficult to sample, the best method being a channel bed trap (Figure 8.12). Worldwide, suspended sediment probably exceeds bedload by 14-fold, but bedload can be important in many rivers and may account for 20–40 per cent or more of all sediment movement (Dunne and Leopold 1978). Yet most measurements relate only to the suspended fraction. Moreover, sediment movement is highly variable in space and time, yet samples taken by hand-held sediment samplers or even automatic time-lapse vacuum samplers normally have a very restricted frequency.

These point samples are usually employed to construct **rating curves**, which can be used to predict suspended solids concentrations from discharge and so calculate point or total annual yields. But this can be a very inexact method, because there is rarely a simple linear relationship between the sediment discharge and the water discharge curves. Both positive and negative *hysteresis* can occur, depending upon the availability of entrainable sediments and the travel times from the main sources of sediment and for river discharge (Figure 8.13). Walling (1974, 1977) tried to overcome some of these problems by calculating seasonal rating curves, whereas Loughran (1976) chose to identify separate flow stages. These problems can now be overcome by continuous monitoring with a light source and photoelectric cell, calibrating concentrations against light attenuation, but such equipment is even rarer.

Viewed in the wider context, suspended sediments are merely the 'present and visible problem'. The overall pattern of sediment movement and deposition can create many secondary sources for pollution. Fine sediments containing adsorbed pollutants can be

Table 8.5 Human impacts upon sediment yield and discharge. Some recent examples from Britain

Type of impact	Source	Processes	Selected reports
Changes in sediment supply	Land management	*Afforestation* drainage ditches increase amount of mobile sediments	Newson (1980) Burt *et al.* (1984) Stott *et al.* (1986)
	Mining and extraction	*Metal ore mining* increases the amount of mobile sediment and concentrations of toxic metals which reduce stabilization of banks and bars by vegetation	Davies and Lewin (1974) Lewin *et al.* (1983) Lewin *et al.* (1988)
		Gravel extraction increases mobile sediment	Leeks *et al.* (1988)
	River regulation/ reservoirs	*Suspended sediments reduced* as reservoirs trap fine sediments	Petts (1980) Lewin *et al.* (1988)
Changes in magnitude/frequency of erosive events	Land management	*Afforestation* drainage ditches increase flood frequency	Robinson (1986a)
		Land drainage reduces flood frequency	Newson and Robinson (1983)
		Decline in floodplain storage due to reclamation for agriculture or construction of channel works	Leeks *et al.* (1988)
	River regulation/ reservoirs	*Reduction in erosive flows* reduced medium to high flows, higher low flows	Petts (1980) Lewin *et al.* (1988)

deposited in backwaters and floodplains for decades or centuries before being eroded and re-entering the river or releasing pollutants into solution. Dunes and bars of polluted bedload can get incorporated in floodplain sediments for extended periods. They can also stimulate local bank erosion as they migrate downstream in a discontinuous 'sediment train' and divert the river currents (Schumm 1977).

8.2.4 Heavy metals and man-made chemicals

Some of the most toxic pollution and the longest-lasting effects come from mining and the products of the chemical industry. Heavy metals, polychlorinated biphenyls (PCBs), pesticides and herbicides are among the worst. Both heavy metals and chlorinated hydrocarbons tend to reduce photosynthetic activity in plants, and cause thinner shells in birds and a variety of genetic disorders in mammals and humans.

Figure 8.12 Institute of Hydrology channel bedload trap. Sediment is periodically removed from the pit by JCB excavator and deposited on a sample divider (right), and one-eighth is weighed in a pig balance

Heavy metals 'Heavy metals' is a loose term generally applied to metals with heavier atomic weights that are most associated with toxicity problems (Alloway and Ayres 1993). Some heavy metals are essential for normal growth in small, natural 'trace' quantities, but the excessive amounts produced by mining and industry tend to be highly toxic.

Metal ore mining really began to be a problem during the nineteenth century and many mining areas long since abandoned still hold the legacy of this mining within a fluvial system that remains very sensitive to disturbance. Abandoned spoil tips are often still too toxic for plants to colonize and remain prone to sheetwash erosion. Stream channels can be overburdened with coarse spoil that forms islets, braided channels and streambanks that remain erodible through lack of plant cover. Acid drainage waters emerging from old mine adits still carry high levels of toxic metals like zinc and cadmium (Figure 8.14) (Fuge *et al.* 1991). And floodplains contain buried deposits laid down during the mining years that can easily be disturbed by excavation work or natural erosion processes (Figure 8.15).

The Maas and Geul basins in the Netherlands contain deposits more than 3.5 m thick created over the last 350 years as a result of ore and coal mining upstream in Germany and Belgium (Rang *et al.* 1986). Contamination reached a peak 100 years ago, but the Maas reworks approximately 500 tonnes of these fossil channel deposits each year (Figure 8.16). This produces an estimated annual input into the river of over 1500 kg of zinc, 330 kg lead, 40 kg copper and 6 kg cadmium. Work by Leenaers (1989) on the Geul shows that the bulk of annual metal transport takes place during just a few major flood events. These events carry two to three times as much sediment as the mean annual flood.

The bulk of this material moves only a relatively short distance before it is redeposited, which means that the sediment train could take centuries to clear naturally. The Maas is now the most polluted river entering the Netherlands and is likely to remain so for a long time. A recent agreement on effluent releases in Belgium should produce better *baseflow* water quality by 2000, but this will have little effect at *high flows,* where sediment entrainment is the main problem. In 1994 plans were unveiled for a long-term solution, in which polluted riverbank materials will be removed and deposited safe from flood erosion behind protective walls.

Monitoring on the River Rheidol in mid-Wales indicates that the legacy of mining operations which ended 50–70 years ago is still critical for water quality at very low and very high discharges (Bradley and Lewin 1982). At low discharges, lack of *dilution* of inputs from adit outfalls causes poorer quality, whereas a *flushing* effect occurs at high flows, with inputs from spoil tips, entrainment of polluted sediments and increased solution aided by an increase in acidity. Toxic levels of

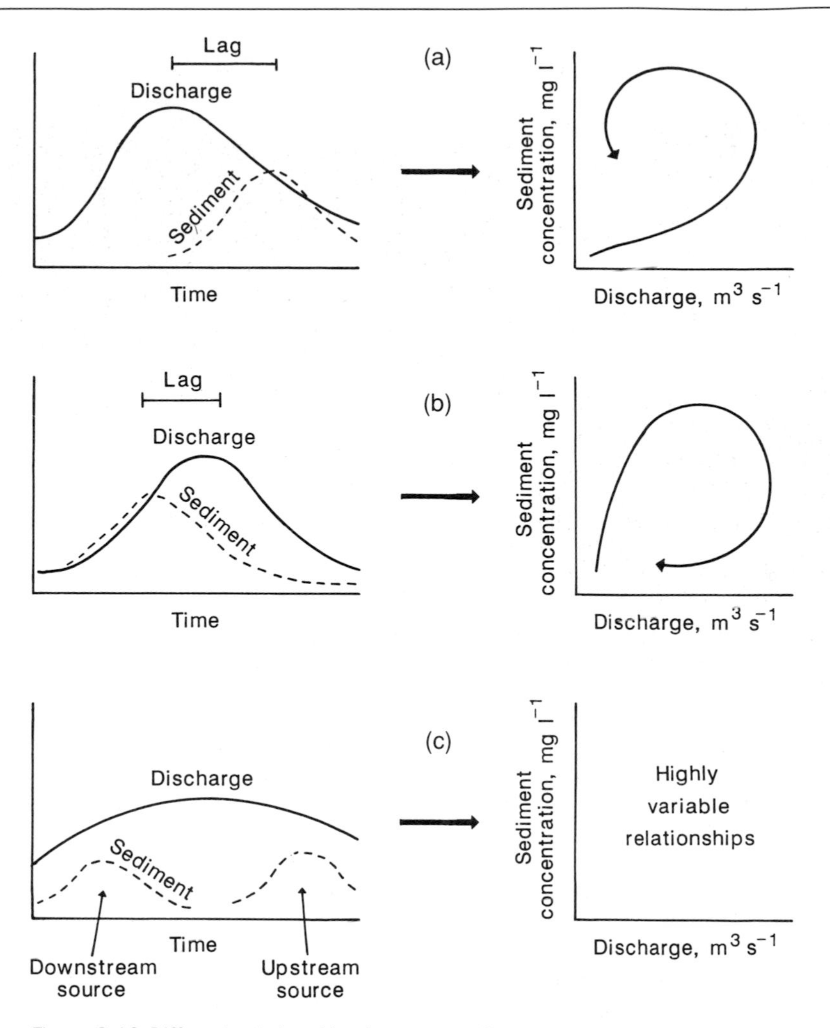

Figure 8.13 Different relationships between sediment concentrations and discharge: (a) positive hysteresis – discharge leads sediment entrainment as rising velocity lifts sediment but concentrations are diluted by high discharge around the peak runoff; (b) negative hysteresis – probably caused by the sediment source becoming exhausted before peak discharge is reached; (c) multiple sources – different tributaries bring sediment with different travel times into the main stream. This can create any pattern. Patterns can vary in space and time depending on the location of the sources of discharge and sediment within the drainage basin

zinc can still reach 0.88 mg l^{-1} and up to 13 tonnes Zn can be shifted in a single month. The metals reside both as ore particles and as coatings deposited on other particles. In such circumstances, the metal need not be dissolved to be poisonous to fish, as they may ingest it on fine particles. Even so, the Rheidol is now generally clean enough for salmon and trout to return, perhaps helped by the river regulation scheme (section 9.3).

The Rhine delta carries a more recent legacy which came principally from manufacturing and chemical industries in the 1960s and 1970s (Table 8.6 and Box 8.1).

Pesticides and man-made organic chemicals There is increasing concern over the proven and suspected effects on human health from man-made organic chemicals. Petroleum, pesticides and herbicides are the main sources. In 1993 the US Congress initiated a review of legislation on the release of new organic chemicals into the environment, with some experts

Figure 8.14 Acid drainage waters from mining adits bring water with high concentrations of toxic heavy metals into the river. Natural dilution in the river has been reduced here by flow diverted into the Rheidol hydroscheme (Figure 7.15). Limestone filterbeds are used to raise the pH and cause metals to precipitate out, but the limestone chippings soon get coated by iron bacteria and become ineffective. Reed beds are being tested for cleaning up mine effluent elsewhere (cf. Figure 8.17)

proposing that they should be tested and licensed like pharmaceutical drugs.

The principal chemicals can be grouped for convenience as: aromatic hydrocarbons, organic solvents, organohalides and other pesticides, although these groupings are not always mutually exclusive. Table 8.7 shows the WHO guidelines for these substances in drinking water. Current medical evidence suggests that some members of these groups are many times more toxic than others.

Aromatic hydrocarbons are associated with coal and oil products. Benzene is not only toxic in itself, but also the source of the extremely toxic polycyclic aromatic hydrocarbons (PAH), formed by the fusion of four or more benzene rings, and it can cause the formation of phenols in the environment (Alloway and Ayres 1993). Its use has increased with the introduction of lead-free petrol.

Organic solvents are also mostly derived from petroleum. Solvents include adhesives, cleaning substances like trichloroethylene and detergents. One of Britain' s worst pollution incidents was due to a leak of 10 000 litres of solvents into the River Roden, a tributary of the Severn, near Wem in Shropshire in 1994 (Figure 9.5) – 250 000 people in the downstream counties of Worcestershire and Gloucestershire either had their water cut off for two days or had to boil it. Ironically, few people were affected in Shropshire itself, because most of their water is abstracted from groundwater (section 9.3).

Organohalides are organic chlorine compounds. They may be introduced into water by pesticides and herbicides or by an ever-increasing range of synthetic polymers used in paint, plastics and electrical insulators. Unfortunately, they can also be created by standard chlorine treatment designed to sterilize drinking water, especially when that water contains high levels of organic matter. Chloroform, a trihalomethane thus formed, is suspected of inducing cancer of the rectum and colon.

DDT (dichlorodiphenyl trichloroethylene) and lindane are important pesticides in this group. Severe limitations, or in the cases of DDT, aldrin and dieldrin complete bans, have been imposed on these polychlorinated agrochemicals throughout most Developed Countries because of their toxicity to fish and other cold-blooded animals, their long persistence (Table 8.8) and accumulation in the animal food chain, and their

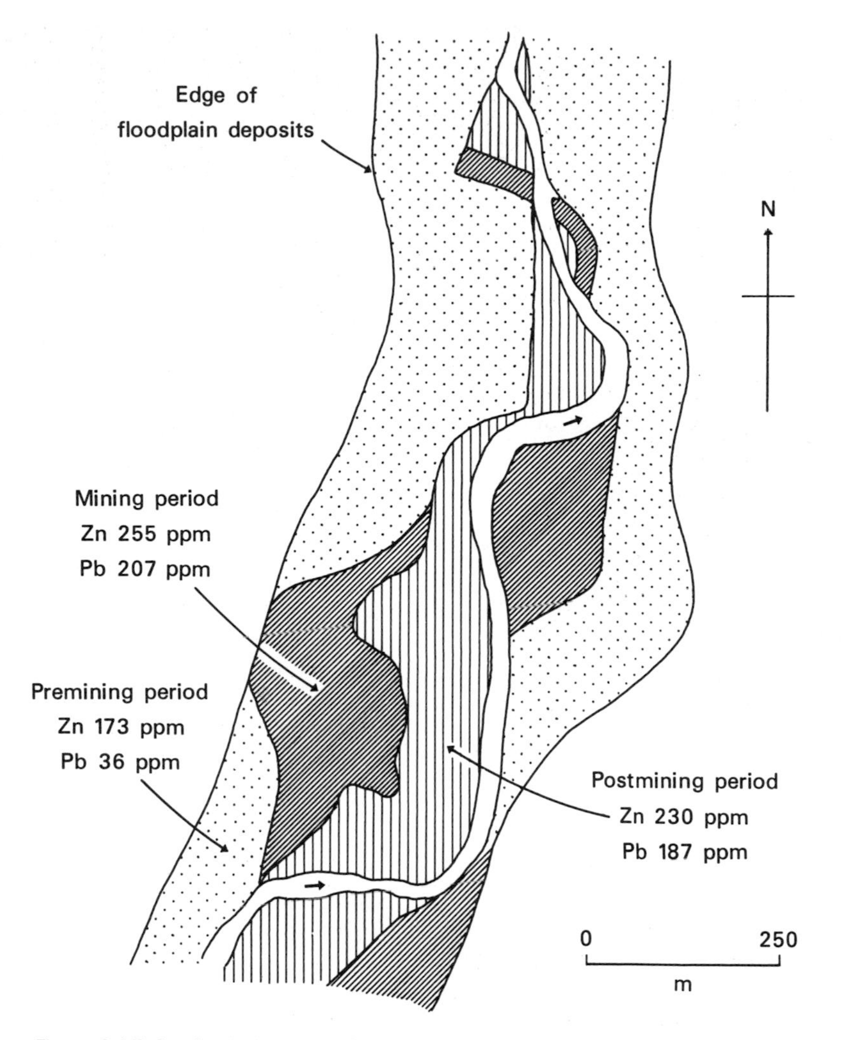

Figure 8.15 Overbank deposits of heavy metals and abandoned channels can provide sources for recontamination of the river when they are eroded or flooded. Map of floodplain deposits on the River Severn at Llandinam by Brewer and Taylor (in press)

frequent degradation into even more toxic by-products. Although worldwide demand has levelled out, their use has increased in the Third World. The insecticide DDT is a common pollutant in rivers draining the pastoral regions of Africa, like the Zambezi. Even in 1995, there was concern at discoveries of DDT in Welsh rivers, where it was officially banned in 1984 and should have largely degraded.

A wide range of organochlorine polymers have been developed since the 1940s, including polyvinylchloride (PVC) and polychlorobiphenyls (PCBs). The decline of the European otter and death and illness in seal and dolphin populations in the North Atlantic have been ascribed in part or all to PCBs (Alloway and Ayres 1993). Highly toxic dioxins have been used in herbicides, notably in Agent Orange, but many dioxins are created by accident as chemical by-products or by the incineration of commercial polymers, and get buried or washed away without the danger being realized. This happened in 1982 in the city dumpsite within the partially dug Love Canal, Niagara, and emerged through the storm sewers (Kamrin and Rodgers 1985). Relatively insoluble pollutants like dioxins may still cause water pollution by adhering to fine suspended particles.

The **organophosphorus** and **carbamate** pesticides and herbicides, like malathion and parathion, introduced progressively since the 1950s, break down far

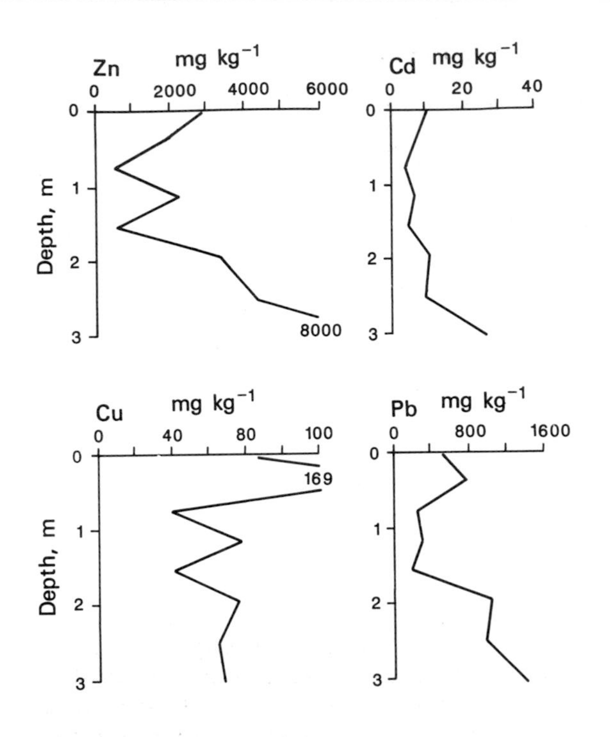

Figure 8.16 Contamination stored in the floodplain deposits of the River Maas, laid down at the peak of mining activity 100 years ago. After Rang *et al.* (1986)

more rapidly in the environment than the organochlorines (Table 8.8). Unfortunately, this advantage may be offset by soil organisms which increase their solubility, so that the agricultural herbicides atrazine and simazine can get washed out of the soil well before they have decayed. Both have polluted water supplies in Britain and are on the Environment Agency's 'red list' of highly dangerous substances. The NRA recently began using satellite imagery to trace maize fields, because maize is one of the few crops for which atrazine is still licensed and maize cultivation in the basin of the Hampshire Avon doubled between 1990 and 1995. Farmers found to be growing maize are visited and given advice.

Bio-accumulation is a major problem with organochlorines and some aromatic hydrocarbons. Early evidence of accumulation in the food chain came from Clear Lake, northern California, in the 1950s (Moriarty 1988). Spraying with DDD, a close relative of DDT, to kill gnats originally produced concentrations of only 20 ppb in the lake, but it accumulated within the food chain so that the birds at the top of the predatory chain accumulated 1600 ppb in their fat.

Much recent concern stems from work in Britain and Florida that links cancers and sexual abnormalities in humans and animals with at least 37 different chemicals that disrupt hormone operations either by acting like the female hormones, oestrogens, or by blocking normal hormonal activity (Sharpe and Skakkebaek 1993; Dibb 1995). An American lake again provides the disturbing evidence. Lake Apopka in Florida has an abnormally high incidence of alligators with trans-sexual characteristics, yet the lake water is apparently 'clean'. There was a major spillage of pesticide in the area in the early 1980s, but neither this nor its decay products can now be found in the water. In fact, the products are there, but they are stored in animal fat and are being recirculated through the predatory cycle. Parallel research in Britain has shown that male fish kept in cages below sewage outfalls rapidly begin to produce female hormones and develop female characteristics.

The problem chemicals include many chlorinated hydrocarbons, DDT and its breakdown product DDE, some PCBs, dioxins and several pesticides and fungicides, many of which are resistant to biodegradation.

Table 8.6 Concentrations of heavy metals and oil in bottom sediments in the Rhine Delta (mg kg^{-1})

	Zn	Cu	Pb	Cd	Oil
River Rhine AD 1500	93	21	31	0.5	0
River Rhine 1984	1500	207	269	50	1000
River Maas/ Meuse	3680	152	524	26	–

Source: Selected from Schouten and Rang (1989).

8.3 Controlling water quality

The foregoing sections have given us ample examples of attempts to rectify problems once those problems have grown to an almost unrectifiable scale. The history of water quality control is not very long and most of it has been dominated by 'end-of-pipe measures', treating the symptoms rather than the causes. The concept of 'polluter pays' legislation was a laudible springboard for action, even though the British legal system at least can sometimes make it very difficult for the enforcers (Kinnersley 1994). The concept ranges from charging for clean-ups after a disaster to simply exacting licence fees for discharges, what the Environment Agency terms 'incentive charging'. But the cost of a clean-up can be extremely high, even beyond the means of the polluter and quite frequently beyond the retribution

Table 8.7 Selected World Health Organization guidelines for manufactured organic compounds in drinking water

Pollutant	Use/source	Guideline limit ($\mu g\ l^{-1}$)
Aromatic hydrocarbons		
benzene	Motor fuel/solvent	10
benzo-(a)-pyrene	diesel exhaust	0.01
Organic solvents	Refrigerants, drugs,	30
chloroform	chlorination of water	
Organohalides		
DDT	Pesticide	1
lindane	Pesticide	3
aldrin, dieldrin	Pesticide	0.03
1,2-dichloroethane	Vinyl chloride for paints, plastics, etc.	10
1,1-dichloroethane	Vinyl chloride for paints, plastics, etc.	0.3

Source: WHO (1984).

Table 8.8 Decay rates for selected pesticides and herbicides

Organic compound	Half-life in environment (approx.)
DDT	8 years
Dieldrin	3 years
BHX	1 year
Glyphosate	60 days
Parathion	2–8 days

Source: Based on Alloway and Ayres (1993)

allowed in law. The Dutch have realized that the State may have to contribute to some of the clean-ups required by the 1986 Soil Protection Act.

Ultimately, the only sustainable way forward lies in the principle that 'prevention is better than cure'. This does not necessarily imply a strict Lvovichian system of isolated recycling units, but it does require setting limits that are based on the best available information about their impacts on the environment and human health, effective monitoring, and a responsive policing and warning procedure. It will also be more effective if it involves a dialogue rather than confrontation between potential polluters and the enforcers. Her Majesty's Inspectorate of Pollution was successful within its policy of 'integrated pollution control' in persuading certain dischargers of cadmium that the value of reclaiming the cadmium and savings in water charges by recycling offset the costs of the new installation. The concept of 'best available technology not entailing excessive cost' (BATNEEC) enshrined in the British 1990 Environmental Protection Act embodies the inevitable balance between conflicting interests. The 'best available' is the most effective current method of prevention, but 'excessive' is qualitatively open-ended. In practice, however, if the environmental argument outweighs the bearable cost, then licences can simply be refused.

8.3.1 Cleaning up the rivers

Legislation and monitoring The 1977 US Clean Water Act and its 1987 extension to nonpoint sources form the backbone of American legislation. This requires states to set local **water quality standards** for their rivers, while the EPA sets national **effluent standards** for industries and municipalities, and both bodies then adjust the permitted effluents to the local water quality requirements. Experience in Colorado shows how important it may be to adjust water quality standards to local conditions and to have a good database: monitoring subsequent to setting the standards revealed that many streams already had background levels which exceeded the limits. This was due to centuries of mining spoil, and the State reviewed standards in light of the mining companies' difficulties. Herein lies a major quandary for legislators: should they aim to retain the total pollution within general ecological or health limits, or should they merely control the additional pollution above local background levels?

The 1987 amendment required states to identify nonpoint sources and produce an action plan that would be funded from $400 million earmarked by the EPA.

Not only was this allocation widely regarded as insufficient, but by the mid-1990s barely half of it has been spent.

In England, the 1989 Water Act that set up the NRA provided for similar **river quality objectives (RQOs)**, set by the Environment Secretary and enforced by the NRA. But these were originally set only to maintain (option 2 in the quandary) rather than improve water quality and were not statutory. The EU Municipal Wastewater Directive should change this, but the costs that government and the private water companies will accept to achieve the objectives remain an open political question (Kinnersley 1994). It is, nevertheless, clear that, despite negotiating opt-outs for some compliances extending into next century, Britain is gradually falling in line with EU environmental legislation after decades of maintaining that islands with short rivers and surrounded by 'self-disinfecting' seas are an exceptional case.

No legislation can be effective if data are not collected appropriately or if the legislation does not take due regard of environmental processes. Even some EU legislation, which in many ways is currently leading the race, suffers from lack of appreciation of the way the environment operates, or more to the point how it may be possible to falsify the apparent situation by carefully selecting the dates, frequencies and locations for taking samples. The Bathing Waters Directive is a publicly visible example where the UK initially nominated only a handful of beaches that seemed to require the legislative controls and where Italy continues to be accused of 'massaging' the data by careful sampling.

The main problem is *temporal variability* in discharge and quality, which affects baselines or background concentrations as well as pollutants. How much is carried by storm events compared with normal flows? There is no universal answer, but there is a general tendency for storms to do most of the work (section 8.1). Typically, 90 per cent of sediment movement occurs in just 20 per cent of the time. Old-style manual monitoring schemes sampling once a week tend to miss these events. Even 'continuous' monitoring with automatic multi-bottle vacuum samplers may miss or under-represent critical events if the sampling frequency is too low. Conversely, if sampling is too frequent then the costs of laboratory analyses can become excessive. It may also give too much statistical weight to baseflow situations.

Sampling proportional to flow is a much better strategy than either regular or random sampling. Statistical tests have suggested that this can improve the accuracy of estimating average values by 15–20 per cent, and it is vital when pollution is concentrated in high flows, as with acid flushes.

General characteristics of river water, like pH, electrical conductivity (a measure of the concentration of dissolved salts) and BOD, can be monitored on a continuous basis by automatic instruments with sampling intervals set at a few minutes to a few hours. The NRA completed deployment of its automatic water quality checking system Cyclops ('the beast with many eyes') in 1994. Cyclops also issues an alert when pollutants exceed a preset limit.

More detailed analyses still require removal of water samples to a laboratory. For example, the British Harmonized Monitoring Scheme specified a common set of 117 determinants to be measured, together with recommended accuracies for determination, in order to satisfy the Control of Pollution Act (COPA) 1974 (Simpson 1980). Not all of these will be measured at every site. In Table 8.9 the most common determinants are grouped under major headings.

Despite major advances in monitoring technology, the number and *spatial distribution* of monitoring sites still frequently present problems for policing pollution. Most monitoring sites are on large rivers, because these tend to be of most interest for water abstraction. However, large rivers are also complex integrators of many sources of discharge and pollution, so chasing the source of contamination can be difficult. Of the 500 or so sites operated under the USGS/EPA NASQAN network about 60 per cent are 'hydrologically independent', i.e. on separate river systems, which means that few large river basins have many sampling sites.

The best solution to this ubiquitous problem at present probably lies in self-monitoring by industries, as embodied in the EPA's effluent standards and the general notion of discharge licences, but this does not cover all eventualities. In Britain in the 1970s the Saline and Freshwater Fisheries Act gave Water Authorities the power to prosecute for pollution incidents. The 1991 Water Resources Act gave the NRA more power to prosecute, but also, significantly, to *prevent pollution.*

This has opened the way to catchment management through water protection zones and nitrate-sensitive areas. In 1995, the NRA proposed the River Dee basin as the first integrated surface and subsurface **water protection zone**. The basin provides water for 5 per cent of the population of England and Wales and the three current monitoring stations record about 40

Table 8.9 Main determinants in the UK Harmonized Water Quality Monitoring Scheme

Determinant	Accuracy
Temperature	0.5°C
pH	0.1 units
Alkalinity (total)	2 mg l^{-1} as $CaCO_3$
Conductivity	5 μS cm^{-1} at 20°C
Oxygen	
Dissolved oxygen	0.1 mg l^{-1} and 1% saturation
Biological oxygen demand (BOD)	not defined
Dissolved anions	
Sulphate, chloride	1 mg l^{-1}
Ammoniacal-N, nitrate-N, nitrite-N, fluoride	0.1 mg l^{-1}
Orthophosphate	0.01 mg l^{-1} as P
Clarity and colour	
Suspended solids	1 mg l^{-1}
Turbidity	2 Formazin turbidity units
Colour	2 Hazen colour units
Carbon	
Dissolved organic carbon	1 mg l^{-1}
COD	4 mg l^{-1}
Macroelements	
Calcium, sodium	1 mg l^{-1}
Magnesium, potassium, silica*	0.1 mg l^{-1}
Phosphorus, manganese, iron	0.01 mg l^{-1}
Detergents	
Anionic and nonionic	0.05 mg l^{-1}
Heavy metals	
Cadmium, chromium, copper, lead, nickel and zinc	0.01 mg l^{-1}
Other rare elements used in manufacturing	
Boron	0.1 mg l^{-1}
Vanadium	0.01 mg l^{-1}
Selenium	0.001 mg l^{-1}
Poisons	
Arsenic and cyanide	0.005 mg l^{-1}
Mercury	0.0001 mg l^{-1}
Man-made organic compounds	
Phenols (monohydric)	0.002 mg l^{-1}
Organohalogens	10 ng l^{-1}

Source: Based on Simpson (1980)
* Silicon dioxide

pollution incidents a year. A pollution incident in 1984 affected two million people. Now about 260 companies will require Environment Agency consent not only to release but also to store named chemicals.

In Holland, the Soil Protection Act gives provincial authorities the power to control fertilizer and manure applications in sensitive areas, even to protect wildlife, although the authorities still have difficulty fighting a well-organized farming lobby (Schouten *et al.* 1988).

Reducing sewage pollution Improved treatment of sewage is a global imperative. Most sewage in the Third World still receives no treatment whatsoever and this is a major source of illness.

Standard methods of treatment operate at three levels (Hardman *et al.* 1994). **Primary treatment** consists of simply filtering solids. **Secondary treatment** involves biochemical treatment to remove organic pollutants, commonly by bacterial biodegradation in trickling filter

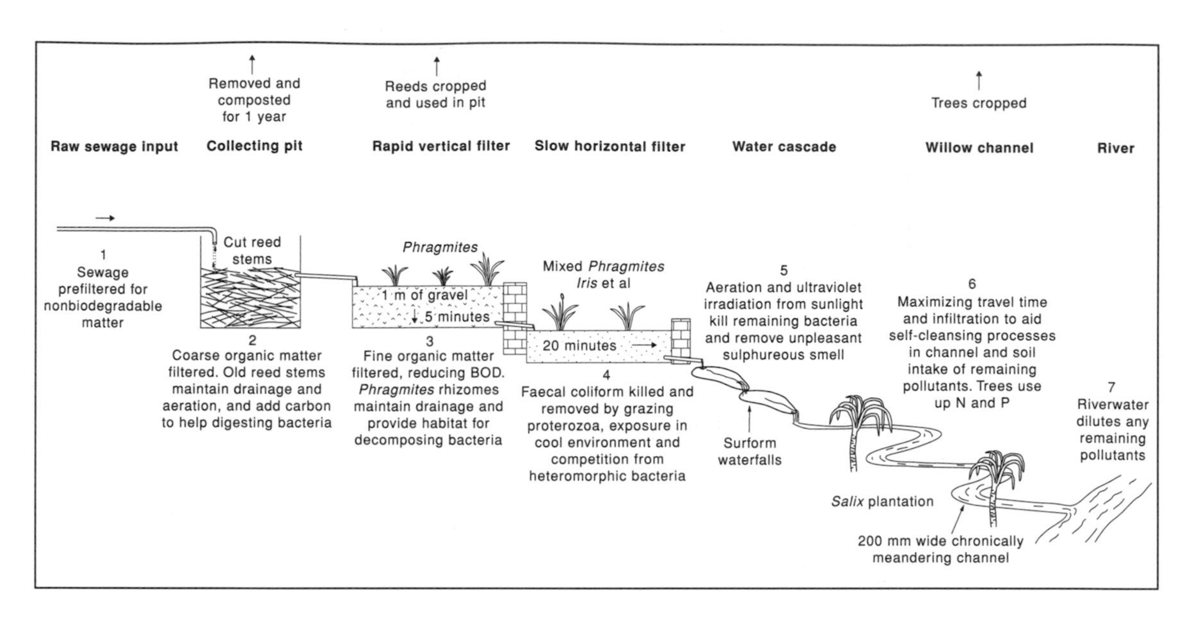

Figure 8.17 Reed beds at the UK Centre for Alternative Technology, which combine both established forms of drainage system (vertical and horizontal), plus some CAT innovations

beds. **Tertiary treatment** also removes the nitrates and phosphates that cause eutrophication. To this the latest technology is adding irradiation with ultraviolet light to eradicate viruses.

Most tertiary treatments are expensive, but a number of cheaper alternatives are being tested. Effluent from secondary treatment can be directed into **holding ponds** where algal blooms, duckweed and water hyacinth scavenge the nitrates and phosphates. The plants can be harvested and used for animal or human consumption.

For smaller communities, **reed beds** can now be designed to provide a complete water treatment based on natural biological processes. Advanced systems have been developed at the UK Centre for Alternative Technology (Figure 8.17) and on the Prince of Wales's estate at Highgrove (Prince of Wales 1993). Most solids are composted and used as fertilizer, while the reeds and other plants are selected to provide high uptake of both nutrients and pathogens.

Yet another process is thermal treatment that will convert the solids into building material (Box 8.1). The Washington Suburban Sanitary Commission has built a complete building with 'biobricks' made by combining sludge with clay and slate.

Some ways of reducing treatment costs Numerous other methods have been suggested or implemented to improve water quality or reduce treatment costs, both for drinking water and protecting riverwater quality. These may be either static or dynamic operations and broadly fall into three categories: dilution with higher quality water, settling basins, or turn-out.

Dilution is achieved by mixing in water from a higher quality source. This may be achieved with groundwater, especially in summer low flows when riverwater quality may be poor (e.g. the Shropshire Groundwater Scheme, section 9.3) or when storm runoff delivers surface water that is 'dirty' or high in aluminium from upland sources.

An interesting policy of **turn-out** has been tested by Yorkshire Water as a cheaper and more effective alternative to expensive chemical treatment of water supplies to solve the 'dirty water' problem (Pattinson *et al.* 1994). The problem has become quite widespread in late summer and autumn in upland source areas in Britain, especially since the great drought of 1976. It appears to be caused by desiccation, aeration and perhaps photochemical processes within organic soils. Since the 'dirtiness' is due to colour or dyes rather than suspended solids, it has proven very resistant to standard treatments like aluminium sulphate to cause flocculation and sedimentation. The Yorkshire turn-out policy identifies the main streams causing the problem and when they must be diverted from the reservoir (Figure 8.18). Real-time modelling, warning and source switching systems, like that on the Bedford Ouse (section 8.3.2), are a variant of this philosophy.

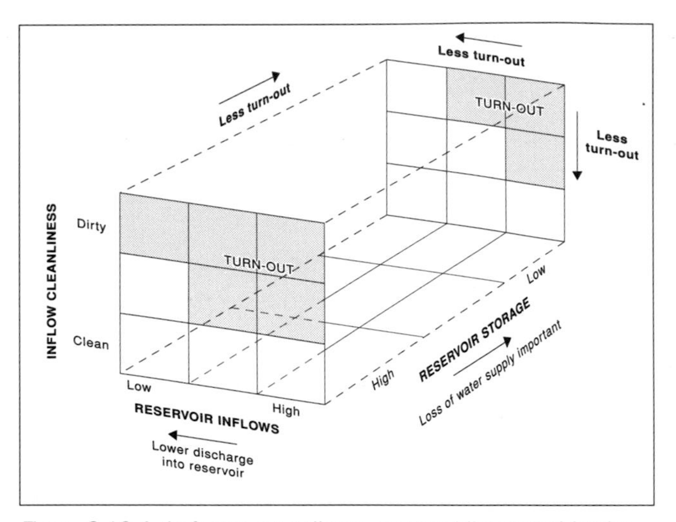

Figure 8.18 A draft turn-out policy to counter 'dirty water' intake in Yorkshire reservoirs. After the principles established by Pattinson *et al.* 1994

Settling basins are a common solution for improving urban and industrial discharges in North America and Europe. Figure 8.19 shows the River Tame downstream from the Birmingham conurbation. In places half its discharge frequently comes from water treatment plants (Figure 7.8). This was one of the most polluted rivers in England, but large settling lakes have proved highly successful and exit concentrations are equal to or better than those upstream of the sewage outfall.

Restoration of the Lower Rhine is currently following the principle laid down in the Rijkswaterstaat's 1989 Third Policy Document on Water Management that the most successful strategy is to *use a combination of measures*. These aim to improve the quality of water and bottom sediments (Box 8.1), as well as undertaking habitat reconstruction (Box 10.1) and a thorough EIA of modified operating rules for the dams and sluices (section 8.3.2).

8.3.2 Modelling water quality

To be truly effective, monitoring, policing and warning systems both require and inform models. The simplest type of model is a **plug flow** or **mixing model**, which calculates a basic mass balance. When two discharges of known concentration mix, the resulting concentration is a simple linear product:

$$\text{Resulting concentration} = \frac{Q_{\text{river}}C_{\text{river}} + Q_{\text{effluent}}C_{\text{effluent}}}{Q_{\text{river}} + Q_{\text{effluent}}}$$

where C is concentration. DOSAG, developed by the Texas Water Development Authority in the 1960s, was of this type.

More elaborate **deterministic steady state models** consider the main chemical and biochemical processes, simulating these until a steady state is reached. The Water Research Council in England produced this type of 'self-purification' model based on 11 processes, including rate of change of BOD, settlement and resuspension rates for solids, changes in dissolved oxygen due to surface aeration and aeration at waterfalls, and the respiration of channel vegetation. The WRC deterministic model divides the channel network up into reaches between inflow points and models each separately. Nonpoint inputs must be entered as point inputs at the start of a new reach. This proved very popular with Water Authorities in the 1970s and 1980s, and the Environment Agency continues to use it, but it is used mostly for planning or impact assessment now.

Most real-time forecasting is now based on **stochastic mass balance models**, which acknowledge the high degree of indeterminacy in both the chemical processes and the data that are entered into the model. River chemistry is often not 'textbook' chemistry and a steady state chemical equilibrium is often not reached before some disturbance, such as a new input, occurs. Hence the trend has been away from deterministic models towards Monte Carlo simulations of mass balances, in which randomized elements are introduced into the processes.

Most advanced stochastic models today are derived from models developed in the USA to meet American laws like the Federal Water Pollution Control Act.

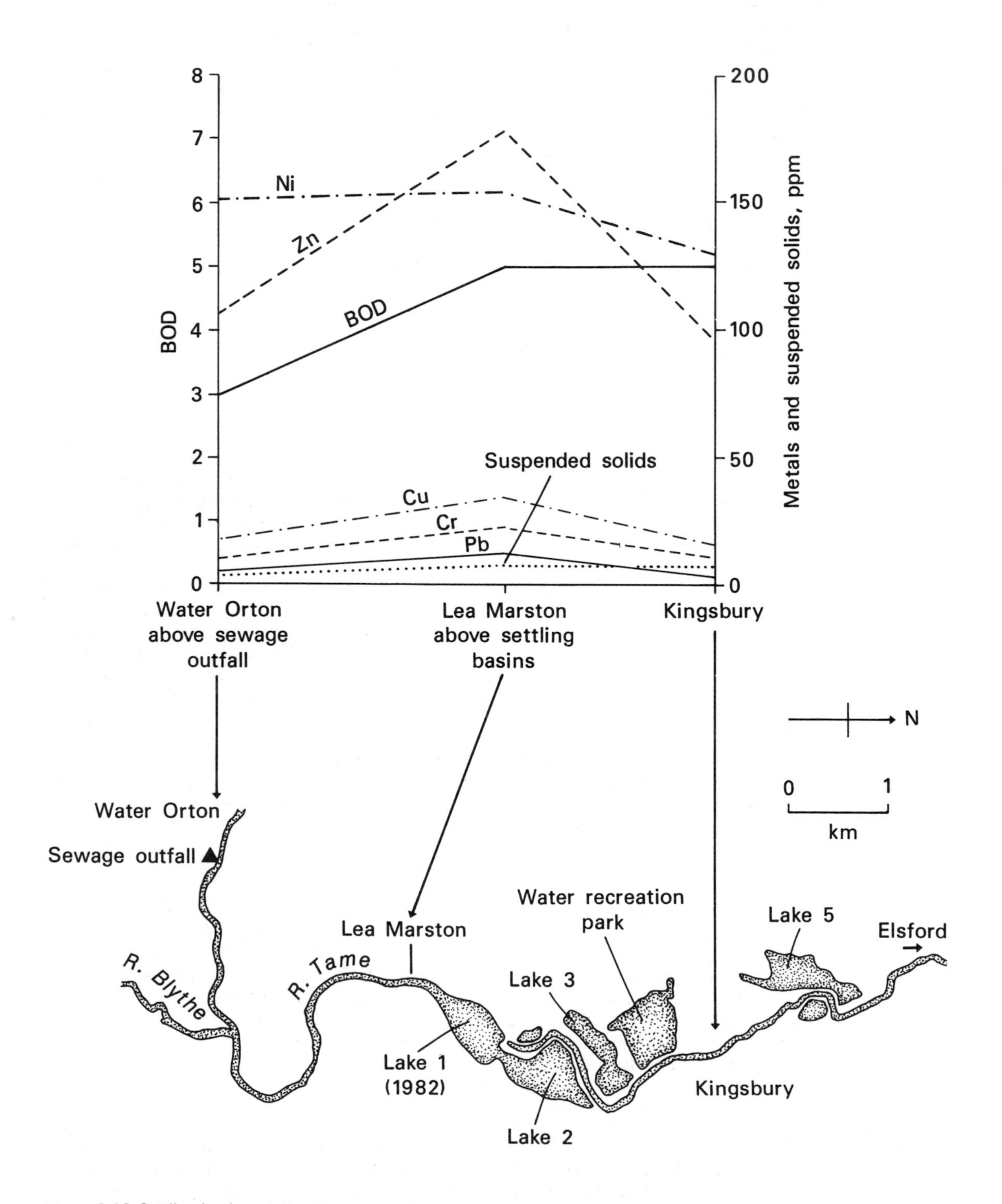

Figure 8.19 Settling basins on the River Tame downstream of Birmingham have successfully improved pollutant levels downstream of the sewage outfalls. (Data for July 1994 from Severn Trent Water)

Box 8.1 Case study: Managing and treating pollution in the Rhine delta

Pollution on the Rhine peaked in the early 1970s, but three significant problems remain: (1) major volumes of polluted sediments within the delta waterways; (2) the risk of pollution incidents, especially from upstream like the Sandos incident (section 8.3.2); and (3) the effect of delta water management upon erosion and deposition patterns and the range of accident response procedures. This is complicated by the needs of navigation on the busiest river in Europe.

The Rhine was once regarded as the 'sewer of Europe' and the low gradients and tides in the delta waterways made them a natural focus for the deposition of large quantities of pollutants. This has been aggravated by flow management. Engineering carried out upstream over the last 200 years to create a swifter, self-flushing flow shifted more sediment downstream. But the Delta Plan has been the most critical (Figure 10.5).

To the delta's misfortune, the obstruction of flows by dams and sluices (Box 10.1) coincided with the peak levels of pollution in the Rhine and Maas in the 1960s and early 1970s. Cadmium and mercury concentrations in the lower river reached 5 and 2–3 mg l^{-1} respectively at their peak in 1971, compared with just 0.1–0.2 mg l^{-1} since 1984. The Grevelingen Dam (1965) and the Haringvliet sluices (1971) trapped millions of m^3 of highly polluted sediment, now often covered by only a thin layer of cleaner sediment.

About 45 million m^3 have to be dredged annually from the waterways. Several million m^3 are strongly contaminated with heavy metals and toxic organic substances. The sludge was used to build dikes, new land or dumped at sea, but these methods are no longer permitted for heavily polluted sludge.

The mud is classified into four categories and treated accordingly. Class 1 has little or no pollution and 10 million $m^3\ a^{-1}$ can be disposed of in the old ways. Classes 2 and 3 are moderately to severely polluted with heavy metals and slightly with toxic organics and oil – 12 million m^3 are removed annually but have not been disposed of at sea since 1985. Their quality is comparable to the material currently transported by the Rhine and Maas. Class 4, generally deposited near factory outfalls, is very severely polluted and may include substantial quantities of PCBs.

Classes 2 and 3 sludge is now *temporarily stored in the 'slufter dam'*, which is designed to last 20 years. Another totally isolated, leak-proof container is established at the Parrot's Beak to hold 1.5 million $m^3\ a^{-1}$ of class 4 sludge until a permanent solution is found. Another storage depot is proposed for the Hollandsch Diep (see below).

The Dutch Institute for Inland Water Management and Waste Water Treatment (RIZA), the environmental consultants CSO and collaborating companies have already made progress towards a solution with a *thermal treatment* that turns sludge into a rock-like building material. The heavy metals are firmly bound in and guaranteed against leaching (Schouten and Rang 1989). A pilot plant is now operating at Dordrecht.

Other approaches aim to reduce the amount of polluted sediment to be dredged. The Rijkswaterstaat and the city of Rotterdam are talking to industry to *reduce present water pollution*; the city can also threaten legal action in the European Court. Since the Sandos emergency, the Rijkswaterstaat recommends that industry builds **calamity basins** to catch accidental releases. Special riverboats pump wastewater directly from barges, and Europort is now served by the largest industrial wastewater plant in the world, the 'tower biology' plant.

The fourth and fifth approaches are to *reduce hydraulic reworking of bottom deposits* and to *reduce recontamination from floodplain deposits*. These involve controlling velocities in the waterways, a floodplain soil sanitation programme (section 8.3.2) and restoration of the *forelands*, the floodplain area in front of the winter dikes (Box 10.2). Velocities need to be managed to reduce entrainment of contaminated deposits in two critical zones, the shallows and the Hollandsch Diep. The shallows tend to have more polluted sediments laid down in the 1960s, but they are relatively accessible for clean-up operations.

The most critical area is the Diep at the junction of the Maas and a major Rhine distributary, which has trapped almost 30 per cent of all the silt from the Rhine and Maas (Figure 8.20); 40 million m^3 of moderate to severely polluted sediment lies in the Diep. It is now proposed to leave this *in situ* and build a protecting island over it.

The 1990 National Policy Plan proposed that the operation of the delta dams could be improved to reduce the risk of pollution, and improve the ecology (section 8.3.2, Box 10.2). Current plans, including removal of sediments to a storage depot, could increase the area of unpolluted deposits in the Haringvliet six-fold by 2035 and reduce that of class 4 sediments from

Box 8.1 (cont.)

from 16 km^2 to just 1 km^2. In many ways *timing* is crucial to the success of the clean-up: there is little point in removing sediments of a grade that is still being replenished by the river nor in removing too much sediment that cannot properly be treated or disposed of with current technology.

QUAL1, developed 25 years ago, began a major family tree of finite difference simulation models. The EPA funded the development of QUAL2, which has been widely adopted in America. Numerous adaptations and refinements have been produced from QUAL2, for example in Britain, Germany and Holland. MODQUAL is a derivative used on the Lower Rhine.

Considerable effort was wasted in the recent past in developing models for specific rivers, many of which took a rather simplistic view of the processes. The QUAL2 family of models now provides a high standard of simulation and a commensurate level of transportability. But it does make heavy demands on data, requiring almost as much data on hydraulic parameters as a flood-routing model (section 6.4) in addition to chemical and biochemical data.

It is important not to confuse complexity with the accuracy of a model. If local data are deficient, then a simpler model, or even maximum and minimum values from the literature, may give as good an indication of the range of possible results for an EIA. Even with good data, where EPA standards aim for 95 per cent explanation during calibration, it expects only 60 per cent explanation during verification because of the difficulties of optimizing parameters (section 6.3.1).

Real-time forecasting Detailed simulation of processes is most needed for early warning systems. A succession of simulation models has been used on the Great Ouse in Bedfordshire to protect water supply

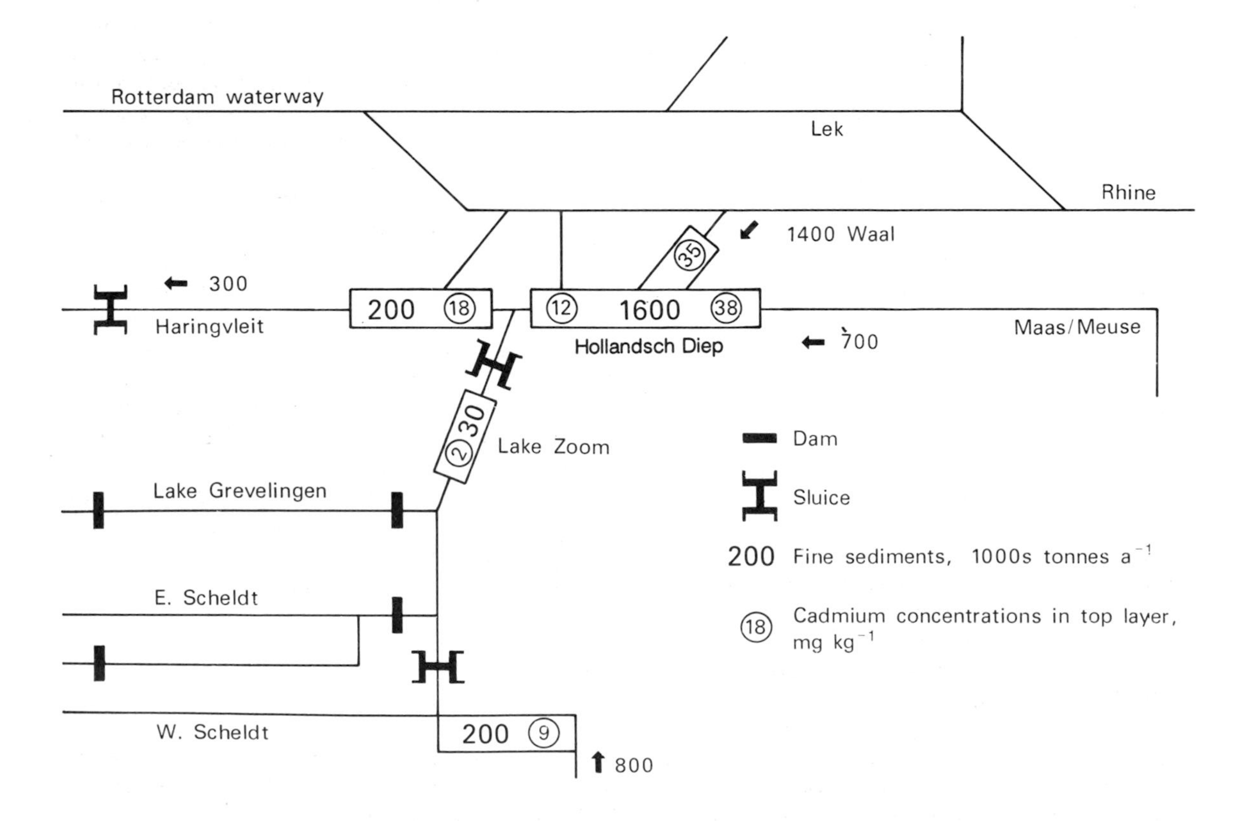

Figure 8.20 Sedimentation of silt and heavy metals in the Rhine delta, 1987. After Bielsma *et al.* (1992)

abstraction points from pollution coming mainly from Bedford, Cambridge and Milton Keynes, particularly the Thames stochastic model TOMCAT. This is linked to an automatic data collection network. There is just 50 km between the sewage outfall for Milton Keynes and the major Bedford water supply intake. Riverwater takes 3–4 days to travel this distance at baseflow, during which self-cleansing is complete. During storms this is reduced to 1–1.5 days and cleansing may be incomplete. The computer program scans river sampling stations at 1–3 h intervals, depending on flow conditions. Closer monitoring is usually needed below sewage outfalls in summer. The program transforms the data to relevant parameters, carries out data reduction and produces daily summaries. It also warns of any monitoring failures or drifts in calibration and issues a telephone alarm signal if predicted concentrations and travel times present a threat to intakes. Operators can then decide when to switch to alternative supplies.

The Sandos chemical plant spillage near Basle in 1986 caused a major improvement of the monitoring and modelling systems on the Rhine. The situation is particularly complex in Holland (Box 8.1). The Rijkswaterstaat has developed a number of models to predict the real-time movement of water pollution and long-term sediment deposition. ZWENDL is a one-dimensional model, i.e. it predicts only horizontal patterns, not vertical. It can predict pollution at numerous points throughout the whole Northern Delta Basin at 10-minute time steps, and requires input on sea levels as well as river properties in order to predict the effect of tides. DISTRO is a two-dimensional model that also predicts the vertical patterns of salt intrusion within the estuaries.

Modelling to plan management strategy The recent review of operating strategy within the Rhine delta works (Box 8.1) has combined water quality with sedimentation modelling to evaluate various operating scenarios. This combined ZWENDL with an empirical sedimentation/channel geometry model, EMPREL. It proved that operating the Haringvliet sluices only as a storm surge barrier, and leaving them open at other times, would dramatically reduce sedimentation, so that only two distributary branches would need dredging. EMPREL also showed that if dredging is stopped, most of the estuary will be silted up within a century and water levels would rise by nearly 1 m. Under the established operating rules, water levels would rise by up to 0.6 m over the next 100 years mainly through sedimentation, whereas the proposed revised rules would halve this. However, sea level rise due to global warming would tend to reduce the need for dredging, because it would raise water levels without raising the channel beds.

8.3.3 Protecting groundwater

Once groundwater is contaminated it is very difficult to rehabilitate, partly because of the slow rate of flow and a retarded microbiological cleansing activity and partly because there is ample opportunity for contaminants to reside in the more slowly permeable parts of the aquifer for very long periods. Cleansing that would take days or weeks in surface waters can take decades in groundwater.

Groundwater protection began in America in the 1980s. The EPA set up an Office of Groundwater Protection in 1984 to oversee its strategy of encouraging individual states to regulate land use and effluents in the vicinity of aquifers. Federal laws passed from 1986 onwards have strengthened groundwater protection, including the EPA's drinking water quality standards, which list 85 regulated substances from industrial chemicals to pesticides. Many states have now banned industrial development above aquifers and most exercise some control over discharges entering groundwater as the most cost-effective way of guaranteeing drinking water quality.

Similar developments have taken place in Europe, especially following the EU Groundwater Directive. Holland is especially vulnerable to groundwater pollution, not only from its own intensive agriculture and industry, but via the rivers from Belgium and Germany, and from the sea. Some groundwater has been polluted by using poor quality riverwater to irrigate farmland. Saltwater incursions are a particular concern for the coastal towns, including The Hague, where water supply is derived from the sand dune aquifers and overexploitation readily encourages sea-water intrusion. It has also been a concern for settlements around the Grevelingen, the artificial lake created by the Delta Project that was belatedly converted to saltwater (Box 10.1). But Holland's 14 million people also share just 40 000 km^2 with 12 million pigs, five million cattle and millions of chickens, which produce 100 million m^3 of manure and vast quantities of ammonia annually. Much of this manure is spread on the land, but the nutrients are often not taken up completely by the crops and the residues leach into groundwater or evaporate into the air and contribute to acid rain. Dutch farmers are forbidden to

Box 8.2 Groundwater clean-up in Rotterdam

Between 1994 and 2000 the city of Rotterdam aims to complete the most ambitious scheme of groundwater and soil rehabilitation yet undertaken anywhere. The area lies under a city housing scheme built on the site of a disused gasworks in the district of Kralingen and covers more than 11 ha. Much of the ground is heavily polluted with cyanides, PAHs like naphthalene, mineral oil and volatile aromatics. Pollution has reached the upper part of the 'first aquifer' of Pleistocene sands and is draining at a rate of 50 m a^{-1} towards the north, threatening groundwater abstraction points 500 m downstream. An emergency water supply system designed to provide drinking water in the event of a pollution incident in the Rhine or severe drought can no longer be used and new applications for groundwater abstraction are being refused.

Full refurbishment according to the letter of the Soil Protection Act proved too expensive, but 'plan B^{+}' adopted in 1993 will cost half as much (£100 million) and aims to remove 97 per cent of the sources of groundwater pollution and 87 per cent of the pollution already in the groundwater. Apartment blocks will be demolished or shored up and 270 000 m^3 of soil removed. The most polluted class C soil will be dumped at a safe site and 54 000 m^3 of soil will be purified and returned. The plan aims to attain an average standard of class B, i.e. 'slightly polluted', the usual class for urban soils, throughout the area.

An impermeable layer of clay will be buried beneath the relaid soil to further protect the groundwater against leachates in rainwater percolating through the soil (Figure 8.21). Beneath this, a layer of gravel will act as a drain for long-term removal of pollution from the Holocene clay/peat layer immediately beneath. Pumping this gravel drain will suck water up from the Holocene layer. Because the clay layer has a low permeability, pumping will have to continue for 100 years to achieve the purity standards required. The low permeability has protected the sand aquifer beneath. So too has the higher permeability of the sand, where the lateral drainage carries pollutants away faster than they enter from above. This means that only pockets of severe pollution exist at the very top of the aquifer at present. However, without any action this situation would deteriorate. Ironically, some of the pollution may be reaching the aquifer through test boreholes sunk during the pollution surveys.

Current action is therefore preceding the expected pollution peak. Five deep wells will be sunk into the sands and pumping is expected to last 35 years. A detailed discussion of pumping and alternative strategies for groundwater clean-up can be found in National Research Council (1994).

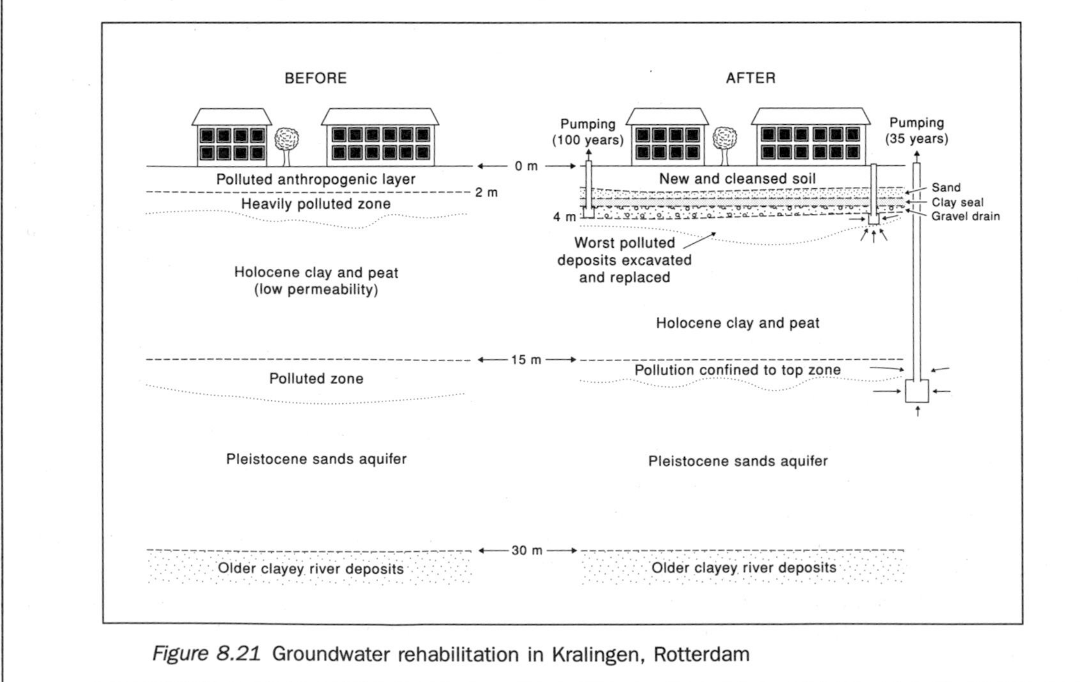

Figure 8.21 Groundwater rehabilitation in Kralingen, Rotterdam

spread manure before April because it is more likely to be washed into the groundwater by the winter rains; the offensive smell that troubled southeast England in April 1995 was tangible proof that the Dutch ammonia can volatilize rapidly.

The Dutch National Environmental Policy Plan published in 1990 aims to bring pollution under control by 2015 at a cost of $26.5 billion. The plan includes designated **groundwater protection zones**, dredging and cleansing channelways (Box 8.2), and **soil sanitation**, whereby heavily polluted soils are removed and replaced to reduce groundwater contamination.

In designated groundwater zones, the number of farm animals and methods of manure disposal are restricted. Related to this plan, the Dutch Ministry of Agriculture, Nature Management and Fisheries is funding research into methods of reducing contamination from manure. The recently introduced technique of injecting manure into narrow slits cut in the topsoil reduces volatilization (and acid rain) but only adds to percolation.

An important part of the problem is caused by phosphorus. Because animals can take up only about one-third of the phosphorus contained in plants, manure is a major source of the phosphorus which permeates through the soil. Research has shown that feeding animals with the enzyme phytase can release the non-utilized phosphorus which is locked up in the compound phytin, so most of it can be metabolized. Pigs no longer need extra phosphorus in their feed, and 33–50 per cent less P is released into the environment.

In England and Wales, the NRA (1991) noted that groundwater has the potential to provide 35 per cent of current demand, but that recent landfill and agricultural practices are presenting a major threat. Many private water companies have introduced groundwater protection policies, but there have been wide differences in approach. The NRA (1991) policy document attempted to outline a common framework and established the working principle of **groundwater vulnerability** to concentrate protection in areas of greatest need. The NRA/Environment Agency plan defines **nitrate-sensitive areas**, using the concept of **buffer zones**, to designate three degrees of protection:

1. **Zone I** is the inner zone around the source of water abstraction, which is most highly protected. This has a minimum radius of 50 m from the source and extends for a distance equivalent to 50 days travel time from the source. The groundwater travel time is determined by the decay time for the biological contaminants that are restricted within the zone.
2. **Zone II** is the outer zone covering groundwater contributing areas with travel times extending up to 400 days, over which more slowly degrading contaminants are controlled.
3. **Zone III** covers the complete catchment area for the source, which is the area needed to support abstraction from long-term annual recharge.

The NRA (1994) proposed 40 groundwater protection schemes, which were increased to 72 **Vulnerable Zones** covering nearly 650 000 ha in 1995. Restrictions are placed on use of manure, fertilizer and sludge.

Physically based models are being developed and used to predict the movement of pollutants into the groundwater. The Danish Hydraulic Institute has added a 3-D groundwater flow module with integrated solute transport for the saturated and unsaturated zones to SHE, linked it with a biological model of root zone uptake, DAISY, and applied it to nitrate movement.

Conclusions

Progress is at last being made in containing pollution, but it is an enormous task. The recent proliferation of nonpoint sources requires new techniques and new legislative powers. The range of pollutants has also been increasing and is constantly outstripping monitoring and legislative controls. Some areas of the world, notably the former Communist states, have a tremendous legacy of environmental pollution that will take decades to clear. Elsewhere, vested interests and political inertia often continue to hamper progress. EU legislation is creating major advances, but there are still instances of inadequate specifications which leave the possibility of misleading or fraudulent data.

Further reading

Technical aspects of pollution are accessibly presented in:

Alloway B J and D C Ayres 1993 *Chemical principles of environmental pollution*. Glasgow, Blackie: 291pp.

The politics and practicalities of pollution control in Britain are analysed in:

Kinnersley D 1994 *Coming clean – the politics of water and the environment*. Harmondsworth, UK, Penguin: 229pp.

Acidification problems are well covered in:

Hornung M and R A Skeffington (eds) 1993 *Critical loads: concepts and applications*. London, HMSO: 134pp.

Steinberg C E W and R F Wright (eds) 1994 *Acidification of freshwater ecosystems – implications for the future*. Chichester, Wiley: 404pp.

Wellburn A 1988 *Air pollution and acid rain: the biological impact*. London, Longman: 274pp.

Biological aspects of pollution are also covered in:

Hardman D, S McEldowney and S Waite 1994 *Pollution ecology and biotreatment*. Harlow, Longman: 322pp.

For a good review of nitrate problems in Britain:

Burt, T P and N E Haycock 1992 Catchment planning and the nitrate issue: a UK perspective. *Progress in Physical Geography* 16(4): 379–404

Discussion topics

1. Investigate the effects of acidification on your local surface waters.
2. Compare the strategies needed to contain point and nonpoint pollution.
3. Consider the pollution potential from either floodplain sediments or groundwater sources in an area you know. How would you deal with it?
4. Should regulators aim to contain the total pollution within general ecological or health limits, or should they merely control the additional pollution above local background levels?
5. How important are floodflows for increasing or decreasing pollution problems?
6. Discuss the problems of modelling pollutant movement.

CHAPTER 9

Managing runoff 1: the design of major water management systems

Topics covered

9.1 Modelling and systems analysis in modern water management

9.2 Large dams and impounding reservoirs

9.3 River regulation and conjunctive schemes

9.4 Large-scale strategic river diversions

There are three ways in which adequate, reliable and economic water supplies can be assured: (1) by conservation and *demand management*; (2) by efficient development of *local resources*; or (3) by importing *strategic resources* from other areas. This chapter will address the resource development options.

The efficient development and control of water utilization schemes requires a delicate balance between physical hydrology, human demands, economic and technical feasibility and environmental impact. Often the most difficult part of a formidable task is taking account of future trends. Computerized techniques of synthesis, simulation and decision making have made an important contribution to both design and management. Even so, major problems can still be caused by lack of hydrological information, poor environmental assessment or an uncertain socio-economic context.

9.1 Modelling and systems analysis in modern water management

Computer modelling techniques are now indispensable in the two broad areas of designing and operating major schemes. At one level, this may mean techniques of streamflow synthesis or simulation (sections 6.2 and 6.3). **Streamflow modelling** may be used in real-time mode for operating flood control or river regulation schemes, or it may be used in the design process to determine the size of reservoirs from the probabilities of floods and droughts, and to devise 'operating rules' for the dams (section 4.2.2).

At other levels, **systems modelling** may help to determine the most effective configuration or mode of operation for a scheme, or else to project demand trends. This approach is commonly termed **systems analysis** or **operations research**. Here, a prime objective is to produce and evaluate alternative solutions. The rise in basin-wide and regional planning, the increasing competition between different types of demand, and rapid developments in computer hardware and software have all accelerated this tendency.

9.1.1 Why systems analysis is necessary

The decision to use systems analysis techniques is driven either by the complexity of modern problems or by the need for greater operating efficiency. Six major factors can be identified:

1. *Large and complex systems* It is usually possible to design and run small-scale schemes with near-optimal efficiency by traditional means. Even so, such schemes are unlikely to respond efficiently to marked and perhaps sudden changes in the objectives or scope of the system, especially expansion of the system. It is much more difficult to predict the responses of the system when a number of water sources interact.
2. *Competing demands for new designs* New schemes usually have to satisfy the requirements of competing users and uses. For example, hydropower needs to maintain high reservoir levels to supply the turbines at a steady rate, but this reduces the efficiency of the dams for containing a flood. Similarly, recreational use of reservoirs may conflict with water quality requirements.
3. *Cost-effectiveness* Modern schemes tend to represent a major monetary investment. Mistakes can cost the regional economy dearly, so a cost-effective design is of paramount importance.
4. *Improving old schemes* It is frequently possible to use these methods to improve the running of existing large schemes. Even quite small improvements in operating methods can be cost-effective. For example, it might be possible to operate a badly designed reservoir system at a nearer to optimum level if the operators can be given reliable and timely flow forecasts or appropriately revised operating rules.
5. *Shift to river regulation and conjunctive schemes* These shifts in approach put greater emphasis on efficient modelling in both design and running. While an impounding reservoir used for direct supply to consumers may be designed to take a nine-month drought, much briefer anomalies can be critical in managing a regulation scheme. **Conjunctive schemes** combine supplies from surface water and groundwater, and they can rely heavily upon modelling for efficient design and operation (section 9.3).
6. *Pollution control* The shift towards river regulation as a method of supply has combined with increased awareness of the environmental dangers of pollution to increase the need for strict control of river pollution (Chapter 8). Modelling now plays a fundamental role in decisions on licensing freshwater abstractions and wastewater discharges, as well as in determining minimum acceptable flows and in devising short-term flushing strategies to dilute accidental pollution. Streamflow or groundwater simulation or synthesis may be sufficient on its own, but as the interacting systems become more complex so this modelling is being treated more as input into broader systems analyses.

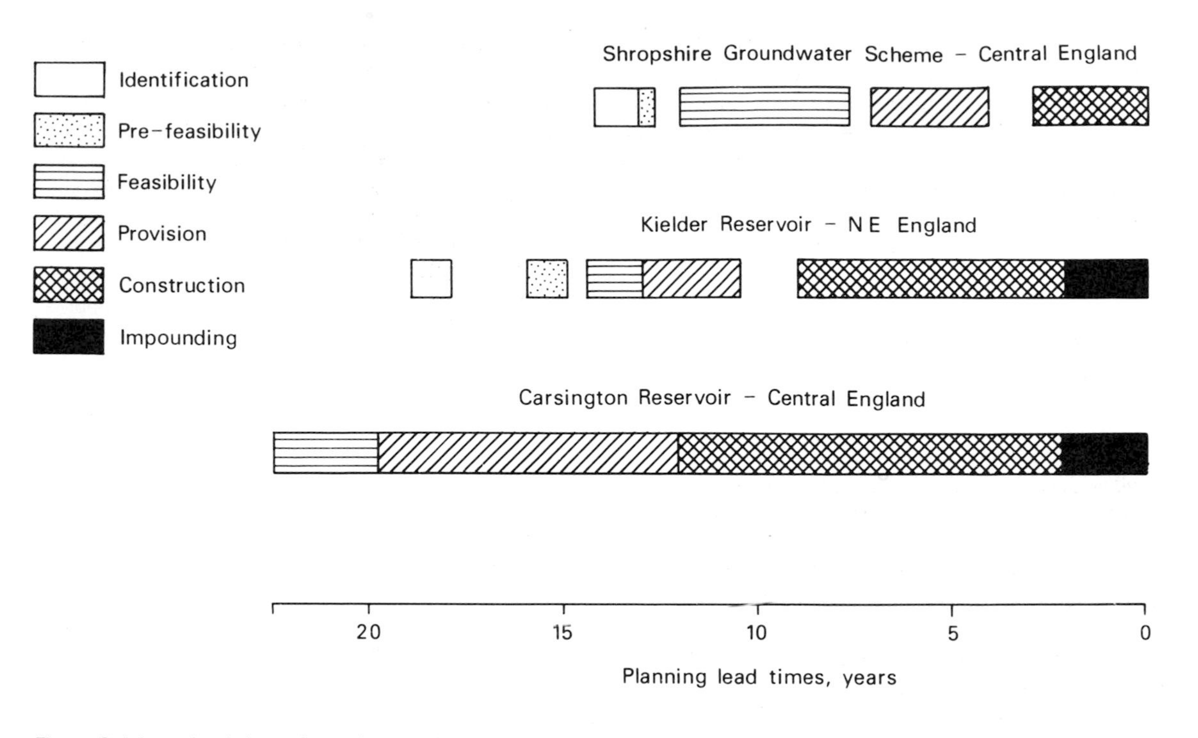

Figure 9.1 Long lead times for major developments. Based on NRA (1994)

The long lead times for major projects are illustrated in Figure 9.1. The case of the Kielder Dam in Northumbria (section 1.2) is an illustration of the problems of forecasting demand over such long periods and the high costs involved. Kielder cost £167 million and was intended as a key element in preparing the regional infrastructure for industrial revival. But after over a decade of planning and construction, the dam was completed in a very different industrial environment. Water consumers in Northumbria faced exceptional increases in charges in the years following its completion, and this has become more of a disincentive to economic revival. By an odd twist of fate, however, Kielder is now seen as a possible source for relieving the water shortages in southeast England (NRA 1992, 1994).

Other examples of the complexity of modern management systems will be discussed in subsequent sections.

9.1.2 Modern principles of system design

Modern systems analytic techniques for network design offer a method of experimenting with different configurations of 'hardware', such as reservoirs, and 'software', such as operating rules, by computer simulation before construction begins. They also force planners to identify quantifiable criteria and, as with all modelling, they tend to give the operators a much deeper understanding of the way the system works and the decisions that have to be made. Modelling frequently has to cope with nonlinear or stochastic aspects of the system, such as the relationships between yields and storages or changes in demand. This can be especially problematic for determining the optimal operating policy in multi-reservoir systems, but rapid optimization techniques are available to deal with these situations.

Before a problem can be analysed, it must be formulated in certain ways. First, it must be *mathematically tractable* and programmable. This requires a clear definition of the problem. It also requires access to data from which the parameters used in the model can be estimated.

Secondly, *appropriate criteria* must be selected or defined that can be used to judge the relative success of each solution. This may be called an 'objective function' and the computer experiments aim to maximize or optimize it (section 6.3.1). Because it can be difficult to compare criteria in different units, they are often formulated in terms of monetary value.

A third and vital element in the formulation is

determining the type and degree of *acceptable risk.* Some degree of risk, i.e. some likelihood of some form of system failure such as flooding or a critical drought, is inevitable. Risk is inherent in the stochastic nature of hydrological events, and is aggravated when records are deficient or local runoff processes poorly understood. Some risks are more acceptable than others. For example, a higher risk of exceedance of capacity is generally acceptable for road drains than for dams, where loss of life or damage to property is likely to be greater. However, it is difficult to assign an acceptable monetary value to loss of human life, and this can frustrate the ideal of a single unit of comparison. This would fall into the category of risk termed a **constraint**, which is essentially a restriction never to be broken at any cost. Other constraints might be specific requirements which must be met but for which there is no advantage in overfulfilling. By definition, over- or under-specification of a dam in terms of capacity, durability or the probability of events that it can withstand would be a breach of such constraints. Some of these constraints might be relaxed if the designer knew more about the processes governing the system. But usually a safe solution has to be sought, and a crude 'safety factor' is added to the design specifications to cover doubt or ignorance.

Less stringent design restrictions are called **objectives**. These are the objectives of optimization. Failure to meet objectives costs but is not critical. The acceptable risk of failure could be defined as the allowable probability of failure to meet some target objective, α. In water supply terms, a criterion of success could then be defined as the probability of supply meeting demand being at least equal to $(1 - \alpha)$.

Resource analysts have identified four basic criteria of performance which incorporate the risk element: **robustness, reliability, resilience** and **vulnerability**. A robust scheme is flexible enough to accommodate a range of new demands at little extra cost. A reliable scheme is one with a low probability of failure, while a resilient one recovers quickly after a failure. Vulnerability measures the consequences of failure. Sometimes not all these criteria can be equally met and compromise may be necessary. Failure might be rare (high reliability), but when it comes it could be expensive (high vulnerability). There will usually be a point at which money is better spent to minimize the consequences of failure rather than to prevent failure altogether.

There comes a point where, in the words of Hashimoto *et al.* (1982a, b), planning a *safe-fail* is better than continuing to search (and pay) for the elusive *fail-safe* solution. This policy applies equally well to protecting supplies as to flood protection. As it gets increasingly and disproportionately expensive to build engineering structures to guard against ever rarer events, and since the worst possible event is an elusive concept (section 4.2), so the fail-safe solution becomes less supportable. The 'hardware' solution must be supplemented or even replaced by 'soft' solutions.

Having defined the criteria and acceptable risks, systems analysis can be used to determine the optimal size and location, for example, of reservoirs, pumping stations or purification plants, their individual operating rules and the rules for operating the system as a whole for multi-source and multiple use systems. *What if* scenarios may then be set up to evaluate the behaviour of the system under extreme conditions, e.g. seeing how well the system copes in different configurations given a period in which demand exceeds supply. Other *what if* experiments might test different techniques of evaluation, changing the criteria, or test the efficacy of alternative data collection networks. They might even change environmental aspects like land use, and rerun the simulation to find the most efficient basin management policy. Johnson and Charoenwatana (1981) described a simulation to determine the most economic and water-efficient planting scheme for certain tropical crops. Ideally, all such experiments should incorporate environmental impact criteria as well as criteria for the hydrological gains.

A wide range of systems analytic techniques exists to generate and evaluate alternative solutions. Examples are: (1) the **Random** method, which can be valuable in producing unexpected solutions; (2) the **Hop, Skip and Jump** method, which is more focused and selects only alternatives that are significantly different yet still satisfy the objectives; or (3) the **Branch and Bound/ Screening** method, which screens the results to produce a small set of solutions that are both feasible and achieve a certain minimal value for the objective function.

All of these techniques should be regarded as aids to decision making, rather than definitive answers, and Chang *et al.* (1982) recommended that the analyst should select from the results of a number of techniques rather than rely on just one. Similarly, Fiering (1982) has argued that when only a small number of design options are available, the resulting selection is likely to be more 'brittle' and sensitive to the optimizing technique than if a larger number of options are combined, leaving a certain amount of redundancy within the system that could be useful when it gets severely tested.

Planners are now aiming to extend the boundaries of acceptable outcomes rather than merely to control the probabilities of unacceptable ones. Failures are likely to be more catastrophic when they are less expected. While accepting the inevitability of failure, they aim either to eliminate surprise or at least to be prepared for it. Better forecasting and warning systems can therefore play an important role.

In flood protection, 'soft' solutions that extend the protection provided by hardware structures could include selective insurance structures that make floodplain locations expensive, **floodplain zoning** which places planning restrictions on the type of land-use activity permitted there, or **floodproofing** of individual buildings, as well as improved warning systems. US floodplain zoning laws normally distinguish: (1) zones flooded regularly, once in 2 years, where no building is allowed; (2) zones flooded once in 10 years, where groundfloor use is restricted unless floodproofed; and (3) a warning zone covering the area of the maximum recorded flood, where hazard drill applies. Floodproofing might be a planning requirement or simply encouraged by favourable insurance premiums. In the UK, the Environment Agency introduced a warning scheme in 1996 which uses the OpenTalk computerized telephoning system. The agency's computer collates weather radar and telemetered river stage data to predict flood levels, and telephones prerecorded warnings to vulnerable properties.

In protecting supplies, 'soft' solutions might include hosepipe bans, restricting supplies to non-essential users, or varying compensation flows.

9.2 Large dams and impounding reservoirs

The concept of building a dam and diverting the water for direct use has been the mainstay of water resources engineering throughout history. Until the mid-twentieth century, schemes were relatively unsophisticated and were commonly operated as single-source, single-use systems. Even where topographic or engineering restrictions required a series of linked reservoirs rather than one single large dam, as in the English Pennines or Wales, they were essentially operated like one large source.

Figure 9.2 shows the dramatic expansion in the construction of large dams since 1960. Globally, the number of large dams nearly doubled in 30 years, but reservoir capacity has more than doubled, indicating a trend towards larger impoundments. Figure 9.2 also shows some reduction in the rate of increase during the 1980s. This probably reflects reduced political enthusiasm and available funding more than environmental concern or inherent limits to exploitable resources.

Over 60 per cent of reservoir capacity is now in the Developing World, where the greatest expansion has occurred since 1960. These developments have often been problematic and controversial (section 1.1.1). Criticism has tended to concentrate initially on the flooding of land and displacement of people. In India, the Narmada River project is being funded by the World Bank, despite protests from many of the one million people due to be displaced and claims of an environmental disaster (Alvares and Billorey 1988). Thirty large dams and thousands of supplementary dams are to be built in the basin, flooding 4000 km^2.

After construction, criticisms are commonly levelled at schemes which have not produced the predicted returns (Box 9.1). To this is now being added concern for public health implications in hot countries (Wooldridge 1991). Diseases like schistosomiasis and malaria have spread and intensified, as reservoirs have caused population migrations and as water surfaces have increased the populations of vectors carrying the diseases, like fluke worms and *Anopheles* mosquitoes (Tsegaye 1991).

China has for long followed the traditional path of building small dams for local use. Between 1950 and 1980, China built 90 000 such dams. But it too is opting for grandiose schemes. The Three Gorges project on the Yangtze River began in 1992 and is planned to take 18 years to complete and cost $30 billion (Box 9.4). It will involve displacing 1.3 million people and flooding 41 000 ha of farmland, but the 185 m dam could provide hydropower equal to 12 per cent of national power requirements, plus water supply and flood control. Half a million people have died in Yangtze floods this century. The greatest fears are that it will silt up rapidly, disrupt navigation and the erosional and ecological balance of the river environment and cause more flooding upstream. Critics argue that the aims could be better achieved by building numerous smaller dams, although 10 smaller schemes are already underway upstream to generate 12 000 MW.

Countries with a coastline may consider offshore reservoirs to overcome political opposition to taking land for reservoirs. In Britain, proposals have existed for decades for **estuarine barrages** to hold freshwater and/or generate hydropower in the Humber, Dee and

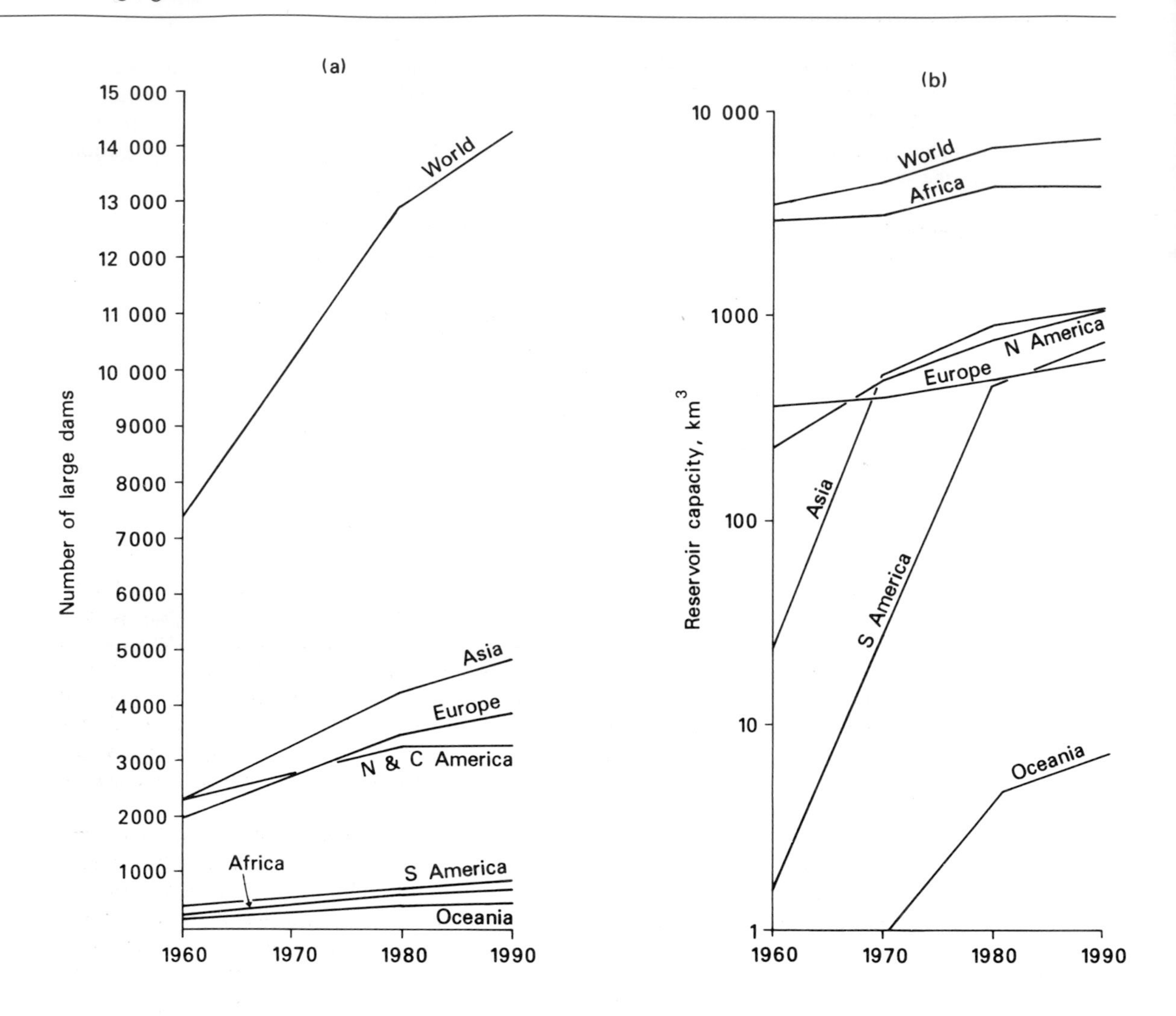

Figure 9.2 Worldwide construction of large dams: (a) number of dams (World Resources Institute 1992); (b) total maximum reservoir capacity (Gleick 1994)

Severn estuaries, the Wash, Morecambe Bay and Solway Firth. But none of these is likely to receive approval given recent environmental impact studies which indicate major disruption of fisheries and coastal processes. A less environmentally disruptive proposal for four floating reservoirs or **drogues** 2 km off the coast of East Anglia was shelved in 1979. A novel proposal in Florida is to harness freshwater from a submarine spring that discharges at 40 $m^3 s^{-1}$ into a 40 m depression in the ocean bed off St Augustine.

9.2.1 Environmental effects of large dams

All large-scale manipulations of water resources are bound to have environmental effects. It is now realized, somewhat belatedly, that a major task of planning and management must be to predict the effects and to minimize them (Box 9.1).

Table 9.1 catalogues the main advantages and disadvantages of large dams, and indicates the breadth of environmental impact. These include impacts on the

Table 9.1 The advantages and disadvantages of large dams

Advantages	Disadvantages
Reliable water source	Large users of land
Flood protection	Often displace large populations
Irrigation	Frequently significant ecological losses
Hydropower	Danger of salinization from irrigation
Support for economic development – attracting industry and jobs	Increase in water-borne and other related diseases in tropical regions
	Economic burden of loans to finance construction
	Social and political conflicts caused by economic stress and disagreement on use or apportionment of water

river and near-river environment created by the normal operation of the dam, indirect impacts caused by activities facilitated by the reservoir such as irrigation, and the impact of dam failure. Problems are by no means confined to the Developing World. Although physical failure of large dams is unlikely in the USA, the Federal Emergency Management Agency estimates that up to 1300 smaller dams are unsafe and threaten populated areas. In 1986 Congress voted $15 million p.a. for improving dam safety (Miller 1992).

Reisner (1993) presents a compelling if polemic account of the dual hegemony of irrigation and large dams in American water resources policy. The USA has some 250 000 dams, of which 50 000 are 'major'. Building dams for irrigation in the drylands west of the 100th meridian began in earnest in the 1880s after droughts had caused a number of agricultural catastrophes. Major John Wesley Powell's Irrigation Survey established both the need and the potential. The Desert Lands Act required proof of irrigation before a farmer could own the land. Irrigation was enshrined in the 1902 Reclamation Act, which set up a Reclamation Fund to finance projects. Thereafter, Reisner claims, the irrigation programme 'became a monster'.

It led directly to the Bureau of Reclamation's policy of **river basin planning and accounting**. Ostensibly, this was a laudable policy. It fitted in well with the 'New Deal' social policy adopted by President Roosevelt to create economic regeneration after the Great Depression, which led to the Tennessee Valley Authority (TVA) programme in 1933 (McDonald and Kay 1988). Under the TVA the river basin was viewed and planned as a whole, both land and water, for the first time. The TVA became world-famous as an example of environmental integration. Indeed, much was achieved, particularly in the realms of land reclamation, soil conservation and farm management, which was aimed specially at reducing siltation in the reservoirs (Newson 1992). The US Soil Conservation Service grew out of the TVA experiment. This provided a vital service, since considerable damage had already resulted from irrigation projects supported by the Reclamation Fund but based on little understanding of soils and drainage processes.

Nevertheless, as Palmer (1986) notes, 'new dams were the cutting edge of the TVA' and this strongly influenced national policy. The TVA dams were built primarily for hydropower, although they also improved navigation facilities and provided flood protection. There were two important consequences: (1) the success of the TVA fortified the engineering view that nature can be completely controlled, that it can be groomed and tidied by engineering structures (Huxley 1943); and (2) it strengthened the bureau's view that integrated basin projects were a means of deriving revenues from more economically viable aspects to fund other parts of the programme (Robinson 1979). The former view has been heavily questioned since the 1993 Mississippi floods (Box 7.5) and in the debate over taming the Brahmaputra (section 11.4.3). The bureau's concept of accounting led to a widespread policy of building hydropower dams to fund uneconomic irrigation schemes.

Reisner (1993) described the result as a 'perversion of a sensible idea' and the 'death sentence for free flowing rivers'. The bureau produced similar basin-wide plans for the Missouri and Bighorn basins in the 1940s. In the 1950s, the Colorado River Storage Project

Box 9.1 Case study: The impact of the Aswan High Dam on the River Nile

The Aswan High Dam was completed in 1964 at a cost of $1 billion and the resettlement of 120 000 people to provide irrigation, hydropower and fish production in Lake Nasser, at a time when scant regard was paid to long-term environmental consequences (Figure 11.7). It was intended to underpin an agricultural and industrial revolution in the country and President Nasser made it a nationalistic symbol for an Egypt recently 'freed' from post-colonial monarchy. In this, it was only partially successful. The agricultural objectives were largely fulfilled during the first 10 years of operation. Year-round irrigation did permit three harvests a year on the floodplain, where from time immemorial the annual Nile flood had supported only one. An additional 405 000 ha of land was reclaimed from the desert, and irrigation saved the rice and cotton crops during the 1972 and 1973 droughts.

But from 1974, Egypt's historical position as a net exporter of food began to falter. This was partly due to high population growth, at 2.5 per cent p.a. It was also due to the severe lack of consideration for environmental effects, many of which rebounded on food production. The dam traps 60 million m^3 of sediment a year and blocks the passage of the vital nutrient-laden silt that for millennia passed down the Nile to fertilize the floodplain. These croplands needed artificial fertilizer to compensate, and much of the newly created hydropower has been absorbed in producing fertilizer costing $100 million p.a. The more continuous, low-velocity irrigation without adequate drainage has caused mineral salts, which used to be flushed out by the flood, to accumulate in the floodplain soils. And salinization has begun in the new croplands. Because of salinization, two-thirds of the land reclaimed over the last 30–40 years is now out of production or marginal. Three-quarters of the gain in food production from the new, but less productive, irrigated land has been offset by salinization.

Fisheries have also suffered in the lower Nile and offshore of the Nile delta. With the loss of the nutrient-rich silt the lobster, shrimp, sardine and mackerel fisheries have almost disappeared, and have yet to be balanced by catfish, carp and bass from Lake Nasser (Miller 1992). More hydropower may be absorbed in producing chemical substitutes to help the fisheries.

Lower sediment loads in the riverwater are also creating erosional problems. Bridges and dams are being undermined. The increased transparency of the riverwater aids eutrophication. Agricultural land is being lost through bank erosion, as riverflow is more competent, and through shore retreat, saltwater intrusion and subsidence in the delta due to lower rates of replenishment combined with a rise in sea level (section 2.1.1). The delta shoreline has been retreating at up to 40 m a^{-1}. A proposed $250 million programme of ten check dams aims to limit bank and bed erosion by reducing the channel gradient.

Water quality and public health are also threatened by reduced riverflows. Shallow or stagnant water encourages malaria and schistosomiasis. Urban wastewater entering the river is less diluted and urban colonization of formerly flooded islands near Cairo has increased the risks.

Water levels could be further affected if an Egyptian Ministry of Irrigation plan to divert water to Sinai were ever implemented (section 11.4.4). Guariso *et al.* (1981) developed a multi-objective programming model to study this proposal, looking at the trade-offs between economic and political objectives and the interactions with water quantity and quality, navigation, crop rotation and irrigation techniques. They concluded that the costs would be high and economic sacrifices would have to be made if the political aims of colonization were to be achieved. They also showed how all decisions are sensitive to the cost of water elsewhere in Egypt.

The final codicil is that a quarter-century after Lake Nasser was due to be full it is still barely two-thirds full, and there is no immediate prospect of a significant rise. With Sudan and Ethiopia now developing their own sections of the Nile, it may never fill (section 11.4.4).

---

produced ten dams that between them store 60 billion m^3, 5–8 times the annual flow of the river, with evaporation losses greater than the storage capacity of most existing US reservoirs. It includes the Hoover Dam and Lake Mead, the largest in the world, from which water is diverted to feed irrigation in the Imperial and Coachella valleys around the Salton Sea in California (environmentally the most damaging aspect) and the discharge is used to generate electricity for Los Angeles (environmentally the best aspect).

9.3 River regulation and conjunctive schemes

There has been a marked tendency in the Developed World over the past 30 years to replace direct supply from impounding reservoirs with planned regulation of riverflows, using the river as an open pipeline. Almost one-fifth of rivers in England and Wales are now regulated. River regulation has a number of technical and environmental advantages. It dispenses with long stretches of pipelines or canals, which are costly and can disrupt the environment. It also allows a diversity of sources from different areas to be linked together without new pipelines, i.e. other regulation dams or groundwater schemes, at lower cost. This means that lowland sources can be incorporated, which reduces the need for such large upland storage reservoirs. The advantage can be seen in two examples described fully later (Figures 9.4 and 9.5). A 50 Gl regulating storage at Clywedog, supplemented by other river and lowland groundwater sources, guarantees 726 Ml day^{-1} at the main abstraction point, whereas 82 Gl at Elan yielded only 341 Ml day^{-1} as a direct supply system.

It also puts a premium on maintaining good water quality standards over longer stretches of river to guarantee the quality at the abstraction point. This in turn encourages good monitoring, legislation and enforcement. The Control of Pollution Acts followed hard on the heels of the first large-scale regulation experiments in Britain (section 8.3.1). River regulation can contribute directly to this enforcement by reducing low flows or by releasing water on demand to dilute accidental spillages. Regulation dams can also be operated to control floods, which may incidentally improve water quality.

Regulation is also philosophically superior in so far as it makes more use of hydrological theory and allows more environmentally sensitive management and flexibility.

Figures 9.3 to 9.5 illustrate some of the complexities of these modern management systems. The Dee regulation scheme was one of the first management schemes designed in Britain (Figure 9.3). It has the dual role of flood control and water supply, with a minor additional hydropower facility. In designing the system during the late 1960s, the Water Resources Board and the River Authority collaborated with the Institute of Hydrology and the Soil Survey (SSEW) to develop a catchment simulation model that could be used to make operating decisions (Jamieson and Wilkinson 1972). The Dee Investigation Simulation Program for Regulating Networks (DISPRIN) model identifies three zones (uplands, hillslopes and valley bottoms), each with nonlinear storages yielding overland flow and quick return flow and feeding a common groundwater storage by linear flow routing (section 6.3). It therefore has seven storages and 21 parameters.

The Meteorological Office tested weather radar here, prior to the establishment of the national network (section 5.2.5), which provided input to the hydrological model, along with remote raingauges and flowgauges that are still interrogated by the central computer at Bala (section 5.1.2). The computer updates the data input for the model and it telephones the duty flood officer when levels are becoming dangerous.

The model allows the controller to ask 'what if' questions about river response. It predicts the downstream migration of natural floodwaves and artificial releases from the four regulating reservoirs, and allows the Bala HQ to make operational decisions to open or close the sluices. This allows minimum flows to be maintained at abstraction points, especially at Chester for Liverpool. In recent years, operation has concentrated on monitored flow levels and on the flow routing part of the model, as experience has shown that the hillslope drainage phase need not be modelled in such detail for this catchment.

The Rheidol scheme is the only one in England and Wales where gravity-fed hydropower generation is the prime purpose (Figure 7.15). It has to balance the demands for power generation with a flood control responsibility, and compensation flows to serve agriculture and fisheries, to support an internationally famous waterfall and to dilute toxic effluent from abandoned lead mines. Conflicts of interest are bound to arise. After privatization of the power industry, Powergen sought a relaxation of NRA rules that restrict the maximum water levels within the reservoirs so that they can accommodate potential flood-generating events. Overall, however, this scheme appears to have been reasonably successful in environmental terms. Flooding problems do appear to have decreased slightly (Figure 7.15). Sediment transport and channel erosion seem markedly lower, there is less heavy metal pollution and migratory fish have returned.

The Elan Valley reservoir complex in mid-Wales is an example of changing functions (Figure 9.4). The complex was planned in Victorian times by Birmingham City Corporation as a set of **direct supply** reservoirs to replace local city sources that had become critically polluted following the Industrial Revolution and the accelerated growth in urban population during the late nineteenth century. Four reservoirs were completed by

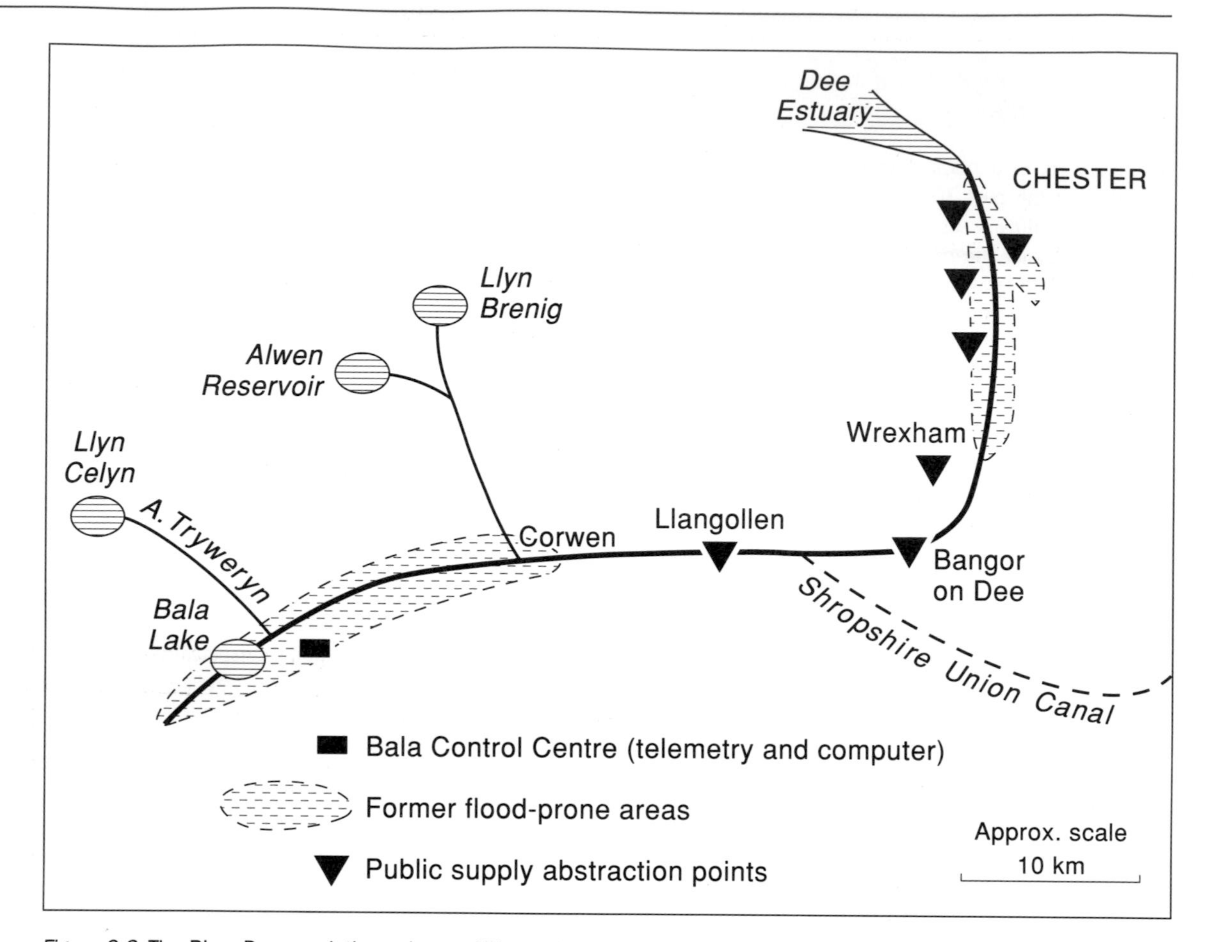

Figure 9.3 The River Dee regulation scheme, UK, combining flood control and water supply

1904 and a fifth in 1952. However, the great drought of 1976 caused a re-evaluation. Southeast Wales suffered particularly from water shortages during the drought, and a plan was subsequently implemented to use part of the capacity of the Elan complex to *regulate* flows in the River Wye and enable water to be pumped out of the Wye near Monmouth and added to the South Wales supply. Supplies for other small towns along the Wye have also benefited. As a purely direct supply system the scheme released 123 Ml day^{-1} in compensation flow and took 341 Ml day^{-1} for water supply. Under the revised rules, compensation flows can be reduced to 68 Ml day^{-1} in favourable conditions and increased up to 232 Ml day^{-1} when water is needed in southeast Wales. The adaptation has worked well and proved effective during the 1984 drought. There are now plans to expand the Craig Goch reservoir and the river regulation role of Elan Valley to support interbasin transfer to the Thames.

The River Severn regulation scheme has also evolved since its inception with the building of Clywedog dam in 1967 (Figure 9.5). As originally conceived, Clywedog's primary role was to supplement the Elan Valley supply for the Birmingham conurbation by ensuring a statutory minimum flow of 726 000 m^3 day^{-1} at the main abstraction point at Bewdley. Other users, particularly thermal power stations, also benefit from the enhanced supply. For environmental purposes it releases 18 160 m^3 day^{-1} in compensation water.

Its secondary role was to act as a flood protection

Figure 9.4 The Elan Valley reservoirs scheme. A Victorian direct supply scheme, now partially operated as a regulation scheme. Proposals to develop the largest man-made lake in Europe around the upper dam, linked by interbasin transfers to the Thames, are officially shelved but reappear as resources in the southeast are perceived to be under pressure (e.g. NRA 1992)

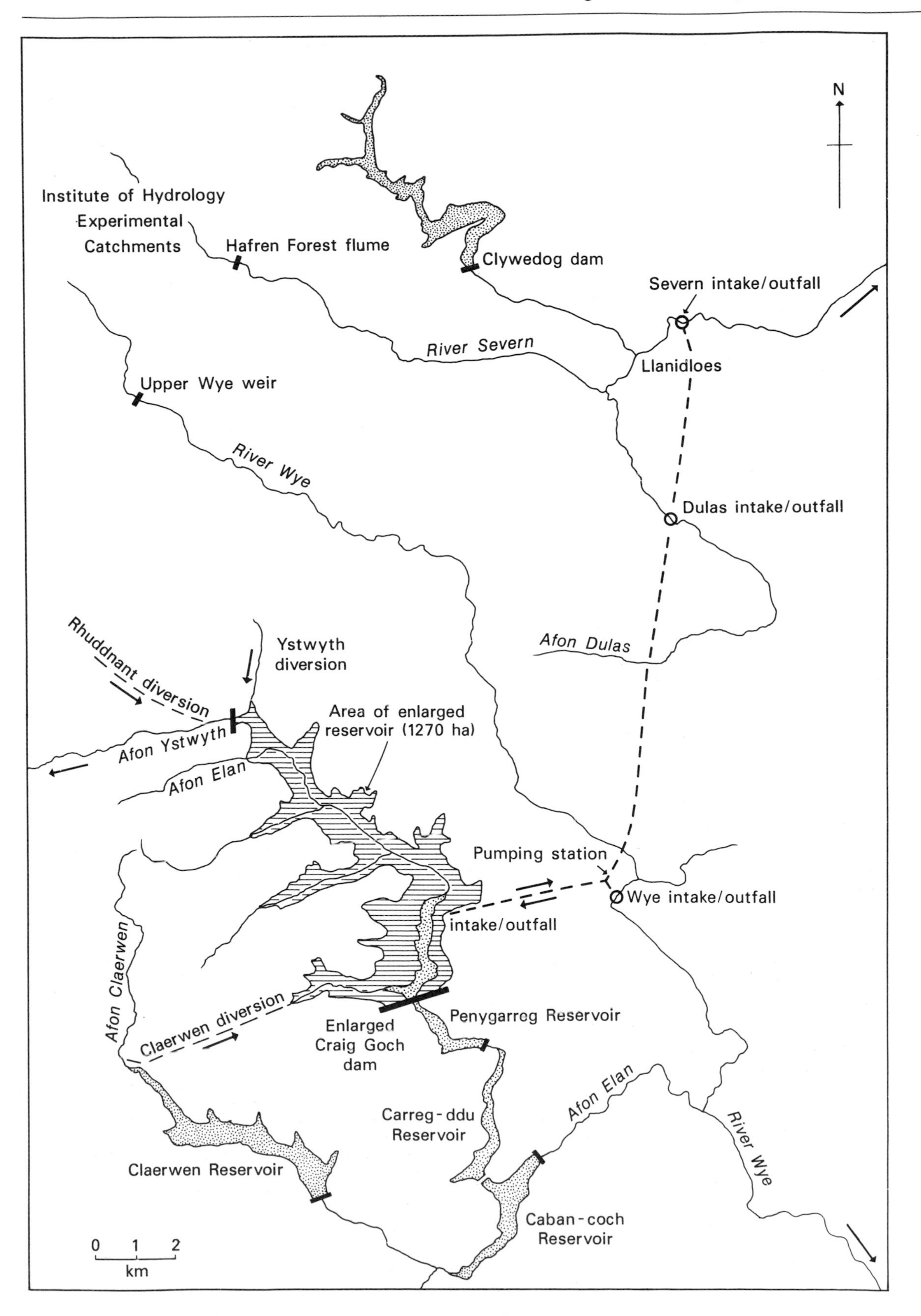
N
Institute of Hydrology
Experimental
Catchments
Hafren Forest flume
Clywedog dam
Severn intake/outfall
River Severn
Llanidloes
Upper Wye weir
River Wye
Dulas intake/outfall
Afon Dulas
Rhuddnant diversion
Ystwyth
diversion
Area of enlarged
reservoir (1270 ha)
Afon Ystwyth
Afon Elan
Pumping station
Wye intake/outfall
intake/outfall
Afon Claerwen
Claerwen diversion
Enlarged
Craig Goch
dam
Penygarreg Reservoir
Afon Elan
River Wye
Carreg-ddu
Reservoir
Claerwen Reservoir
Caban-coch
Reservoir
0 1 2
km

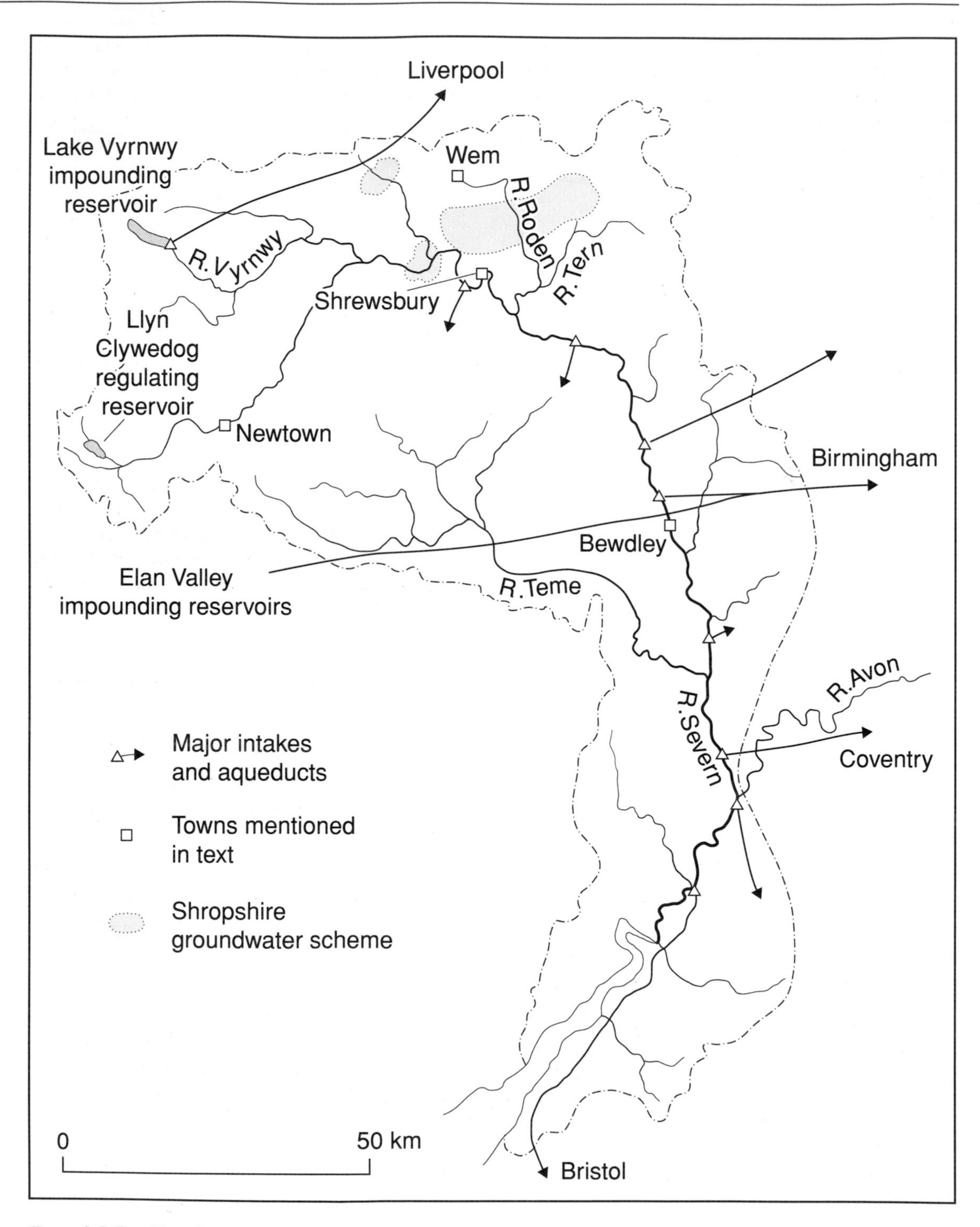

Figure 9.5 The River Severn conjunctive scheme of river regulation and groundwater abstraction, with a subsidiary flood control role

measure for a section of the Severn valley centred on Newtown. Newtown suffered major flooding in the early 1960s, but has avoided major problems since, partly because of the dam, partly through supplementary works at Newtown and partly perhaps through chance (Figure 9.6). The flood protection around vulnerable urban sites is supplemented by a zone further downstream where controlled overbank flooding is allowed on the agricultural floodplain.

During the main winter flood period and through the summer period of heavy convective storms, the reservoir level should be some way below maximum. Reservoir control rules aim for the level to reach maximum by the beginning of April to allow maximum storage of the winter surplus for subsequent regulating releases over the summer. April tends to be a period of low flood risk, but provision is also made for short-term releases to increase storage ahead of potential flood storms (section 5.2.5). Clywedog has since been supplemented by the Shropshire Groundwater Scheme and by adding a regulatory role to the former direct supply reservoir at Vyrnwy, since Liverpool now has regulated supplies from the Dee. It is now a conjunctive scheme, in which water is pumped out of the Triassic sandstone aquifer into small tributaries of the Severn in Shropshire at times when surface water regulation alone is insufficient to meet demand. Because pumping is expensive, the groundwater option is used sparingly. It is also limited by environmental considerations. If too much deoxygenated water of a different temperature is released into small streams aquatic life could be harmed. Both pumping and reservoir releases are initiated on orders from the Birmingham control centre, based on a continuous-flow simulation model.

A third role for Clywedog is hydropower. Although only 500 kW, this has been used more under the financial incentive of the Nonfossil Fuel Obligation in the 1989 Electricity Act. In April 1994, Clywedog performed yet another role. Britain's worst chemical spill for a decade occurred when solvents entered a tributary at Wem (section 8.2.4). Clywedog released water to dilute the pollution and to support subsequent additional abstractions from the Severn while reservoir clean-up operations took place (Figure 9.5). Fortunately for consumers in east Shropshire supplied directly from the aquifer, groundwater remained unaffected, but the event cost Severn Trent Water £1 million.

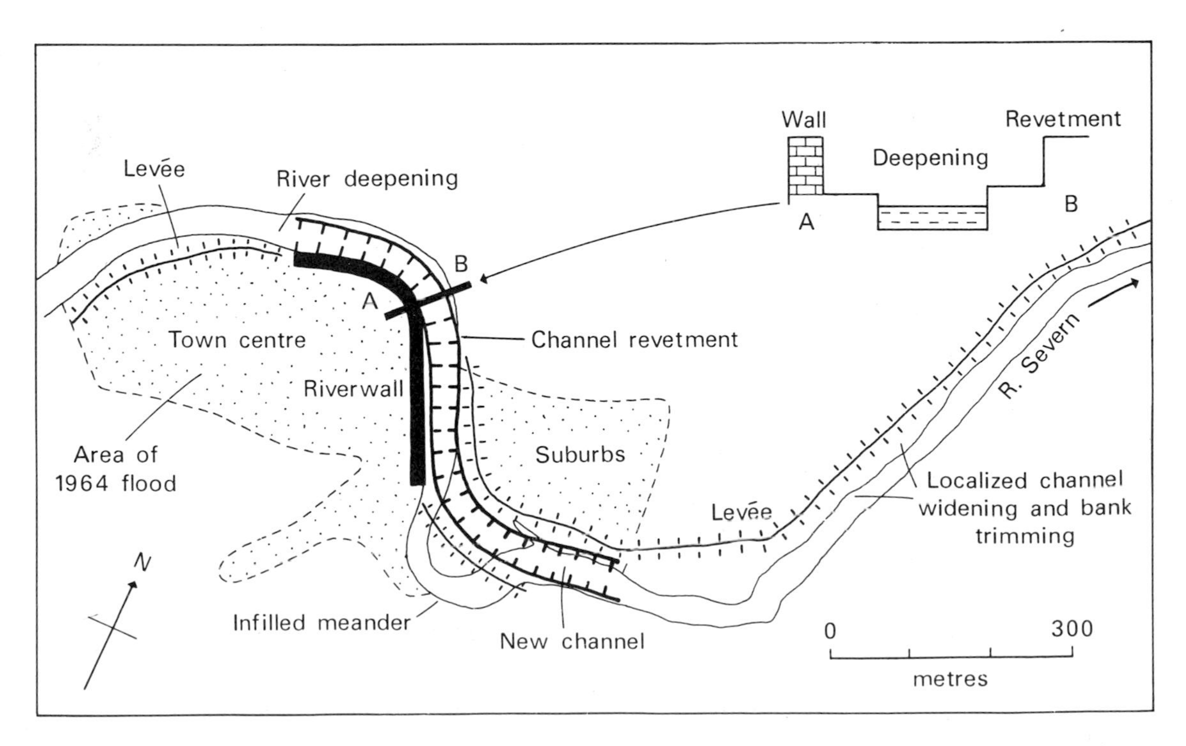

Figure 9.6 Engineering measures for flood protection on the River Severn at Newtown to supplement the protection given by river regulation, the channel is regularly dredged upstream of the town

9.4 Large-scale strategic river diversions

A number of countries have proposed countrywide or even continental-scale diversions to solve their regional water deficiencies. As long ago as 1830, Alexander Shrenk proposed reversing the flow of the Pechora River into the Kolva tributary of the Volga. This very proposal formed part of a grand diversion scheme that was briefly adopted by the Soviet authorities in the 1980s (Figure 9.7). The reversal of the Russian Arctic rivers remains the one scheme that most nearly came to fruition (Box 9.2). A similar North American proposal was stillborn (Box 9.3). So were plans to divert the Congo to Lake Chad and Amazon headwaters across the Andes. Australian plans to divert water from the east coast across the Great Dividing Range to irrigate the drylands to the west remain more likely to succeed. So too do the strategic plans in the UK, which could involve diverting water to southeast England from the Kielder Dam (section 1.2) or from an expanded Craig Goch scheme (section 9.3). But both the Australian and British plans are of a lesser order. The largest of the schemes currently likely to reach fruition are in China (Figure 9.9, Box 9.4).

There have been vociferous environmental lobbies against all of the more grandiose plans, particularly in North America. Objections have ranged from concern over the climatic effects of reducing the flow of freshwater into the Arctic Ocean to the ecological and aesthetic impacts of dams and canals. In the final analysis, perhaps the most persuasive environmental argument has been that provision of vast quantities of 'new' water will only encourage further profligate usage and support the expansion of irrigated agriculture in arid lands, which has had such a disastrous effect on both soils and agriculture throughout history (section 1.2.3). Despite these environmental objections, however, the main reasons why most of these schemes have not materialized are political or economic. They often require international collaboration. Yet they tend to favour one nation more than the others and they stretch international goodwill to the limit. They are also high-cost solutions.

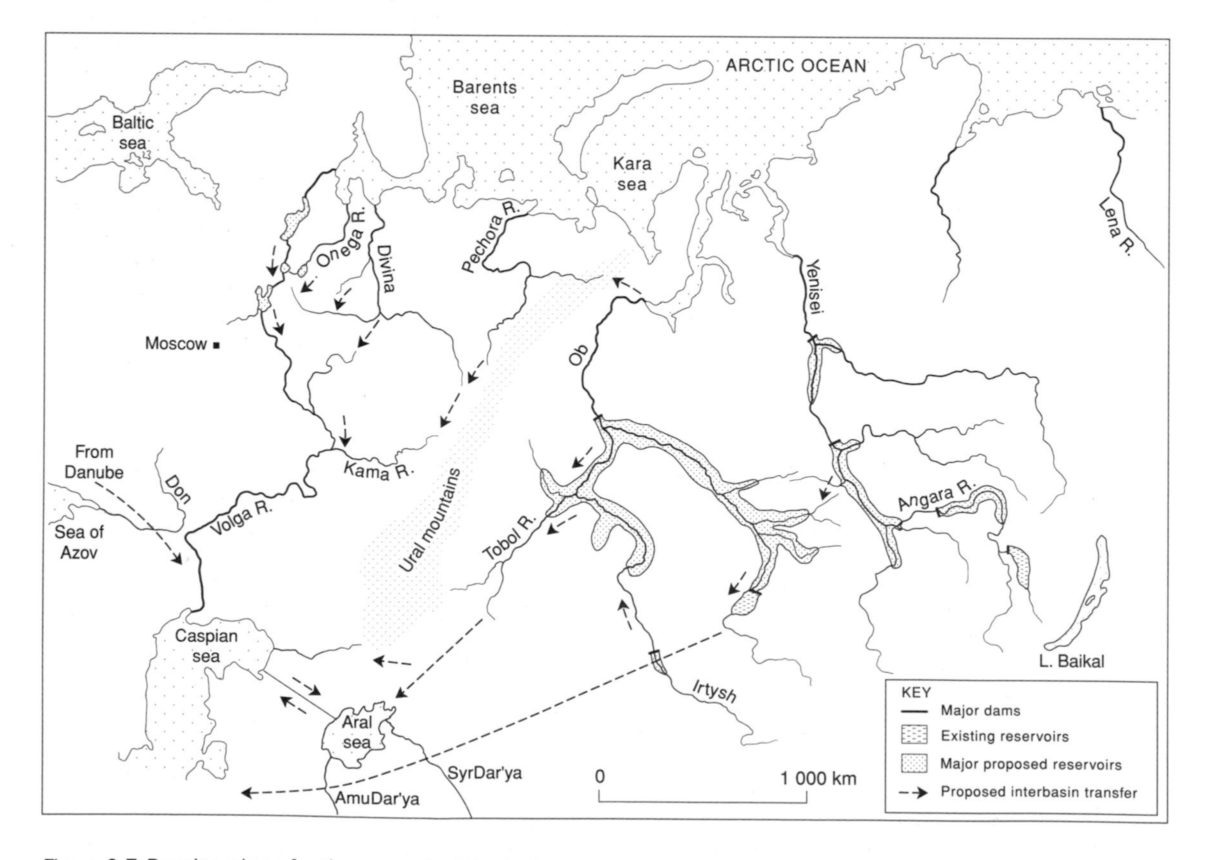

Figure 9.7 Russian plans for the reversal of Arctic rivers, approved and then cancelled in the 1980s

Box 9.2 Case study: Reversal of Russian Arctic rivers

The Soviet plan to divert most of the major rivers in the Russian Arctic basin to supply the grain lands of the Ukraine and Central Asian Republics could have been the largest single human interference in global hydrology (Figure 9.7). The largest country on Earth until 1990, the Soviet Union was unique in having within its own national boundaries both the huge, virtually untapped discharges of the great Arctic rivers and one of the world's top grain lands, which was suffering from a chronic combination of natural drought and mismanagement (Holt 1983). It also had a dictatorial Communist régime with a centrally planned economy. If river diversion were to be achieved anywhere on this scale, then this superpower was likely to achieve it. In contrast, the other superpower, the USA, has had comparable plans thwarted by Canada (Box 9.3).

The former USSR was a prime example of imbalance between the distribution of water resources and population and industry, with 80 per cent of the water in Asia and 70 per cent of the population in Europe. Even in the European sector, the water resources were mainly in the Arctic basin, whereas the population was mainly in the Caspian and Black Sea basins.

The Supreme Soviet approved an operational plan in 1982 and work was due to begin in the European sector in 1986 and in Asia in 1988. At the last moment, Gorbachev won a vote to shelve the scheme in 1986, as part of his reforms. His argument was that falling yields were largely caused by the Communist agricultural system and that a grand technological 'fix' would only delay making the system more efficient and introducing incentives for the workers.

The scheme would have taken over 50 years to complete and eventually involved diverting 60 km^3 of water a year (Micklin 1986). By 1990, the first 6 $km^3\ a^{-1}$ were to have been diverted to the Volga from the N Dvina and Pechora Rivers. By 2000, 20 $km^3\ a^{-1}$ would be diverted to the Caspian, Sea of Azov and Black Sea basins, and a beginning made on diverting 27 $km^3\ a^{-1}$ from western Siberia to Kazakhstan, the other Central Asian Republics and the Aral Sea. Much of the Central Asian diversion would be via a very inefficient new open and unlined canal. This would be shadowed by 'leak collector' channels that would return about half of the 3.5 km^3 lost through leakage by collecting groundwater seepage and local surface water.

The ultimate aim was to create 2.3 million ha of new irrigated land and to relieve falling water levels in the Caspian and Aral Seas (Box 7.1). Both seas would be constricted by dikes in order to reduce the amount of water needed.

Environmental impacts

Most of the perceived impacts would have been negative. Undoubtedly, the sequences of large reservoirs needed to reverse flows would have caused population displacements and loss of agricultural land and commercial forests. Some new waterlogging would result and areas of northern agriculture that rely on water from snowmelt floods would be adversely affected. But this would be partly offset by better drainage and less spring meltwater flooding elsewhere.

Migratory fish like salmon would certainly be affected by dams blocking their migration, especially the proposed dam in the Onega Gulf of the White Sea, unless special arrangements were made. However, the effects of greatest concern, but also some controversy, would have been on sea ice. Scientific controversy has been due partly to the complexity of the interactions that might be involved, partly to lack of knowledge, and partly through the paucity of available computer modelling facilities in the USSR. But it was no doubt encouraged by a lack of official interest in negative environmental arguments.

The most established view is that reducing the input of freshwater into the Arctic Ocean would reduce sea ice. The lower density freshwater floats on top of the seawater and freezes more readily than saltwater. Reducing the extent of sea ice could also affect the winter position of the Ferrel jetstream over the North Atlantic (section 2.3.2) and aggravate rainfall shortages in southern Europe (section 2.6.2).

The alternative view holds that reducing freshwater discharges from the colder continent will cause warmer sea surfaces and higher evaporation. This will encourage snowstorms along the coast, which will chill the water in sheltered bays and provide freezing nuclei that stimulate ice growth. But the ice could be much thinner and melt earlier, because more snowcover and cloudcover would reduce radiative cooling. A similar controversy exists over the effects of global warming on sea ice. The route nature might take seems to depend mainly upon the critical balance between sea surface temperatures and river discharge, with windiness playing a supporting role in mixing the surface layers.

Ultimately, however, the decision to stop the scheme was purely politico-economic.

Box 9.3 Case study: American plans for continental-scale river diversion

American plans for the diversion of Arctic rivers were published in the 1960s. The main aim was to provide water for the states west of the Mississippi, particularly states suffering from the inadequacy of supplies from the Colorado River (section 4.3) or from the rivers and aquifers of the Great Plains (section 7.5). To this was added what was seen as the major political, if not environmental, objective of being able to restore the volume and quality of water in the Colorado as it crosses the border into Mexico: a grievance still of active concern to Mexico (Reisner 1993). It could also improve flows in the similarly depleted Rio Grande.

The **North American Water and Power Alliance (NAWAPA)** plan was expected to take 30 years and $100 billion to complete (Kelley 1967). It would transfer water from Alaska, the Rockies, the Mackenzie and other Arctic basins to the southwest via the Southwest Reservoir Canal, to the Mississippi via the Dakota Canal, and even to New York via the St Lawrence using nearly 14 600 km of canals and tunnels and numerous major reservoirs. The largest reservoir would fill 800 km of the American Rocky Mountain Trench with water pumped to an altitude of 900 m to provide a head for gravity-feeding water to the southwestern states (Figure 9.8). A total drainage area of 3 328 000 km^2 was involved in the diversion schemes, yielding a mean annual runoff of 817 798 million m^3. Provision was made to use

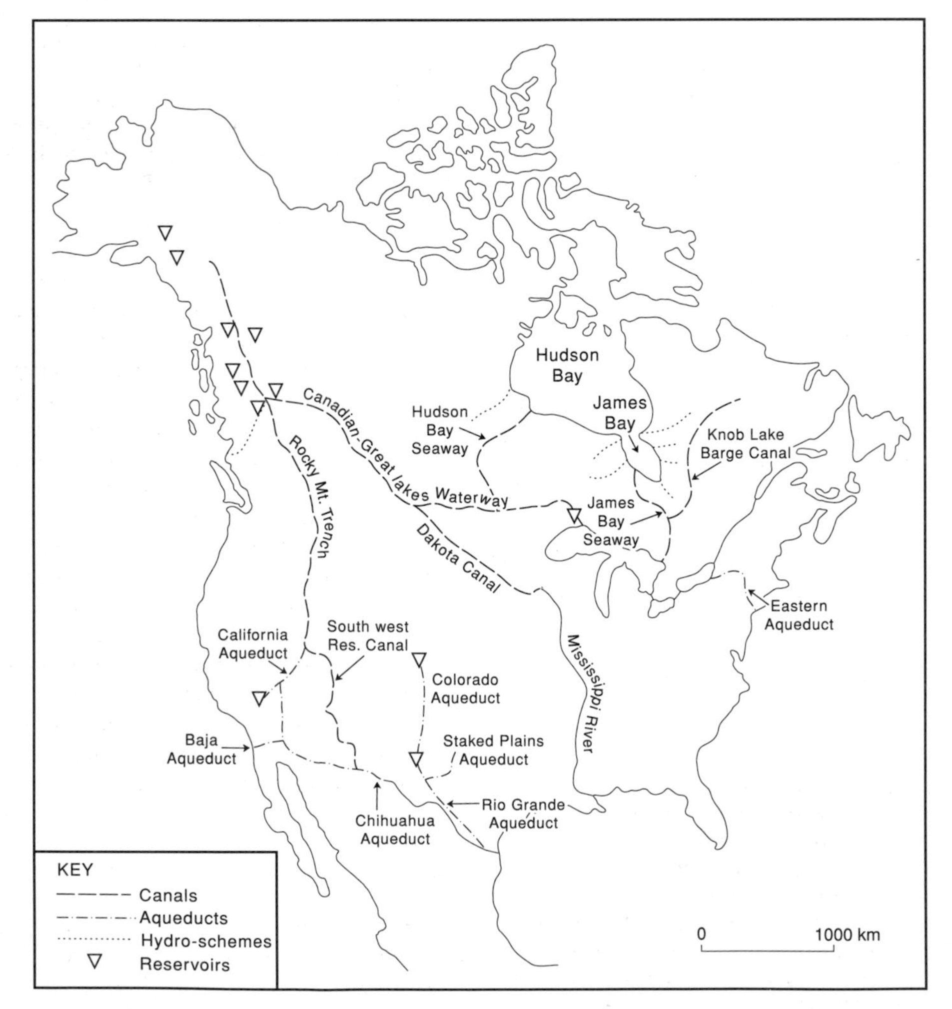

Figure 9.8 The North American Power and Water Alliance plans, proposed in the 1960s and recently the subject of revived interest

Box 9.3 (cont.)

up to 40 per cent of this runoff, although only half of this would normally be used.

Unfortunately, most of the Arctic water lies in Canada, and a land link with Alaska must pass through Canada. Canada was wooed by the prospect of hydropower, by improved water supply for the drought-ridden prairies, by topping up falling water levels in the Great Lakes and by new north–south navigation routes. It was estimated in 1967 that the hydropower could generate 50 per cent of Canada's total power requirements by 1980, as well as 15 per cent of Mexico's and just 4 per cent of the US domestic requirement.

But Canada was not persuaded and has since identified its own requirements for the runoff from the Arctic and Rocky Mountain drainage basins for water supply, especially for prairie agriculture, for hydropower and for environmental protection in the Mackenzie basin. Improving the water quality of the Great Lakes with unpolluted water from the western mountains, with a typical dissolved salts content of under 50 ppm, has been supplanted by direct action since 1972 under the joint US–Canadian pollution control programme to control wastewater discharges into the lakes. Canada has managed thus far to do without many of the navigational provisions of the scheme, although river modifications in the Mackenzie basin aided a sharp increase in river freight tonnage during the late 1970s and 1980s.

A 1600 per cent increase in oil prices in the 1970s, the Three Mile Island nuclear disaster and the worst drought in Californian history led the USA to reconsider the plans in the early 1980s. But by then an already expensive scheme had doubled in cost (Reisner 1993).

Meanwhile, Canada has developed more hydropower than it needs domestically, by selectively modifying and implementing aspects of the NAWAPA scheme and by adding others. This includes harnessing the vast potential of the low-head schemes on James Bay (Box 10.3) recognized by NAWAPA, as well as the high-head sites in the western mountains.

Box 9.4 Case study: China's national plans for interbasin transfer

China suffers from the triple problems of: (1) a burgeoning population; (2) a major imbalance between the distribution of water and agricultural resources; and (3) an erratic climate that often causes the north to suffer crippling drought while the south is fighting catastrophic floods. Despite draconian family-planning measures, the population of China is rising at 1.3 per cent p.a. and may not peak till 2050. Most water resources are in the south, yet most of the cultivable land lies in the north. Water tables have been falling 1 m a^{-1} in northern China and twice this around Beijing. By the early 2000s farmers could lose 30–40 per cent of their current water as Beijing's demands increase 50 per cent and farm wells dry out (Postel 1992). Moreover, there seems little possibility at present of reducing the high evaporative losses from the flood-irrigated paddy fields that grow the staple rice crop.

Climatic variability is displayed by the behaviour of the Hwang Ho, the main river serving the North China Plain, which in 1995 reached a maximum discharge of only 1000 $m^3 s^{-1}$ compared with its usual level of 4500 $m^3 s^{-1}$ in its middle reaches and for over five months ran virtually dry below Zhangzhou. In contrast, the Yangtze River is a frequent source of widespread flooding.

In planning industrial and economic modernization, China intends to use more hydropower and the major rivers that rise in the western mountains and plateaux have large potential, especially the Brahmaputra, Mekong and Yangtze in the south. Unfortunately, developments on the first two will affect downstream states (section 11.4) and the optimum site for development on the Yangtze, the Three Gorges at the point where it emerges from the eastern uplands, lies in a highly populated area (section 9.2).

A large number of transfer schemes have been advanced to meet the dual requirements of water supply and energy. Three main routes are currently under construction or discussion (Figure 9.9). The **western route** would transfer water from the Yangtze to the Hwang Ho in its upper reaches. This could be supplemented by a recent proposal to divert water from the Brahmaputra to the Yangtze. The annual discharge of the Brahmaputra or Tsangpo before it crosses into Assam exceeds 100 billion m^3. Some of the hydropower generated by a large dam at this point could be used to pump water 2000 m up over the divide and regenerated by hydropower stations set on the downward slope towards the Yangtze. It is

Box 9.4 (cont.)

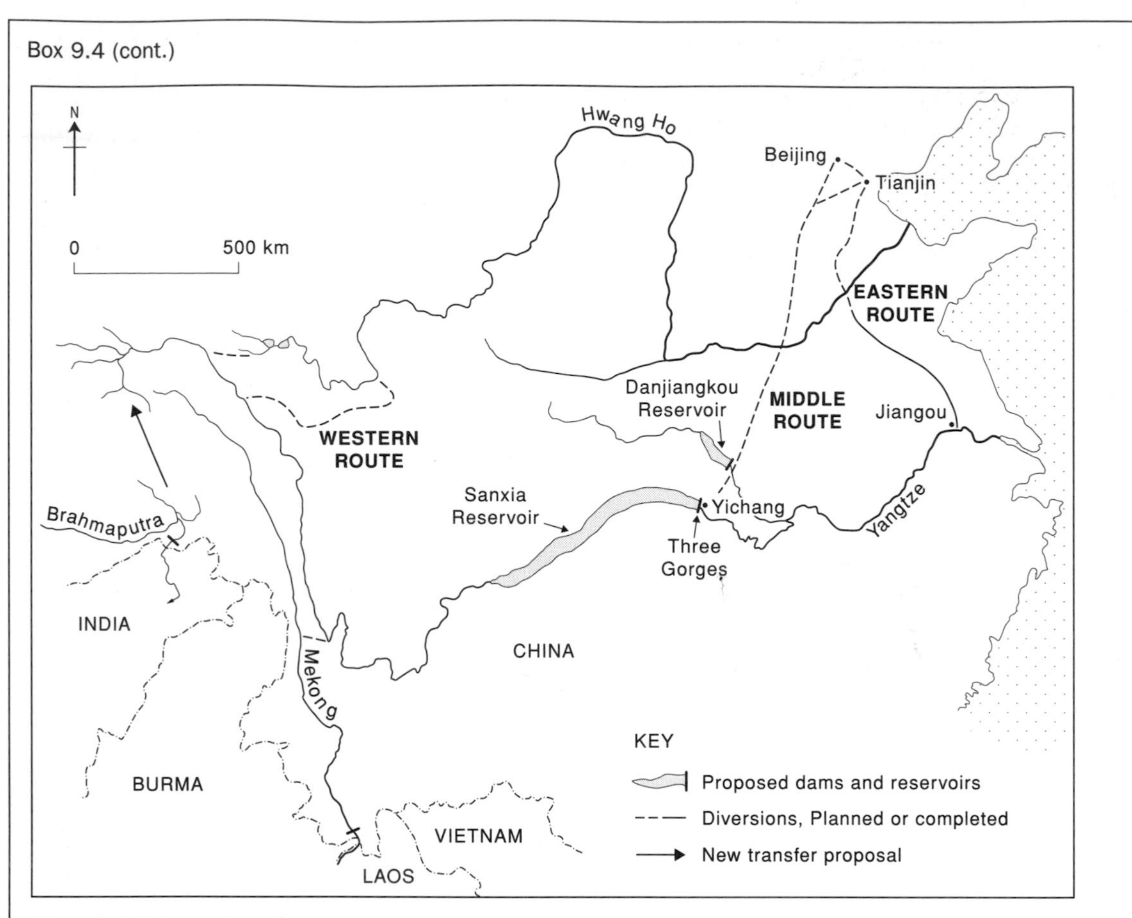

Figure 9.9 Chinese river diversion plans

claimed that this could help to reduce flood problems in India and Bangladesh, but it could also aggravate water shortages (section 11.4.3). Similar flood protection claims are made for a 1.2 million kW hydroscheme planned for the Mekong (section 11.4.2).

The **middle route** would run from the Three Gorges Scheme northwards to Beijing and construction could begin soon after the year 2000. The **eastern route** will run from the lower reaches of the Yangtze towards Beijing and is already half completed. Both these routes would be supported by some water transfer from the Mekong.

The environmental impact of these schemes could be as large outside China as within, but none of these arguments is likely to hold sway in a centrally planned economy with a pressing need to feed its people and develop its industry. The predicted effects of global warming will add to the need for water transfer, with a drying trend in the north and stronger monsoon and typhoon development in the south, trends which may have already begun (Liu and Fu 1996).

Conclusions

Most of the grand schemes for interbasin transfer proposed over the last four decades have not reached fruition. The only major exception to this effective moratorium appears to be China. The environmental impact of such schemes is bound to be enormous. Nevertheless, continued growth in population, urbanization and regional economies is continuing to aggravate the geographical incongruences between consumers and resources.

Management techniques now offer the prospect of

operating smaller-scale transfers with environmental sensitivity. Systems analysis enables an endless variety of system configurations to be explored, so that the end product can be more reliable, more flexible and more environmentally friendly. Through computer-aided design and control, it is possible to link complex multi-source, conjunctive and multi-use systems in order to maximize utilization and minimize impacts at the same time.

Further reading

Goldsmith E and N Hildyard (eds) 1986 *The social and environmental effects of large dams. Vol. II case studies.* Cornwall, Wadebridge Ecological Centre. Detailed case studies.

McDonald A T and D Kay 1988 *Water resources issues and strategies.* London, Longman: 284pp. A general textbook introduction.

Newson M D 1992 *Land, water and development.* Routledge: 351pp. Similar to the last, but more environmentally oriented.

Reisner M 1993 *Cadillac Desert.* New York, Penguin: 528pp. An excellent read.

Shikloanov I A 1985 Large scale water transfers. In J C Rodda (ed.) *Facets of hydrology II.* Chichester, Wiley: 345–87. Provides a detailed discussion of river diversion plans.

Wooldridge R (ed.) 1991 *Techniques for environmentally sound water resources development.* London, Pentech: 332pp. A collection of papers, mainly from Developing Countries.

Discussion topics

1. Analyse an example of multi-source or conjunctive resource management known to you.
2. Analyse the environmental impact of one or more large dams or interbasin transfer schemes.
3. Discuss the advantages and disadvantages of river regulation.

CHAPTER 10

Managing runoff 2: unconventional and environmentally sound solutions

Topics covered

10.1 Human-scale developments: a principle from the past

10.2 Environmentally sound management of channels and river basins

10.3 Controlling the atmospheric components: the potential for extended management options

10.4 Controlling the surface cryosphere

The rise in interest in new environmentally sound solutions over the last quarter of the twentieth century has coincided with a re-evaluation of some older approaches that have been overlooked or undervalued in the drive for major structural solutions. These include small-scale local water resource development, which may be particularly relevant in the LDCs, and a more rational and integrated use of rainmaking technology.

10.1 Human-scale developments: a principle from the past

In centuries past, most developments were 'human scale', that is, relatively small-scale and local developments undertaken by local communities. When modern engineering technology gained the ascendancy, two vital aspects were lost. One was regard for the local environment. The other was the feeling of involvement among the local people. Solutions imposed by remote planning bodies too often destroyed both, and resulted in a lack of understanding, a lack of care, more waste, and a feeling of helplessness or dependency. This has led to many wasteful projects in the LDCs. But the principle is also applicable to some aspects of developments in developed nations.

Proposals for small-scale hydropower dams in Britain satisfy this principle (section 1.2.2). So in some respects do small-scale groundwater developments undertaken in Wales during the 1980s, on the Rivers Monnow, Rheidol and Clwyd. These do not provide adequate supplies in themselves, but when linked to existing surface sources provide increased capacity without enlarging the reservoirs. They also tap a resource with a different sensitivity to rainfall. The Rheidol scheme allowed Aberystwyth to avoid drought orders during the severe 1984 drought, which has since become the 'design drought' for most resource planning in Wales. It also yields high-quality water that has been filtered by the fluvial gravel aquifer, which can be used to dilute surface water during periods of high metal pollution, acidic or 'dirty' water (section 8.3.1).

It is neither practical nor desirable to dismantle existing large-scale schemes in developed countries. But revision of established philosophy can often be beneficial and small-scale elements might usefully supplement existing systems. The NRA/Environment Agency's declared policies of consultation (NRA 1992) and of environmentally sound, sustainable development (NRA 1994) are attempts to reverse the dehumanizing trend, without necessarily returning to the old human scale.

In Developing Countries, however, the situation may be different. The human-scale or the village-scale approach may well be more sustainable than big schemes financed by foreign aid and entailing crippling debt repayments.

Rainwater harvesting has ancient roots, yet it is the subject of renewed interest in many parts of the arid realm. It uses a range of methods to collect rainwater by redirecting overland flow or using impermeable surfaces. Three-quarters of the farmland of the Hopi Indians in the Mojave Desert was fed by runoff harvesting, draining surface runoff from a wide area of hillslope into the fields. Gilbertson *et al.* (1994) describe the 'bunds' (walls), terraces and spillways used in ancient Libya to collect floodwaters.

Small-scale abstraction of groundwater still has a role today. Rodda (1994) describes the Kachogo Kakola area project near Lake Victoria, where the Kenya Water for Health Organization has developed a network of shallow wells and **handpumps**, operated by 60 women's groups, to replace the surface water sources that are polluted by agrichemicals and human excrement used on the irrigated rice fields. She notes that the large traditional role of women in collecting 70 per cent of the water in LDCs has been disregarded by modern commercial irrigation, with severe environmental and social effects. Large-scale irrigation of sugar cane from groundwater in India has caused village wells to dry up. In some Asian examples, the water table has fallen by up to 80 m since the 'Green Revolution' brought two to three crops a year of high-yielding rice, which consumed vast quantities of water. Similarly, draining of wetlands for commercial agriculture has caused some villagers to lose valuable fish resources.

10.2 Environmentally sound management of channels and river basins

Attitudes to river management changed markedly in the Western world during the 1980s and 1990s. In part, this has been due to increasing conflicts brought about by the general level of development and competition for the use of resources. In part, it is due to 'increasing public clamour for wildlife and landscape protection, and for recreation' (Hockin *et al.* 1988). Hockin *et al.* (1988) note the importance of a flexible approach to finding solutions as well as the important contribution that the educational policy of environmental agencies can make (e.g. Lewis and Williams 1984).

Improvements in scientific knowledge and our ability to predict the effects on wildlife and landscape are allowing planners and engineers to work *with* the natural processes rather than against them.

10.2.1 Channel engineering

Channel engineering is beginning to take account of recent advances in geomorphology and 'botanical engineering'. The days of the 'ideal' straight, trapezoidal channel, as designed by British engineers in Victorian India, are numbered as river engineers consider the aesthetic, ecological and erosional advantages of a wandering channel. Brookes (1988) describes many of the advances that have been made in this field in recent years. In Holland and Denmark straightened streams have been re-engineered to meander, to reduce velocities and allow fish to breed in the shallows on the inside of the bends.

Working with a meandering river can be difficult, whatever the objective. Lewin (1978) describes a case where repeated attempts to straighten a section of the River Ystwyth were followed by deposition and renewed meandering. Newson and Leeks (1987) describe the case on another Welsh river, the Trannon, where, although the meandering channel form was left in place, the gabions placed to stop the natural migrating tendency of the channel were ineffective. 'Hardware' solutions like this may be ineffective for two main reasons: (1) they generate more rapid, eddying flows that cause increased erosion immediately downstream; or (2) they are placed inappropriately, so that they do not protect the most vulnerable part of the bank. The second problem can be more difficult to deal with than might appear, because the point of maximum erosive stress shifts as the depth and velocity of flow change. Advances in **modelling channel erosion** in meanders are enabling more effective design that is in harmony with natural processes (Thorne and Osman 1988). The solution shown in Figure 10.1 looks incongruous and it remains to be seem whether it will stabilize the meanders. Indeed, for maintaining a diverse ecology there is every justification for allowing localized bank erosion.

Equally important may be the recognition by river engineers of the **basin-wide geomorphic context** for individual sections of channel (Figure 10.2). In order to understand the processes of erosion and deposition in a specific section, and so to engineer in sympathy with long-term and basin-wide trends, it may well be necessary to take the broader view (section 8.2.3). Local erosional processes can be affected by the supply of sediment to the channel upstream and the long-term staged migration of sedimentary deposits downstream.

Natural harmony can also be preserved by using **botanical engineering**, especially using plants to protect the vulnerable toeslopes of the riverbanks against scouring and undermining, which is the normal prelude to bank slumping. Work in the Rhine delta has successfully investigated this 'soft' form of protection, using reed beds, which create a wildlife habitat at the same time as they reduce bank erosion.

Figure 10.1 Recent grading of banks on the upper Wye in Wales attempts to reduce erosion

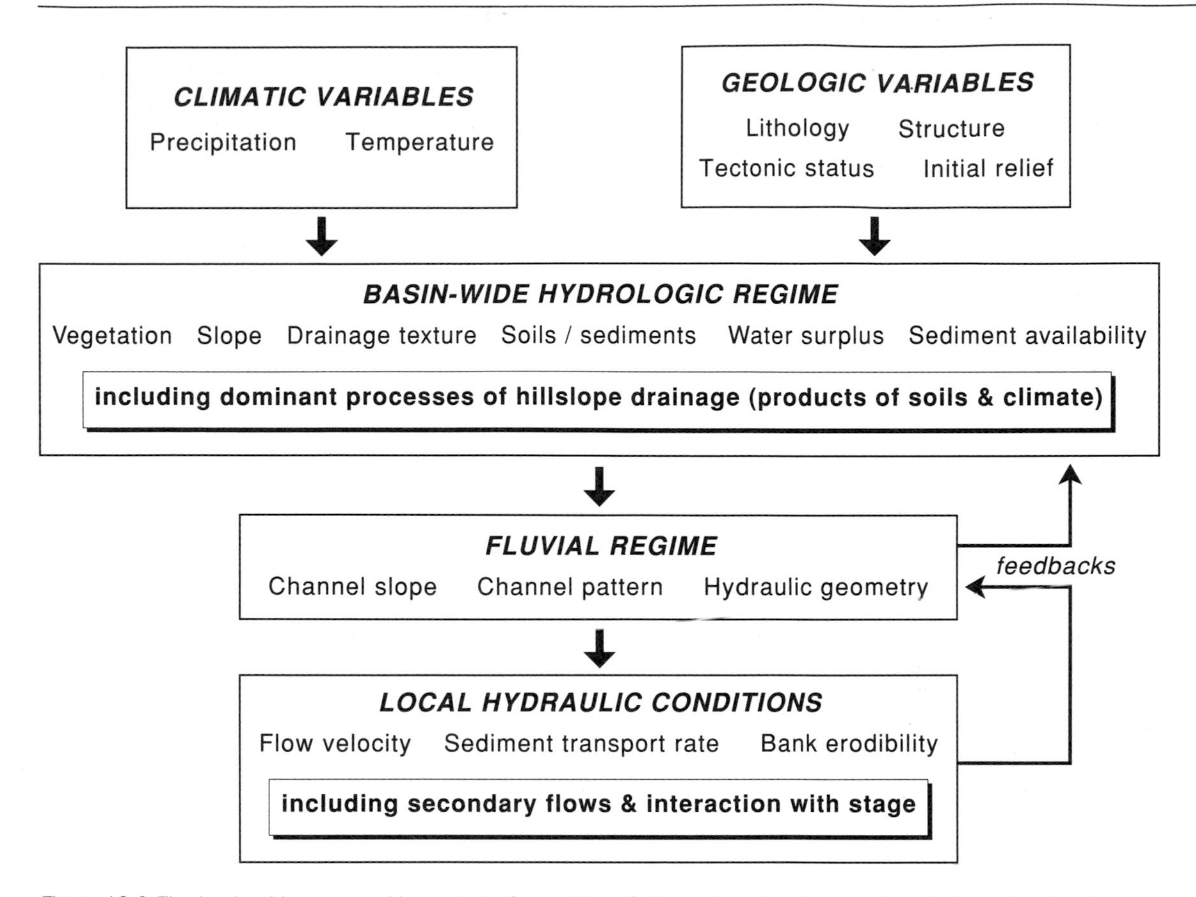

Figure 10.2 The basin-wide geomorphic context of a section of channel creates important controls on local patterns of erosion and deposition

10.2.2 Flood protection

Modern flood management is also tending to pay more regard to natural *out-of-channel* processes on the floodplain and in the wider basin.

In its natural state, the floodplain becomes a channel during heavy storms. The concept of the **controlled flood** recognizes that some return to this situation may be cost-effective and may even be ecologically valuable. The practice of floodplain zoning has long recognized that flooding can rarely be completely prevented (section 9.1.2). A positive extension of this idea was to allow controlled flooding on agricultural areas. Attempts are now being made to go one step further and to re-create floodplain **wildscapes**, where both ecological processes and flooding will operate with as little human interference as possible.

Figures 10.3 and 9.6 show two examples where planned overbank storage is allowed on agricultural land either side of a town. At Yarm, the town defences are built to withstand a 40-year flood, but surrounding agricultural land is only protected against the 10-year event. This allows overbank storage to attenuate the flood wave. The scheme is a compromise between the 5-year flood protection on agricultural land provided up to the early 1990s and the need to provide significantly more attenuation for flood waves of 20-year return period and above, at which the Gumbel curve shows a dramatic increase in flood levels. The effectiveness of the overbank storage is illustrated by a 34-year flood in January 1982 which was reduced from 580 m^3 s^{-1} upstream at Broken Scar to 460 m^3 s^{-1} at Low Moor. Planners judged that this compromise was more cost-efficient than either increasing the cross-sectional area of the channel or providing new flood storage upstream of Yarm.

The latest trend makes greater use of the floodplains for **landscape ecology**. This creates a more natural

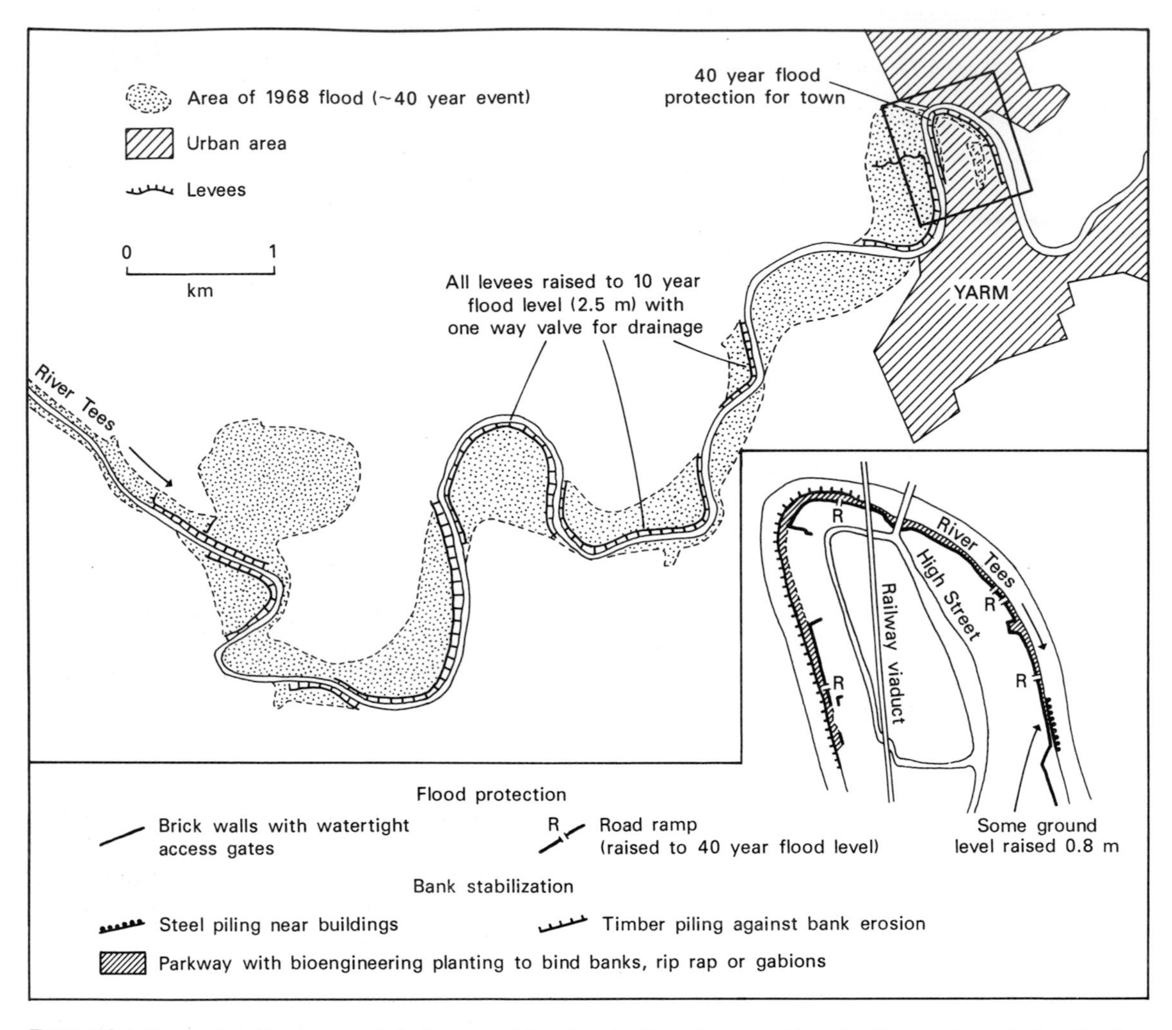

Figure 10.3 Planned overbank storage in the recent flood protection scheme at Yarm in Cleveland. Controlled flooding is allowed by lower levées upstream of town

environment that looks better and benefits wildlife at the same time as it restores the floodplain's hydrological function to delay flow and attenuate the floodwave (Figure 10.4). New **linear parks** along rivers in the suburbs of Birmingham 'absorb' floodwaters. Booth (1991) described this as planting a **flooding fringe**. The established idea of using **flood storage reservoirs** to reduce flooding from urban stormwater is being supplemented by excavating new **lakes for overbank storage** along open stretches of river. In Milton Keynes new town, large lakes have been created in floodplain areas, some with a rural milieu, others manicured and urban with fountains. In the Netherlands, excavations in rural floodplains are being allowed to go wild (Box 10.2). Such nature reserves recognize the important role of rivers as **wildlife corridors**, which allow migration, dispersal and cross-fertilization and so ensure the natural survival of species.

More natural methods of flood control are an important element in the 'new' approach known as **strategic catchment planning**. It may owe something to TVA-style planning (section 9.2.1), but this approach specifically aims to dispense with large-scale physical flood prevention structures. The approach developed by the Thames Region of the NRA not only employs a wide range of disciplines, but it also integrates them and aims for maximum public consultation (Gardiner 1991). Indeed, planning for a large basin in a democratic country nowadays requires extensive

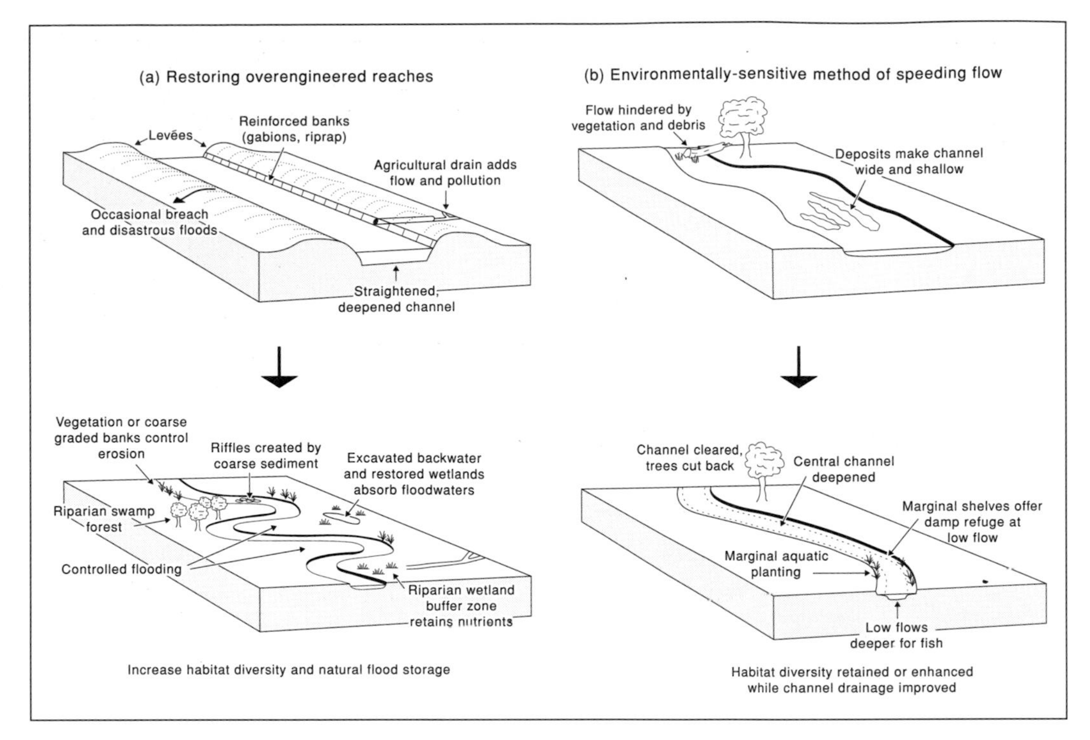

Figure 10.4 Restoration of floodplain landscapes

consultation and persuasion to satisfy both the environmental and the recreation lobbies. The days of imposing top-down solutions seem limited.

The River Rhine and its delta are a good example of the change in management philosophy that has occurred over the last 20 years (Figure 10.5). With 60 per cent of the country potentially vulnerable to flooding from the sea and sinking at a rate of 200 mm a century, the Netherlands has always been a flood-conscious nation. The Dutch are now among the most skilled in melding engineering and environmental interests.

The final fillip that launched the ambitious Delta Project was the disastrous coastal floods of February 1953, which killed 1835 people, displaced 72 000 and flooded 200 000 ha. Whereas the solutions of the 1950s and 1960s were based on solid engineering structures, the 1980s saw a shift towards working *with* the environment, regulating rather than impounding (Box 10.1), and the 1990s has seen the balance move very rapidly in favour of restoring nature and utilizing natural checks where possible (Box 10.2). With this have come the first steps towards basin-wide planning and collaboration through the International Commission for the Hydrology of the Rhine Basin (CHR).

The Rhine–Maas floods of January/February 1995, in which over 250 000 people were evacuated in Holland and one critical stretch of levée at Ochten on the Waal only narrowly avoided collapse, renewed calls to redouble the 'return to nature', just as the 1993 Mississippi floods did in the USA (section 7.6.2). For Europe, as for the USA, these were the costliest river floods of the century, causing nearly $4 billion worth of damage – 28 people died, 16 of them in France. It followed another severe flood only 13 months before.

The trigger was exceptional rains from Britanny to the Rhine; Belgium received three times its normal January rainfall. But the underlying causes identified by the French Environment Minister and the President of the French Office of Geological and Mineral Research include expanded urban paving and drainage, and the destruction of hedgerows to give large expanses of parallel ploughing: 'the frequency of flooding is going to increase and occur farther and farther upstream', said the geologist (Walsh 1995).

Box 10.2 looks at the methods of rehabilitation now being employed on the Rhine.

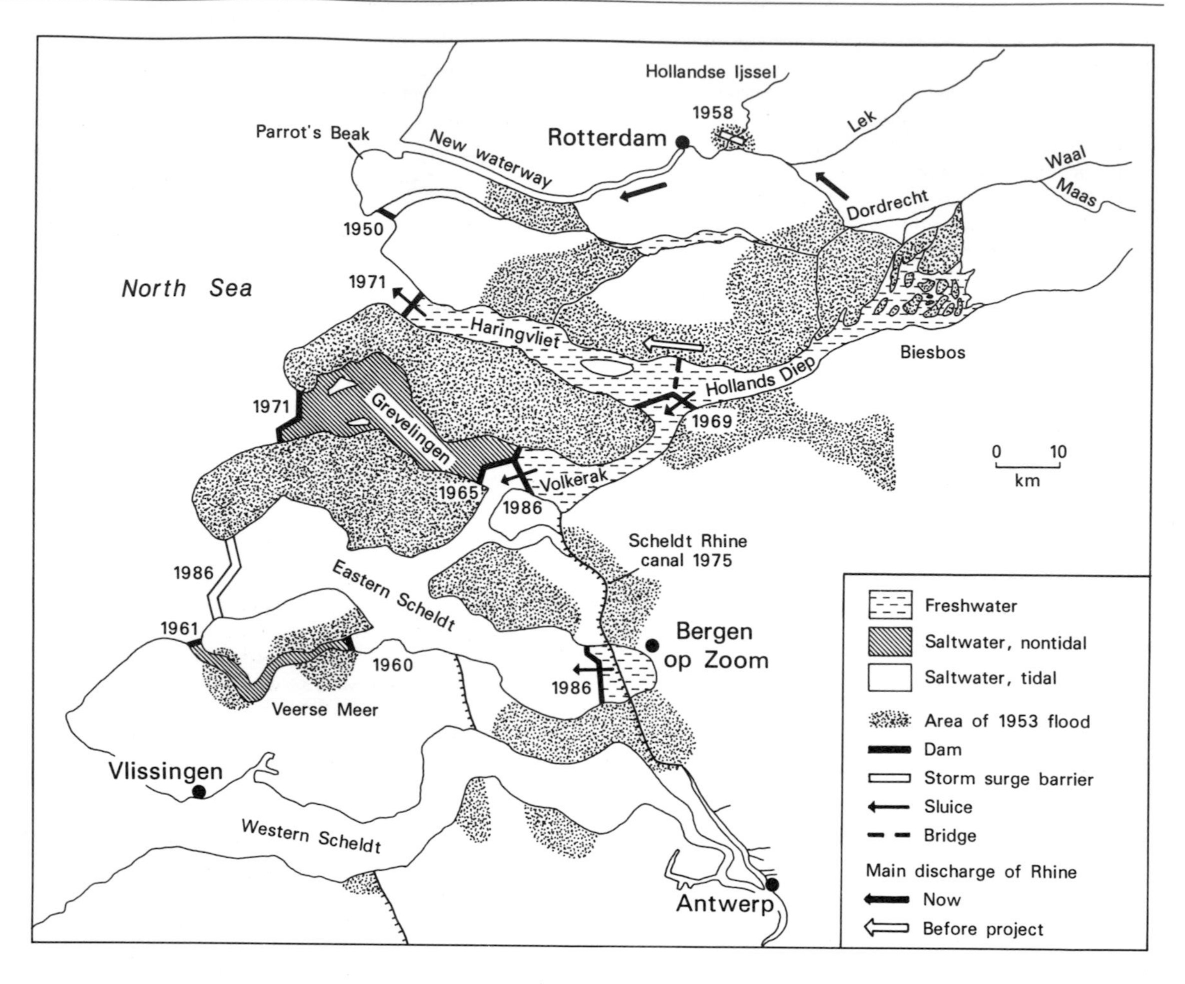

Figure 10.5 The Rhine Delta Project

Box 10.1 Case study: Environmental protection and rehabilitation in the Rhine delta.
Part 1: Environmental evolution of the Delta Project

Initially, the project aimed at flood protection with little concern for nature, but from the late 1960s, as the project progressed, public concern for the environment increased.

Damming the estuaries created four main environmental problems:

1. *Excluding saltwater* and reducing *tidal oscillations* threaten estuarine ecosystems through changes in water quality and habitat. Wave erosion concentrates on a narrow range of shoreline and many tidal flats have been eroded away and heavily polluted sediments reworked.
2. *Raising water surface levels* affects groundwater levels and can also destroy ecosystems such as mudflats and salt marshes, and the water birds that use them.
3. Sea barrages might increase the threat of *river floods* behind them. Hence only a few can be solid and even sluice dams like the Haringvliet need to be operated in relation to river discharges.
4. *Water pollution* may be increased: (i) by altering river currents so that they erode shoals that contain highly toxic chemicals; (ii) by dams hindering the 'safer' discharge of

Box 10.1 (cont.)

pollution to the sea; and (iii) by salt incursions from the artificially maintained saltwater lake at Grevelingen. The dams have been a major cause of the estimated 100 million m^3 of polluted estuary sediments.

Increasing environmental awareness led to re-evaluation of the construction and operation of existing structures and to major changes of plan for parts of the scheme. Most of the early work consisted of **solid dams** designed to reduce the length of coastline that needed to be protected by dikes. Some estuaries, however, had to remain open to discharge riverwater and to allow access for shipping, and for this reason a **storm surge barrier**, which is kept open until a flood threatens, was built on the Hollandse IJssel. The 1970s saw completion of the **regulating sluices** at Haringvliet, whose prime purpose is to regulate water levels and to discharge excess water from the Maas and Rhine.

By the 1980s, environmental concerns were becoming more important. In 1981, 10 years after Lake Grevelingen had been created by the solid Brouwers Dam, a sluice was added to allow saltwater to re-enter and permit the re-establishment of the valuable oyster industry that was being decimated in the Eastern Scheldt by pollution and disease. Unfortunately, with the water level also raised for boating, this has threatened some local agricultural water supplies through saltwater incursions into the groundwater.

More radical changes took place on the Eastern Scheldt, where the original plan was for a solid dam. After prolonged protests and a special environmental study, this was replaced by the least intrusive compromise, a storm surge barrier. The Eastern Scheldt is particularly rich in wildlife. It plays a crucial role in North Sea and Atlantic fisheries as a nursery for plaice, cod, sole and herring, and as a breeding ground for anchovy. It is internationally renowned as a sanctuary for migrating and breeding birds, including oystercatchers and avocets. Much of this has been maintained, although the barrier inevitably affects tidal range. Some plants like sea purslane are dying out, but in general the lower tidal line results in *relocation of habitats* rather than their destruction.

The concept of **compartmentalization** was employed to restrict the extent of any flooding failures with internal 'compartmentation dams', but it also helps maintain the tidal range by restricting the area subject to tides (Figure 10.5). The internal dams also separate freshwater from saltwater and divert freshwater for navigation and for water supply in the **polders**, reclaimed land below sea level.

Flow through the compartments is regulated according to the discharge of the Rhine as it enters Holland at Lobith. When the discharge is *low*, under 2200 $m^3\ s^{-1}$, the Lower Rhine, Lek and Maas are closed off, diverting water into the IJssel, Biesbosch and Hollandsch Diep. The IJsselmeer, Hollandsch Diep and Haringvliet then act as reservoirs. Reserves of freshwater stored during high discharge may be released from these reservoirs to supplement natural riverflow or to supply the polders with freshwater. Water quality is improved by oxygenation and settlement as it stands in these reservoirs, so that Haringvliet water is of better quality than the Rhine at Lobith. Closure of the Haringvliet also diverts freshwater into the New Waterway and halts advancing saltwater.

When discharge is *average*, the IJsselmeer, Lower Rhine and Lek sluices are open and the Haringvliet sluices are opened just enough to allow flow in the New Waterway to prevent saltwater advance. Freshwater reservoirs are topped up with riverwater, but the Maas sluices are closed. At *high* discharges, over 2200 $m^3\ s^{-1}$, all sluices remain open at low tide, but the Haringvliet closes at high tide to limit salt incursion. The polders have a water surplus and all pumps are working to capacity.

The scheme is now designed to withstand the 1 in 4000 year event. The final stages involve another storm surge barrier for Rotterdam, completed in 1995, and a 'waterdefense' dike around Europort, for final completion in 2000. Meanwhile, final decisions on the changes to the operating rules of the delta sluices recommended in the 1994 impact statement (Box 10.2) are expected by 1997.

Box 10.2 Case study: Environmental protection and rehabilitation in the Rhine delta. Part 2: Restoring habitats

From the early 1980s four approaches have been taken to restoration work:

1. Preserving and restoring marshes and mudflats
2. Improving water quality and removing heavily polluted sediments (Box 8.1)
3. Restoring the connection between the estuary and the sea (Box 10.1)
4. Restoring floodplain wetlands.

The 1994 Environmental Impact Statement initiated under the 1990 National Policy Plan recommended improved rules for operating the sluices, which include allowing fish through and increasing tidal ranges to re-create lost tidal flats. Migratory fish and fish-eating animals like the otter and cormorant have almost disappeared, but leaving the sluices open longer could reduce the velocity of freshwater outflows and restore a more balanced connection to the sea.

Tidal flats are being re-created in the lagoons behind new rubble walls set in the shallows at low-tide level, 25–250 m offshore. When the sluices are open, these walls will help prevent increased tidal erosion in the shallows, where polluted sediments are often covered by only a thin layer of recent, cleaner deposits. Elsewhere, artificial sand shoals are being built to create feeding and breeding grounds for birds and fish.

During the 1990s the Rijkswaterstaat and conservation groups have been re-evaluating flood control along the rivers. There are clear parallels between the Rhine and the Mississippi (section 7.6.2). Over 90 per cent of its wetlands have been drained. River straightening means it is now 80 km shorter than in 1830, and Alpine meltwater that used to take 60 hours to reach Karlsruhe now takes just 30 h. The Dutch began experiments to re-create wetland overbank storage areas in 1986 initially as wildlife refuges under the Stork Plan, but their value for natural flood alleviation was appreciated in the 1990s. De Blauwe Kamer Nature Reserve, on the Neder Rijn near Wageningen, is a good example (Figure 10.6), where 32 000 m^3 of earth has been excavated to re-create a floodplain storage area and a wetland habitat for wildlife. In 1992 a breach was dug in the 'summer dike' to allow summer floods as well as winter floods into the wetland. And in the February floods of 1995, the Lower Rhine presented no problem downstream of the reserve, although De Blauwe Kamer itself was under 4.5 m of water. Ecologically, such reserves are designed to provide: (1) variable water levels, which provide habitat diversity; and (2) slow-flowing waters, which favour species like the mayfly. The **alluvial forests** of willows and black poplars and marshy backwaters are slowly returning and with them birds like the stork and cormorant.

Figure 10.6 The artificial backwater at De Blauwe Kamer in the early stages of naturalization designed to provide overbank storage and a habitat for migrating birds

Box 10.2 (cont.)

Numerous similar projects are in the making. Although each is only a few kilometres long, they represent a significant move in the right direction. Restoration in Holland is helped by the state owning all land in front of the summer dikes and about 15 per cent of the area between the winter and summer dikes, known as the **forelands**. As farming tenancies lapse, the state aims to hand the land over to conservation groups for restoration. The forelands are also mainly used for dairying and EU dairy surpluses mean they are largely redundant. Within South Holland the government plans to transform up to 2500 ha of farmland adjacent to lakes and waterways into wetland wildlife refuges between 1990 and 2010. By 2040 all the 12 000 ha of forelands in Holland could be restored. If the Haringvliet reverts to an estuary under new operating rules (Box 8.1, section 8.3.2), the agricultural margins, which could be affected by salt incursion, may also be converted into wetlands. Other wetlands could be created on land reclaimed in the past from the sea by reflooding polders. The De Dood polder is now a wetland. After an accidental dike breach in 1990 the Selena polder rapidly reverted to nature. Similar proposals have been advanced to improve the ecology of the Eastern Scheldt.

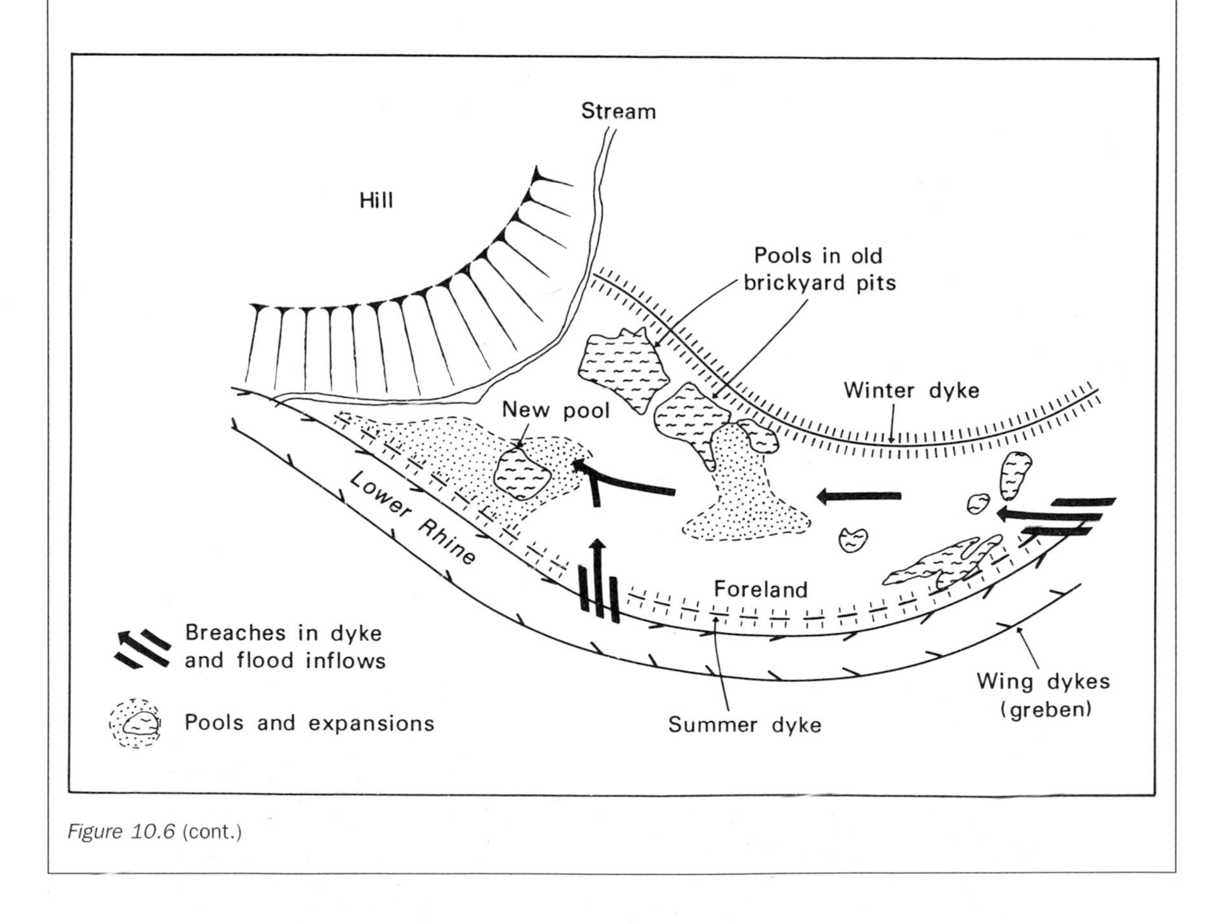

Figure 10.6 (cont.)

The increase in environmental awareness is also apparent in the contrast between Canada's James Bay hydropower development, a product of 1960s planning (Figure 10.7), and the Mackenzie basin management scheme developed a decade later (Figure 10.8). Boxes 10.3 and 10.4 outline the different approaches.

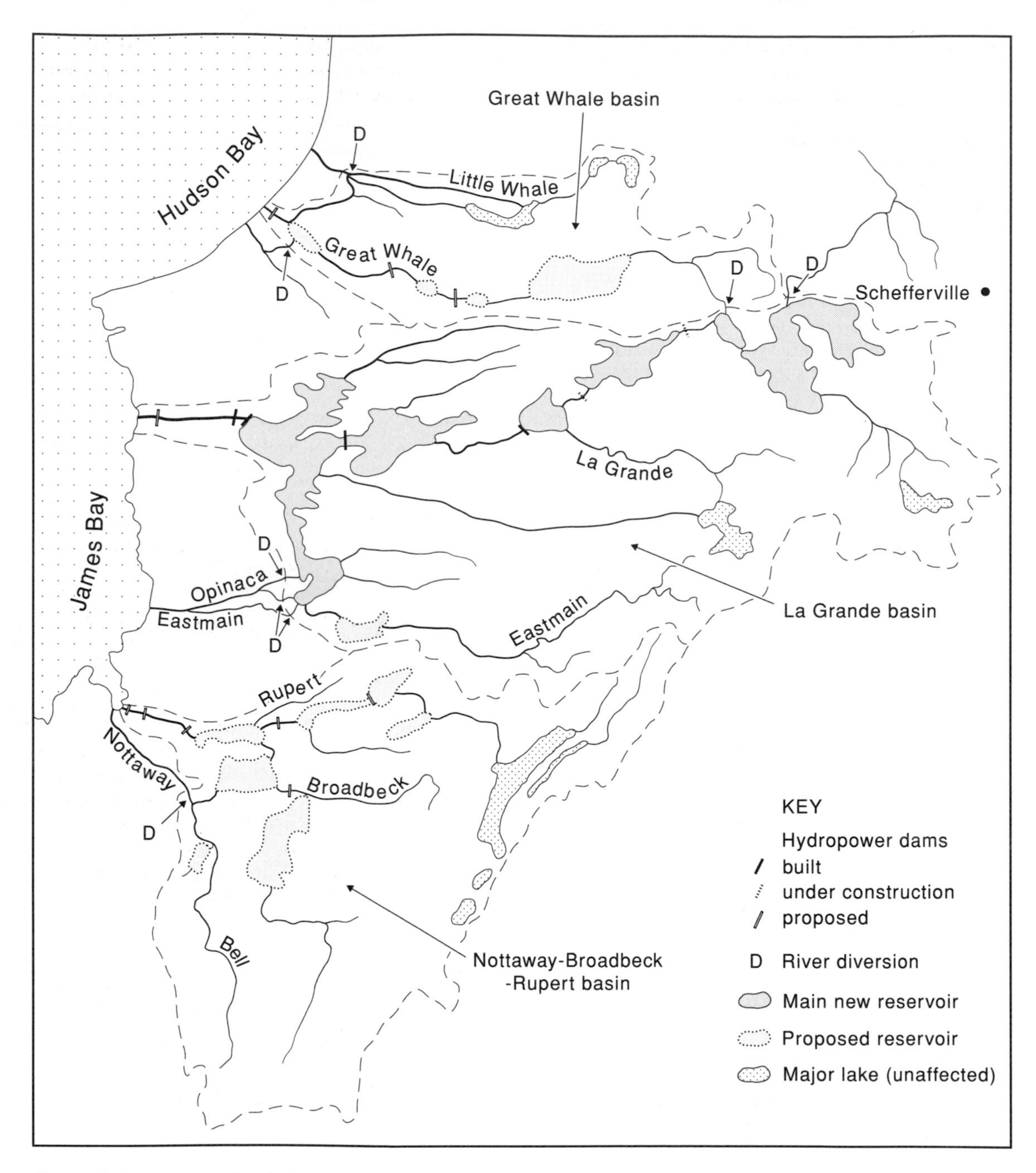

Figure 10.7 The James Bay hydropower schemes in northern Quebec

Box 10.3 Case study: Contrasting environmental awareness in Canada.
Part 1: The James Bay hydroschemes

A 50-year long programme to develop 26 000 MW from the rivers of the James and Hudson Bays began in 1971 without a proper environmental assessment (Figure 10.7). Hydro-Quebec's $50 billion 'project of the century' is a low-head scheme, which capitalizes upon the large quantities of snow that accumulate in northern Quebec and the ease of river diversion in a glaciated lowland with subdued catchment divides. The main technical detraction is that the rivers are highly seasonal, with flows dominated by the spring snowmelt. Consequently, the reservoirs need to be exceptionally large, and this is a major source of the environmental impact.

The scheme was spurred on by sales of electricity to the USA, following Newfoundland's successful exportation of power from Churchill Falls through Quebec to New York, and by Quebec's own separatist politics. 'James Bay One' on the La Grande River involved building 215 dams including three enormous ones, diverting 19 billion m^3 of annual discharge from seven rivers and flooding 10 000 km^2. The even larger 'James Bay Two' plans to divert another 10 rivers in the Great Whale and Nottaway–Broadbeck–Rupert basins. By completion, the project will have reversed or drastically altered 19 major rivers, flooding 176 000 km^2 of boreal forest and tundra. The hydrology and the general environment of an area the size of France will be redesigned and one-fifth of Quebec's surface water, or more than 1.5 per cent of the world's liquid freshwater, will be harnessed.

Immediate environmental impacts and long-term fears abound. A major forest fire raged after an accident during construction work on the La Grande in the early 1980s. Fears have been expressed for the valuable lobster population of James Bay as sediment yields have risen and for the *Beluga* whale due to salinity changes in Hudson Bay. Concern has also been expressed over possible geological shifts due to compression from the weight of the impounded water, especially for the La Grande-3 reservoir, which covers the glacial deposits to a depth of 90 m. Reservoirs and diversions can also disrupt migratory mammals. Interference with flows on the Caniapiscau River caused an exceptional flood in the spring of 1984 in which 10 000 migrating caribou died, despite an emergency helicopter airlift. The extension of Lake Bienville in Phase II could disrupt another caribou calving area.

The Cree Indians have borne the brunt of the development and are complaining at the destruction of their environment – 2500 Cree were advised to leave their traditional township on Fort George Island in the mouth of the La Grande because hydraulic engineers feared, apparently needlessly, that the La Grande scheme would cause the sandy island to erode. As feared on the Mackenzie (Box 10.4), higher winter discharges have destabilized the river ice, so that it no longer forms a bridge across the river mouth for the Indian hunters, while 150 km inland the Eastman River is now reduced to a trickle by the river diversion. Worse, decomposition of flooded vegetation has caused a toxic build-up of mercury in the food chain and the Cree no longer eat fish from the reservoirs. Problems caused by decomposition could be long-lasting because of the slow rate of decay in subarctic temperatures. One reservoir built 80 years ago in the Russian taiga is still suffering from decomposition problems. Ecologists also fear that the large surface area of stagnant reservoirs will accumulate toxic pollutants from the atmosphere.

Compensation of $90 million paid to the Cree and Inuit nations under the James Bay and Northern Quebec Agreement has helped, and the agreement gave the Cree unprecedented powers of self-government for an aboriginal 'First Nation'. Not surprisingly the Cree have been vociferous in their opposition to Phase II, particularly the damming of the Great Whale River. Using the law to protect their traditional hunting and fishing grounds, they obtained a federal court order for a full EIA and public hearings in 1990.

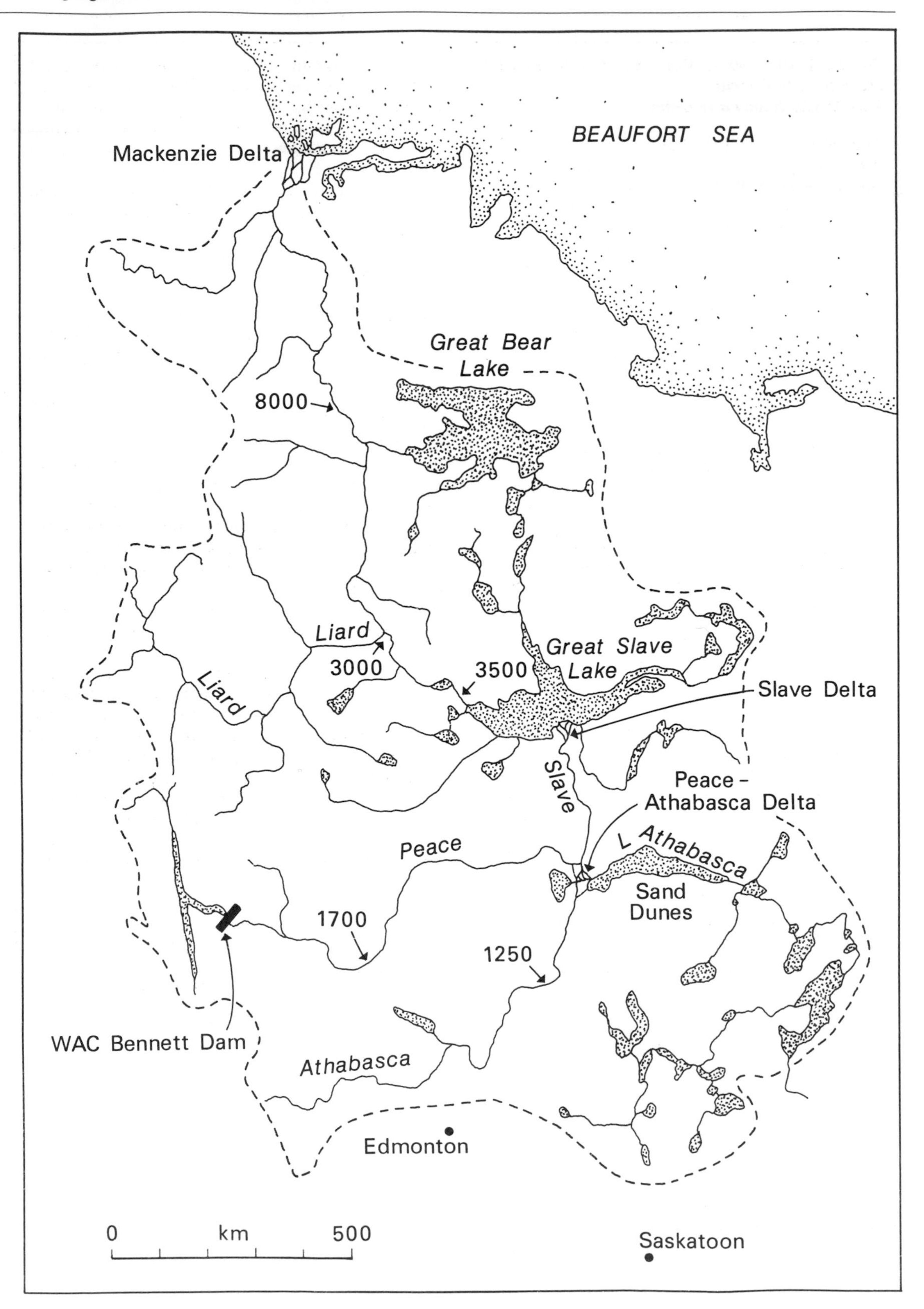
BEAUFORT SEA
Mackenzie Delta
Great Bear Lake
8000
Liard
3000
3500
Great Slave Lake
Slave Delta
Liard
Slave
Peace - Athabasca Delta
L Athabasca
Peace
Sand Dunes
1700
1250
WAC Bennett Dam
Athabasca
Edmonton
0
km
500
Saskatoon

Box 10.4 Case study: Contrasting environmental awareness in Canada.
Part 2: The Mackenzie basin

The Mackenzie River Basin Committee (MRBC) was set up in 1977 to lay the foundations for coordinated development in Canada's largest river basin, comprising 1 787 000 km^2, 6095 km of channel and 65 550 km^2 of lakes (Figure 10.8). The Mackenzie–Peace is the eleventh longest river in the world, and the management scheme is one of the largest and most environmentally friendly. The MRBC was set up not long after Canada rejected American proposals to integrate the Mackenzie into a continental water grid (Box 9.3).

The committee's 10-volume report (MRBC 1981) established a framework based on:

1. Agreements on **transboundary water management,** MAFs, water quality and river regulation between five provincial/territorial governments and the federal government.
2. A permanent **integrated network of stations** for hydrometric and meteorological observations, including snowcover, sediment and water quality.
3. A suite of **computer models** to predict responses in all parts of the water budget, including a daily routing model for flow in the main river network.
4. A series of **special basin-wide and local area studies** to cover key environments and processes. These included spring ice break-up and sediment transport, and the ecology of alluvial areas and the Athabasca sand dunes.
5. Every project should be judged according to an **Environmental Impact Assessment.**
6. Information should be readily shared through the **national water document database,** WATDOC.

Spring break-up is the most important annual event. During the spring floods of 1974, 24 million tonnes of sediment were carried daily into the sea. Ice jams are a major cause of erosion and flooding, and the regular destruction of bankside vegetation by floods and ice floes is vital in maintaining habitat diversity. The Mackenzie, Slave and Peace–Athabasca deltas are internationally important as staging posts for hundreds of thousands of migrating birds. They provide **breeding and feeding grounds** not found in the surrounding boreal forest. Arctic char, grayling and trout spawn in the gravel and thrive on a period of low sedimentation during incubation. Conversely, the nutrient-rich sediment carried from the mountains during spring floods is the life-blood of the diverse and highly productive ecology along the waterways and in the Beaufort Sea.

Abnormal patterns of **sediment mobilization** caused by construction or mining could be disastrous. Similarly, changes in lake or river levels could cause **changes in local groundwater** which would be disastrous for the unique and sensitive ecosystem of the Athabasca sand dunes. Figure 10.9 is an example of retrodictive modelling, using the daily routing model to assess the impact of the WAC Bennett Dam on downstream flow.

Over the past decade, settlement and resource development have proceeded in broad sympathy with this sensitive environment as a result of the MRBC management system. Although the Mackenzie River could generate 14 000 MW in hydropower, development has been limited by the realization that large reservoirs could cause severe degradation of the alluvial ecosystems by reducing the spring freshet, and reduce the stability of winter ice covers by increasing winter flow levels.

Figure 10.8 The Mackenzie River basin. Figures indicate mean annual discharge in $m^3\ s^{-1}$

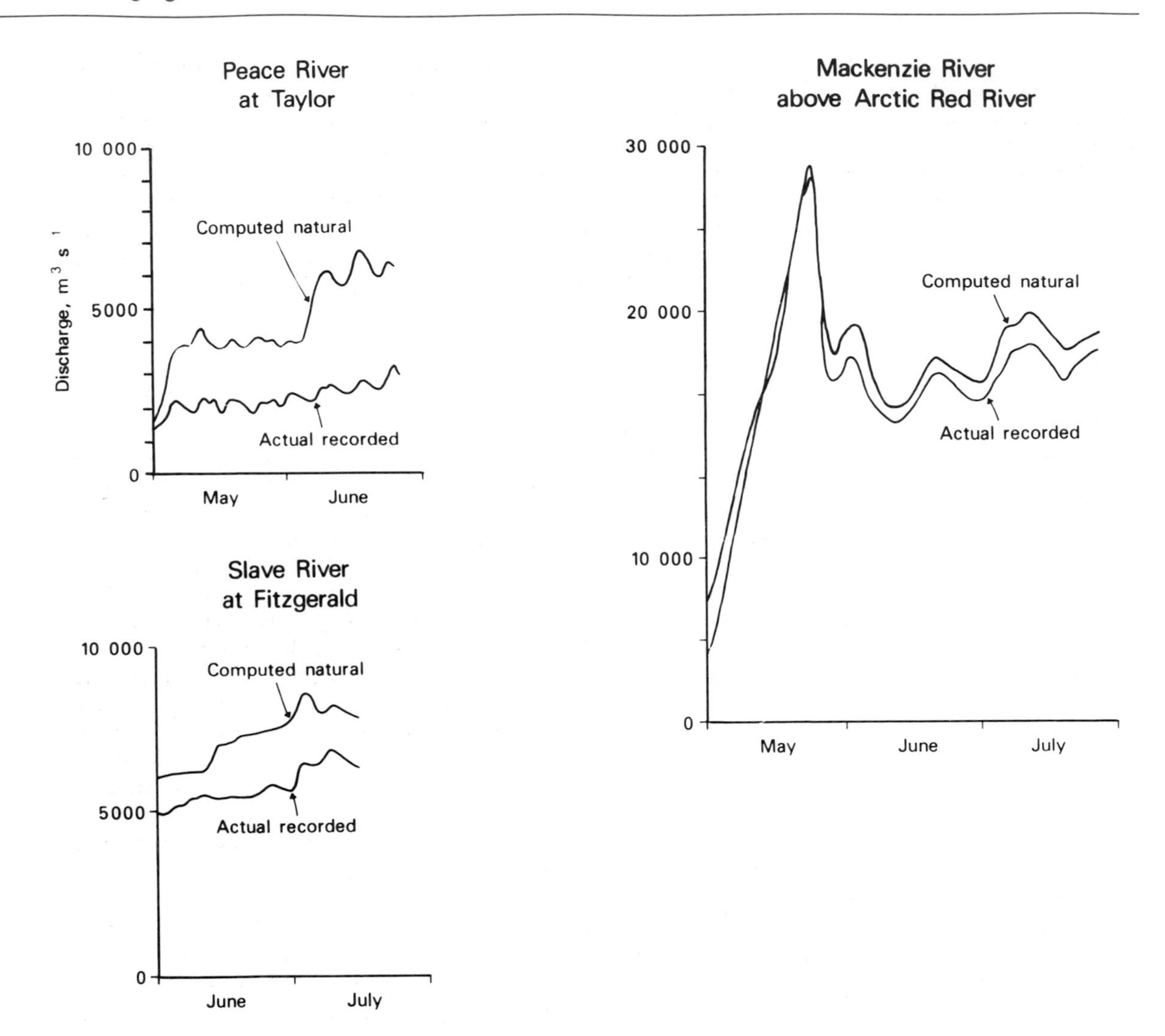

Figure 10.9 Retrodictive modelling assesses the impact of the WAC Bennett Dam on downstream flows within the Mackenzie basin. Based on MRBC (1981)

10.3 Controlling the atmospheric components: the potential for extended management options

Water managers have traditionally limited their attention to streamflow. Yet advances in our understanding of precipitation and evapotranspiration processes, plus a growing need for alternative options in some parts of the world, have made this traditional restriction less logical. Some 'atmospheric' options may actually interfere less with the visible landscape than river engineering, be less capital intensive and be capable of more immediate impact. They may be particularly attractive in temporary problems of drought or flood, where the hazard does not persist long enough or occur frequently enough to merit a permanent engineering solution. Or they may provide a temporary stop-gap while permanent solutions are being sought.

Increasing rainfall or reducing evapotranspiration may be alternatives to irrigation. Triggering rainfall before it reaches flood-sensitive areas has already been used in the USA as an alternative method of flood management. The spread of weather radar and of satellite–radar link-ups make this an increasingly viable option (section 5.1.2). One of the most impressive routine storm-dissipation programmes is operated by the US NWS in the Caribbean, where long-distance airborne 'seeding' of hurricanes is carried out from Florida. Hurricane-generated floods are a major hazard in many areas on the edge of the tropical world (Figure 2.5). Hurricanes typically combine intense rainfall with high winds and coastal storm surges, and all three elements can be reduced by standard rainmaking methods (next section). The amount of potential damage merits a regular hurricane watch with US Air Force planes on alert in Florida during the summer hurricane season.

The **meteorological option** is now well established in the USA and has been used on a more or less *ad hoc* basis in a number of countries. Ideally, it should be treated as one in a set of alternatives, which includes the option of modifying human activities as well. Hence, weather modification as a means of reducing flood hazard should be judged in cost–benefit terms in relation to other options, such as flood control dams or improved early-warning systems, which will allow farmers to evacuate cattle. Such cost–benefit assessments should consider the *environmental* costs and benefits, balancing for example the possible ecological effects of chemicals used in rainmaking or inhibiting evapotranspiration against the potential damage to the soil caused by irrigation. Not only is this rarely attempted, but, as is so often the case, our ability to modify the environment is running ahead of our ability to assess the environmental costs.

Weather modification and improved forecasting may be both alternatives and complements. Forecasting is at its least reliable with localized convective storms and this is precisely where rainmaking can be most effective. On the other hand, improved forecasting systems can enable weather modification to be more effective, helping in the initial decision to employ weather modification and in selecting the clouds for seeding.

Repeated weather modication amounts to willful climate modification and needs to be evaluated against similar long-term options like altering the crops grown or installing interbasin transfers. Both traditional plant and animal breeding and modern genetic engineering are actively seeking varieties that are more resistant to summer drought.

10.3.1 Rainmaking

Making rain is one of humanity's most ancient aspirations, yet the technical ability to achieve it is only half a century old, some 6000 years behind river control.

Early experiments concentrated on a **macrophysical** approach, which tries to augment convective turbulence and so stimulate collision and coalescence. The Russians have experimented with this approach since the 1930s, using dive-bombing aircraft or throwing bags of cement off transport aircraft (Haman 1976). But this approach has not been particularly successful and energy costs are disproportionate. Indeed, the amounts of energy that mankind can muster are small compared with natural rain systems and unlikely to make much difference. A single thunderstorm is equivalent to 1/10 000th of mankind's total daily energy consumption. A mid-latitude depression contains the energy of 1000 large nuclear bombs.

A more effective method relies on the **microphysical** approach, in which small amounts of 'seed' substances (rather than energy) are introduced into the clouds to stimulate condensation or sublimation. Notable success has been achieved in stimulating the *Bergeron process* in clouds that contain or can be induced to produce supercooled water droplets, either by introducing freezing nuclei or by chilling the cloud with 'dry ice' (section 2.3.1). Dry ice is solid carbon dioxide and has a temperature of –78°C. It causes instant *homogeneous nucleation* (freezing) of some liquid droplets, which then act as ice nuclei. Because CO_2 is a gas at normal atmospheric pressures, the dry ice pellets begin to volatilize as they are discharged into the clouds from aircraft. This causes cooling as it takes the latent heat of volatilization from the surrounding air.

Shortly after Vincent J Schaefer and Irving Langmuir of the General Electric Company, New York, published their experiments with dry ice in 1946, Bernard Vonnegut published the results of his experiments using silver iodide (AgI) as a freezing nucleus. AgI has many advantages over dry ice. It is easier to handle. It can be supplied more cheaply from burners on the ground as well as from burners on the wings of aircraft. It also has a hexagonal crystal structure similar to ice, which makes it an efficient nucleus for sublimation, beginning in the critical temperature range from –4°C where supercooled water tends to be present in sufficient amount. It causes freezing as it touches the supercooled droplets and provides a crystalline base for the sublimation of water vapour. Other seeds have been used, like lead iodide (PbI) and cupric sulphide (CuS), but AgI has become the dominant cloud-seeding substance.

Cloud seeding has been applied widely throughout the world, especially in North America, Europe, Russia, Japan and Australia. American and French planes carried out seeding for the Niger government to help local agriculture at the height of the Sahel drought in 1973. The Russians use it to prevent rain disrupting the annual May Day parade, and the 1995 VE Day celebrations, by inducing rainfall around but not in the centre of Moscow. American power companies began experiments in the 1960s and concluded that seeding could boost hydropower resources sufficiently to avoid increasing output from less efficient thermal power stations in a dry year. As little as 2 per cent more runoff could cover the costs of seeding in basins over 1000 km^2, compared with a 10 per cent increase in basins one-tenth of the size.

It is not, however, without its detractions. The first of these are the *amount* of rain that can be made and the *reliability* of the methods. After his experiments in the 1940s, Langmuir claimed that rainmaking could increase rain over most of the USA east of New Mexico by 3–10 times. In the heady days of the early 1950s, when up to $5 million was being spent annually on cloud-seeding exercises covering about 10 per cent of the USA, many US rainmaking companies claimed increases of several hundred per cent. However, investigations carried out by the US Defense Department under Project Cirrus and the President's Advisory Committee on Weather Control report of 1958 reduced the figure to only 10–15 per cent. Some experiments showed no increase or even a decrease.

A persistent problem for sceptics and converts alike is proving whether seeding has had an effect, especially since the best results tend to be achieved from clouds that are about to rain naturally. The problem is to assess accurately both how much would have fallen anyway, and how much did fall? Three main approaches have been taken:

1. **Randomized seeding experiments** select clouds for seeding at random and compare the rainfall from seeded and unseeded groups. This ensures that both groups contain an equally representative collection of the physical range of cloud characteristics.

2. **Target–control regression** takes an identical area as a control, establishes a long-term relationship between rainfall in the control and the target areas prior to seeding, and then measures the degree of change after a period of seeding. Unfortunately, this can suffer from the same problem as sequential experiments in vegetation change (section 7.1), that weather patterns can change from one period to the next and confound the effects. It is also possible to falsify the results by using weather forecasts to select the best occasions for seeding and so bias the results.

3. **Target–control crossover procedure** tries to combine the best elements of the others by randomizing the choice of area for seeding on each occasion. This overcomes the problem of shifts in weather patterns and it reduces the probability of a biased sample, while identifying a fixed area that can be instrumented for rainfall measurement.

Even so, a 10 per cent increase is perilously close to the accuracy of many rainfall measurements (section 5.2).

It is now clear that many physical factors affect the results. These include the temperature profiles of the clouds and the surrounding air, the amount of seed, the type and concentration of natural nuclei, the timing of seeding relative to the life history of the cloud, the location of seeding within the cloud's structure, and therefore the mode of injection, whether from above or below.

Overseeding can increase competition with more ice crystals competing for the same amount of water vapour, so that fewer large crystals are formed. There is a potential use for deliberate overseeding in controlling flood-producing rainfall. But it is often very difficult to judge the amount of seed that would be needed, especially since data on nuclei concentrations in clouds are not readily available.

Premature seeding can have a similar effect, causing too many nuclei to become active too early. Alternatively, it might cause enough latent heat to be released to enable buoyancy to overcome a temperature inversion and cause an explosive vertical growth of the cloud.

Ground-based seeding runs the danger of introducing nuclei into the heart of the convective updraught and stimulating 'runaway' growth. One of the more spectacular failures of seeding is reported from Argentina, where ground-based burners used to reduce hail damage by premature/overseeding were successful in reducing it by 70 per cent in frontal storms, but dramatically *increased* it by 200 per cent in convective storms.

Clearly, a lot of questions still remain to be resolved. And even if adequate simulation models were available, collecting the data on location in order to feed real-time models presents major difficulties.

Generally less success has been achieved in augmenting *warm-cloud* rainfall. Seeding experiments have either introduced large water droplets to set off the coalescence process or they have injected hygroscopic nuclei. The former seems economically questionable. Classic experiments by the University of Chicago in the

Caribbean required vast quantities of water, with aircraft spraying nearly 1300 litres km^{-1} on to the clouds, and still only half the seeded clouds rained. In contrast, pulverized salt is cheap and effective as a hygroscopic nucleus. Delivery by aircraft in the West Indies and ground sources in India have both proved effective.

For those techniques which use nuclei there may be concern over *environmental pollution*. The amount of common salt used in rainmaking should be much less than may occur naturally in coastal districts. Lead iodide, however, is likely to cause more legitimate concern. Silver iodide is also a pollutant and even Vonnegut has admitted that our understanding of the environmental effects is inadequate. In this respect, dry ice seems a safer choice.

Yet more concerns arise in the *social and legal* field. It is generally extremely difficult to target the rainfall within a few kilometres. It could easily fall on a neighbour's farm or in another catchment. Even if it is correctly targeted, it is likely to mean less rain downwind, as seen in the urban-induced rainfall example in Figure 7.8. There is also the danger of creating unexpectedly large storms. The tragic flash flood that swept away much of Rapid City, South Dakota, and killed 237 in June 1972 followed cloud seeding in the area and lawsuits were filed (Dunne and Leopold 1978).

In the USA, citizens' rights to the natural water on their land extend to the water in the atmosphere above. Not surprisingly, private litigation and public legislation to control rainmaking activities have become widespread in America. Roberts and Lansford (1979) recount the case of a public hearing in San Luis valley, southern Colorado, brought under the 1972 Colorado Weather Modification Act. The case centred on rainmaking to support the barley crop to be sold to a brewing company. The 'beer-barley' farmers hired Atmospherics Inc. of California to maintain the optimum 190 mm of rainfall for the main growing season, suppress damaging hail and provide a dry period for ripening and harvest. However, neighbouring ranchers and feed-crop barley growers were not so fussy about the amount, form or timing of precipitation and they objected to activities that could reduce their own economic returns. The case was finally decided by scientific evidence which suggested that overseeding, whether deliberate or acccidental, could reduce precipitation over thousands of square kilometres in the southwestern USA.

The Colorado Weather Modification Act remains one of the best of its type. Under the Act, all weather modifiers are licensed annually by the state. If they prove incompetent or troublesome, their licence can be revoked.

Another now classic case of litigation followed a widespread programme of snowmaking in the western USA during the latter part of winter 1976–77. Seeding was commissioned in a number of states, especially Colorado and Washington, after water agencies predicted a 50 per cent shortfall in winter snowfall from the Coast Ranges to the Rockies. Washington state spent $400 000 on seeding in the Cascades to prevent a springtime drought caused by reduced snowmelt, despite threats of legal action by Idaho and Montana to prevent it 'stealing' their water.

In the event, the measures were probably largely unnecessary. Certainly this appears to have been so in the Yakima valley, where farmers using riverwater for irrigation were advised by the Bureau of Reclamation to take precautionary measures (Glanz 1982). Many sank new wells or even transplanted sensitive crops, like mint, to adjacent valleys. Statisticians and scientists were ambivalent about the claims that snowmaking had been successful. But there was no shortage of water in the spring, largely because of an error in the bureau's water management model which failed to account for the amount of return water reaching the river from the irrigation schemes. The farmers subsequently filed complaints against the bureau for the costs needlessly incurred.

Despite the problems, cloud seeding remains a feasible alternative to interbasin transfer or to modifying agricultural systems. Interest was reborn in the 1970s by fears of global food shortages, out of which came the US National Weather Modification Policy Act of 1976, which empowered the Secretary of Commerce to establish a national policy and to direct research funding. Subsequent research by the NOAA in Florida has suggested that as much as 20–70 per cent more rainfall can be obtained from thunderclouds, and seeding has become commonplace in parts of the High Plains, where groundwater resources are in marked decline (section 7.5).

In the 1990s, fears that the High Plains may be entering a cyclical drought like the 1930s and 1950s have been aggravated by fears of global warming. A 30 per cent shortfall has been predicted for the wheat harvest of the Saskatchewan prairies in a double-CO_2 scenario. Since Saskatchewan alone currently accounts for 18 per cent of the wheat traded on the world market, this could have important global consequences.

10.3.2 Controlling evapotranspiration losses

Less spectacular, but in many ways more direct and reliable, methods of improving water supply focus on reducing losses rather than increasing inputs. Four basic types of approach have been adopted: (1) control of the surface heat budget; (2) control of vegetation species; (3) barriers to restrain water losses; or (4) use of chemical antitranspirants.

Early methods used in the 1950s concentrated on reducing direct evaporation from reservoirs by spreading **monomolecular layers** of alcohols, like hexadecanol and octadecanol, over the surface. These long-chain molecules act like a net to hold down the hotter molecules in the water and so hinder evaporation. They were particularly popular in America and in Australia, where losses from some reservoirs were excessively high. Unfortunately, such layers are easily broken by waves in winds over 2 m s^{-1}. In low-wind situations, Australian tests have shown savings of up to 50 per cent in water losses in lakes of less than 11 km^2, but in the windy environment of Lake Heffner (11 km^2), Oklahoma, savings of only 9–14 per cent have been achieved. Retaining the hotter water molecules also raises the water temperature and increases the chances of water vapour eventually breaking through the barrier. Hence the Bureau of Reclamation originally overestimated the effectiveness by nearly 14 per cent. Nevertheless, costs per litre of water saved are comparable with many alternative methods of increasing water supplies.

Raising the albedo of the reservoir with white wax, rubber or styrofoam blocks became popular in the 1960s. These **high-albedo covers** are very effective in reducing water temperatures and are essentially non-polluting. Where **suitable aquifers** are available, an attractive alternative is to pump or direct water into the aquifer using the overburden as a thermal insulator. This has been successfully tested in the Thames Valley chalk and is especially attractive in hot climates, where surface reservoirs are very inefficient.

Both barriers and higher-albedo covers are used to reduce evaporation from soils. The time-honoured technique of **mulching** is designed to insulate the soil. It offers a barrier to heat rather than vapour, by reducing heat transmission to the soil with a cover of material such as leaves, wood chips or gravel that has a low thermal conductivity. White paper also raises the albedo (Oke 1992). **Dry farming** is another traditional technique. This uses the soil itself as an insulator by lightly tilling the surface to increase the total porosity, so that the air-filled pores reduce the thermal conductivity.

Even **monomolecular films** can be sprayed on to the soil, where they do not suffer from break-up in high wind. They can also be sprayed on vegetation as a form of antitranspirant. In this, they are less effective than active chemicals, because spraying is usually from above and the stomata are mainly on the underside of the leaves, and they run the danger of causing damage to the plants from overheating. This problem can be avoided if white substances are used, but there remains the big question of whether the environmental pollution is acceptable.

Active **chemical antitranspirants** are designed to encourage the stomata to close. Phenyl mercuric acetate has been shown to be an effective agent in commercial forests, delaying the pattern of water use by up to six weeks in aspen forests in Utah. However, there are legitimate concerns over potential damage to the crop or the environment. Application at the wrong growth stage or in the wrong amount could cause irreversible damage to the plants' metabolism. Wider concern could be expressed at the prospect of highly toxic mercury getting into the wildlife foodchain.

The potential for saving water by **manipulating the vegetation composition** has been raised in section 7.1. Most of the observed hydrological effects of inadvertent changes in vegetation can be used as the basis for purposeful water management, despite the inexactitude of predictive methods.

The US Forest Service estimated that manipulation of species in the upper Colorado basin could save more water than could be generated by cloud seeding. While the USFS estimates that 1850 million m^3 could be saved this way, the USNWS estimates that 1233.5 million m^3 could be created by cloud seeding (Stetson 1980). If these two approaches were combined, they could halve the calculated shortfall in supplies to Arizona and California (section 1.1.4).

10.4 Controlling the surface cryosphere

Snowmaking has been used operationally to manipulate snow resources since the second Project Skywater demonstrated its effectiveness in Nevada in the early 1970s. The potential for using cloud seeding to alter the hydrological parameters of the snowpack is illustrated in Table 10.1. Increments of up to 20 per cent in the water equivalent of snowpacks may well be possible, especially in mountainous regions, though it is expensive and there is no guarantee of success.

Table 10.1 The potential of cloud seeding for manipulating snowcover, according to Howell (1972). Figures in percentages

	To increase			To reduce		
Year	1972	1980	2000	1972	1980	2000
Seasonal snow accumulation	15	20	25	0	0	–5
Individual susceptible storms	100	120	150	–5	–17	–30
Specific environments						
Rocky Mountains Colorado (cyclonic + orographic)	20	25	30	0	0	–5
Great Plains (cyclonic)	10	15	20	0	0	–10
New England (cyclonic)	15	20	25	0	0	–15

10.4.1 Controlling snowcover and snowmelt

Although snowfall and rainfall are closely related meteorologically, the opportunities for manipulation once they have reached the ground are very different. By not infiltrating the soil, snowfall offers more options. It can be a more efficient form of storage than liquid water. Evaporation losses from snow are minimal. And at melt-time less tends to be lost by infiltration, either because the ground is frozen or because the large amounts of meltwater rapidly saturate the soil.

Snow reservoirs can be created with minimal structural work by capitalizing on the processes of drifting and blowing, and manipulating near-ground windspeeds with permeable barriers. Barrow, Alaska, has been supplied from a snow reservoir created by snow fences since the early 1970s. Outcalt *et al.* (1975) calculated that the reservoir need be little more than 4 m deep. Snow reservoir design has been extensively researched by the USDA in Wyoming and the University of Arizona School of Renewable Natural Resources, with particular emphasis on manipulating the species and spacing of trees as permeable windbreaks.

Mixed conifer forest offers the ideal high initial storage followed by slow release. Observations in Russia have shown that the melt period lasts only 5–20 days on the open plain, but 20–30 days under forest. Relatively low-density or 'honeycomb' forests located in low-insolation and high-altitude environments provide optimal conditions, with a maximum *storage-duration index.* A honeycomb structure allows the maximum accumulation to spread over a larger area, but the trees must also be dense enough to provide sufficient shade to spread or slow down snowmelt, so that the water is not wasted in a snowmelt flood that cannot be harnessed.

Simpler forms of snow reservoir may be created where the objective is to maximize the direct supply of water to agricultural fields rather than runoff. Steppuhn (1981) illustrated the principle of **stubble management**, in which crop stubble is left standing over winter to reduce deflation of the snow and increase soil moisture content at the beginning of the new season in Saskatchewan. He calculated that if just 0.3 m were retained over the northern plains and prairies it would provide 213.5 billion m^3 of meltwater for agriculture, an amount comparable with the entire annual discharge of the Great Lakes into the St Lawrence River.

A novel form of snowpack manipulation initially tried in the Colorado Rockies is **artificial avalanching**, whereby explosive changes are used to create deeper, denser bodies of snow in order to extend the melt period (Obled and Harder 1979).

Although these management methods appear to be relatively 'environmentally friendly', concern has been expressed by some ecologists that deeper snowcovers could adversely affect wildlife habitats and limit the movements of the larger mammals, like elk, confining them to lower altitudes and possibly creating a food crisis. Ecological studies by the Bureau of Reclamation in Colorado and Wyoming, fortunately, have not revealed any problems. Indeed, reducing the stress on wildlife caused by floods and droughts might outweigh such effects.

10.4.2 Managing icemelt

Much more serious controversy surrounds certain aspects of artificial icemelting. The discharge from land glaciation may be utilized either as meltwater or as icebergs calved into the sea. Both occur naturally, but the idea of *inducing* glacier icemelt and proposals for towing icebergs to irrigate distant desert areas have met with considerable opposition.

In all, four types of icemelt need to be considered: (1) harnessing natural glacial meltwaters for power or water supply; (2) inducing glacier icemelt mainly to increase water supply; (3) inducing icemelt on lakes and rivers to reduce flood hazard or improve navigation; and (4) harnessing icebergs for remote use in water supply and power generation.

Harnessing **natural meltwaters** is relatively commonplace. Snowmelt and icemelt combine to provide 40 per cent of the annual flow of the Canadian prairie rivers, much of which is harnessed for irrigation and public supply. And many deserts are irrigated largely from meltwater: the Thar from the Indus, the Ulan Buh from the Hwang Ho, and the Kyzl-Kum from the Amu Darya. Meltwater provides most of the flow in the Indus prior to the summer monsoons. Significant hydropower schemes rely heavily on glacial meltwater in the Swiss Alps, as at Val de Dix (Lang and Dayer 1985), in Norway at Bondhusbreen or the 14 hydroplants along the Columbia River in Canada (Power 1985). Harnessing natural glacial meltwater is essentially no more environmentally hazardous than harnessing any other natural runoff.

Inducing glacier icemelt, however, is very different. Canada rejected the idea in its national survey, despite fears of increasing drought frequency in the prairie provinces and predictions of serious problems in regional supplies by 2000. The rejection is based on the technical problem of how to 'turn the tap off', plus the environmental pollution caused by materials used to induce melting, and on the recognition that it is really a form of mining. The only method of inducing melt that has been proven is to lower the albedo over some part of the ablation zone by spraying the surface with dark material, like carbon black or more rarely black plastic discs. Removing this material once it has induced sufficient meltwater is fundamentally impractical: the 'tap' remains on unless and until the meltwaters can flush the material away.

Even if the process works in the short term, it is drawing upon a store that is being replenished at a far slower rate (section 2.1.2): the store is being mined and future supplies and the environment are both being imperilled.

Nevertheless, the CIS continues to take the prospect seriously. Krenke and Kravchenko (1996) have calculated the potential for controlled inducement in the Aral Sea basin under present conditions and under global warming. Currently, glaciers provide 15 per cent of annual runoff and 40 per cent of summer runoff in the basin. They estimated that inducement could increase current runoff by a much-needed 4.12 km^3, but that this is likely to be reduced to 3.93 km^3 in a double-CO_2 scenario because of reduced snowfall and higher equilibrium lines.

Inducing freshwater icemelt is normally undertaken for quite different purposes. Ice jams are a common source of flooding in Canada and Russia, and may be relieved by spreading carbon black, by cutting leads or by designing channels to eliminate jamming points and trapping ice floes in booms (section 4.1.1).

Harnessing **icebergs** raises much more important questions, both technically and environmentally. Technically, iceberg farming or *ice tec* has been considered feasible since the early 1970s when it was expected to become an operational procedure by the mid-1980s (Weeks and Campbell 1973).

About 5000 icebergs issue from Antarctica annually, containing 1000 km^3 of ice in total. Three times more issue from the snouts of glaciers in Greenland, but they are generally smaller to start with and they tend to get trapped by the shallow seaward thresholds of the Arctic fjords, perhaps for up to 10 years, until they have melted enough to float out into the open sea. At this stage they weigh just a few million tonnes compared with perhaps 250 million tonnes for an Antarctic iceberg. They are also more pyramidal in shape and therefore more difficult to tow. Antarctic bergs tend to calve directly into the open sea from the broad edges of the ice shelves, like Ross, Ronne and Amery, and have a tabular shape. Antarctic icebergs are typically 200–250 m thick and range up to several kilometres across. They represent 3000 years of snowfall, but this is a natural discharge and exploitation would not harm the icesheet in the way induced melting harms glaciers.

It is suggested that suitable icebergs might be

identified from satellites and perhaps marked by radio beacons that could be tracked until the ocean currents carry them to a suitable harnessing point. Oceangoing tugs would then haul them to their destination. The South Atlantic offers the best source, because here the east-flowing Antarctic Circumpolar Current (West Wind Drift), which carries bergs from the Ross Ice Shelf, is diverted equatorwards by the Antarctic Peninsula, while an ocean gyre in the lee of the peninsula brings bergs up from the Ronne Ice Shelf. Icebergs may drift up to 26°S and Namibia and South Africa are well placed to take advantage. The 'mother of all icebergs', Iceberg A24, which drifted from the Ronne Ice Shelf to threaten South Atlantic shipping in October 1991, contained 1000 billion m^3 of water.

Australia is less well-favoured and is 'upcurrent' of the main sources of icebergs, but it could well become one of the first substantial users of ice tec. In any event, natural drifting is slow and even towing speeds are limited because the drag increases with the square of the velocity. Icebergs therefore lose a lot of their useful mass in transportation. Nevertheless, the southern hemisphere icebergs are so large that there have even been proposals to transport them across the equator to the Middle East or India.

Since these icebergs have a large draught of 200 m or more, another important consideration is to get the ice close enough inshore. In this respect, the deep Peru–Chile Trench offers an advantage to Chile that might aid irrigation in the Atacama Desert. In most cases, however, the bergs could be blasted apart 10–30 km offshore and the smaller pieces towed inshore to an artificial lagoon. Once there, they would be allowed to melt. The cold meltwater could be used to drive gas turbine electricity generators, before being piped off for water supply (Figure 10.10).

A million-tonne berg could drive a 100 MW station for three years, supplying electricity worth over £100 million and water worth about £75 million. According to another estimate, just one typical delivery could irrigate 1600 km^2 and cost only 1–2 cents for 10 m^3 of water in regions where 8 cents is the present norm. This could indeed make substantial parts of the deserts on the western sides of the southern continents 'green and fertile', as well as provide a massive pollution-free source of power: one somewhat fatuous projection holds that it could satisfy the present energy requirements of the USA for 1000 years. One 'reasonable' estimate of the global potential for water supply is 1000 billion $m^3\ a^{-1}$, or 50 per cent of the annual discharge of the Mississippi.

The main environmental concern must be whether the long-term consequences of such massive irrigation would be as disastrous as the historical record suggests (section 1.2.3). The extra evaporation is unlikely to increase continental rainfall downwind, because these

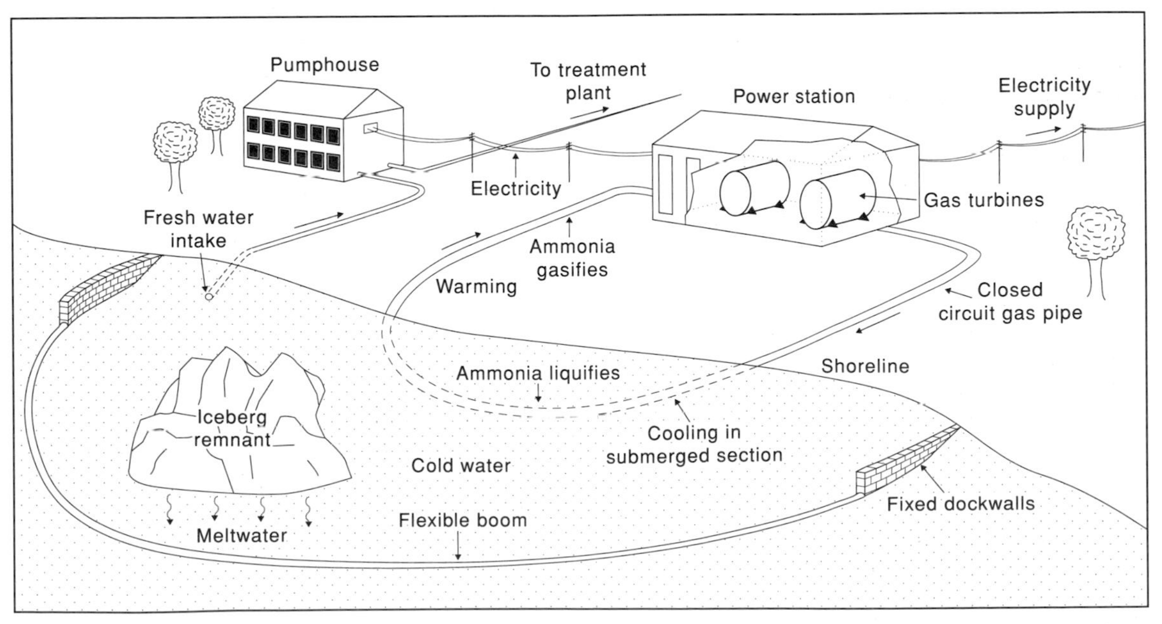

Figure 10.10 A proposal to generate electricity from melting icebergs, based on liquefacation and revaporization of ammonia gas. The ammonia is liquefied by passing through pipes immersed in the cold meltwater, and then revaporized by passing through warm seawater before entering the turbines

latitudes are dominated by the subtropical high-pressure cells.

An alternative strategy is to utilize natural meltwater from the ice shelves. This has a more limited capability and is more suited to public supply than irrigation. But the George VI Ice Shelf is estimated to yield 40 $km^3\ a^{-1}$, which forms a freshwater pool beneath the ice. This could be tapped using supertankers or rubber pods.

Conclusions

There is clearly a wide variety of possible approaches to water management that are more environmentally friendly than many modern established methods. They range from reverting to 'human-scale' methods from the past to applying recent scientific advances in understanding hydrological processes. The most successful are likely to be those which work in sympathy with nature. In this respect, rainmaking, some methods of controlling evapotranspiration and some of the latest proposals to utilize frozen resources and induce meltwater still need to be tested or used with care.

Further reading

Ecologically friendly channel management methods are covered fully in:

Boon P J, P Calow and G E Petts (eds) 1992 *River conservation and management*. Chichester, Wiley: 470pp.

Brookes A 1988 *Channelized rivers: perspectives for environmental management*. Chichester, Wiley: 326pp.

Newson M D 1994 *Hydrology and the river environment*. Oxford, Clarendon Press: 221pp.

A good review of the techniques and the impacts of weather modification is available in:

Williams M C and R D Elliot 1985 Weather modification. In J C Rodda (ed.) *Facets of hydrology II*, Chichester, Wiley: 99–129.

A standard, if ageing text on snow and ice resources is:

Gray D M and D H Male (eds) 1981 *Handbook of snow: principles, processes, management and use*. Oxford, Pergamon: 776pp.

Discussion topics

1. Discuss the need for a thorough understanding of hydrological and environmental processes in order to manipulate the hydrosphere in the most environmentally friendly way.
2. What are the merits and demerits of rainmaking?
3. How much can we learn from history or the pre-industrial era about effective water management today?
4. How dangerous is it to interfere with surface snow and ice covers? Could they provide a sound source in some regions?

CHAPTER 11 Towards a sustainable future?

Topics covered

11.1 Desalination – a source of 'new' water?

11.2 Controlling demand and waste

11.3 Protecting the environment

11.4 Hydropolitics

11.5 The threat of global warming

11.6 Integrated water resource management

The ultimate aim of water management must be sustainable consumption. The need for sustainability advocated by Lvovich in 1970 is even clearer now that some parts of the world are approaching the limits of exploitation, and the impacts of unrestrained development on water quality and the environment are so obvious. Agenda 21 of the UN Conference on Environment and Development in Rio in 1992 specified the need to protect the quality and supply of freshwater resources by an integrated approach to the development, management and use of water in a sustainable way.

In a report that laid many of the foundations for the new perception of a sustainable future, the UN World Commission on Environment and Development (1987) defined a sustainable society as one that satisfies its needs without compromising the ability of future generations to meet theirs. According to the textbook definition of Camp (1994) a sustainable world 'means that human activity would not degrade the planet's carrying capacity forever for other humans'. More simply, it is a matter of living within our resources.

For water resources, sustainability may be defined at three different levels: (1) the narrow viewpoint of maintaining the physical water resources; (2) the broader aim of maintaining basin ecosystems; or (3) the all-encompassing aim of sustaining a balance between social and physical components, between economic returns, social equity, and ecological and hydrological needs (Dixon and Fallon 1989). Adopting the latter goal, sustainable water resource management is defined for UNESCO as 'a set of activities that ensures that the value of the services provided by a given water resource system will satisfy present objectives of society without compromising the ability of the system to satisfy the objectives of future generations' (Hufschmidt and Tejwani 1993). 'Services' are here taken to include the protection of ecosystems.

Our assessment begins by questioning whether the calls for sustainable use of water resources might not be premature given the technology that already enables many areas of the globe to utilize the vast resources of the oceans. After hopefully dismissing this diversion, we will look at prospects for curbing demand and for sustainably protecting the environment. The chapter concludes by looking at the twin threats to sustainability from international conflict and global warming.

11.1 Desalination – a source of 'new' water?

Creating more water than the local environment provides naturally has been the water engineer's role from time immemorial. Interbasin transfers and conjunctive management schemes are simply among the latest and most sophisticated extensions of traditional methods for redistributing terrestrial water (Chapter 9). Rainmaking is a novel, but rather limited, way of increasing terrestrial water (section 10.3.1). So is controlling evapotranspiration (section 10.3.2). In contrast, obtaining large quantities of water from an alien environment like the polar cryosphere (section 10.4) or the oceans represents a major departure.

Not only are the oceans the greatest single store of water on Earth, they are also the nearest source to many of the world's major deserts. This offers a potential source of water that for the foreseeable future will be limited more by technology and costs than by the amount available or the ecological implications. In addition, large amounts of groundwater and even surface water are brackish or saline. **Brackish water** has a typical salt content of 0.5 per cent, against 3.5 per cent in standard seawater.

For a number of purposes even unprocessed seawater is adequate and it can release better quality water for uses with more stringent demands. For 30 years, Britain and America have sited power stations, especially nuclear plants, in coastal locations in order to use seawater as a coolant. Many tens of cubic kilometres are used this way in the USA each year, although the UNESCO (1978) estimate of 290 $km^3 a^{-1}$ by 2000 is likely to be too high. In Japan, nearly 30 per cent of industrial requirements were met by seawater as long ago as 1965.

Alternatively, a variety of techniques are available for the purification of saline or brackish waters. The choice is largely determined by the volumes of freshwater required, the type of salinity and thus the unit costs of processing. For chemical treatment or **electrolysis** the costs in chemicals or electricity are proportional to the salt content of the water, so they are best used for lower levels of dissolved salts, e.g. brackish water. Chemical treatment with **water softeners**, such as washing soda (sodium carbonate) or sodium polyphosphates like Calgon, which form insoluble complexes, or sodium aluminium silicate (Permutit), which operates by **ion exchange**, is generally restricted to removing excess calcium and magnesium from alkaline terrestrial water.

Distillation is more universally applicable and is one of the main methods of desalination. But it can require substantial capital and running costs, depending on the method used. *Multistage flash distillation* is a popular form, which artificially reduces the pressure so that boiling occurs at 80–85°C to reduce fuel costs. *Vapour compression distillation* economizes on fuel and wastewater by pumping steam through a mixture of fresh seawater and recycled seawater (brine) that has already passed through the system at least once. Another popular method is **reverse osmosis**. In this, water is forced through thin filter membranes which have pores designed to allow water molecules to pass through but not the larger dissolved salts and minerals.

Refrigeration is less common. It uses the fact that impurities are expelled during ice formation. In *vacuum freezing* the boiling point of water is lowered to the freezing point at a pressure of only 400 pascals (4 mbar). Because of the different salt contents, the brine boils but the seawater freezes, and freshwater is obtained from both the vapour and the ice. *Secondary refrigeration* uses evaporating butane gas to freeze the salt water.

There are now over 1800 desalination plants worldwide. The largest desalination plant in the world at al-Jubail in Saudi Arabia produces nearly 2.3 million litres a day, or 0.8 $km^3 a^{-1}$, mainly for Riyadh, and was probably the main target for Iraqi oil pollution in the Gulf during the Gulf War of 1990. For all intents and purposes, Saudi Arabia has no rivers, receives less than 100 mm of rainfall a year, and its groundwater is largely connate and in critical decline through overexploitation. It therefore matters little that the oil-fired desalination plants produce water that is more costly than refined petrol.

Even when the plentiful supply of cheap oil is gone, however, the Gulf states will be left with endless sunshine. Solar power could transform the economics of desalination for large areas of the world. The main problem is harnessing power in sufficient quantities for large-scale desalination. In its simplest form, it need involve nothing more than a seawater pool covered by a glasshouse that replicates the natural ocean evaporation–rainfall cycle in miniature. The water condenses on the glass and trickles down into collecting gullies. Such a pool was constructed on the Greek island of Patmos some 25 years ago: though producing only 27 m^3 a day, it doubled the local water supply. There is no need for expensive and inefficient conversion of solar energy into electricity for such relatively small-scale applications.

Small islands in general might benefit from desalination, especially as the demands of tourism increase. It is important for Key West in Florida and for Malta. Long Island, New York, uses a refuse incinerator to power desalination. In 1992, the Scilly Islands became the first part of the UK to have a public desalination plant. Although the NRA's strategic plan still saw desalination as generally inappropriate for Britain, for the Scillies it is much preferable to the old system of importing water by ship.

The economics of desalination are continually changing. Distillation is commonly thought of as attractive only where there is a critical need for water and where large volumes are to be processed. This is not entirely correct, but cheaper, more efficient solar-powered methods are needed for it to become important in global terms. Even then, its distribution is likely to be circumscribed by the availability of cheaper solutions or by insolation receipts. Costs could be reduced by combining desalination with power generation or treatment of saline groundwater, but again opportunities are localized.

Britain is certainly an area where the likelihood of desalination becoming a competitive solution has actually decreased since 1970 despite rising demand, as other solutions have advanced (National Water Council 1978; NRA 1994). It is unlikely to prove worthwhile for irrigated agriculture, either because of production costs (currently 4–10 times more than conventional sources like dams) or because the areas in need of water are remote from the sea, as in the USA.

For public water supply in the drylands and islands, however, it is already advancing apace. Two-thirds of world production occurs in the Middle East and North Africa, where WMO/UNESCO (1991) identify an approaching water crisis. The USA now has over 100 large plants, producing more than 1250 million litres daily, though this is still only 0.001 per cent of the national supply. The American plants serve mainly islands, tourist areas or communities that have expanded well beyond the support of local resources in recent decades, especially in Florida, California and Texas, and even parts of the Northeast. A number of cities in southern California may soon turn to desalination to tackle the advancing regional water shortage. Los Angeles has opted for a $500 million plant powered by nuclear power and diesel.

Environmental impact of desalination Concern has been expressed that desalination increases the ability of mankind to exploit environments that are fragile where there is a real potential for ecological damage. Unique wildlife and coral reefs are threatened by the recent expansion of settlements in the Florida Keys, which is partly supported by desalination (Chiras 1994). Throughout mainland Florida, population growth is putting great strain on the environment from increased sewage and other water pollution.

Conversely, desalination is taking some of the strain off the local terrestrial environment to provide the water, and it might be viewed as assisting the recent programme to limit exploitation and re-establish an ecological balance in the Everglades. If only it could stay that way, and not follow the seeming 'law' of water provision that providing more water only provides a higher ceiling for exploitation!

There is also a potentially more direct environmental effect. The salt that is a by-product of desalination must be disposed of. It may be of commercial value. But if large quantities are disposed of in the sea nearby, which is the cheapest and easiest solution, marine and estuarine wildlife could be threatened (Miller 1992).

11.2 Controlling demand and waste

For the present, the spread of desalination or 'ice tec' is limited by costs, technology and geography. Even if this were not so, there are strong reasons for choosing a different pathway to the future, one of frugality and conservation rather than for ever increasing supplies.

More water is wasted in Britain than is likely to be lost in the worst global warming scenario. On average, nearly 25 per cent is lost through leaks in the supply system. **Reducing leaks** is now a high priority for most water companies in Britain, by laying new mains or relining the old. South West Water aimed to reduce leakage rates from 32 per cent in 1989 to 20 per cent in 1995. The NRA (1994) *Strategy* states that 'water companies must be required to achieve economic levels of leakage and metering before new abstraction licences are granted for strategic developments'. The consumer protection agency, OFWAT, also requires water companies to provide annual data on leakages and officially encourages improvements, although OFWAT's remit is cost rather than technical efficiency.

Britain lags well behind Germany in leakage reduction. In Bielefeld only 5 per cent is lost through leakage and 32 teams are replacing old piping at the rate of 2 per cent a year, four times the British average. Even so, Britain is developing and applying some of the most advanced techniques in the world, and in October 1994 the National Leakage Control Initiative published a document entitled *Managing Leakage* to publicize the latest methods of detecting, measuring and controlling leaks.

In cities like London the sheer age and complexity of the pipe networks present major problems for detection and control. One of the aims of the new London ring main completed in 1994 is to reduce leakage in the older, smaller pipes by reducing the water pressure within them (Figure 11.1). Elsewhere, the high cost of replacing old mains has focused attention on *management* of leakage rather than eradicating leakage. This includes better monitoring of flows and rapid response to major leaks.

Domestic meters are another proven means of reducing wastage. Meters are required throughout Germany and demand has been static for the past decade. A private experiment by a Japanese engineer suggested that the average domestic consumption in Tokyo could be reduced by about 40 per cent and that total domestic consumption might easily be reduced from 910 million m^3 to 800 million m^3 a year, markedly reducing the need for extra supplies (section 1.2.1). An official experiment at Normanton, Yorkshire, reported in the early 1990s that water bills for 700 metered homes were 29 per cent lower than for unmetered homes. However, there was no charge for installing the meters and other modifications in tariffs during the trial.

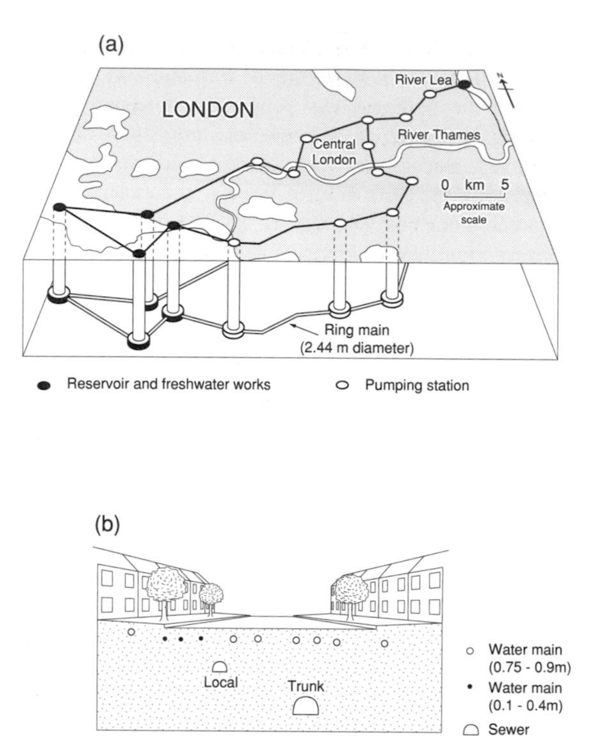

Figure 11.1 The new London ring main for water supply (a), designed to relieve pressure on old mains (b) and reduce leakage losses

A 3-year experiment in 100 per cent metering of domestic users on the Isle of Wight reported a 20 per cent overall saving in 1992. The Isle of Wight experiment probably overestimates the likely savings on the mainland because there is little industry on the island, whereas industry is already metered, and the social structure is more middle class. Nevertheless, for the island at least, the experiment showed that metering makes the newly completed water pipeline under the Solent superfluous.

Southern Water has also calculated that installing meters in the mid-Kent region at a cost of £30–£40 million is a cheaper and better solution to the local water shortage problem, which was aggravated by the 1988–93 drought, than spending £70 million on the proposed Broad Oak Reservoir. Broad Oak would flood 250 ha of prime agricultural land and has been the subject of vociferous environmental objections.

A graphic illustration of the degree of public economy that is possible occurred during the San Francisco water crisis of 1976–78. The city administration requested a 25 per cent reduction, but a 40 per cent economy was actually achieved. Unfortunately, the administrative system proceeded to work against the savings, because the Public Utilities Commission had granted the Water Department a 43 per cent rate increase in order to maintain its income with the requested reduction. The 40 per cent fall led the Department to request an additional 22 per cent rate increase and then, at the height of the drought, to suggest that people should consume more to prevent rates going higher!

There is still considerable room for economy of use. UK consumption has risen 70 per cent over the last 50 years with little official inducement to economize. The main incentives for industry to economize have come from elsewhere: three major droughts in the last 25 years and international commercial competition. The Environment Agency (NRA 1994) is now advocating a proactive role for itself in promoting water use efficiency in industry, commerce, agriculture and the home.

More efficient water-based processes, using less and recycling more, serve the economy better as well as the environment. Miele washing machines now use 60 per cent less water than 10 years ago. Major advances have been made in introducing less water-intensive industrial processes in Britain since the traumatic drought of 1976 (Kirby 1984).

Nevertheless, considerable amounts of water can still be consumed by centralized sewage systems. Roughly one-third of domestic water in Britain is flushed down the toilet. England and Wales flush 3000 Ml of potable water down the toilet every day. This seems a gross waste of high-quality water. It made sense as part of the programme to improve environmental health in nineteenth-century cities, when ample pure water was available not too far away. But the practice is increasingly questioned as resources are becoming scarcer. While the 13.5-litre cistern is still normal in Scotland, East Anglia encourages 4.5-litre tanks and variable volume or vacuum flushes are being introduced elsewhere.

It is possible to do without water altogether for sewage processing on a small scale. Totally dry technology has found a certain niche with the 'bioloo', particularly for temporary buildings, and the UK Centre for Alternative Technology is engaged in research on dry recycling of toilet waste as fertilizer for vegetable plots. Many large cities in China have 'honey trucks' that collect organic wastes from buildings not connected to central sewage systems and deposit it on the surrounding farmland.

It is more difficult for large Western settlements to abandon central sewage. However, Melbourne, Australia, has achieved notable success by using raw sewage to grow cattle forage; while not dispensing with flushing systems, this does limit processing costs and the return of polluted water into the rivers. A thousand or more townships in the USA deposit their sewage on farmland or parkland. Many have adopted the procedure in recent decades in preference to expensive reprocessing plants. Arizona manages this with careful statutory controls on the quality of slurry that can be spread on the land. But while this may be feasible for townships of 50 000 or less, large metropolitan cities tend to suffer from high accumulations of human pathogens and industrial toxins and tertiary treatment becomes necessary. Singer (1973) made the interesting calculation that if the population of the USA were perfectly distributed according to the available water resources, 250 million people could be accommodated without requiring any sewage treatment.

Unfortunately, global urbanization is spreading apace (section 1.2.1). For urban living the flush toilet system has obvious advantages, and the main thrust is likely to be towards smaller cisterns and perhaps some recycling of higher grade washing water within buildings.

Recycling could clearly make a valuable contribution towards reducing demand. Much of this recycling could be achieved without reprocessing, i.e. using the poorer grade *non-potable water* for non-critical processes. Currently, only a few towns with a dire shortage of water, like Windhoek, Namibia, operate recycling on an urban scale. But the extension of recycling on this scale is more a question of current economics than of available technology.

11.3 Protecting the environment

True sustainability requires more than just containing demand. At the very least, it requires careful protection of resources. At best, it requires an active interest in rehabilitation and improvement directed not only at water resources themselves, but at the broader environment that sustains them and that they nourish.

11.3.1 Conserving ecosystems

Nature conservation and ecologically planned exploitation of land resources cannot be divorced from sound management of water.

The worldwide destruction of **wetland habitats** must be halted. These range from riverside and hillside bogs, marshes and swamps to coastal mangroves and salt marshes. All act as important reservoirs for wildlife, and many play significant hydrological roles, reducing the effects of floods and droughts, filtering pollutants (section 8.1), limiting erosion and stimulating accretion (Gilman 1994). River engineering, agricultural drainage, urbanization, pollution, and logging all threaten the wetlands. Nearly half the wetlands of the USA have been destroyed since 1780 (Gleick 1993). Since 1950, the main OECD countries have lost nearly 13 per cent of their wetlands, and Germany and the Netherlands have lost over 50 per cent, with important consequences for the Rhine.

The Florida Everglades are an example of both success and failure in wetland conservation. Land reclamation, drainage canals, floodgates and the water demands of tourism and a burgeoning resident population have lowered the water table, reduced surface flows to critical levels, increased water pollution and allowed saltwater incursion into surface and groundwater. A series of canals now diverts water that formerly flowed from the Kissimee River into the Everglade wetlands towards coastal towns and the sea. Flood alleviation works on the Kissimee River between 1964 and 1970, which reduced the river to 40 per cent of its former length and drained three-quarters of its 16 000 ha of wetland, had such an adverse effect that within two years fish kills occurred in Lake Okeechobee, flood problems increased downstream and calls began for restoration. Loss of the natural cleansing effect of the swamps, combined with pollution from agriculture and low water levels, reduced wildfowl populations by 90 per cent. In 1984 the Kissimee Restoration Scheme was launched to reflood the marshes, to buy back land and naturalize it. In 1992 a further 15-year programme began, aiming to restore the environment of 100 years ago.

More success was achieved further south with the creation of protected landscapes in the Everglades National Park (1947) and Big Cypress National Preserve (1974). The South Florida Water Management District seeks to maintain more constant water levels by elaborate flow management there and in adjacent parts of the Everglades, to prevent surface floods and droughts and to maintain the head in the important Biscayne aquifer. In 1971 Congress prescribed minimum flows into the National Park that must be maintained.

Belated recognition of the loss of these vital habitats occurred in the 1970s with the Convention on Wetlands of International Importance at Ramsar, Iran, in 1971 and the subsequent UNESCO network of Biosphere Reserves providing a framework for protecting major sites. The Lands Directorate of Environment Canada produced its first national wetlands inventory map in 1981 as a prelude to protection. In Britain, all drainage activity is now controlled under the 1981 Wildlife and Countryside Act and additional government guidelines from 1985 onwards. Nevertheless, barely 2 per cent of the world's wetlands have been protected under the Ramsar Convention.

Even more belated has been the recognition of the importance of wetlands as flood storage areas (Box 7.5). In Britain, recent collaboration between water companies, the NRA and environmental agencies like the Countryside Councils has helped in protecting and even restoring wetlands for both ecological and hydrological purposes, but the tide of losses is far from stemmed. Some of the more elaborate restoration work is being undertaken in Holland (Box 10.2).

Deforestation also threatens both ecological and hydrological resources. The floods that created havoc in Khartoum and towns along the Nile throughout northern Sudan in August 1988 were triggered by a 1000-year rainfall event in headwaters of the Blue and White Niles, but deforestation clearly increased the amount and velocity of runoff. Over the last 100 years, forest cover in Ethiopia has been reduced from 40 to just 4 per cent of the land area. Flooding is further aggravated by silt from the deforested areas accumulating in dams and riverbeds (section 11.4.3).

The main remedies now being undertaken in the Developing Countries include reafforestation, management of forests in a sustainable way, i.e. replanting at least as much as is felled, and agroforestry, in which a proportion of tree cover is retained over the agricultural crops. Both Nepal and Pakistan have centralized responsibilities for forestry, soil conservation and watershed management in response to these needs (Hufschmidt and Tejwani 1993). Watershed rehabilitation is now a key strategy in the Himalayas (Tejwani 1993).

In Developed Countries the emphasis is now upon restoring and protecting the near-river environment (Gore 1985; Boon *et al.* 1992). Petts (1990) extols the ecological value of the lost **forested river corridors** for wildlife protection, but it is also increasingly appreciated that the benefits can work both ways, i.e. that hydrology can profit from a 'return to nature' in the riparian zone (section 10.2.2). The new *Rivers and Wildlife Handbook* published by the Royal Society for the Protection of Birds, the NRA and the Wildlife Trusts (1994) and Harper and Ferguson's (1995) *Ecological Basis for River Management* represent a valuable integrated approach to river conservation. The USA established its **wild and scenic river system** as early as 1968. Under the system, free-flowing rivers are classified as wild, scenic or recreational and protected accordingly. A scenic river may be upgraded to a **wild river** by cutting off road access and removing human constructions just as 'wilderness areas' are created within national parks.

11.3.2 Pollution monitoring, environmental audit and EIA

Water pollution is largely 'under control' in Western countries, although controls on many dangerous organic substances are currently lacking or inadequate (section 8.2.4). Even where water is not controlled by a publicly accountable body, pressure from regulating bodies, competitors, public relations and shareholders has raised environmental protection high on the agenda. Institutional shareholders are particularly keen to invest only in ethically sound companies. British companies in general see 'toxicity assents' coming in, and have been carrying out **environmental audits** and waste reduction for some time. The electronics group GPT (GEC–Plessey) claims to have made a 62 per cent reduction in water use following a recent environmental audit. Establishing agreed standards for environmental audits can, however, be problematic when environmental impacts cannot always be measured on a single scale (section 9.1). Nevertheless, there seems a real possibility that standards for environmental accounting comparable to money audits will be available within a few years. At the same time, industry, agriculture and forestry in the EU are all now well acquainted with requirements for **environmental impact assessments** as part of planning procedures (Wathern 1990).

Many Third World and former Communist nations, however, carry chronic legacies, face worsening prospects for the future, and have little in the way of ethical or legal controls. The opportunities for financial gain from developments are often too attractive to be superseded by antipollution legislation (Gladwell and Low Kwai Sim 1993).

UNESCO (1992), the International Hydrological Programme and the Man and the Biosphere Programme are all emphasizing the need for active participation from local communities in protecting their own environment and water supplies. But Western assistance is clearly needed. Major rivers like the lower Niger still have not a single water quality monitoring station. The UNESCO/WMO GEMS monitoring network (section 1.2.4) is now being supplemented by a worldwide real-time data collection system for water quantity and quality, WHYCOS. This aims to create a data collection system comparable to the World Weather Watch system for meteorology. WHYCOS will be satellite-linked (Figure 11.2). A scheme for the Mediterranean has been agreed with World Bank funding. Britain and France are committed to supporting a scheme for the Aral Sea basin, and progress is being made in Africa.

11.4 Hydropolitics

One of the most worrying developments in recent years has been the politicizing of water resources on an international scale. A report by the Center for Strategic and International Studies in 1988 predicted that water is likely to take over from oil as a cause of regional tensions in the Middle East, although this report came rather ironically just before the outbreak of the Gulf War, which was dominated by questions of oil. The 1992 Dublin Conference, *Water and the Environment – development in the 21st century*, and the Rio Earth Summit both reaffirmed the seriousness of regional disagreements over water.

11.4.1 Conflicts, treaties and international law

Three geographic scenarios tend to foster hydropolitical problems:

1. International drainage basins, in which upstream states have control over resources vital for downstream neighbours.

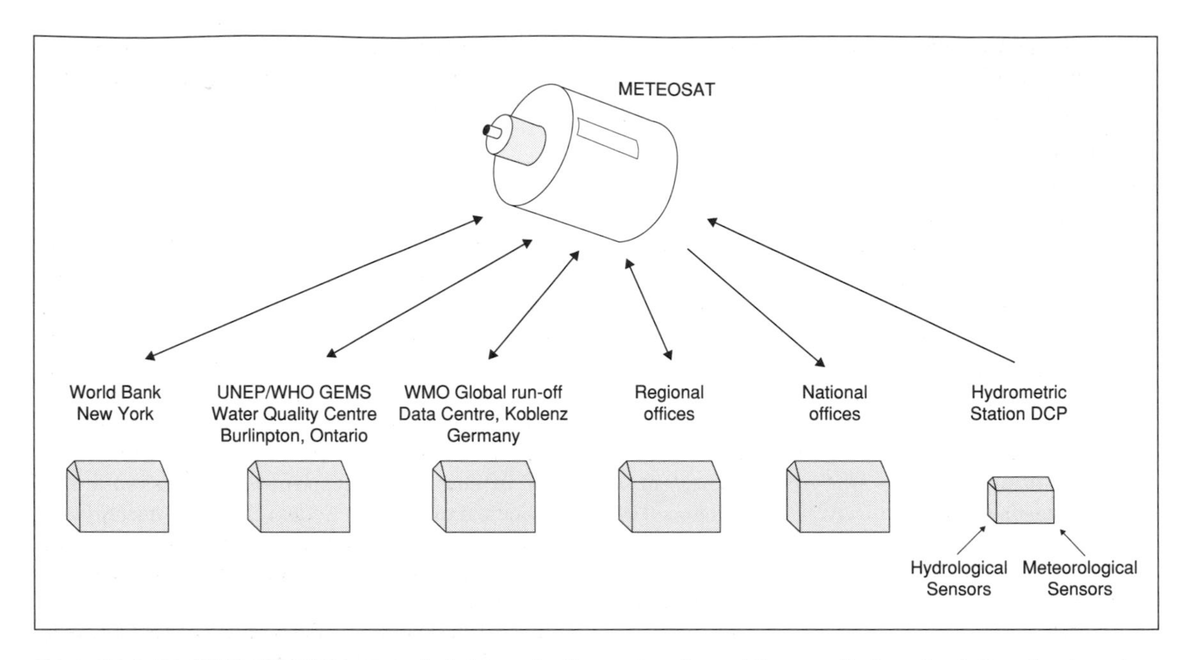

Figure 11.2 The WMO WHYCOS hydrological data collection system for real-time monitoring of pollution, floods and droughts

2. International aquifers, in which overpumping by one state can reduce resources for its neighbour.
3. Strong contrasts in water resources between neighbouring states.

The first two of these are more easily justifiable sources of conflict in the modern world, but the third scenario has certainly played its role in historical conflict, probably at least since the invasion of the Nile Delta by the Hyksos from Canaan in the eighteenth century BC. The NAWAPA scheme might be viewed as a more subtle modern approach to the age-old scenario (Box 9.3), although the USA and Canada are not likely to engage in serious conflict over what must be a mutually agreeable scheme to proceed.

More serious matters of international politics are arising in the CIS. Water was at the heart of popular rejection of Communism in the southern republics. Many felt they were puppets of central government planning that compelled them to grow irrigated cotton and grain for a 'foreign' market and offered continental interbasin transfers which would make them even more dependent. The damage done to local water resources and the environment in these newly independent republics is major and a continuing source of political conflict with Russia, which includes non-cooperation and calls for compensation.

The potential for armed conflict, true 'water wars', is greatest where water resources naturally straddle international frontiers. Forty per cent of the world population live in international basins, many of which were created by treaties between former colonial powers that eschewed natural topographic boundaries. Despite the UN recognizing access to sufficient and safe water as a basic human right, no international law covers this situation.

International law merely offers four principles that fall rather short of being mutually supportive, and there is no effective mechanism for resolving the conflicts that can arise:

1 **The Principle of Absolute Territorial *Sovereignty*** or the **Harmon Doctrine** proclaims that states have unlimited rights to use the resources within their territory, and therefore provides an argument that favours upstream users. In contrast,

2 **The Principle of Absolute Territorial *Integrity*** holds that no state may use its resources to the detriment of a downstream state. This is supported and extended by

3 **The Principle of Condominium** or **Common Jurisdiction** holds that the rights of a state are strictly limited and that prior consent from other interested states is needed before water resource

developments can take place. This is fundamental to **Integrated Drainage Basin Development**.

The most commonly invoked principle falls slightly short of requiring mutual agreement. This is:

4 **The Principle of Equitable Utilization** or **Limited Territorial Sovereignty**, which holds that developments are permissible if they do not harm the resources of a neighbour. The principle was established by the International Law Commission in 1966 and forms the core of the Helsinki rules on the use of water in international rivers. It supports a reasonable and equitable sharing of resources.

In Europe and North America there are generally written agreements that recognize the rights of downstream users. The Rhine and Columbia Rivers are reasonably good examples of this, especially since the 1986 Basel chemical spillage forced a reappraisal of agreements and procedures on the Rhine. Unfortunately, the Danube continues to be a source of argument, especially between Hungary and Slovakia over the latter's continuation of the Gabcikovo hydropower dam on the frontier, nor has Mexico been entirely satisfied with American use of the Colorado (Box 9.3).

In Africa and Asia, however, written agreements are the exception, and yet 60 per cent of Africa and 65 per cent of Asia are covered by shared basins. Matters are aggravated by looming water shortages in these regions. Eleven of the 20 countries in the Middle East and North Africa use over 50 per cent of their renewable resources. Libya and most of the Arabian peninsula already use over 100 per cent with the support of desalination and fossil groundwater. And population is due to double or even quadruple before stabilizing (Falkenmark 1989). South Asia and China are somewhat better served, but with approaching 20 per cent of renewable resources in use they do have regional problems.

In the past, South Africa has had disputes with many of its neighbours over shared water resources, especially Mozambique and Swaziland. Now, the Mandela government has made major progress with a cooperative agreement on water resources signed by the 12-nation Southern African Development Community, plus a hydropower compact with Zaire. The cooperative agreement signed in August 1995 aims to mitigate the effects of the recurrent droughts in the region. The accord includes a new joint water administration, agreement to follow the Principle of Equitable Utilization, and exchange of expertise and information. This should ensure harmonious relations as Lesotho undertakes a large resource development and plans are afoot for an international water grid linking the Zambezi with other rivers.

11.4.2 The Mekong basin

The Mekong could also become a peaceful exception, if the agreement signed by Laos, Vietnam, Cambodia and Thailand in December 1994 leads to successful collaboration. The four states have been trying to cooperate with varying degrees of enthusiasm throughout four decades of civil and international wars in the region. A water-sharing treaty is important to all four states, since one-third of their total population lives within the basin, largely subsisting on fish and rice production. Vietnam's economy is particularly vulnerable to upstream developments, since 60 per cent of its agricultural production comes from the Mekong delta (Figure 11.3).

The UN has played a significant role in achieving cooperation. It set up the Committee for Co-ordination of Investigations of the Lower Mekong Basin in 1957, which produced a plan in the early 1960s for 100 dams to regulate the 30-fold seasonal range in discharges and serve irrigation, hydropower and navigation. The committee has been accused of favouring Thailand, but, although the six main projects that have been completed primarily serve Thailand, in reality Thailand is the only country that has remained free from war (Brady 1993). No major works have been completed since 1970, partly because of the military and political disturbances and partly because in the 1980s the committee bent to environmentalist pressure and favoured small-scale and non-structural solutions.

During the 1990s, plans have been relaid for a series of major hydropower dams that will transform the wild river and the fertilizing annual flood cycles that agriculture has traditionally depended upon (Lohmann 1991). Laos is keen because exporting hydropower is already a major source of income. But reducing the monsoon discharge of the Mekong could have major environmental impacts, not least upon the unique Tonle Sap or Great Lake of Cambodia. Floodwaters from the Mekong reverse the flow of the Tonle Sap River each year to create a 13 000 km^2 lake, which is vital to local wildlife and the national economy based on fisheries and irrigated agriculture. Cambodia has proposed it as a UNESCO World Heritage Site (Heywood 1994).

Quite apart from the environmental consequences, two important questions remain: (1) will the new agreement brokered by the UN Development Programme

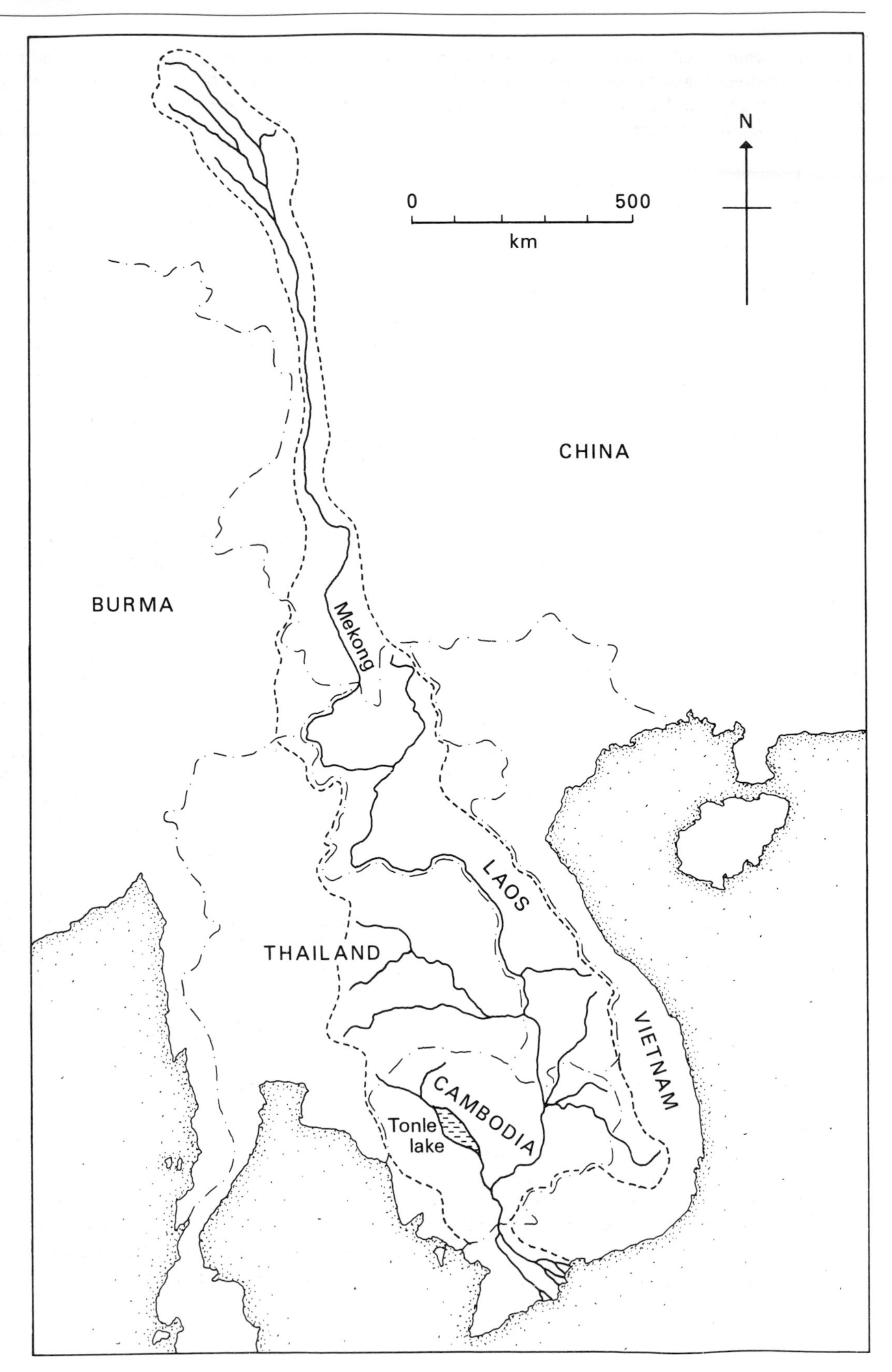
N
0
500
km
CHINA
BURMA
Mekong
LAOS
THAILAND
VIETNAM
CAMBODIA
Tonle
lake

protect the interests of Vietnam, which must remain primarily agricultural; and (2) can it succeed without the involvement of China and Burma, which control the upper basin? China already has diversion plans (Box 9.4).

11.4.3 The Ganges–Brahmaputra

Bangladesh is a classic downstream user: 80 per cent of the country is formed by the delta floodplains of the Ganges, Meghna and Brahmaputra Rivers, but barely 10 per cent of the total catchment lies under Bangladeshi control (Figure 11.4). Frequent floods created by snowmelt or by monsoon rains are vital to its agriculturally based economy and normally 20–50 per cent of the land is flooded each year with fertile silt. But flood frequencies have been changing on both rivers as a result of activities in the states upstream. During the 1990s, Bangladesh has been suffering from the unlikely combination of floods that are both too few and too severe. All water-sharing agreements with India had lapsed and until 1996 India was refusing to renegotiate.

The first problem that Bangladesh faced was an increased frequency of disastrous floods linked with deforestation in the Himalayas – 93 per cent of Nepal's energy comes from woodburning and the country's forests are being depleted by 400 000 ha a^{-1}. As a result, soil is being lost at rates of 75–200 t ha^{-1} a^{-1}. Helped by high rainfall intensities, like the 760 mm h^{-1} reported by Starkel (1972) in Sikkim, and highly erodible soils in Assam, the Brahmaputra has the third highest annual sediment yield in the world. The delta rivers carry up to 2.5 billion t a^{-1}. Deposition has raised channel beds, which has increased channel migration, levée erosion and destruction of the silt mounds called 'chars' that the farmers settle and cultivate. Deposition has combined with faster runoff due to the deforestation to cause more disastrous floods.

The severe floods of 1987 and 1988, which covered most of the country, led to the 1990 Five-year Action Plan for Flood Control with funds from the World Bank and 15 donor countries; $150 million is being spent on research and modelling studies, and it could cost over $10 billion to 'tame' the river. The plan originally comprised 'river training' and bank protection schemes, plus improvements in forecasting and early warning. The proposed engineering aims to reduce the numerous channelways or *thalwegs* to one or two narrower but deeper channels over 15–20 years. It includes dams to regulate flow, wing-dikes to collect sediment and divert flow towards the main channel, concrete-reinforced islands, and up to 8000 km of levées. The plan has been criticized as too reliant on structural engineering, especially since the 1993 Mississippi disaster (section 7.6.2). Planners have responded to this and to views that it is neither practicable nor desirable to 'tame' such a powerful river, and have partly adopted the principle of 'controlled flooding' (section 10.2.2; Hosh 1994).

Improvements to satellite surveillance systems may help monitor not only heavy rain and snowmelt in the mountains, but also tropical cyclones in the Bay of Bengal, which cause heavy rainfall in the poorly drained delta and generate coastal storm surges. However, experience in the 1991 cyclone disaster, in which 200 000 may have died, showed that the greatest problem lies in disseminating information, establishing flood drills and providing safe concrete retreats for the rural population.

The second problem facing Bangladesh is caused by Indian exploitation of river resources. Whereas the flood problems are accidental, the low-flow problems are due to deliberate interference. India is taking more and more from the Ganges and has built scores of dams to regulate dry season flows to supply its rising population. In 1993 only 260 m^3 s^{-1} entered Bangladesh, whereas nearly 2000 $m^3 s^{-1}$ was diverted at the Farakka Barrage, 18 km upstream in India, in a river diversion scheme designed to reduce siltation and improve navigation in and upstream of the port of Calcutta. Farakka was constructed in 1975 without any long-term water-sharing accord. In contrast, India has an accord with Nepal in which some of the regulation needed to supply the 120 million people of Uttar Pradesh province is achieved by damming in Nepal.

By 1995 one-third of Bangladesh was threatened with economic and environmental catastrophe because end-of-winter low flows had been reduced by 80%. The largest irrigation scheme in southwest Bangladesh, the Ganges–Kobadak scheme covering more than 121 000 ha, is forced to close completely in some dry seasons through lack of water. Crops worth some £2 billion may have been lost since Farakka was completed. Desertification is reported. Thousands of fishermen have lost their livelihood through low flows and saltwater encroachment. With lower river and groundwater levels, seawater now reaches 220 km inland and over 25 000 km^2 of farmland has been salinated. River salinity at Khulna has increased 58-fold since Farakka was built. The natural environment is also suffering. Parts of the world's largest mangrove forest, the Sunderbans along the coast of the Ganges delta, are dying because of lower river levels.

Figure 11.3 The Mekong basin

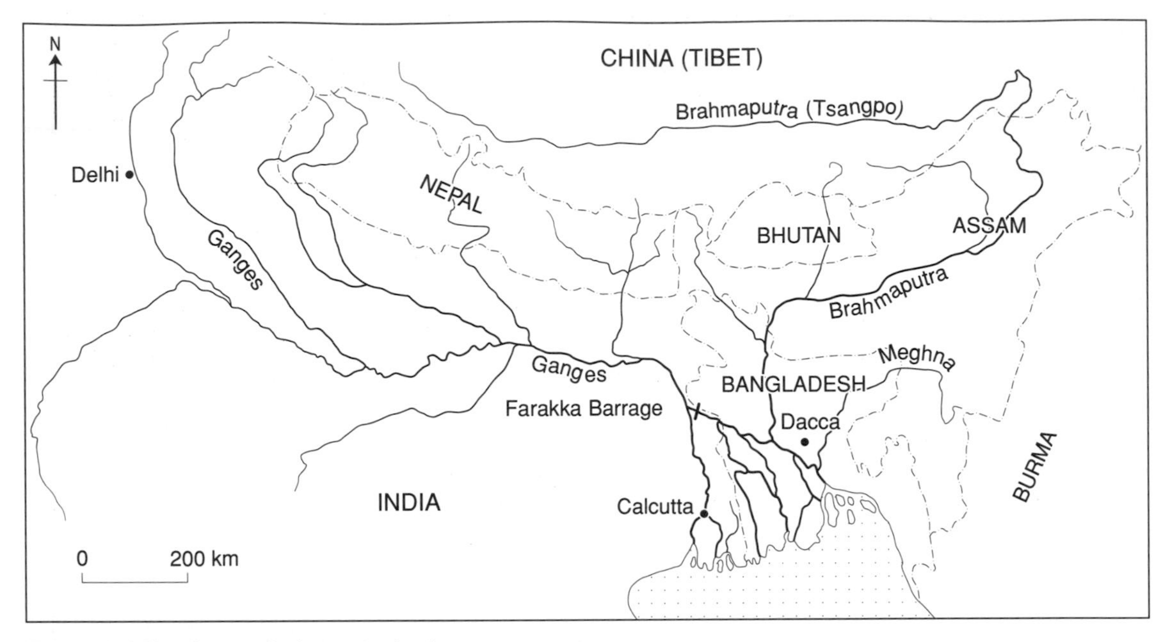

Figure 11.4 The Ganges–Brahmaputra basin

Thankfully a 30-year agreement was signed in December 1996, which guarantees Bangladesh 50 per cent of discharge during the critical March-May period, rising to 80 per cent in severe drought. This follows a series of political and technical pressures: Bangladesh blocked a plan to divert the Brahmaputra to meet the Ganges in India, the World Bank refused to fund an Indian-sponsored hydropower scheme in Nepal because of the Ganges dispute, the diversion has not stopped siltation in Calcutta, although the water is still needed for irrigation and public supply, and Bangladesh realizes that effective flood control requires Indian cooperation.

11.4.4 The Arab world

One-quarter of the Arab world has no surface water, yet it has the fastest growing population outside tropical Africa. Population in the Middle East could increase by 34 million over the next 30 years and water demand is likely to rise to 470 billion m^3 a^{-1}, 132 billion more than the total available even with dramatic conservation measures (Anderson 1991c). Worse still, most of the significant water resources are shared between Arab and non-Arab nations that have a history of enmity (Rowley 1993).

Local rainfall is both sparse and highly variable. Cairo has a mean annual rainfall of 22 mm, which ranges from 1.5 to 64 mm, yet at least 400 mm is needed for successful cropping. The Arabian peninsula is most vulnerable of all, with less than 100 mm a^{-1} and no major rivers. Saudi Arabia has only one-third of the estimated minimum rainfall resources needed for producing successful crops. It has become the seventh largest grain exporter only by 'mining' its aquifers, which will become exhausted in 20–100 years. Despite this paucity of freshwater resources, however, at least Saudi Arabia and the Gulf states are not (as yet) dependent on international water sources, and they do have coastlines and cheap oil to power desalination plants. Iraq, Jordan and Egypt are in more difficult positions.

The Tigris–Euphrates basin The headwaters of the two great rivers of ancient Mesopotamia lie in modern-day Turkey (Figure 11.5). Iraq and Syria both depend heavily on storm and meltwater runoff from the mountains of eastern Turkey. Yet this is also the main area of water surplus for Turkey. Turkey began the Southeast or Greater Anatolia Project (GAP) in the 1960s, aiming to develop up to 26 billion kW of hydropower, equivalent to about 60 per cent of national requirements, and to irrigate up to 1.6 million ha, capable of growing half the national production of rice and vegetables and allowing the introduction of a commercial cotton crop (Hellier 1990). Turkey aims to make the area the 'bread basket of the Middle East' by 2005.

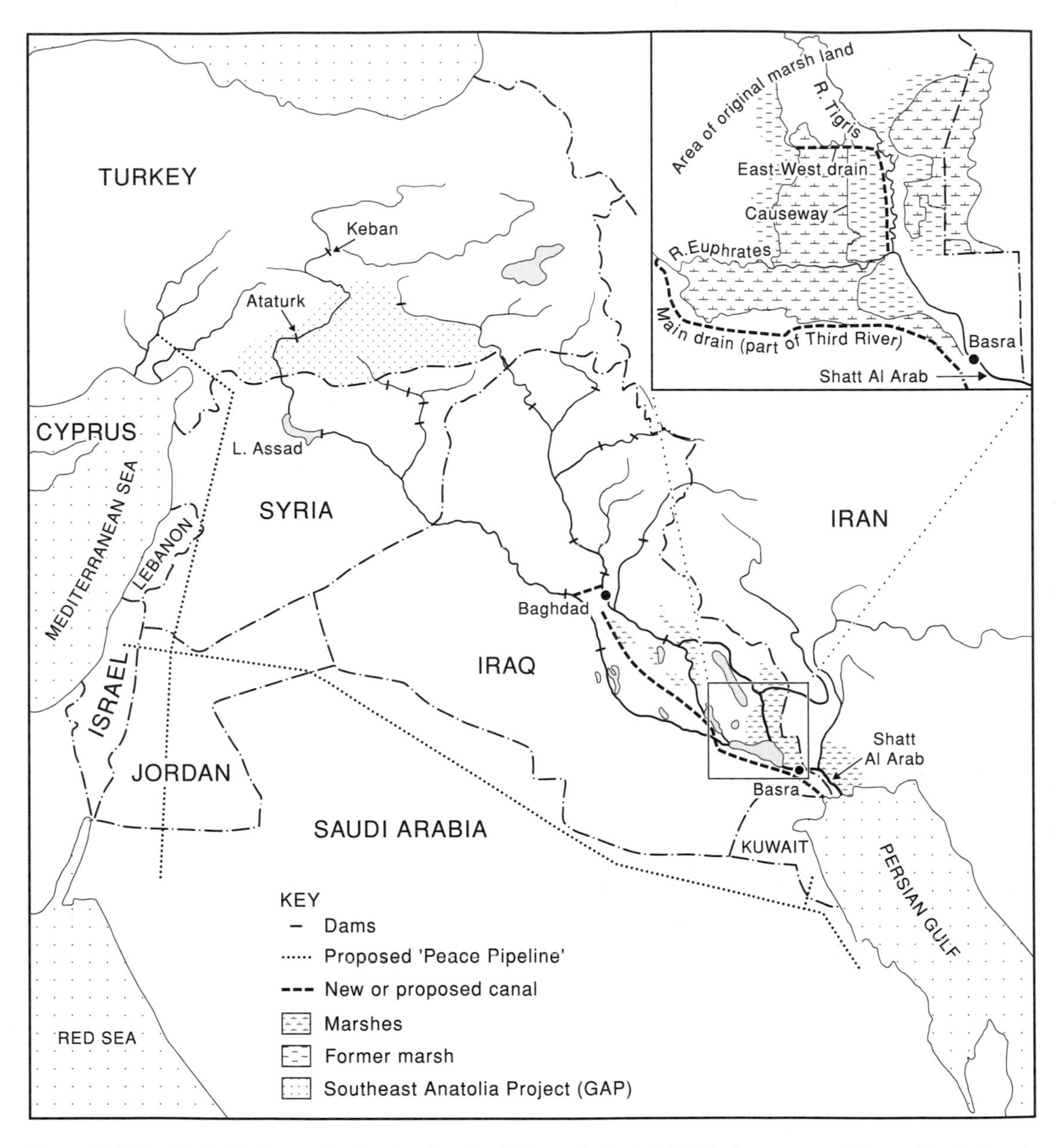

Figure 11.5 The Tigris–Euphrates basin, showing the SE Anatolia Project (GAP), the proposed 'Peace Pipeline' and the draining of the Tigris–Euphrates marshlands. Sources: North (1993), Anderson (1991c) and others

Wary of political ramifications, the World Bank and other major sources of foreign aid refused to finance the scheme. Three decades later, the scheme is still little more than half complete, and technical problems have been caused by poor management of soil erosion on the new arable land. The Keban Dam accumulated 2 million m^3 of sediment in its first 10 years. Even so, the three main dams hold the equivalent of 3 years discharge for the Euphrates. On occasion, flow has ceased entirely, and the economic and environmental impacts are matched by strong political antagonism. International politics are further aggravated by the fact that southeast Anatolia is the homeland of the Kurds, who seek independence and wish to join with the Kurdish peoples of eastern Syria and northern Iraq. Tens of thousands of Kurds have already been displaced by the scheme, and more will be displaced and dispossessed when the scheme extends to the Tigris, giving the PKK terrorists further motivation.

Control of water resources has always been a major source of political power in the Middle East. The Allies considered asking Turkey to cut off supplies to Iraq during the Gulf War in 1991, but this would have affected Syria as well. Turkey did just this in 1990, when the Ataturk Dam, the fifth largest rock-filled dam in the world, was completed. Syria and Iraq laid plans for joint military retaliation. Flow was restored a month later after a secret agreement in which Syria asked for a guaranteed minimum flow of 500 $m^3 s^{-1}$ and Turkey asked for Syria to control and return Kurdish rebels. Some 15 years earlier, Iraq had threatened military action against Syria over the Assad Reservoir. Designed to irrigate 640 000 ha of near desert, Syria regarded it as vital to its national economy, but Iraq accused it of threatening the livelihood of 3 million Iraqi peasant farmers.

In the mid-1980s, Turkey publicized plans for a 'Peace Pipeline', ostensibly to earn money for the GAP by exporting water up to 3000 km, as far as the Arabian peninsula (Figure 11.5). Covertly, the scheme would have extended Turkish political influence throughout all of Iraq's western and southern neighbours, while pointedly cutting off Iraq. None of the states could accept the political implications. If politics were not an issue, however, this could have been a viable solution to many of the region's problems. The pipeline could have transferred up to 8 billion $m^3 a^{-1}$ and Saudi Arabia could have obtained water at one-third of the cost of desalination.

With or without the pipeline, the impact upon the Euphrates River in Iraq is dire. Turkey has argued that the flow regulation assists Syria and Iraq by evening out seasonal flows, which used to range from 7000 $m^3 s^{-1}$ during snowmelt from March to May down to 100 $m^3 s^{-1}$ in the dry season. But this is outweighed by a drastic reduction in annual discharge. Since 1990, discharge in the Euphrates entering Syria has decreased by 50 per cent. Discharge entering Iraq is down nearly 80 per cent, as a result of Syria's own al-Thawrah dam and other schemes. Furthermore, Syria and Iraq have complained of poorer water quality due to less dilution of sewage and other pollutants and return waters for irrigated land with higher salinity.

The Marsh Arabs and the military use of water The Shi'ite Muslims who occupy the marshlands around the confluence of the Tigris and Euphrates are suffering near-genocide from the strategy of the Iraqi military dictatorship, which is compounding the effects of the GAP on water levels (North 1993). In the 1970s, about 220 000 Marsh Arabs occupied these unique marshlands the size of Wales (Figure 11.5). Now perhaps only 10 000 remain, trying to maintain their 5000-year-old lifestyle based on fishing and reed production. The rest have emigrated or died. Their livelihood is being destroyed by the systematic draining of the marshes, 60 per cent of which have entirely disappeared.

Saddam Hussein's government maintains this is part of a long-term plan for agricultural improvement, which has become more vital since the UN sanctions were imposed following the Gulf War. The plan's prime aim is to use the water to rehabilitate millions of hectares of former irrigated land destroyed by salinization by flushing out the salts. The problem is severe, 80 per cent of all the land that has been irrigated in the Tigris–Euphrates basin is now suffering salinization and 33 per cent has been totally abandoned. The Marshes scheme began in 1984. The Third or Leader River was completed in December 1992 and aimed to cleanse 1.5 million ha of salinated land. Tunnels have been constructed under the marshes to keep the saline return water separate.

There are serious doubts that the scheme is technically adequate for its overt purpose, but graver fears that the side-effects will be very damaging. Although some international consultants believe the rehabilitation could be successful, the World Conservation Union predicts that most of the *drained* area will become desert within 10–20 years and claims it is the worst ecological disaster of modern times. Much of the local topsoil has been used to build the dikes and the newly exposed land is already suffering salinization. A unique local ecology is being destroyed. The Indian porcupine, smooth-coated otter and grey wolf are already extinct and many species that evolved in the local habitat, like the Basrah reed warbler, are threatened. But the ecological impact will be much wider. The marshes were a winter haven for two-thirds of all the wildfowl in the Middle East as well as for many others from as far away as western Siberia. The effects on discharges, shown on computer models developed at the University of Exeter, predict that up to 40 per cent of Kuwait's shrimp stocks will be destroyed through loss of spawning grounds.

Whatever the agricultural merits, the military value of the scheme is not in doubt. After the failed Shi'ite uprising following President Bush's exhaltation to revolt in 1991 and the Allied imposition of a no-fly zone in August 1992, the Iraqi army could only proceed with the elimination of the Shi'ites by ground operations. The marshes provided an impenetrable refuge, but operations began with poisoning the water. Subsequent draining has enabled artillery to be moved in. During 1992 and 1993, 40 rivers that used to flow into the al-Amarah Marsh were diverted and the Mother of All Battles Project completed the Fourth River.

The Jordan basin Water has unusual strategic significance for Israel, which now draws 40 per cent of its water supply from outside the internationally recognized pre-1967 borders. High levels of immigration, especially from the former Soviet Union, increased its population by 25 per cent in the early 1990s, and securing its water resources continues to be an important element in military and political decisions.

Plans unveiled at the 1964 Arab Summit in which Syria and Jordan proposed to divert water from the Yarmuk River, a headwater of the River Jordan, were dropped only after Israeli air strikes. More than 60 per cent of Israel's water and 70 per cent of Jordan's come from the Jordan River system. Israel's capture of the Golan Heights in the Six Day War of 1967 secured strategically important high ground, but it also secured control of the Yarmuk Dam site (Figure 11.6). The military buffer zone established in 1978 in southern Lebanon similarly secured the Hasbani River and brought the Litani River under Israeli control. A report to the UN Economic and Social Commission for Western Asia in 1994 claimed that Israel has been diverting over 200 000 $m^3 a^{-1}$ from the Litani and Wazzani Rivers, and that large areas of southern Lebanon are short of water for irrigation and drinking.

Meanwhile, an important element in negotiations over the return of the West Bank to the Palestinians has been the large aquifer that straddles the border. Before 1967, Israeli took 60 per cent of the water abstracted from this aquifer; it now takes 80 per cent. Loss of control over this aquifer would make Israel strategically vulnerable. The only aquifer within the state of Israel as set up by the UN in 1948 runs along the Mediterranean coast and is suffering from saltwater incursion caused by overexploitation.

Israel cannot be accused of wasting water; it is one of the most efficient water users in the world. At least one-quarter of Israel's water is reused and it aims to recycle 80 per cent. By 2010, Israel hopes to recycle 430 million $m^3 a^{-1}$. Its National Water Commission has unrivalled legal control over water abstraction and use (Hillel 1992). Its irrigation systems are already the most efficient in Asia (section 1.2.3), yet Israel is still aiming to recycle 35 per cent of irrigation water. Some orchards are traditionally watered by 'dew farming', with rocks piled around the trees to cool and cause condensation at night. Even so, the higher level of economic development in Israel means that its per capita consumption is 300 litres a day against Jordan's 80.

One of the greatest sources of concern for Jordan has been the diversion of tributaries of the River Jordan around the Sea of Galilee by Israel's National Water Carrier (Figure 11.6). Israel completed this in 1964 to redress the imbalance in resources between the north and the south: 93 per cent of water is in the north, but only 35 per cent of arable land (Hillel 1992). Israel also diverted unwanted saline springs into the Jordan, so that the river is now virtually useless to Jordan because of its low discharge and high salinity.

For Jordan, Israel's river diversion is being compounded by Syrian exploitation of the Yarmuk River. Unable to secure World Bank support for a second joint dam project with Jordan at another site till all riparian parties agree, Syria has proposed a 20-dam scheme on the Yarmuk totally within Syrian territory. As Turkey squeezes Syria on the Euphrates, Syria is turning to the Yarmuk and in turn squeezes Jordan, which has the bad luck to be last in line. Both Turkey and Syria are exercising the Principle of Absolute Sovereignty.

Despite Israel importing 400 million $m^3 a^{-1}$ in containers from Turkey, Israel and Jordan between them are using 120 per cent of the 'safe yield' from regional resources (Anderson 1991a). Water tables are falling in Jordan and the West Bank, springs are drying up and groundwater around Amman is suffering salination.

Economic development in Jordan and an autonomous Palestine, and a sustained future for Israel's economy, require more than improved efficiency in water use. They require political agreement and joint husbandry of water resources. The Israeli–Palestinian accords between 1993 and 1995 were major advances. They appear to make water wars less likely and open the way for more balanced exploitation of resources, barring a resurgence of inter-Arab conflicts. After the 1993 settlement, Turkey even relaunched its Peace Pipeline proposal.

The Nile basin The Nile river system flows through nine countries (Figure 11.7) and is another major water resource that is suffering from a combination of political and economic pressures (Agnew and Anderson 1992; Howell and Allan 1994). The experience of the last 30 years has demonstrated that the resource cannot reliably meet even current levels of demand, and demand is rising rapidly with the 'demographic explosion' both in Egypt and the headwater states (section 9.2.1). The downstream user, Egypt, derives over 90 per cent of its water supply from the river, but the 1959 Nile Waters Agreement binds only Egypt and Sudan.

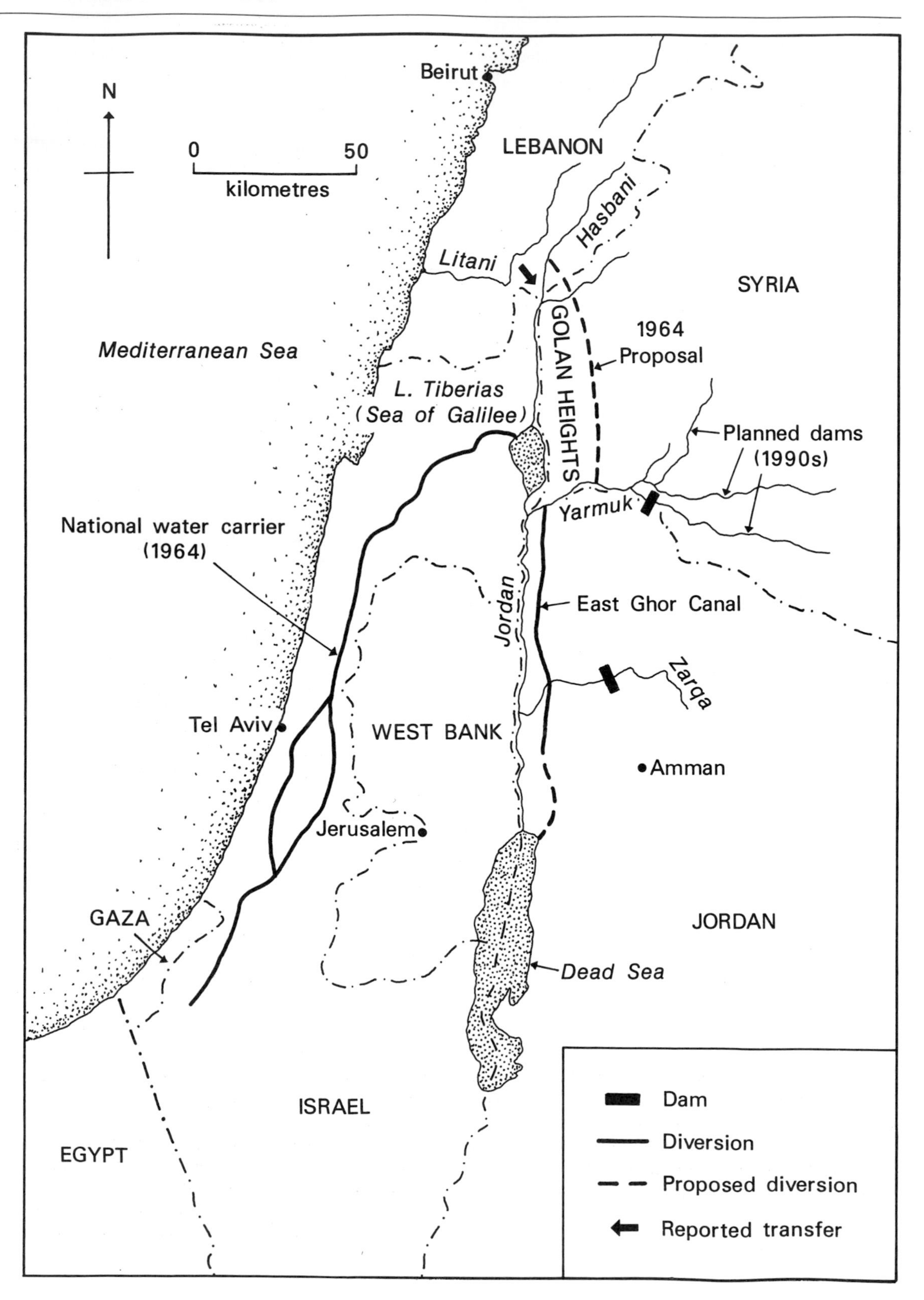

N
0
50
kilometres
Beirut
LEBANON
Hasbani
Litani
SYRIA
GOLAN HEIGHTS
1964
Proposal
Mediterranean Sea
L. Tiberias
(Sea of Galilee)
Planned dams
(1990s)
National water carrier
(1964)
Yarmuk
Jordan
East Ghor Canal
Zarqa
Tel Aviv
WEST BANK
Amman
Jerusalem
GAZA
JORDAN
Dead Sea
ISRAEL
EGYPT
Dam
Diversion
Proposed diversion
Reported transfer

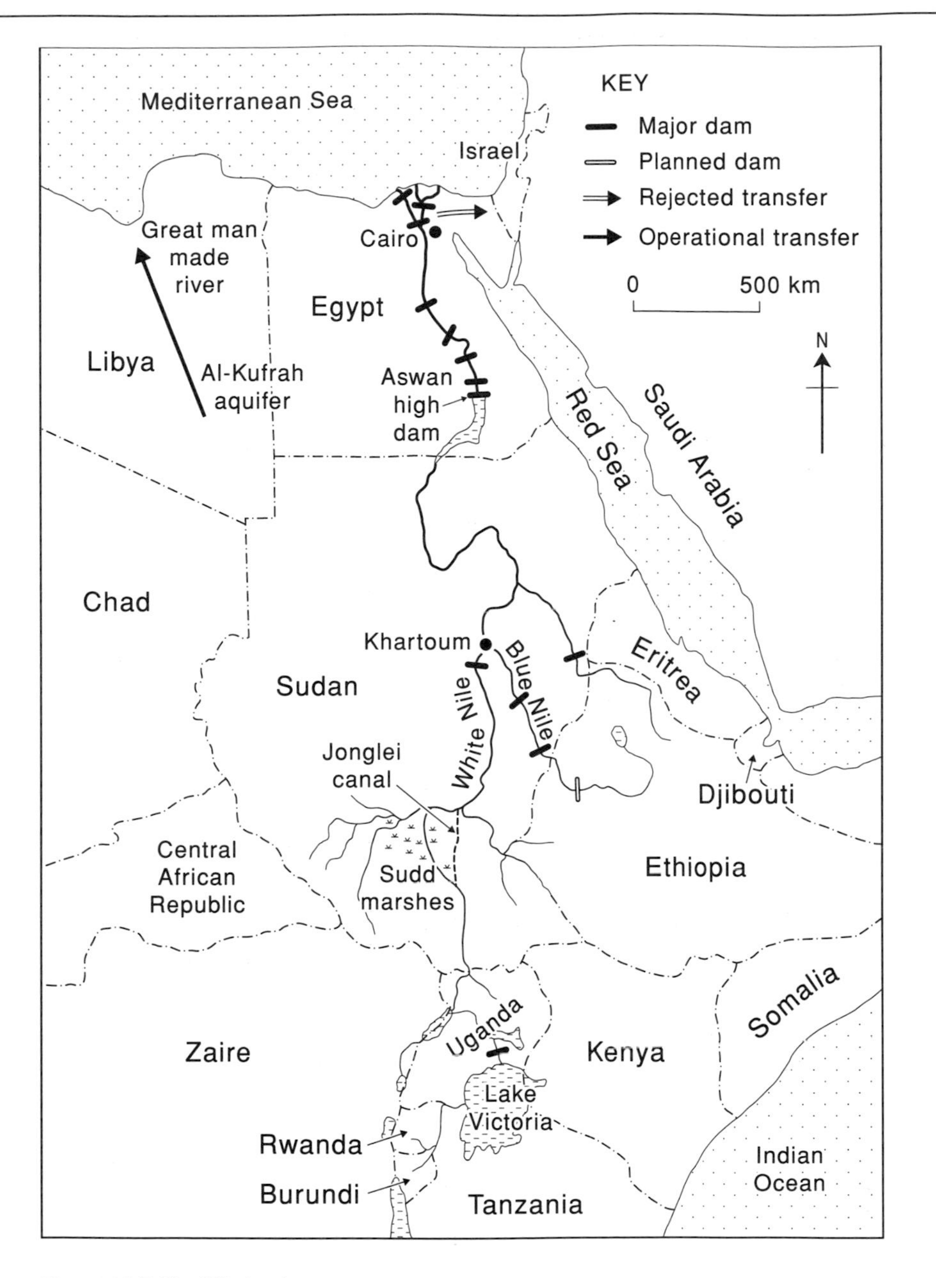

Figure 11.7 The Nile basin

After the Peace Treaty with Israel following the Camp David accord in 1979, the Egyptian government was keen to use Nile water to reclaim land from the Sinai Desert (Guariso *et al.* 1981) and even to accede to Israel's request for 1 per cent of the flow of the Nile to be diverted into the Negev. Even though President Sadat had proposed that this could be extended to Lebanon and Jordan and form the basis for regional cooperation, a deal to share the Nile with Israel caused widespread concern in Egypt and was stopped by threats to President Sadat's life. Turkey was soon to offer its alternative (see page 332).

Even without this diversion, Egypt is already exceeding the consumption of 55.5 billion m^3 a^{-1}

Figure 11.6 Israel and the Jordan basin

agreed under the 1959 treaty. It regularly borrows from Sudan and is still planning more irrigation (Hellier 1990). Ironically, civil unrest in Ethiopia, Uganda and Sudan has delayed their national plans to use more Nile water, but as these national problems are solved upstream demands will increase and so will the potential for international conflict. As its neighbours assert their rights to the water resources in their own territories, Egypt is becoming more wary. Ethiopia has extensive plans to dam the Blue Nile, the source of three-quarters of the discharge in the Egyptian Nile, to help reconstruct its post-Civil War economy. Sudan's plans to drain the Sudd Marshes with the Jonglei Canal have been delayed for a decade by civil war. The plan would reduce the major evapotranspiration losses and so increase water resources for new irrigation projects in southern Sudan.

Some Egyptian hydrogeologists even fear that Colonel Gaddafi's 'Great Man-Made River' in Libya could reduce the Nile's discharge. The Man-Made River is a pipeline fed by 120 wells in the Kufrah aquifer and capable of carrying 2 million m^3 day^{-1} 1900 km northwards to serve the arid but populated Mediterranean coastlands. The aquifer extends into the Nile basin and could draw off water from the river.

As the shortage of water intensifies in Egypt, so water quality and environmental problems increase (section 9.2.1). The sustainability of recycling irrigation water through the aquifer is causing particular concern, since without sufficient local rainfall to flush them through fertilizers are accumulating, especially nitrates. Small wonder that Egypt is reportedly training military units for jungle warfare on the upper Nile and desert warfare in Libya in order to protect its water interests (Bulloch and Darwish 1993). The UN Secretary-General, Boutros Boutros-Ghali, has emphasized the need for a regional agreement: 'In the next few years the demographic explosion in Egypt, in Kenya and in Uganda will lead to all those countries using more water; unless we can agree on the management of water resources we may have international or inter-African disputes.'

11.5 The threat of global warming

The incidence of drought in Africa may have been on the increase in recent decades and there are many Third World countries whose economies will be highly sensitive to climatic shifts predicted for the near future. But ironically parts of the Developed World may have to adapt more under global warming. Resources are likely to change most in some mid-latitude regions, as subtropical high-pressure belts shift northwards. And many of the sophisticated water supply systems in these regions are attuned to exploit *current* resources to the full. Table 11.1 summarizes recent hydrological changes in Europe, many of which are consistent with a warming trend.

11.5.1 Some global predictions

Predictions of rainfall patterns from different GCMs do not show the same degree of convergence as predicted temperatures, and can vary widely at regional scales. Evapotranspiration losses, soil moisture and runoff are even more difficult to predict at present (Jones *et al.* 1996).

Recent reviews of GCM output suggest a 4–12 per cent increase in precipitation and evaporation globally associated with double-CO_2, but the distribution is likely to be uneven. Reduced runoff and soil moisture could affect many agricultural areas in the northern hemisphere. Of particular concern were early American predictions of reduced rainfall in two of the three major world 'breadbaskets', the American Great Plains–Prairies and the Ukraine. Only India and Southeast Asia might benefit, with a more reliable monsoon season (Figure 11.8a). Increased rainfall in some desert areas will be of little practical value. Large increases in precipitation of up to 40 per cent north of 50° N and south of 40° S, predicted by Manabe and Stouffer (1979) and most recent models, could benefit water resources in northwest Europe and the margins of the polar ice sheets.

There is evidence that polar ice has already increased by *c.* 40 000 km^3 this century. Warrick and Farmer (1990) calculate that accumulation in Antarctica could reduce sea level *rise* by up to 200 mm. However, the polar high-pressure belt is likely to block the intrusion of snow-laden depressions far into the continental interior and so limit accumulation to the continental margins. At slightly lower latitudes, milder winters would reduce snowfall amounts and snowmelt would come earlier, decreasing meltwater flooding and reducing soil moisture levels during the growing season.

Initial attempts have been made recently to put a monetary value on the effects on water resources. Fankhauser (1995) estimates the global 'welfare loss in the water sector' at nearly $47 billion, of which the USA and the EU each accounts for about $14 billion. These estimates cover only the cost of lost water. However, Titus (1992) also looked at the costs of

Table 11.1 Trends in hydrological components in Europe during the twentieth century

Precipitation	Evaporation, soil moisture and groundwater	Streamflow
England & Wales More in winter (1900–85) More in spring (1960–90) Less in summer (1960–90)	*England & Wales* Higher soil moisture deficits (1927–77)	
Scandinavia +13%/century (55–70° N) More in winter (1950–85) – Denmark	*Scandinavia* More evapotranspiration (1945–87) Lower soil moisture (1890–1990) Higher water tables (1964–83) in Denmark	*Scandinavia* Increased (1980–90) – Finland More river ice break-up floods (1980–90)
Western Europe -5%/century (37–55° N)		*Western Europe* Increased in Rhine (1970–90)
Southern & Central Europe Less (1960–90) – Italy Desertification in Sicily 1980+ Less snowfall (in warmer 1930–40 & 1970–90) – Austria	*Southern & Central Europe* Droughts (1980–90) – Hungary	
NO TRENDS Seasonal/annual (1931–88) – Scotland & N Ireland	Groundwater (1962–89) – Finland	

Based on Jones (1996).

increased water pollution in the USA and estimated these alone to amount to $33 billion under a 4°C warming.

Detailed regional coverage is available in Jones *et al.* (1996), including East Asia (Liu and Woo 1996), and Australia (Chiew *et al.* 1996), but the next two sections will look briefly at Europe and North America.

11.5.2 The impact in Europe

Recent output from GCMs tends to indicate reduced rainfall over much of southern Europe and increased rainfall in the north and west (Warrick *et al.* 1990). These changes would be consistent with a northward shift in the wind belts and depression tracks, plus increased evaporation in the North Atlantic (Jones 1996). The shifts are clearly indicated in the Hadley Centre's UKHI model (Figure 11.9).

The GCMs agree less over the likely changes in soil moisture (Bach 1989). This presents a more difficult problem for modelling since it requires accurate representation not only of rainfall and evapotranspiration, but also of 'surface hydrology', that is of soil water movement and land-surface drainage. Surface hydrology is an area where rapid improvements are now being made in a number of GCMs (Jones 1996). However, the general indication is of further desiccation in Mediterranean Europe, with possible summer extensions into Western Europe and southeast England. The implications are already being considered by EU policy-makers.

Desertification is presently spreading in southern Spain, Sicily and the Italian mezzogiorno as a result of encroachment by the subtropical high-pressure belt. Italian rainfall has been decreasing since 1900, especially in the last 30 years (Conte *et al.* 1989). Winter droughts in the Alps, Pyrenees and Apennines, due to lack of frontal snowfall, have important consequences for summer supplies, as in Italy in 1988–89. In the summer of 1992 the UK National Rivers Authority designated 40 rivers in southeast England as 'endangered', many of them actually dry after the worst multi-year drought this century. Whether such extreme events are harbingers of change or not, both floods and droughts can be expected more frequently under global warming (Warrick and Farmer 1990).

The combined effect of a northward shift in rainfall resources and temperature shifts, which should cause a marked increase in the length of growing season in

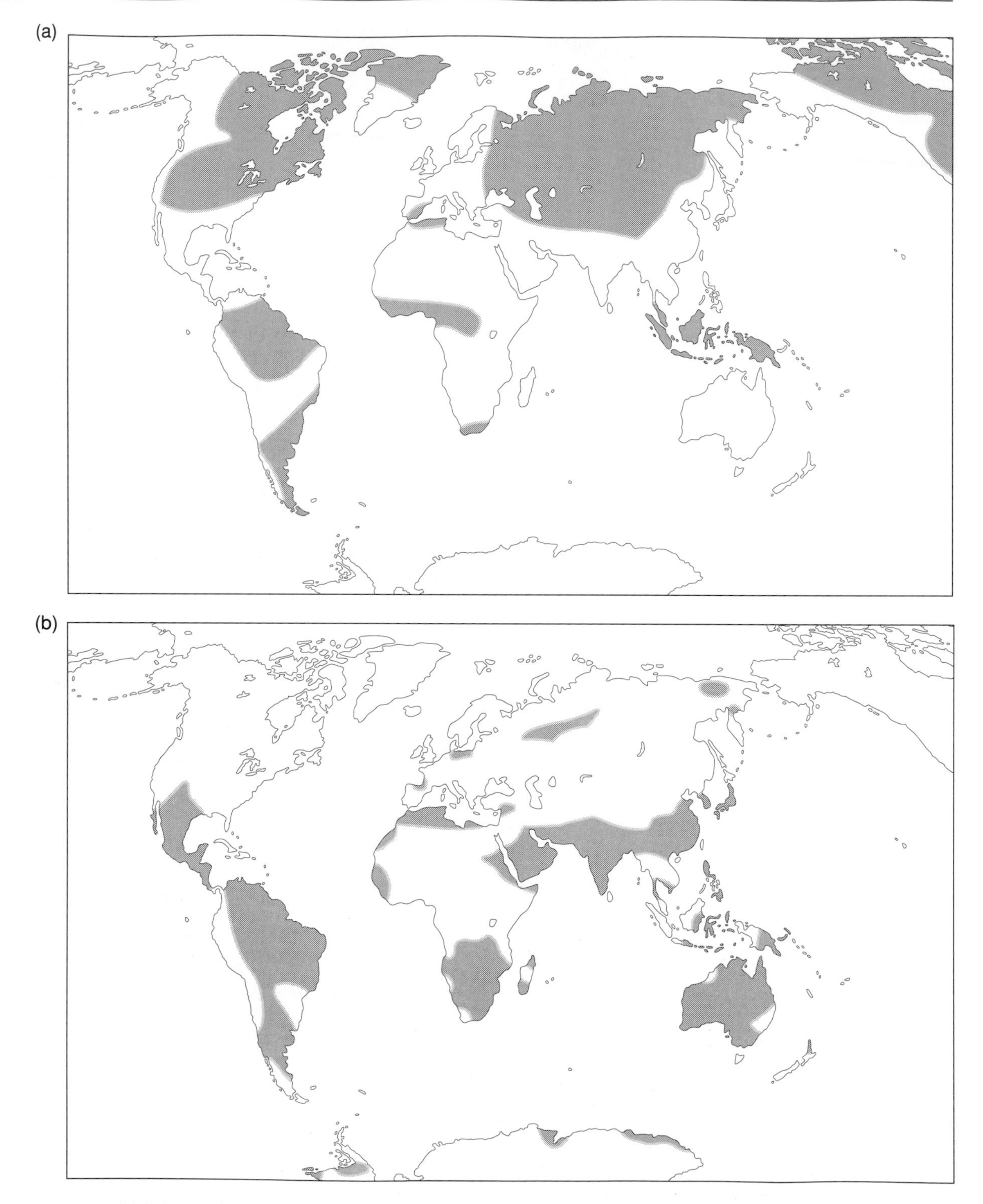

Figure 11.8 Areas of the world predicted to get drier with global warming: (a) according to early American models; (b) according to the Hadley Centre (1995)

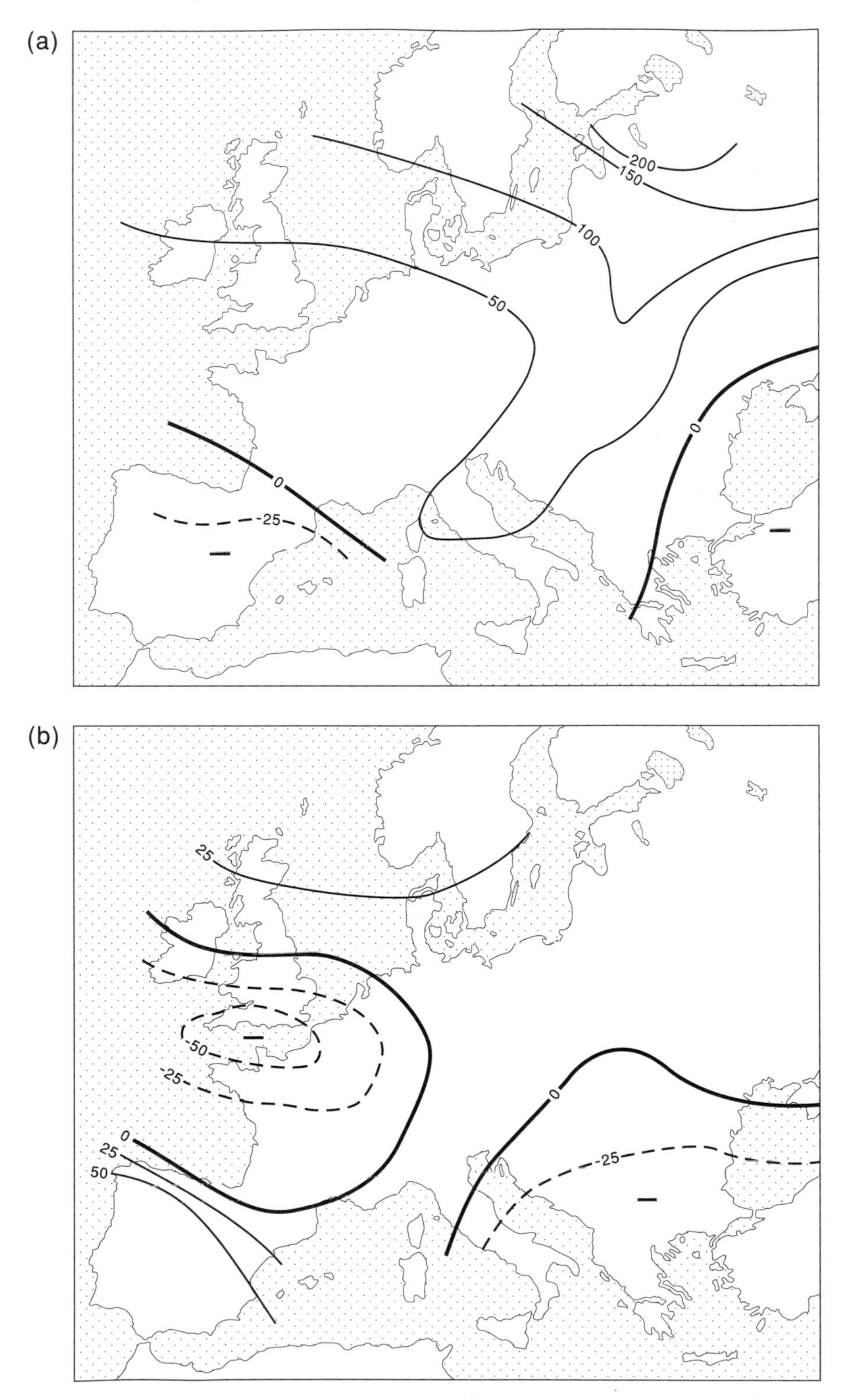

Figure 11.9 Shifts in European rainfall predicted by the first equilibrium experiment with the UKHI. Percentage change for (a) winter and (b) summer

northwest Europe, will be to cause a shift in agricultural potential towards the north and west. Since this is likely to be accompanied by increased risk of summer drought, the already rising demand for summer irrigation in this area will increase.

Increased summer evapotranspiration uncountered by increased rainfall will also cause problems of water quality, owing to lack of dilution for effluents and loss of oxygen as degassing occurs in the warmer water, as evidenced by widespread fish kills in southeast England in the hot July of 1995. Higher temperatures combined with higher solute concentrations will encourage algal blooms on both fresh and saltwater bodies, as in the Adriatic in 1990.

Nor will increased winter rain necessarily maintain summer water supplies in Britain, since most reservoirs are generally already full at the end of winter. Extra rain will only mean extra spill without constructional work or modified operating rules (Law 1989). Studies by the UK Institute of Hydrology suggest that the present gradient in water resources between the northwest and the southeast will intensify, increasing pressure for water transfer schemes (e.g. Arnell 1992; Arnell and Reynard 1993). Arnell found that changes in runoff amplify rainfall changes by up to threefold in southeast England. The only salvation is that basins with greater groundwater storage (more likely in the southeast than northwest) may benefit during summer from higher winter rains. Bultot *et al.* (1988) found a similar situation in Belgium (Figure 11.10). Holt and Jones (1996) compared equilibrium and transient model scenarios for Wales and concluded that shortfalls are likely to start as an autumn problem and move progressively earlier into the summer in later years as the oceanic system warms up.

Annual and seasonal averages give only a partial view of likely changes. Variability and the frequency of extreme events may also change. One of the harbingers of change may be *increased storminess*, as frontal storms are fed by increased evaporation in the North Atlantic, plus a poleward shift in the westerlies and perhaps a temporary steepening of the latitudinal temperature gradients in mid-latitudes as warmer waters push closer to snow and ice surfaces. Convective storms should also increase in frequency and intensity in northwest Europe because of warmer summer

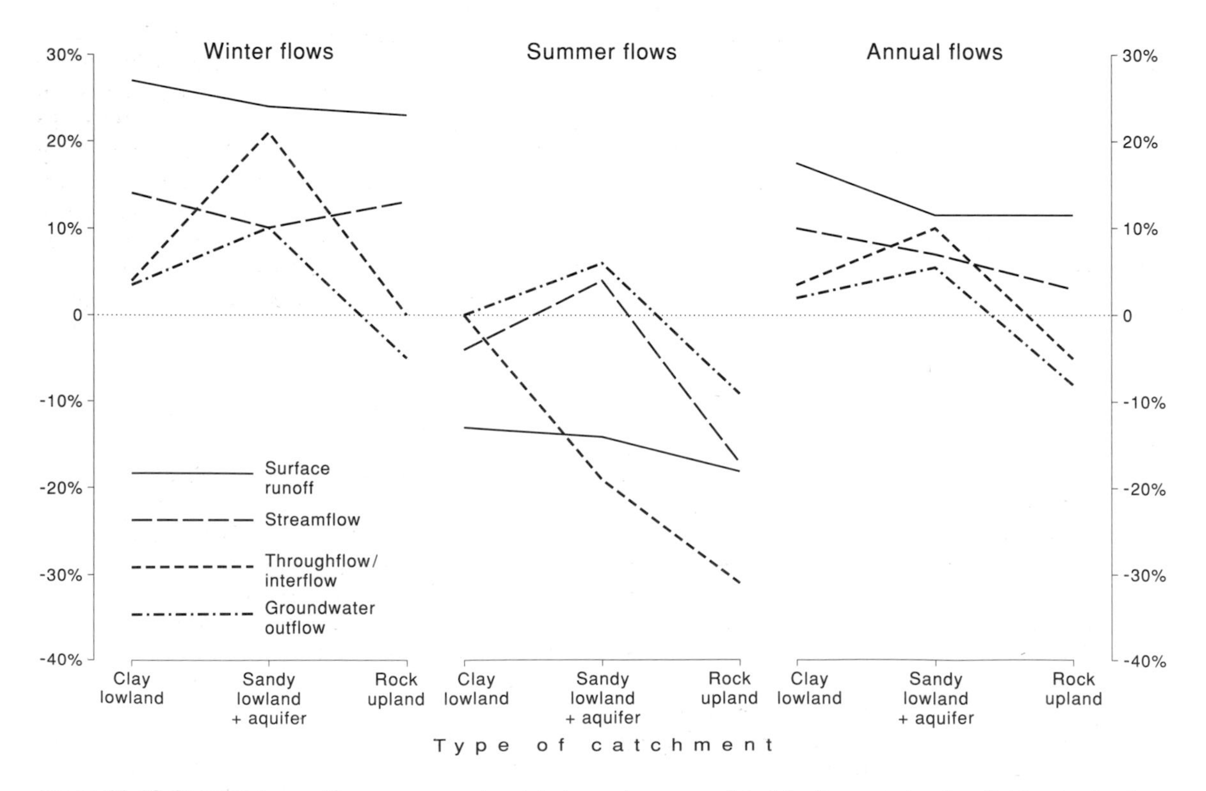

Figure 11.10 Changes in runoff processes under global warming as predicted for three contrasting Belgian basins by Bultot *et al.* (1988)

surfaces, increasing the probability of local thermal low-pressure cells developing over the eastern North Atlantic. The probability of severe storms developing from the remains of a tropical hurricane also increases as the area of the North Atlantic over 26°C extends further north. At the same time, the length and frequency of summer droughts are likely to increase throughout much of Europe, aggravating pollution events and the 'dirty' water problem (section 8.3.1).

Ironically, northwest Europe could experience more floods as a result, even during drier summers, partly because of possible increases in rainfall intensity and partly because many soils may crack and develop hydrophobic properties which encourage more rapid surface runoff. In Belgium, Gellens (1991) has shown how feeding likely climatic scenarios into a conceptual hydrological model increases the frequency of winter floods and summer droughts. In Britain, Cole *et al.* (1991) have demonstrated that the increased variability is likely to require even greater increases in reservoir capacity in order to hold the risk of failure of supplies at current levels.

The storminess could interact with elevated sea levels to increase the risk of coastal and estuarine flooding. Strong onshore winds generate destructive waves, pile water up on the coast and cause riverflow to back-up in estuaries, and lower atmospheric pressure itself raises sea level, as was seen in the North Wales floods of February 1990 or more spectacularly in the North Sea floods of 1953 in Holland and East Anglia, in which nearly 1500 people died. The Rhine Delta Project set in train after these floods was finally completed in 1986 and the Thames Barrier in 1982.

Adequate safety margins are built into these schemes for some decades ahead. But it is worth noting that increased storminess could make a seemingly innocuous sea level rise of 300 mm by 2030 (Warrick and Farmer 1990; UK Climate Change Impacts Review Group 1991) more critical. The Environment Agency currently regards all land less than 5 m OD in Britain as prone to coastal flooding. Nonlinear increases in risk factors could extend this to 6–7 m OD by 2100. However, if warming progresses beyond the point at which winter pack ice retreats from the Atlantic, there is likely to be a reduction in the frequency and intensity of depressions affecting most of western Europe. Depression tracks often follow the edge of the winter pack ice quite closely, in part because of the marked contrast in air mass temperatures astride this edge. The demise of the sea ice will reduce thermal gradients.

Global warming will also have a marked effect on land-based snow and ice in Europe. Collins (1989) has shown that sensitivity to climate change in the French Alps varies between low glacierized basins, where runoff is sensitive to precipitation input, and heavily glacierized basins, which are more sensitive to energy inputs. Data from the Austrian Alps indicate reduced snowfall during the warm period of the 1930s and a similar tendency beginning in 1975–80 apparently linked with the recent period of warming (Mohnl 1991). A reconstructed 450-year flow record for the Rhine shows the sensitivity of flow in the upper basin to snowmelt, but also suggests that land-use changes in the lower basin could be of comparable importance, particularly as they affect evapotranspiration (Schadler 1989). In 1989, the International Commission for the Hydrology of the Rhine Basin (CHR) initiated a collaborative research programme to develop a basin-wide hydrological model to simulate these effects (Parmet and Mann 1993). It seems that the recent EC policy of 'set-aside', designed to reduce agricultural overproduction by subsidizing afforestation, could have as much effect on riverflow as global warming on meltwaters. As hydrologists are increasingly aware, physical models of the hydrological system have to take account of such human inputs.

The impact on demand for water is likely to be no less important than changes in discharge, although it is more difficult to study. The 125 per cent increase in demand in southwest England during July 1995, the third hottest on record, is a clear indication of the likely problem. Arnell *et al.* (1994) calculate a 4 per cent increase in annual demand in southern and eastern England by 2021, but summertime peak demands are likely to be the main problem. Demand for spray irrigation in England could increase by 50 per cent by 2021.

11.5.3 The impact in North America

North America spans a wider latitudinal range than Europe and will be more immediately affected by any strengthening of the Hadley cell in the south and by higher rates of warming in the subarctic and Arctic. Many areas of the south and east could receive less precipitation due to expansion of the Hadley circulation and an intensification of the subtropical jet. This could shift rainfall currently affecting the eastern seaboard out into the Atlantic (Rind 1988). In contrast, there is likely to be an increase in precipitation receipts along the borders of the North Pacific. This could mean more winter snowfall around the Gulf of Alaska. However, higher summer temperatures are likely to cause a net reduction in annual accumulation. Elsewhere in the north and west the expected shift in favour of rainfall rather than snowfall could have greater effect than a slight increase in annual precipitation. This is because snowmelt dominates the annual hydrograph over much of the area and is critically timed in relation to demand, particularly from agriculture.

A significant reduction is likely in snow accumulation in the subarctic and in the temporal and spatial extent of frozen ground (Woo 1996). This would alter the seasonality of runoff and affect the operating procedures for the James Bay hydroschemes (Box 10.3), although increased annual precipitation should protect the overall viability of these schemes. As in Europe, winter temperatures are likely to rise more than summer, further reducing snow accumulation. This could prove critical for the Canadian prairies, where snowmelt is an important source of soil moisture in the early growing season: it could cause summer droughts, which are a worrying aspect of the present climate, to become more troublesome.

Further south, Lins *et al.* (1990) predict reduced precipitation and higher evapotranspiration in the Great Lakes basin. This could reduce the effectiveness of a plan to divert water to the High Plains to help stem the water shortages that have so affected the Ogallala aquifer. Cohen (1986) calculated an 11 per cent reduction in runoff under a GISS scenario suggesting a temperature rise of around 4.5°C. Runoff seems likely to decrease over much of west central USA, including the Colorado River, the Great Basin and the Rio Grande. The only area likely to show marked improvement in surface water resources is the Pacific Northwest and the coast of British Columbia, where higher sea surface temperatures will provide more moisture. California will also benefit from this in winter, but summer droughts are likely to be more intense (Lins *et al.* 1990). All 10 scenarios tested by Gleick (1987), some with up to 20 per cent increase in annual precipitation, showed large decreases in summer runoff and soil moisture in the Sacramento–San Jaoquin basin. Estimates have suggested that a 3°C rise could reduce local supplies by up to 15 per cent at 1990 demand levels, and demand is projected to increase by 30 per cent by 2010 (Lins *et al.* 1990). Based on the Californian estimates, Cline (1992) predicts a 10 per cent reduction in US water resources.

Studies in the east, on the Delaware River and in the Tennessee valley, again show a tendency to increased seasonality with springs tending to be wetter and summers drier. Miller (1989) calculated that it would be possible to operate the larger reservoirs in the TVA scheme and to maintain navigation reasonably effectively even in the driest scenario with over 30 per cent less precipitation, but at the expense of the peripheral reservoirs. McCabe and Ayers (1989) estimated that a 2°C rise would cause a doubling of drought frequency on the Delaware, and that a 10 per cent reduction in precipitation would raise drought frequency five-fold.

Despite obvious geographical differences, the studies in North America and Europe show a strong consistency, and they reveal that even areas where annual precipitation is likely to increase may not be free of water supply problems.

11.6 Integrated water resource management

Integrated water resource management (IWRM) is being actively promoted by the UN as a means of linking the development and protection of both natural and human resources within a basin under one overarching management structure. Problems caused by piecemeal developments and fragmented institutional responsibilities have been all too common, especially in Developing Countries (Hufschmidt and Tejwani 1993).

Figure 11.11 presents a systems view of the framework for IWRM based on the proposals of Hufschmidt (1993). IWRM combines a scientific focus on the surface and subsurface water resources, and their links with the soil and biota, with careful consideration of their value to society and its economic and social objectives (Table 11.2). The basic aim is to avoid developments in one field, like urban expansion or forest clearance, that might have important detrimental impacts upon another. Human occupancy of floodplains is one such case that has increased hazards for inhabitants downstream as well as the risk of local flood damage (Table 11.3).

Most cities in the humid tropics currently suffer from severe lack of integrated planning, combined with grossly inadequate databases, and lack of scientific analysis of urban hydrology or concern for wastewater control (Bonell *et al.* 1993). But this is not solely a problem for Developing Countries. The northern Italian floods of January 1994 were widely blamed on a combination of lack of planning control for building development on the floodplains and continuing environmental degradation and deforestation in the Alpine headwaters. Table 11.3 illustrates the ways in which IWRM may help.

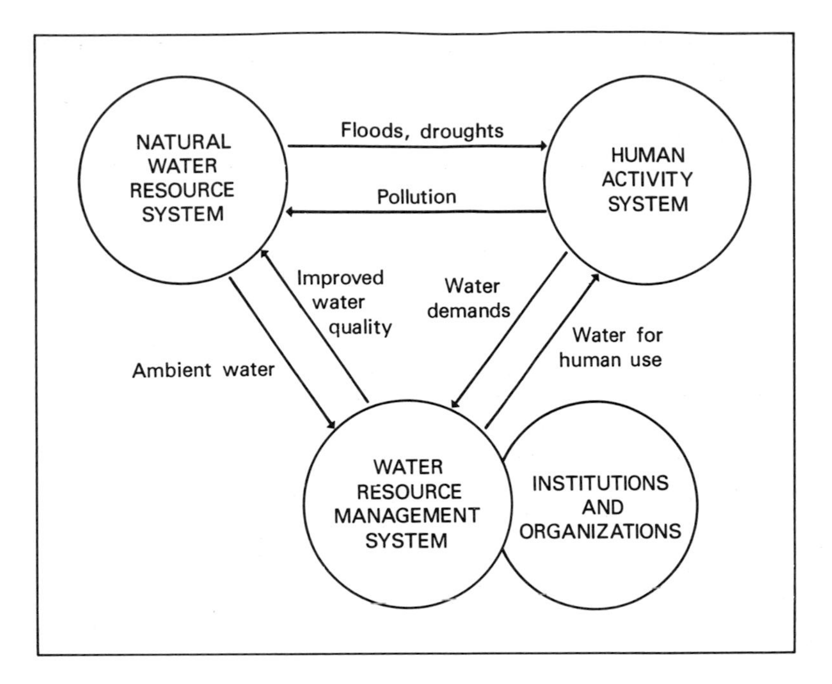

Figure 11.11 The framework for integrated water resource management according to Hufschmidt (1993)

Table 11.2 Integrated watershed management: rationale, implementation and problems

Rationale major aims	Implementation stages of development	Problems common causes of failure
1. Treating watershed as a functional region with interrelationships between water and land management	1. Establishing watershed management objectives	1. Lack of local participation
2. Evaluating the bio-physical linkages of upstream and downstream activities	2. Formulating and evaluating alternative resource management actions, involving institutional structures and tools for implementation	2. Inadequate technical support and guidance
3. Enabling planners/managers to consider all relevant facets of development	3. Choosing and implementing a preferred course of action	3. Inadequate management practices
4. Strong economic logic in integration	4. Evaluating performance, i.e. degree of achievement of specific objectives, by monitoring activities and outcomes	4. Delays in key inputs, especially financial
5. Allows ready assessment of environmental impacts		5. Fragmented governmental control structure
6. Can be integrated into other programmes, e.g. forestry and soil conservation		6. Ignoring downstream interests
		7. Inappropriate institutional arrangements
		8. Political boundaries unrelated to catchment boundary

Source: Adapted from Hufschmidt and Tejwani (1993).

Table 11.3 Examples of value of integrated water resource management, especially in the Developing Countries

Problem	Major causes	Implications for integrated management
Groundwater:		
Aquifer depletion, increasing pumping costs, saltwater intrusions on coast	Unsustainable groundwater pumping rates	*Need for conjunctive groundwater–surface water management* Link timing of pumping with status of surface resources Use surplus surface water for artificial recharge
Land subsidence in urban areas	Excessive pumping of groundwater	*Need aquifer-wide regulation of pumping rates* Establish pumping permits, metering and monitoring
Floods:		
Flood risks and damages increased	Rapid urbanization and human occupancy of floodplains	*Need effective land-use planning* Floodplain zoning, flood proofing, evacuation programmes, redevelopment of flood-prone areas
Structural flood controls too expensive		*Emphasize non-structural and low-cost solutions* Use existing field or paved areas as temporary flood storage Create new low-lying areas to do this
Ineffective flood control programmes	Lack of high-level management	*Establish appropriate institutions and integrated flood management for basin*
Urban water supply:		
Many, especially poor, areas not served	Bias towards high-cost technology, shortage of investment funds, unequal distribution of infrastructure	Increase charges to present users Adopt low-cost technology Promote community involvement in management
Unreliable distribution system and high losses	Poor operation and maintenance Faulty metering and billing	Increase water charges to invest in improvements Improve organization and training Community involvement
Urban water pollution:		
Domestic sewage enters water bodies uncollected	No centralized system through lack of funds or perception of need	Low-cost sanitation techniques Community involvement in building and operating sanitation systems
Domestic sewage collected but not treated	Cost of standard treatment	Low-cost treatment like detention basins Wastewater reuse and cost-sharing
Industrial wastewater discharged untreated	Lack of regulatory controls	Recycling Incentives like subsidies for pollution control, effluent charges or fines
Major nonpoint pollution	Lack of regulatory controls Limited community understanding or commitment	Watershed management programme Community involvement in sanitation and clean-up
Centralized sewage collection and treatment too costly	Lack of funds	Low-cost technology Community involvement Cost-sharing, e.g. sewer charges
Existing sewage treatment systems poorly maintained	Lack of funds	Community involvement in operation and maintenance Improving financing, e.g. by taxes

Source: Adapted from Hufschmidt (1993)

Box 11.1 Some general conclusions

It is rarely possible to undertake developments without any deterioration of environment or resource, but it should be possible to minimize these side-effects and it must be possible to avoid irreversible damage. Integrated and environmentally aware development combined with demand management and waste reduction offer the safest way forward.

Importing strategic resources from other areas need not be antipathetic to these aims, but it needs to be carefully considered in terms of its likely environmental impact, the durability of the solution and a range of alternatives, including socio-economic adjustments. The USA and Canada are reportedly reviewing the NAWAPA plan (Box 9.3) in light of global warming (Evans 1994). No price may be too high for water if global warming scenarios materialize, but currently the plan makes no economic sense and the 240 large dams and 30 years of construction could be very damaging to the environment.

Sustainability is the keyword for future development. Stable sustainability will probably only be achieved by controlling demand – by recycling, by reducing pollution and by cutting overall water use. However, it may be difficult to achieve true stability within natural, technical and economic environments that are constantly changing. Technological developments, cheaper methods or changing economic priorities may enable new approaches, such as desalination or importing icebergs, which could expand local resources in some areas. History suggests, however, that engineering new resources only 'buys time' until demand catches up, and there are still many legitimate environmental and technological reservations over widespread use of these technologies.

We should also be careful to avoid the narrow anthropocentric view of sustainability, as the sustainable management of resources for humanity. The scale of human interference with the hydrological cycle now demands a more altruistic view of the impacts upon the rest of the living world. In the eloquent words of US Vice-President Al Gore (1992):

> The rains bring us trees and flowers; the droughts bring gaping cracks in the world. The lakes and rivers sustain us; they flow through the veins of the earth and into our own. But we must take care to let them flow back out as pure as they came, not poison and waste them without thought for the future.

Gore (1992) identified five strategic threats to the global water system: the redistribution of freshwater; rising sea levels; changes in land use, especially deforestation; worldwide contamination of water resources; and rapid population growth. Water is at last on the international political agenda and thankfully on a higher plane than 'water wars'.

Environmental protection is still, nevertheless, too often regarded as the preserve of Developed Nations. And even in these countries many aspects of understanding and enforcement need further development. Transferring concepts of environmental protection to different environments and to nations that understandably regard them as a secondary consideration to economic development can be problematic. But it will be necessary.

Increased international cooperation in science and technology will hopefully play an important role in achieving sustainability. Natural systems need to be thoroughly understood before they can be safely managed. Too often in the recent past, technological solutions devised in one environment have been transported to less well understood environments where they have proved inappropriate or even damaging. In this, the work of the IAHS, the WMO and UNESCO is of paramount importance, particularly as vehicles for communicating methods of 'best practice' and for identifying regional scientific and data needs.

The Dutch government has actively embraced IWRM to ensure that all water uses operate harmoniously, especially with a view to restoring natural habitats and restricting 'overengineered' solutions. Waterways containing heavily polluted sediments are now dredged only where it is absolutely necessary for navigation. Otherwise they are best left until better treatment methods are available. In addition to legal protection of groundwaters (section 8.3.3), farmers are encouraged not to dredge ditches and ponds so regularly, e.g. under the Cooperation Waterland scheme north of Amsterdam. The Biesbosch National Park, created in the early 1990s, is successfully integrating the increasing demand for leisure use of the water with sound ecological management of a deltaic wetland.

IWRM has three guiding principles: *multiple purpose, multiple objectives and multiple means.* Integrated plans need to balance a wide range of water uses and management purposes. They need to include objectives that balance economic productivity with environmental quality, health and social considerations. And they should achieve these ends by combining physical structures with regulations and economic incentives. This may involve sacrificing some economic advantage to preserve resources for the next generation or to maintain and enhance the value of water to society.

IWRM is commonly seen as relating to the river basin unit (Kirby and White 1994), but this is increasingly inadequate. The 'hydrological unit' could be an aquifer or an artificial unit created by interbasin transfer. Challenging as this is on a national scale, it can be forbidding when the units are international. The 1993 UNESCO report suggested that 'detailed integrated planning for such large river basins is probably administratively infeasible', and noted that even the Lower Mekong Commission had to date only concerned itself with framework planning and data collection (Hufschmidt and Tejwani 1993). Smaller basin units in the Philippines or Caribbean offer greater chance of success for IWRM.

Further reading

General coverage of many issues raised in this chapter can be found in:

Biswas A K and M A H Abu-Zeid (eds) 1993 *Water for sustainable development in the 21st century*. Oxford, Oxford University Press.

Canter L 1988 *Environmental impact of water resources projects*. Chelsea, Michigan, Lewis Pub.: 352pp.

Postel S 1992 *The last oasis*. London, Earthscan: 239pp.

Aspects of regional interest, including hydropolitics and climate change, are covered in:

Agnew C and E W Anderson 1992 *Water in the arid realm*. London, Routledge: 329pp.

Bulloch J and A Darwish 1993 *Water wars: coming conflict in the Middle East*. London, Gollancz: 224pp.

Jones J A A, C M Liu, M K Woo and H T Kung (eds) 1996 *Regional hydrological response to climate change*. Dordrecht, Kluwer: 425pp.

NRA 1994 *Water, nature's precious resource – an environmentally sustainable water resources development strategy for England and Wales*. London, HMSO: 93pp.

Thomas C and D Howlett (eds) 1993 *Resource politics, freshwater and regional relations*. Milton Keynes, Open University: 210pp.

Rodda J C and N C Matalas (eds) 1987 *Water for the future: hydrology in perspective*. International Association of Hydrological Sciences Pub. No. 164: 517pp.

Young G J, J C Dooge and J C Rodda (eds) 1994 *Global water resources issues*. Cambridge, Cambridge University Press: 194pp.

Discussion topics

1. Is demand management a feasible option without compulsion?
2. Is there any real evidence that countries are curbing their demands?
3. How can the Developed Nations best help the LDCs to cope with water shortage?
4. Is the natural environment doomed always to be the loser?
5. Is global warming a real threat?
6. Are water shortages only a temporary aberration that improved technology will overcome?

References

Abbott M B, J C Bathurst, J A Cunge, P E O'Connell and J Rasmussen 1986 An introduction to the European Hydrological System – Système Hydrologique Europée̒n "SHE" 2. Structure of the physically based, distributed modelling system. *Journal of Hydrology*, 87: 61–77

Adams W P 1981 Snow and ice on lakes. In D M Gray and D H Male (eds) *Handbook of snow: principles, processes, management and use*. Oxford, Pergamon: 437–74

Adler J 1993 Troubled waters. *Newsweek* 122(4): 20–5

Agnew C and E W Anderson 1992 *Water in the arid realm.* London, Routledge: 329pp

Ahrens C D 1991 *Meteorology today*, 4th edition. New York, West Publishing Co.: 576pp

Akan A O 1984 Simulation of runoff from snow-covered hillslopes. *Water Resources Research*, 20: 703–13

Albertson M L 1983 The impact of hydropower on society. In UNESCO Impact of Science on Society No 1, *Managing our freshwater resources*, Paris, UNESCO: 69–81

Alloway B J and D C Ayres 1993 *Chemical principles of environmental pollution.* Glasgow, Blackie: 291pp

Alvares C and R Billorey 1988 *Damming the Narmada: India's greatest planned environmental disaster*. Malaysia, Third World Network/Asia-Pacific People's Environment Network

American Society of Civil Engineers (ASCE) 1970 *Design and construction of sanitory sewers*. ASCE Manuals and Reports on Engineering Practice No 37

Anderson D G 1970 *Effect of urban development on floods in northern Virginia*. US Geological Survey Water Supply Paper 2001–C

Anderson E W 1991a White oil. *Geographical Magazine*, 63(2): 10–14

Anderson E W 1991b Making waves on the Nile. *Geographical Magazine*, 63(4): 10–13

Anderson E W 1991c The violence of thirst. *Geographical Magazine*, 63(5): 31–4

Anderson M G and T P Burt 1977a Automatic monitoring of soil moisture conditions in a hillslope spur and hollow. *Journal of Hydrology*, 33: 27–36

Anderson M G and T P Burt 1977b A laboratory model to investigate the soil moisture conditions on a draining slope. *Journal of Hydrology*, 33: 383–90

Anderson M G and T P Burt 1978 The role of topography in controlling throughflow generation. *Earth Surface Processes*, 3: 331–4

Anderson M G and T P Burt (eds) 1985 *Hydrological forecasting*. Chichester, Wiley: 604pp

Anderson M G and T P Burt (eds) 1990 *Process studies in hillslope hydrology*. Chichester, Wiley: 539pp

Anderson M G, D Bosworth and P E Kneale 1984 Controls on overland flow generation. In T P Burt and D E Walling (eds) *Field experiments in fluvial geomorphology*. Norwich, GeoBooks: 21–34

Archer D R 1981 Severe snowmelt runoff in north-east England and its implications. *Proceedings of Institution of Civil Engineers*, part 2, 71: 1047–60

Archer D R 1983 Computer modelling of snowmelt flood runoff in north-east England. *Proceedings of Institution of Civil Engineers*, 75: 155–73

Arnell N 1989 *Human influences on hydrological behaviour: an international literature survey*. Paris, UNESCO

Arnell N W 1992 Impacts of climatic change on river flow regimes in the UK. *Journal of Institution of Water and Environmental Management*, 6(4): 432–42

Arnell N W and N Reynard 1993 *Impact of climate change on river flow regimes in the United Kingdom*. Wallingford, Institute of Hydrology: 129pp

Arnell N W, A Jenkins and D G George 1994 *The implications of climate change for the National Rivers Authority*. Bristol, National Rivers Authority, R & D Report 12: 94pp

Atkinson B W 1979 Urban influences on precipitation in London. In G E Hollis (ed.) *Man's influence on the hydrological cycle in the United Kingdom*. Norwich, GeoBooks: 123–33

Atkinson T C 1978 Techniques for measuring subsurface flow on hillslopes. In M J Kirkby (ed.) *Hillslope hydrology*. Chichester, Wiley: 73–120

Bach W 1989 Projected climatic changes and impacts in Europe due to increased CO_2. In *Conference on Climate and Water*. Helsinki, Academy of Finland, vol. 1: 31–50

Baker D, H Escher-Vetter, H Moser, H Oerter and O Reinwarth 1982 A glacier discharge model based on results from field studies of energy balance, water storage, and flow. In J W Glen (ed.) *Hydrological aspects of alpine and mountain areas*. International Association of Hydrological Sciences Pub. No 138: 103–12

Balek J, J Cermak, J Kucera, M Palous and A Prax 1986 Regional transpiration assessment by remote sensing. In A I Johnson (ed.) *Hydrologic applications of space technology*. International Association of Hydrological Sciences Pub. No 160: 141–8

Bardossy A, W E Kelly and I Bogardi 1986 Network design of geoelectric surveys for estimating aquifer properties. In M E Moss (ed.) *Integrated design of hydrological networks*. International Association of Hydrological Sciences Pub. No 158: 85–96

Barrett E C and R W Herschy 1986 A European perspective on satellite remote sensing for hydrology and water management. In A I Johnson (ed.) *Hydrologic applications of space technology*. International Association of Hydrological Sciences Pub. No 160: 3–12

Barrett E C and D W Martin 1981 *The use of satellite data in rainfall monitoring*. Academic Press: 340pp

Barry R G and R J Chorley 1992 *Atmosphere, weather and climate*, 6th edition. London, Routledge: 392pp

Barth H (ed.) 1987 *Reversibility of acidification*. Amsterdam, Elsevier: 175pp

Bastin G, B Lovert, C Duque and M Gevers 1984 Optimum estimation of the average areal rainfall and optimum selection of rain gauge locations. *Water Resources Research*, 20: 463–70

Bathurst J C 1988 Flow processes and data provision for channel flow models. In M G Anderson (ed.) *Modelling geomorphological systems*. Chichester, Wiley: 127–52

Battan L J 1973 *Radar observation of the atmosphere*. Chicago, University of Chicago Press: 324pp

Batterbee R W 1990 The causes of lake acidification with special reference to the role of acid deposition. *Philosophical Transactions of the Royal Society London Series B* 327: 339–47

Batterbee R W, T E H Allott, A M Kreiser and S Juggins 1993 Setting critical loads for UK surface waters: the diatom model. In M Hornung and R A Skeffington (eds) *Critical loads: concepts and applications*. London, HMSO: 99–103

Baumgartner A 1967 Energetic bases for differential vaporization from forest and agricultural land. In W E Sopper and H W Lull (eds) *Forest hydrology*, Oxford, Pergamon: 381–9

Baumgartner M F, K Seidel, H Haefner, K I Itten and J Martinec 1986 Snow cover mapping for runoff simulations based on Landsat-MSS data in an Alpine basin. In A I Johnson (ed.) *Hydrologic applications of space technology*. International Association of Hydrological Sciences Pub. No 160: 191–200

Beaumont C D 1979 Stochastic models in hydrology. *Progress in Physical Geography*, 3: 363–91

Beaumont C D 1982 The analysis of hydrological time series. *Progress in Physical Geography*, 6: 60–99

Becker A and Z Kundzewicz 1987 No nlinear flood routing with multilinear models. *Water Resources Research*, 23(6): 1043–8

Becker A and P Serban 1990 *Hydrological models for water-resources system design and operation*. Geneva, WMO Operational Hydrology Report No 34: 80pp

Bellon A, S Lovejoy and G L Austin 1980 Combined satellite and radar data for short-range forecasting of precipitation. *Monthly Weather Review*, 108: 1554–66

Beltaos S 1981 *Ice freeze up and breakup in the Lower Thames River: 1979–80 observations*. Burlington, Ontario, National Water Research Institute, Canada Centre for Inland Waters: 34pp

Betson R P 1964 What is watershed runoff? *Journal of Geophysical Research*, 69: 1541–52

Betson R P and J B Marius 1969 Source areas of storm runoff. *Water Resources Research*, 5: 574–82

Betton C, B W Webb and D E Walling 1991 Recent trends in $NO_3 - N$ concentration and loads in British rivers. In *Sediment and stream water quality in a changing environment: trends and explanation*, International Association of Hydrological Sciences Publication No 203, 169–80

Beven K J and J A Binley 1992 The future of distributed models: model calibration and uncertainty prediction. *Hydrological Processes* 6(3): 279–98

Beven K J, A Calver and E M Morris 1987 The Institute of Hydrology distributed model. Wallingford, Institute of Hydrology Report No 98: 33pp

Beven K J and P F Germann 1982 Macropores and water flow in soils. *Water Resources Research*, 18(5): 1311–25

Bielsma L and J Kuijpers 1989 River water and the quality of the Delta waters. In J C Hooghart and C W S Posthumus (eds) *Hydro-ecological relations in the Delta waters of south-west Netherlands*, The Hague, Rijkswaterstaat: 3–26

Binns W O 1986 Forestry and fresh water: problems and remedies. In J F de L G Solbé (ed.) *Effects of land use on fresh waters*, Chichester, Ellis Horwood: 364–77

Biswas A K and M A H Abu-Zeid (eds) 1993 *Water for sustainable development in the 21st century*. Oxford, Oxford University Press: 273pp

Blackie J R and C W O Eeles 1985 Lumped catchment models. In M G Anderson and T P Burt (eds) *Hydrological forecasting*. Chichester, Wiley: 311–45

Blackie J R and M D Newson 1986 The effects of forestry on the quantity and quality of runoff in upland Britain. In J F de L G Solbé (ed.) *Effects of land use on fresh waters*. Chichester, Ellis Horwood: 389–412

Bleasdale A 1965 Rain gauge networks development and design with special reference to the United Kingdom. *Proceedings WMO/IASH Symposium on the design of hydrological networks*. International Association of Scientific Hydrology Pub No 67: 46–54

Bloschl G, R Kinbauer and D Gutknecht 1991 A spatially distributed snowmelt model for application in alpine terrain. In H Bergmann, H Lang, W Frey, D Issler and B Salm (eds) *Snow, hydrology and forests in high alpine areas*, International Association of Hydrological Sciences Pub. No 205: 51–60

Bonell M, M M Hufschmidt and J S Gladwell (eds) 1993 *Hydrology and water management in the humid tropics*. Cambridge, Cambridge University Press: 590pp

Bonell M, M R Hendricks, A C Imeson and L Hazelhoff 1984 The generation of storm runoff in a forested clayey drainage basin in Luxembourg. *Journal of Hydrology*, 71: 53–77

Boon P J, P Calow and G E Petts (eds) 1992 *River conservation and management*. Chichester, Wiley: 470pp

Boorman D B, J M Hollis and A Lilly 1995 Hydrology of soil types: a hydrologically based classification of the soils of the United Kingdom, Institute of Hydrology Report No 126, Wallingford: 137pp

Booth D B 1991 Urbanization and the natural drainage system – impacts, solutions and prognoses. *Northwest Environmental Journal*, 7(1): 93–118

Bosch J M and J D Hewlett 1982 A review of catchment experiments to determine the effect of vegetation changes on water yield and evapotranspiration. *Journal of Hydrology*, 55: 3–23

Box G E P and G M Jenkins 1970 *Time series analysis, forecasting and control*. San Francisco, Holden-Day: 553pp

Bradley S B and J L Lewin 1982 Transport of heavy metals in suspended sediments under high flow conditions in a mineralised region of Wales. *Environmental Pollution, Series B*, 4: 257–67

Brady C 1993 South-East Asia: the Mekong River. In C Thomas and D Howlett (eds) *Resource politics, freshwater and regional relations*. Buckingham and Bristol, Open University Press: 86–109

Bras R L and R Colon 1978 Time-averaged areal mean of precipitation: estimation and network design. *Water Resources Research*, 14(5): 878–88

Brewer P and M P Taylor in press. The spatial distribution of heavy metal contaminated sediment across terraced floodplains *Catena*

Brewer R and J R Sleeman 1963 Pedotubules: their definition, classification, and interpretation. *Journal of Soil Science*, 14: 156–66

BSI 1981 *Methods of measurement of liquid flow in open channels, BS 3680*. British Standards Institution: 66pp

Broecker W S 1989 Greenhouse surprises. In D E Abrahamson (ed.) *The challenge of global warming*, Washington, D C, Island Press: 196–208

Brookes A 1988 *Channelized rivers: perspectives for environmental management*. Chichester, Wiley: 326pp

Bruce J P and R H Clark 1980 *Introduction to hydrometeorology*. Toronto, Pergamon: 320pp

Bryan R B and I A Campbell 1986 Runoff and sediment discharge in a semiarid drainage basin. *Zeitschrift für Geomorphologie, Supplement Band* 58: 121–43

Bryan R B, A Yair and W K Hodges 1978 Factors controlling the initiation of runoff and piping in Dinosaur Provincial Park Badlands, Alberta, Canada. *Zeitschrift für Geomorphologie, Supplement Band* 29: 151–68

Bryden H L, D Roemmich and J A Church 1991 Ocean heat transport across 24°N in the Pacific. *Deep Sea Research*, 38(3A): 297–324

Bryson R A 1973 Drought in the Sahel: who or what is to blame? *The Ecologist*, 3(10): 366–71

Bryson R A and T J Murray 1977 *Climates of hunger*. Madison, University of Wisconsin Press: 171pp

Bryson W 1994 Riding out the worst of times. *National Geographic*, 185(1): 82–6

Budyko M I 1970 *The water balance of the oceans*. International Association of Scientific Hydrology Pub. No 92: 24–34

Budyko M I 1989 Climatic conditions of the future. In *Conference on Climate and Water*. Helsinki, Academy of Finland, vol. 1: 9–30

Budyko M I, N A Yefimova, L I Zubenok and L A Strokina 1962 The heat balance of the earth. *Soviet Geography Review and Translation*, 3(5): 3–16

Bull K B 1991 The critical loads/levels approach to gaseous pollutant emissions control. *Environmental Pollution*, 69: 105–23

Bulloch J and A Darwish 1993 *Water wars: coming conflict in the Middle East*. London, Gollancz: 224pp

Bultot F, A Coppens, G L Dupriez, D Gellens and F Meulenberghs 1988 Repercussions of a CO_2-doubling on the water cycle and on the water balance: a case study from Belgium. *Journal of Hydrology*, 99: 319–47

Bunting B T 1961 The role of seepage moisture in soil formation, slope development and stream initiation. *American Journal of Science*, 259: 503–18

Burt T P 1992 The hydrology of headwater catchments. In P Calow and G E Petts (eds) *The river handbook*, vol. 1. Oxford, Blackwell: 3–28

Burt T P and D P Butcher 1986 Development of topographic indices for use in semi-distributed hillslope runoff models. *Zeitschrift für Geomorphologie, Supplement Band* 58: 1–19

Burt T P, M A Donohoe and A R Vann 1984 A comparison of suspended sediment yields from two small upland catchments following open ditching for forestry drainage. *Zeitschrift für Geomorphologie*, NF 51: 51–62

Burt T P and N E Haycock 1992 Catchment planning and the nitrate issue: a UK perspective. *Progress in Physical Geography*, 16(4), 379–404

Calder I R 1985 What are the limits on forest evaporation? Comment *Journal of Hydrology*, 82: 179–84

Calder I R 1986 A stochastic model of rainfall interception. *Journal of Hydrology*, 89(1/2): 65–71

Calder I R 1990 *Evaporation in the uplands*. Chichester, Wiley: 224pp

Calow P and G E Petts (eds) 1992 *The river handbook: hydrological and ecological principles*, vol. 1. Oxford, Blackwell: 526pp

Camp W G 1994 *Environmental science for agriculture and the life sciences*. Albany, New York, Delmar: 439pp

Canter L W 1985 *Environmental impact of water resources projects* Chelsea, Michigan Lewis Pub.: 352pp

Carsel R F and R S Parrish 1988 Developing joint probability distributions of soil water retention characteristics. *Water Resources Research*, 24: 755–69

Carson M A and M J Kirkby 1972 *Hillslope form and process*. Cambridge, Cambridge University Press: 475pp

Carson R 1962 *Silent spring*. London, Hamish Hamilton: 304pp

Carter R W 1961 Magnitude and frequency of floods in suburban areas. US Geological Survey Professional Paper 424-B: B9–11

Carter R W and R G Godfrey 1960 Storage and flood routing. US Geological Survey Water Supply Paper 1543–B: 93

Cebeci T and P Bradshaw 1977 *Momentum transfer in boundary layers*. New York, McGraw-Hill: 391pp

Chahine M T 1992 The hydrological cycle and its influence on climate. *Nature*, 359: 373–80

Chang S-Y, F Downey Brill and D Hopkins 1982 Use of mathematical models to generate alternative solutions to water resource planning problems. *Water Resources Research*, 18(1): 58–64

Changnon S A 1968 The La Porte weather anomaly – fact or fiction? *Bulletin of American Meteorological Society*, 49: 4–11

Chidley T R E and R S Drayton 1986 Visual interpretation of standard satellite images for the design of water resources schemes. In Johnson A I (ed.) *Hydrologic applications of space technology*, International Association of Hydrological Sciences Pub. No 160: 239–56

Chiew F H S, Q J Wang, T A McMahon, B C Bates and P H Whetton 1996 Potential hydrological responses to climate change in Australia. In J A A Jones, C M Liu, M K Woo and H T Kung (eds) *Regional hydrological response to climate change*, Dordrecht, Kluwer: 337–50

Childs E C 1967 Soil moisture theory. *Advances in Hydroscience*, 4: 73–117

Chiras D D 1994 *Environmental sciences: action for a sustainable future*. 4th edition. Redwood City, California, Benjamin/Cummings Publishing Company: 611pp

Chorley R J, D E G Malm and H A Pogorzelski 1957 A new standard for estimating drainage basin shape. *American Journal of Science*, 255: 138–41

Chow V T 1951 A general formula for hydrologic frequency analysis. *Transactions of the American Geophysical Union*, 32: 231–7

Clarke R T and M D Newson 1979 Land-use and upland water resources in Britain: a strategic look. *Water Resources Bulletin*, 15(6): 1628–39

Clarke R T, M N Leese and A J Newson 1973 Analysis of data from Plynlimon raingauge networks, April 1971–March 1973. Wallingford, Institute of Hydrology Report No 27: 75pp

Cline W R 1992 *The economics of global warming*. Washington, Institute for International Economics

Cohen S J 1986 Impacts of CO_2-induced climate change on water resources in the Great Lakes basin. *Climate Change*, 8: 135–53

Cole J A, S Slade, P D Jones and J M Gregory 1991 Reliable yield of reservoirs and possible effects of climate change. *Hydrological Sciences Journal*, 36(6): 579–98

Collier C G, P R Larke and B R May 1983 A weather radar correction procedure for real-time estimation of surface rainfall. *Quarterly Journal of Royal Meteorological Society*, 109: 589–608

Collinge V K and C Kirby (eds) 1987 *Weather radar and flood forecasting*. Chichester, Wiley: 296pp

Collins D N 1989 Influence of glacierisation on the response of runoff from Alpine basins to climate variability. In *Conference on Climate and Water*. Helsinki, Academy of Finland, vol. 1: 319–28

Conacher A and J B Dalrymple 1977 The nine unit landsurface model: an approach to pedogeomorphic research. *Geoderma*, 18(1/2): 1–154

Confederación Hidrografíca del Ebro 1976 *Memoria 1946–1975*. Zaragoza, Ministerio de Obras Públicas: 474pp

Conte M, A Giuffrida and S Tedesco 1989 The Mediterranean oscillation: impact on precipitation and hydrology in Italy. In *Conference on Climate and Water*. Helsinki, Academy of Finland, vol. 1: 121–37

Cook H L 1946 The infiltration approach to the calculation of surface runoff. *Transactions of the American Geophysical Union*, 27: 726–43

Crawford N H and R K Linsley 1966 *Digital simulation in hydrology. Stanford watershed model IV*. Stanford, Department of Civil Engineering, University of California, Technical Report No 39

Currie R G and D P O'Brian 1992 Deterministic signals in USA precipitation records: part I. *Journal of Climatology*, 10: 795–818

Davies B E and J Lewin 1974 Chronosequences in alluvial soils with special reference to historical lead pollution in Cardiganshire, Wales. *Environmental Pollution*, 6: 49–57

Davies T D, P W Abrahams, M Tranter, I Blackwood, P Brimblecombe and C E Vincent 1984 Black acid snow in the remote Scottish Highlands. *Nature*, 312(5989): 58–61

Dawdy D R and T O'Donnell 1965 Mathematical models of catchment behavior. *Proceedings of American Society of Civil Engineers*, HY4, 91: 123–37

Department of Environment (DoE) 1986 *Nitrate in water*. Nitrate Coordination Group, Pollution Papers No 26, London, HMSO

Department of Environment/Department of Health (DoE/DoH) 1990 *Cryptosporidium in water supplies: report of the group of experts*. J Badenoch, Chairman, London, HMSO: 230pp

Dhar O N, A K Kulkarni and R B Sangam 1975 A study of extreme point rainfall over flash flood prone regions of the Himalayan foothills of north India. *Hydrological Sciences Bulletin*, 20(1): 61–7

Dibb S 1995 Swimming in a sea of oestrogens – chemical hormone disrupters. *The Ecologist*, 25(1): 27–31

Dingman S L 1994 *Physical hydrology*. New York, Macmillan: 575pp

Dixon J A and L A Fallon 1989 The concept of sustainability: origins, extensions, and usefulness for policy. *Society and Natural Resources*, 2(2): 73–84

Dooge J C I 1959 A general theory of the unit hydrograph. *Journal of Geophysical Research*, 64(2): 241–56

Doornkamp J C, K J Gregory and A S Burn (eds) 1980 *Atlas of drought in Britain 1975–6*. London, Institute of British Geographers: 87pp

Dunne T 1978 Field studies of hillslope processes. In M J Kirkby (ed.) *Hillslope hydrology*. Chichester, Wiley: 227–94

Dunne T 1980 Formation and controls of channel networks. *Progress in Physical Geography*, (2): 211–39

Dunne T and R D Black 1970 Partial area contributions to storm runoff in a small New England watershed. *Water Resources Research*, 6: 1296–311

Dunne T and R D Black 1971 Runoff processes during snowmelt. *Water Resources Research*, 7: 1160–72

Dunne T and L B Leopold 1978 *Water in environmental planning*. San Francisco, Freeman: 818pp

Eagleson P S 1967 Optimum density of rainfall networks. *Water Resources Research*, 3(4): 104–33

Edwards R W, A S Gee and J H Stoner 1990 *Acid waters in Wales*. Dordrecht, Kluwer: 337pp

Elsom D 1992 *Atmospheric pollution: a global problem*, 2nd edition. Oxford, Blackwell: 320pp

Emmett W W 1978 Overland flow. In M J Kirkby (ed.) *Hillslope hydrology*. Chichester, Wiley: 145–76

Evans R 1994 Run river run. *Geographical Magazine*, 66(7): 17–20

Falkenmark M 1989 The massive water scarcity now threatening Africa – why isn't it being addressed? *Ambio*, 18(2): 112–18

Fankhauser S 1995 *Valuing climate change*. London, Earthscan: 180pp

Farquharson F A K, M J Lowing and J V Sutcliffe 1975 Some aspects of design flood estimation. Symposium of inspection, operation and improvement of existing dams, Newcastle, paper 4.7

Farquharson F A K, D Mackney, M D Newson and A J Thomasson 1978 Estimation of runoff potential of river catchments from soil survey. Soil Survey of England and Wales Special Survey No 11, Harpenden, UK: 29pp

Fawcett K R, M G Anderson, P D Bates, J–P Jordan and J C Bathurst 1995 The importance of internal validation in assessment of physically based distributed models. *Transactions of the Institute of British Geographers*, 20(2): 248–65

Fels E and R Keller 1973 World register of man-made lakes. In W C Ackerman, G F White and E B Worthington (eds) *Man-made lakes, their problems and environmental effects*. American Geophysical Union Monograph No 17

Ferguson R I 1985 Runoff from glacierized mountains: a model for annual variation and its forecasting. *Water Resources Research*, 21: 702–8

Ferguson R I and E M Morris 1987 Snowmelt modelling in the Cairngorms, NE Scotland. *Transactions of Royal Society of Edinburgh: Earth Sciences*, 78: 261–7

Fetter C W 1994 *Applied hydrogeology*, 3rd edition. New York, Macmillan: 691pp

Fiering M B 1982 A screening model to quantify resilience. *Water Resources Research*, 18(1): 27–32

Folland C K, T R Karl, N Nicholls, B S Nyenzi, D E Parker and K Y Vinnikov 1992 Observed climate variability and change. In J T Houghton, B A Callander and S K Varney (eds) *Climate change 1992. The supplementary report to the IPCC scientific assessment*. Cambridge, Cambridge University Press: 135–70

Forestry Commission 1988 *Forest and water guidelines*. Edinburgh, Forestry Commision: 28pp

Foster E E 1948 *Rainfall and runoff*. New York, Macmillan: 487pp

Foster J L, A Rango, D K Hall, A T C Chang, L J Allison and B C Diesen 1980 Snowpack monitoring in North America and Eurasia using passive microwave satellite data. *Remote Sensing of the Environment*, 10: 285–98

Fountain A G and W Tangborn 1985 Overview of contemporary techniques. In G J Young (ed.) *Techniques for prediction of runoff from glacierized areas*. International Association of Hydrological Sciences Pub. No 149: 27–41

Francis I 1988 Afforestation: what really goes down the drain. *Ecos*, 9: 23–7

Fuge R, I M S Laidlaw, W Perkins and K Rodgers 1991 The influence of acidic mine and spoil drainage on water quality of the mid-Wales area. *Environmental Geochemistry and Health*, 13(2): 70–5

Gaiser R N 1952 Root channels and roots in forest soils. *Proceedings of the Soil Science Society of America*, 16: 62–5

Gardiner J L (ed.) 1991 *River projects and conservation: a manual for holistic appraisal*. Chichester, Wiley: 272pp

Gardiner V and C C Park 1978 Drainage basin morphometry – review and assessment. *Progress in Physical Geography*, 2(1): 1–35

Gash J H C 1979 An analytical model of rainfall interception by forests. *Quarterly Journal of Royal Meteorological Society*, 105: 43–55

Gee A S and J H Stoner 1989 A review of the causes and effects of acidification of surface waters in Wales and potential mitigation techniques. *Archives of Environmental Contamination and Toxicology*, 18: 121–30

Geiger R 1957 *The climate near the ground*. Boston, Harvard University Press: 611pp

Gellens D 1991 Impacts of a CO_2-induced climatic change on river flow variability in three rivers in Belgium. *Earth Surface Processes and Landforms*, 16: 619–25

Gerits J J P, J L M P De Lima and T M W Van Den Broek 1990 Overland flow and erosion. In M G Anderson and T P Burt (eds) *Process studies in hillslope hydrology* Chichester, Wiley: 173–214

Germann P F 1990 Macropores and hydrologic hillslope processes. In M G Anderson and T P Burt (eds) *Process studies in hillslope hydrology*. Chichester, Wiley: 327–63

Gilbertson D D, C O Hunt, N R J Fieller and G W W Barker 1994 The enviromental consequences and context of ancient floodwater farming in the Tripolitanian Pre-desert. In A C Millington (ed.) *Environmental change in drylands*. Chichester, Wiley: 229–51

Gilman K 1994 *Hydrology and wetland conservation*. Chichester, Wiley: 101pp

Gilman K and M D Newson 1980 *Soil pipes and pipeflow – a hydrological study in upland Wales*. Norwich, GeoBooks, British Geomorphological Research Group Monograph No 1: 114pp

Gladwell J S and K S Low 1993 *Tropical cities: managing their water*. Paris, UNESCO: 25pp

Glanz M H 1982 Consequences and responsibilities in drought forecasting: the case of Yakima, 1977. *Water Resources Research*, 18(1): 3–13

Glazovsky, N F 1990 *The Aral Crisis: the source, the current situation and the ways to solving it*. Moscow, Nauka: 136pp

Gleick J 1988 *Chaos: making a new science*. London, Heinemann: 352pp

Gleick P H 1987 Regional hydrologic consequences of increases in atmospheric CO_2 and other trace gases. *Climatic Change*, 10: 137–61

Gleick P H (ed.) 1993 *Water in crisis: a guide to the world's fresh water resources*. Oxford, Oxford University Press: 473pp

Goldsmith E, N Hildyard and D Trussell (eds) 1986 *The social and environmental effects of large dams. Vol. II case studies*. Cornwall, Wadebridge Ecological Centre: 243pp

Goldstein R A, S A Gherini, C W Chen, L Mak and R J M Hudson 1984 Integrated acidification study (ILWAS): a mechanistic ecosystem analysis. *Philosophical Transactions of the Royal Society London*, Series B 305: 409–25

Gore A 1992 *Earth in the balance*. London, Earthscan: 407pp

Gore J A 1985 *The restoration of rivers and streams*. Boston, Butterworth: 280pp

Grasty R L 1981 Direct snow-water equivalent measurement by airborne gamma-ray spectrometry. *Journal of Hydrology*, 55: 213–36

Gravelius H 1914 *Flusskunde*. Berlin, Goschensche Verlagshandlung: 176pp

Gray D M and D H Male (eds) 1981 *Handbook of snow: principles, processes, management and use*. Toronto, Pergamon: 776pp

Gray N F 1994 *Drinking water quality: problems and solutions*. Chichester, Wiley: 315pp

Green F H W 1970 Some isopleth maps based on lysimeter observations in the British Isles, in 1965, 1966 and 1967. *Journal of Hydrology*, 10: 127–40

Green F H W 1979 Field under-drainage and the hydrological cycle. In G E Hollis (ed.) *Man's influence on the hydrological cycle in the United Kingdom*, Norwich, GeoBooks: 9–17

Gregory K J 1974 Streamflow and building activity. In K J Gregory and D E Walling (eds) *Fluvial processes in instrumented watersheds*, Institute of British Geographers Special Pub. No 6: 107–22

Gregory K J, L Starkel and V R Baker 1995 *Global Continental palaeohydrology*. Chichester, Wiley: 352pp

Gribbin J and S Plagemann 1977 *The Jupiter effect*. Glasgow, Fontana/Collins: 178pp

Guariso G, D Whittington, B S Zikri and K H Mancy 1981 Nile water off Sinai: framework for analysis. *Water Resources Research*, 17(6): 1585–93

Gumbel E J 1958 *Statistics of extremes*. New York, Columbia University Press: 375pp

Gurnell A M 1976 A note on the contribution of fog drip to streamflow. *Weather*, 31(4): 121–6

Hadley Centre 1992 *The Hadley Centre transient climate change experiment*. Bracknell, Hadley Centre for Climate Prediction and Research, UK Meteorological Office: 20pp

Hadley Centre 1993 *Progress report 1990–1992 and future research programme.* Bracknell, Hadley Centre for Climate Prediction and Research, UK Meteorological Office: 72pp

Hall A J 1986 Surface-water networks in Australia. In M E Moss (ed.) *Integrated design of hydrological networks.* International Association of Hydrological Sciences Pub. No 158: 11–21

Hall P (ed.) 1966 *The isolated state*, by J H von Thünen. Oxford, Pergamon: 304pp

Haman K E 1976 Physical problems of weather modification. *Hydrological Sciences Bulletin*, 21(4): 587–602

Hamer M 1990 The year the taps ran dry. *New Scientist* 127(1730): 20–1

Hamilton K, 1988 A detailed examination of the extratropical response to El Niño/Southern Oscillation events. *Journal of Climatology*, 8: 67–86

Hardman D J, S McEldowney and S Waite 1994 *Pollution ecology and biotreatment.* Harlow, Longman: 322pp

Hardy J H and H Gucinski 1989 Stratospheric ozone depletion: implications for marine ecosystems. *Oceanography* 2(2): 18–21

Harper D M and A J D Ferguson 1995 *The ecological basis for river management.* Chichester, Wiley: 614pp

Harryman M B M 1989 Water source protection and protection zones. *Journal of the Institution of Water and Environmental Management*, 3(6): 548–50

Hashimoto T, J R Stedinger and D P Loucks 1982a Reliability and resiliency, and vulnerability criteria for water resource system performance evaluation. *Water Resources Research*, 18(1): 14–20

Hashimoto T, D P Loucks and J R Stedinger 1982b Robustness of water resource systems. *Water Resources Research*, 18(1): 21–26

Hata T and M G Anderson 1990 The application of a simple lumped river flow forecasting model to hillslope soil water storage estimation. *Catena*, 17, 249–59

Hauhs M 1994 Does reduction in SO_2 and NO_x emissions lead to recovery? In C E W Steinberg and R F Wright (eds) *Acidification of freshwater ecosystems – implications for the future.* Chichester, Wiley: 313–23

Hebson C and E F Wood 1982 A derived flood frequency distribution using Horton order ratios. *Water Resources Research*, 18(5): 1509–18

Hellier C 1990 Draining the rivers dry. *Geographical Magazine*, 62(7): 32–5

Hemond H F 1994 Role of organic acids in acidification of fresh waters. In C E W Steinberg and R F Wright (eds) *Acidification of freshwater ecosystems – implications for the future.* Chichester, Wiley: 103–15

Henderson G S and D L Golding 1987 The effect of slash and burning on the water repellency of forest soils at Vancouver, British Columbia. *Canadian Journal of Forest Research*, 13(2): 353–5

Hendrick R L and G H Comer 1970 Space variations of precipitation and implications for raingauge network design. *Journal of Hydrology* 10: 151–63

Hendrickx J M H 1990 Determination of hydraulic soil properties. In M G Anderson and T P Burt (eds) *Process studies in hillslope hydrology.* Chichester, Wiley: 43–92

Herbst P H, D B Bredenkamp and H M G Barker 1966 A technique for evaluation of drought from rainfall data. *Journal of Hydrology*, 4: 264–72

Herschy R W (ed.) 1978 *Hydrometry, principles and practices.* Chichester, Wiley: 511pp

Hershfield D M 1961 Estimating the probable maximum precipitation. *Journal of Hydraulics Division, American Society of Civil Engineers*, No HY5: 99–116

Hershfield D M 1975 Some small-scale characteristics of extreme storm rainfalls in small basins. *Hydrological Sciences Bulletin*, 20(1): 77–85

Hewitt K 1982 Natural dams and outbursts of the Karakoram Himalaya. In J W Glen (ed.) *Hydrological aspects of alpine and mountain areas.* International Association of Hydrological Sciences Pub. No 138: 259–69

Hewlett J D 1961 Soil moisture as a source of base flow from steep mountain watersheds. US Department of Agriculture Forest Service, Southeastern Forest Experimental Station, Asheville, North Carolina, Station Paper No 132: 11pp

Hewlett J D and J D Helvey 1970 Effects of forest clear–felling on the storm hydrograph. *Water Resources Research*, 6: 768–82

Hewlett J D and A R Hibbert 1967 Factors affecting the response of small watersheds to precipitation in humid areas. In *Proceedings of the International Symposium on Forest Hydrology (1965), Pennsylvania State University*, Pergamon: 275–90

Hewlett J D and W L Nutter 1970 The varying source area of streamflow from upland basins. In *Proceedings of the Symposium on Interdisciplinary Aspects of Watershed Management, Montana State University, Bozeman*, New York, American Society of Civil Engineers: 65–83

Hey R D 1986 River response to inter-basin water transfers: Craig Goch feasibility study. *Journal of Hydrology*, 85: 407–21

Heywood D 1994 Reversal of fortune. *Geographical Magazine*, 66(11): 26–28, 45

Hibbert A R 1967 Forest treatment effects on water yield. In W E Sopper and H W Lull (eds) *Forest hydrology.* Oxford, Pergamon: 527–43

Hibbert A R 1971 Increase in streamflow after converting chaparral to grass. *Water Resources Research*, 7(1): 71–80

Higgs G 1987 Environmental change and hydrological response: flooding in the Upper Severn catchment. In K J Gregory, J Lewin and J B Thornes (eds) *Palaeohydrology in practice.* Chichester, Wiley: 133–61

Hillel D 1992 *Civilization and the life in the soil.* London, Anrom Books: 321pp

Hindley D R 1973 The definition of dry weather flow in river flow measurement. *Journal of the Institution of Water Engineers*, 27: 438–40

Hockin D L, I R Whittle and R A Bailey 1988 Managing and engineering rivers for the benefit of Man. *Journal of the Institution of Water and Environmental Management*, 2(2): 151–8

Hodge S M 1979, Direct measurement of basal water pressures: progress and problems. *Journal of Glaciology*, 23(89): 309–19

Hollis G E 1977 Water yield changes after urbanization of the Canon's Brook catchment, Harlow, England. *Hydrological Sciences Bulletin*, 22(1): 61–75

Hollis G E (ed.) 1979 *Man's influence on the hydrological cycle in the United Kingdom*. Norwich, GeoBooks: 278pp

Holt C P and J A A Jones 1996 Equilibrium and transient global warming scenario implications for water resources in Wales – a comparison. *Water Resources Bulletin*, 32(4): 711–21

Holt T 1983 Soviet river diversions – a social need, a climatic hazard? *Climate Monitor*, 12(3): 91–5

Hornung M and R A Skeffington (eds) 1993 *Critical loads: concept and applications*. London, HMSO: 134pp

Horton R E 1933 The role of infiltration in the hydrological cycle. *Transactions of the American Geophysical Union*, 14: 446–60

Horton R E 1945 Erosional development of streams and their drainage basins: hydrophysical approach to quantitative morphology. *Bulletin of the Geological Society of America*, 56: 275–370

Hosh J 1994 Living with the landscape. *Geographical Magazine*, 60(7): 24–7

Houghton J T, G J Jenkins and J J Ephraums (eds) 1990 *Climate change. The IPCC scientific assessment*. Cambridge, Cambridge University Press: 364pp

Houghton J T, B A Callander and S K Varney (eds) 1992 *Climate change 1992. The supplementary report to the IPCC scientific assessment*. Cambridge, Cambridge University Press: 200pp

Howe G M, H O Slaymaker and D M Harding 1967 Some aspects of the flood hydrology of the upper catchments of the Severn and Wye. *Transactions of Institute of British Geographers*, 41: 33–58

Howell P P and J A Allan (eds) 1994 *The Nile: sharing a scarce resource*. Cambridge, Cambridge University Press: 416pp

Howell W E 1972 Impact of snowpack management on snow and ice hydrology. In *The role of snow and ice in hydrology*. International Association of Hydrological Sciences Pub. No 107: 1464–72

Hudson J A 1988 The contribution of soil moisture storage to the water balances of upland forested and grassland catchments. *Hydrological Sciences Journal*, 33(3): 289–309

Hudson J A and K Gilman 1993 Long-term variability in the water balances of the Plynlimon catchments. *Journal of Hydrology*, 143: 355–80

Huff F A 1977 Urban effects on storm rainfall in Midwestern United States. In *Effects of urbanization and industrialization on the hydrological regime and on water quality*, International Association of Hydrological Sciences, Pub. No 123: 12–19

Hufschmidt M M 1993 Water policies for sustainable development. In A K Biswas and M A H Abu-Zeid (eds) *Water for sustainable development in the 21st century*. Oxford, Oxford University Press

Hufschmidt M M and K G Tejwani 1993 *Integrated water resource management – meeting the sustainability challenge*. Paris, UNESCO, IHP Humid Tropics Programme Series No 5: 37pp

Huntley B 1990 Lessons from climates of the past. In J Leggett (ed.) *Global warming – the Greenpeace report*, Oxford, Oxford University Press: 133–48

Hurst H E, R P Black and Y M Simaike 1965 *Long-term storage: an experimental study*. London, Constable: 145pp

Huxley J 1943 *TVA: adventure in planning*. London, Architectural Press: 142pp

Institution of Civil Engineers (ICE) 1960 *Floods in relation to reservoir practice*. London, Institution of Civil Engineers: 66pp

Institution of Civil Engineers (ICE) 1980 *Flood studies report – five years on*. London, Thomas Telford: 159pp

Independent Commission on International Issues (Brandt Commission) 1980 *North–south: a programme for survival*. London, Pan

Institute of Hydrology (IH) 1980 *Low flow studies report*. Wallingford, Institute of Hydrology

International Association of Hydrological Sciences (IAHS) 1977 *Effects of urbanization and industrialization on the hydrological regime and on water quality*. International Association of Hydrological Sciences Pub. No 123: 572pp

International Water Power and Dam Construction 1992 *Water power and dam construction handbook 1992*. Sutton, Reed Enterprise

Ives J D 1957 Glaciation of the Torngat Mountains, Northern Labrador. *Arctic*, 10: 67–88

Jack L and A O Lambert 1992 Operational yield. In M N Parr, J A Charles and S Walker (eds) *Water Resources and Reservoir Engineering*. London, Thomas Telford: 65–72

James L D 1973 Hydrologic modelling, parameter estimation, and watershed characteristics. *Journal of Hydrology*, 17: 283–307

Jamieson D G and J C Wilkinson 1972 River Dee research programme. *Water Resources Research*, 8(4): 899–920

Johnson A I (ed.) 1986 *Hydrologic applications of space technology*. Wallingford, International Association of Hydrological Sciences Pub. No 160: 488pp

Johnson S H, III and T Charoenwatana 1981 Economics of rainfed cropping systems: Northeast Thailand. *Water Resources Research*, 17(3): 462–8

Jones H G and J Stein 1990 Hydrogeochemistry of snow and snowmelt in catchment hydrology. In M G Anderson and

T P Burt (eds) *Process studies in hillslope hydrology*. Chichester, Wiley: 255–97

Jones J A A 1969 The growth and significance of white ice at Knob Lake, Quebec. *Canadian Geographer*, 13(4): 354–72

Jones J A A 1971 Soil piping and stream channel initiation. *Water Resources Research*, 7(3): 602–10

Jones J A A 1986 Some limitations to the a/s index for predicting basin-wide patterns of soil water drainage. *Zeitschrift für Geomorphologie Supplement Band* 60: 7–20

Jones J A A 1987a The initiation of natural drainage networks. *Progress in Physical Geography*, 11(2): 207–45

Jones J A A 1987b The effects of soil piping on contributing areas and erosion patterns. *Earth Surface Processes and Landforms*, 12(3): 229–48

Jones J A A 1990 Piping effects in humid lands. In C G Higgins and D R Coates (eds) *Groundwater geomorphology: the role of subsurface water in earth-surface processes and landforms*, Boulder, Geological Society of America, Special Paper 252: 111–38

Jones J A A 1996 Current evidence on the likely impact of global warming on hydrological regimes in Europe. In J A A Jones, C M Liu, M K Woo and H T Kung (eds) *Regional hydrological response to climate change*. Dordrecht, Kluwer: 87–131

Jones J A A 1997a Subsurface flow and subsurface erosion: further evidence on forms and processes. In D R Stoddart (ed.) *Process and form in geomorphology*. London, Routledge: 74–120

Jones J A A 1997b Pipeflow contributing areas and runoff response. *Hydrological Processes*, 11(1): 35–41

Jones J A A and F G Crane 1984 Pipeflow and pipe erosion in the Maesnant experimental catchment. In T P Burt and D E Walling (eds) *Field experiments in fluvial geomorphology*. GeoBooks, Norwich: 55–72

Jones J A A and J A Taylor 1983 Climate *National atlas of Wales*. Cardiff, University of Wales Press: Section 1.4

Jones J A A, P Wathern, L J Connelly and J M Richardson 1991 Modelling flow in natural soil pipes and its impact on plant ecology in mountain wetlands. In P Nachtnebel (ed.) *Hydrological basis of ecologically sound management of soil and groundwater*, International Association of Hydrological Sciences Pub. No 202: 131–42

Jones J A A, C M Liu, M K Woo and H T Kung (eds) 1996 *Regional hydrological response to climate change*. Dordrecht, Kluwer: 425pp

Jones J A A, J M Richardson and H J Jacob (in press) Factors controlling the distribution of piping in Britain: a reconnaissance. In J A A Jones and R B Bryan (eds) *Geomorphology – special edition on piping* Amsterdam, Elsevier

Jones P D, K R Briffa and J R Picher 1984 Riverflow reconstruction from tree rings in southern Britain. *Journal of Climatology*, 4: 461–72

Kahya E and J A Dracup 1994 The influence of Type 1 El Niño and La Niña events on streamflow in the Pacific southwest of the United States. *Journal of Climate*, 7(6): 965–76

Kaminskii V S 1977 Closed water-use cycles in urban industrial communities. In *Effects of urbanization and industrialization on the hydrological regime and water quality*. International Association of Hydrological Sciences, Pub. No 123: 448–54

Kamrin M A and P W Rodgers 1985 *Dioxins in the environment*. Washington, DC, McGraw-Hill: 328pp

Kang E 1991 Relationship between runoff and meteorological factors and its simulation in a Tianshan glacierized basin. In H Bergmann, H Lang, W Frey, D Issler and B Salm (eds) *Snow, hydrology and forests in high alpine areas*, International Association of Hydrological Sciences Pub. No 205: 189–202

Kayane I 1996 An introduction to global water dynamics. In J A A Jones, C M Liu, M K Woo and H T Kung (eds) *Regional hydrological response to climate change*, Dordrecht, Kluwer: 25–38

Kelley R P 1967 Can we use water and power from Alaska? It's costly, but feasible. *Power Engineering*, 71(1): 34–7

Kinnersley D 1994 *Coming clean – the politics of water and the environment*. Harmondsworth, UK, Penguin: 229pp

Kirby C 1984 *Water in Great Britain*, revised edition. Harmondsworth, Penguin: 156pp

Kirby C and W R White 1994 *Integrated river basin development*. Chichester, Wiley: 584pp

Kirkby M J (ed.) 1978 *Hillslope hydrology*. Chichester, Wiley: 389pp

Kirkby M J 1985 Hillslope hydrology. In M G Anderson and T P Burt (eds) *Hydrological forecasting*. Chichester, Wiley: 37–75

Kirkby M J and R J Chorley 1967 Throughflow, overland flow and erosion. *Bulletin of the International Association of Scientific Hydrology*, 12: 5–21

Kirkham D 1947 Studies of hillslope seepage in the Iowan drift area. *Proceedings of the Soil Science Society of America*, 12: 73–80

Kittredge J 1948 *Forest influences*. New York, McGraw-Hill: 394pp (reprint 1973)

Klemes V 1985 Sensitivity of water resources systems to climate variations. Geneva, WMO, World Climate Programme Report No 98

Knapp B J 1978 Infiltration and storage of soil water. In M J Kirkby (ed.) *Hillslope hydrology*. Chichester, Wiley: 43–72

Knight C 1979 Urbanization and natural stream channel morphology: the case of two English new towns. In G E Hollis (ed.) *Man's influence on the hydrological cycle in the United Kingdom*. Norwich, GeoBooks: 181–98

Kohler M A, T J Nordensen and W E Fox 1955 *Evaporation from pans and lakes*. Washington, US Weather Bureau, Technical Paper No 38

Krenke A N and G N Kravchenko 1996 Impact of eventual climatic change on glacier runoff and the possibilities of artificial increases in the Aral Sea basin. In J A A Jones, C M Liu, M K Woo and H T Kung (eds) *Regional hydrological response to climate change*. Dordrecht, Kluwer: 259–67

Kuittinen R 1986 Determination of areal snow-water equivalent values using satellite imagery and aircraft gamma-ray spectrometry. In A I Johnson (ed.) *Hydrologic applications of space technology*. International Association of Hydrological Sciences Pub. No 160: 181–90

Kuprianov V V 1977 Urban influences on the water balance of an environment. In *Effects of urbanization and industrialization on the hydrological regime and on water quality*, International Association of Hydrological Sciences, Pub. No 123: 41–7

Kutzbach J E 1983 Monsoon rains of the late Pleistocene and early Holocene: pattern, intensity and possible causes of changes. In Street-Perrott A, M Beran and R Radcliffe (eds) *Variations in the global water budget*. Dordrecht, Reidel: 371–401

Lamb H H 1977 *Climate: present, past and future*, vol. 2. London, Methuen: 835pp

Lamb H H 1994 *Climate, history and the modern world*, 2nd edition. London, Routledge: 387pp

Landsberg H E 1981 *The urban climate*. New York Academic Press: 275pp

Lang H and G Dayer 1985 Switzerland case study: water supply. In G J Young (ed.) *Techniques for prediction of runoff from glacierized areas*. International Association of Hydrological Sciences Pub. No 149: 45–57

Langan S J 1987 Episodic acidification of streams at Loch Dee, SW Scotland. *Transactions of the Royal Society of Edinburgh – Earth Sciences*, 78: 393–7

Laurenson E M 1964 A catchment storage model for runoff routing. *Journal of Hydrology*, 2: 141–63

Law F 1956 The effect of afforestation upon the yield of water catchment areas. *Journal of the British Waterworks Association*, 38: 489–94

Law F M 1989 Identifying the climate-sensitive segment of British reservoir yield. In *Conference on Climate and Water*, Helsinki, Academy of Finland, vol. 2: 177–90

Lawrance A J and N T Kottegoda 1977 Stochastic modelling of riverflow time series. *Journal of Royal Statistical Society*, Series A, 140: 1–27

Leeks G J, J Lewin and M D Newson 1988 Channel change, fluvial geomorphology and river engineering: the case of the Afon Trannon, mid-Wales. *Earth Surface Processes and Landforms*, 13(3): 207–23

Leenaers H 1989 The transport of heavy metals during flood events in the polluted River Geul (The Netherlands). *Hydrological Processes*, 3: 325–38

Leopold L B 1968 Hydrology for uban land planning – a guidebook on the hydrologic effects of urban land use. US Geological Survey, Circular 554

Lewin J 1978 Meander development and floodplain sedimentation: a case-study from mid-Wales. *Geological Journal*, 13: 25–36

Lewin J, S B Bradley and M Macklin 1983 Historical valley alluviation in mid-Wales. *Geological Journal*, 18: 331–50

Lewin J, M G Macklin and M D Newson 1988 Regime theory and environmental change – irreconcilable concepts? In W R White (ed.) *International Conference on River Regimes*. Chichester, Wiley: 431–45

Lewis D C 1968 Annual hydrologic response to watershed conversion from oak woodland to annual grassland. *Water Resources Research*, 4(1): 59–72

Lewis G and G Williams 1984 *Rivers and wildlife handbook: a guide to practices which further the conservation of wildlife on rivers*. Sandy, Bedfordshire, UK, Royal Society for the Protection of Birds/Royal Society for Nature Conservation: 295pp

Liebscher H J 1987 Paleohydrologic studies using proxy data and observations. In *The influence of climate change and climatic variability on the hydrologic regime and water resources*. International Association of Hydrological Sciences Pub. No 168: 111–21

Lins H, I Shiklomanov and E Stakliv 1990 Hydrology and water resources. In W J McG Tegart, G W Sheldon and D C Griffiths (eds) *Climate change: the IPCC impacts assessment*. Canberra, Australian Government Publishing Service: 4.1–4.42

Linsley R K, M A Kohler and J L H Paulhus 1982 *Hydrology for engineers*, 3rd edition. New York, McGraw-Hill: 492pp (SI/metric edition 1988 by J S Wallace)

Liu C M and M K Woo 1996 A method to assess the effects of climatic warming on the water balance of mountainous regions. In J A A Jones, C M Liu, M K Woo and H T Kung (eds) *Regional hydrological response to climate change*, Dordrecht, Kluwer: 301–15

Lockwood J G and P J Sellers 1983 Some simulation model results of the effect of vegetation change on the near-surface hydro-climate. In A Street-Perrott, M Beran and R Radcliffe (eds) *Variations in the global water budget*. Dordrecht, Reidel: 463–77

Lohmann L 1991 Engineers move in on the Mekong Basin. *New Scientist*, 131 (1777): 44–7

Loughran R J 1976 The calculation of suspended sediment transport from concentration versus discharge curves: Chandler River, N.S.W. *Catena* 3: 45–61

Lovelock J E 1979 *Gaia: a new look at life on Earth*. Oxford, Oxford University Press: 157pp

Lükewille A 1994 Billion dollar problem, billion dollar solution? Transboundary air pollution calls for transboundary solutions. In C E W Steinberg and R F Wright (eds) *Acidification of freshwater ecosystems – implications for the future*. Chichester, Wiley: 17–31

Lull H W 1964 Ecological and silvicultural aspects. In V T Chow (ed.) *Handbook of applied hydrology*. New York, McGraw-Hill: 6.1–6.30

Lundquist D 1982 Modelling of runoff from a glacierized basin. In J W Glen (ed.) *Hydrological aspects of alpine and mountain areas.* International Association of Hydrological Sciences Pub. No 138: 131–6

Lusby G C, G F Turner, J R Thompson and V H Reid 1963 Hydrologic and biotic characteristics of grazed and ungrazed watersheds of the Badger Wash Basin in western Colorado, 1953–58. US Geological Survey Water Supply Paper 1533–B

Lvovich M I 1970 *Water resources of the world and their future.* International Association of Scientific Hydrology Pub. No 92: 317–22

Lvovich M I 1977 World water resources present and future. *Ambio* 6(1): 13–21

Lvovich M I 1979 *World water resources and their future.* Washington, DC, American Geophysical Union: 415pp

Lvovich M I and E P Chernishov 1977 Experimental studies of changes in the water balance of an urban area. In *Effects of urbanization and industrialization on the hydrological regime and on water quality.* International Association of Hydrological Sciences Pub. No 123: 63–7

Mackenzie River Basin Committee 1981 *Mackenzie River Basin Study Report.* Governments of Canada, Alberta, British Columbia, Saskatchewan, North West Territories and Yukon Territory: 231pp

Mairson A 1994 The great flood of '93. *National Geographic,* 185(1): 42–81

Maizels J K 1983 Palaeovelocity and palaeodischarge determination for coarse gravel deposits. In K J Gregory (ed.) *Background to palaeohydrology.* Chichester, Wiley: 101–39

Manabe S and R J Stouffer 1979 CO_2 – climate sensitivity study with a mathematical model of the global climate. *Nature,* 282(5738): 491–3

Mandelbrot B B and J R Wallis 1968 Noah, Joseph, and operational hydrology. *Water Resources Research,* 4: 909–18

Mandelbrot B B and J W van Ness 1968 Fractional Brownian motions, fractional noises, and their applications. *SIAM Review,* 10: 422–37

Manley R E 1975 A hydrological model with physically realistic parameters. International Association of Hydrological Sciences, Pub. No 115: 154–61

Mannion A M 1991 *Global environmental change.* London, Longman: 404pp

Marker M E 1976 Soil erosion 1955 to 1974: a review of the incidence of soil erosion in the Dundas Tableland area of Western Victoria, Australia. *Proceedings of Royal Society of Victoria,* 88(1/2): 15–22

Marsalek J 1977 Runoff control in urbanizing catchments. In *Effects of urbanization and industrialization on the hydrological regime and on water quality.* International Association of Hydrological Sciences, Pub. No 123: 153–61

Marsh P and M-K Woo 1984 Wetting front advance and freezing of meltwater within a snow cover. Observations in the Canadian Arctic. *Water Resources Research,* 20: 1853–64

Mason B J 1962 *Clouds, rain and rainmaking.* Cambridge, Cambridge University Press: 145pp

McCabe G and M Ayers 1989 Hydrologic effects of climate change in the Delaware River basin. *Water Resources Bulletin,* 25: 1231–42

McCaig, M 1984 The pattern of wash erosion around an upland streamhead. In T P Burt and D E Walling (eds) *Catchment experiments in fluvial geomorphology.* Norwich, GeoBooks, 87–114

McCalla T M and T J Army 1961 Stubble mulch farming. *Advances in Agronomy* 13: 125–96

McDonald A T and D Kay 1988 *Water resources issues and strategies.* London, Longman: 284pp

McKibben W 1990 *The end of nature.* Harmondsworth, Penguin Books: 212pp

Meadows D H, D L Meadows, J Randers and W W Behrens III 1972 *The limits to growth.* New York, Universe Books: 205pp

Meybeck M, D Chapman and R Helmer (eds) 1990 *Global freshwater quality: a first assessment.* Oxford, Blackwell: 306pp

Micklin P P 1986 Soviet river diversion plans: their possible environmental impact. In E. Goldsmith and N Hildyard (eds) *The social and environmental effects of large dams. Vol. II case studies.* Cornwall, Wadebridge Ecological Centre: 91–108

Middleton N 1991 *Desertification.* Oxford, Oxford University Press: 48pp

Miller D H 1977 *Water at the surface of the Earth.* New York, Academic Press: 557pp

Miller G T 1992 *Living in the environment,* 7th edition. Belmont, California, Wadsworth: 703pp

Miller R A 1989 Impact of warming on TVA lakes. *Forum for Applied Research and Public Policy*: 37–42

Miller V C 1953 A quantitative geomorphic study of drainage basin characteristics in the Clinch Mountain area, Virginia and Tennessee. *Columbia University, Office of Naval Research Project NR 389–042,* Technical Report 3

Ministry of Transport, Public Works and Water Management (Netherlands) 1994 *Plan for environmental improvement of the River Maas.* Emmeloord, Bibliotheek Waterloopkundig Laboratorium: 15 volumes (in Dutch)

Mitchell J F B 1991 The equilibrium response to doubling atmospheric CO_2. In M E Schlesinger (ed.) *Greenhouse-gas-induced climatic change: a critical appraisal of simulations and observations.* Amsterdam, Elsevier: 49–61

Mohnl H 1991 Fluctuations of snow parameters in the mountainous region of Austria within the last 90 years. *Extended abstracts, IAHS Symposium H5, XX General Assembly of IUGG,* Vienna: 264–8

Monteith J L 1985 Evaporation from land surfaces: progress in analysis and prediction since 1948. In *Advances in evaporation*. American Society of Agricultural Engineers: 4–12

Moriarty F 1988 *Ecotoxicology: the study of pollutants in ecosystems*. London, Academic Press: 289pp

Moses J and E C Barrett 1986 Interactive procedures for estimating precipitation from satellite imagery. In A I Johnson (ed.) *Hydrologic applications of space technology*. International Association of Hydrological Sciences Pub. No 160: 25–40

Moss A J, P Green and J Hutka 1982 Small channels: their experimental formation, nature and significance. *Earth Surface Processes and Landforms*, 7: 404–15

Moss M E (ed.) 1986 *Integrated design of hydrological networks*. Wallingford, International Association of Hydrological Sciences Pub. No 158: 405pp

Muniz I 1991 Freshwater acidification; its effects on species and communities of freshwater microbes, plants and animals. *Proceedings of the Royal Society Edinburgh*, 97B: 227–54

Murphy J M and J F B Mitchell 1995 Transient response of the Hadley Centre coupled ocean–atmosphere model to increasing carbon dioxide. Part II Spatial and temporal structure of response. *Journal of Climate*, 8: 57–80

Musgrave G W and H N Holtan 1964 Infiltration. In V T Chow (ed.) *Handbook of applied hydrology*. New York, McGraw-Hill: 12.1–12.30

Nace R L 1970 World hydrology: status and prospects. In *Symposium on the world water balance*, vol. 1. International Association of Scientific Hydrology, Pub. No 92: 1–10

Nagel J F 1956 Fog precipitation on Table Mountain. *Quarterly Journal of Royal Meteorological Society*, 82: 452–60

Nash J E 1957 The form of the instantaneous unit hydrograph. International Association of Scientific Hydrology, Pub. No 45(3): 114–21

National Geographic 1993 *Water – the power, promise, and turmoil of North America's fresh water*. National Geographic special edition: 120pp

National Research Council 1994 *Alternatives for ground water cleanup*. Oxford, National Academy Press: 336pp

National Rivers Authority (NRA) 1991 Policy and practice for the protection of groundwater

National Rivers Authority (NRA) 1992a *Water resources development strategy: a discussion document*. Bristol, National Rivers Authority: 12pp

National Rivers Authority (NRA) 1992b *Groundwater protection policy*. Bristol, National Rivers Authority

National Rivers Authority (NRA) 1993 Water pollution incidents in England and Wales 1992. Water Quality Series No 13

National Rivers Authority (NRA) 1994 *Water – nature's precious resource. An environmentally sustainable water resources development strategy for England and Wales*. London, HMSO: 93pp

National Water Council 1978 *Water industry review 1978*. London, National Water Council: 99pp

National Water Council 1981 River quality: the 1980 survey and future outlook. London, National Water Council: 39pp

Neal C, N Christophersen, R. Neale, C J Smith and P G Whitehead 1988 Chloride in precipitation and streamwater for the upland catchment of River Servern, mid-Wales: some consequences for hydrological models. *Hydrological Processes* 2: 155–65

Neal C, B Reynolds, P A Stevens, M Hornung and S J Brown 1990 Dissolved inorganic aluminium in acidic streams and soil waters in Wales. In R W Edwards, A S Gee and J H Stoner (eds) *Acid waters in Wales*. Dordrecht, Kluwer: 173–88

Nemec J and A J Askew 1986 Mean and variance in network design philosophies. In M E Moss (ed.) *Integrated design of hydrological networks*. International Association of Hydrological Sciences Pub. No 158: 123–32

NERC 1975 *Flood studies report*. London, Natural Environment Research Council, 5 volumes

Newson M D 1978 Drainage basin characteristics, their selection, derivation and analysis for a flood study of the British Isles. *Earth Surface Processes*, 3: 277–93

Newson M D 1980 The erosion of drainage ditches and its effect on bed-load yields in mid-Wales: reconnaissance case studies. *Earth Surface Processes*, 5: 275–90

Newson M D 1981 Mountain streams. In J Lewin (ed.) *British Rivers*. London, Allen and Unwin: 59–89

Newson M D 1986 River basin engineering – fluvial geomorphology. *Journal of the Institution of Water Engineers and Scientists*, 40(4): 307–24

Newson M D 1992 *Land, water and development*. London, Routledge: 351pp

Newson M D 1994 *Hydrology and the river environment*. Oxford, Clarendon Press: 221pp

Newson M D and I R Calder 1989 Forests and water resources: problems of prediction on a regional scale. *Philosophical Transactions of Royal Society*, B324: 283–98

Newson M D and G J Leeks 1987 Transport processes at the catchment scale. In C R Thorne, J C Bathurst and R D Hey (eds) *Sediment transport in gravel-bed rivers*. Chichester, Wiley: 187–223

Newson M D and J Lewin 1990 Climatic change, river flow extremes and fluvial erosion – scenarios for England and Wales. *Progress in Physical Geography*. 15(1): 1–17

Newson M D and M Robinson 1983 Effects of agricultural drainage on upland streamflow: case studies in mid-Wales. *Journal of Environmental Management*, 17: 333–48

Nicholass C A, P E O'Connell and M R Senior (1981) Rainguage network rationalization and its advantages. *Meteorological Magazine*, 110: 92–102

Nicholson A E 1989 African drought: characteristics, causal theories and global teleconnections. In A Berger, R E Dickinson and S W Kidson (eds) *Understanding climatic change*, American Geophysical Union Geophysical Monograph 52: 79–100

Nicholson G 1969 Wet Thursdays again. *Weather*, 24: 117–19

Nieuwolt S 1977 Tropical climatology – an introduction to the climates of the low latitudes. London, Wiley: 207pp

Nisbet T R 1990 *Forests and surface water acidification.* Forestry Commision Bulletin 86, London, HMSO: 8pp

NOAA 1994 *The great flood of 1993*. US Department of Commerce, National Disaster Report

North A 1993 Saddam's water war. *Geographical Magazine*, 65(7): 10–14

NRA, see National Rivers Authority

Nyberg L, K Bishop and A Rodhe 1993 Importance of hydrology in the reversal of acidification in till soils, Gårdsjön, Sweden. In B Hitchon and R Fuge (eds) *Environmental Geochemistry, Applied Geochemistry*, Supplementary Issue No 2: 61–6

Nye J F 1976 Water flow in glaciers: jökulhlaups, tunnels and veins. *Journal of Glaciology*, 17(76): 181–207

Obled C H and H Harder 1979 A review of snow melt in the mountain environment. In S C Colbeck, M Ray (eds) *Modeling of snow cover runoff.* Hanover, New Hampshire, US Army Corps of Engineers, Cold Regions Research and Engineering Laboratory: 179–204

O'Connell P E 1974 A simple stochastic modelling of Hurst's Law. In *Mathematical models in hydrology*. International Association of Hydrological Sciences, Pub. No 100: 169–87

O'Connell P E, M A Beran, R J Gurney, D A Jones and R J Moore 1977 *Methods for evaluating the UK rain gauge network.* Wallingford, Institute of Hydrology Report No 40: 262pp

Oke T R 1992 *Boundary layer climates*, 2nd edition reprint. London, Methuen: 435pp

Ommanney C S L 1980 The inventory of Canadian glaciers: procedures, techniques, progress and applications. In *World glacier inventory*. International Association of Hydrological Sciences Pub. No 126: 35–44

Ormerod S J and A Jenkins 1994 The biological effects of acid episodes. In C E W Steinberg and R F Wright (eds) *Acidification of freshwater ecosystems – implications for the future*. Chichester, Wiley: 259–72

Outcalt S I, C Goodwin, G Weller and J Brown 1975 Computer simulation of the snowmelt and soil thermal regime at Barrow, Alaska. *Water Resources Research*, 11(5): 709–15

Packman J C 1979 The effect of urbanization on flood magnitude and frequency. In G E Hollis (ed.) *Man's influence on the hydrological cycle in the United Kingdom.* Norwich, GeoBooks: 153–72

Palmer T 1986 *Endangered rivers and the conservation movement*. Berkeley, University of California Press: 316pp

Parmet B W A H and M A M Mann 1993 Influence of climate change on the discharge of the River Rhine – a model for the lowland area. In *Exchange processes at the land surface for a range of space and time scales.* International Association of Hydrological Sciences Publication No 212: 469–77

Paterson W S B 1994 *The physics of glaciers*. 3rd edition Oxford, Pergamon: 480pp

Pattinson V A, D P Butcher and J C Labadz 1994 The management of water colour in peatland catchments. *Journal of the Institution of Water and Environmental Management*, 8: 298–307

Paulhus J L H 1965 Indian Ocean and Taiwan rainfall set new records. *Monthly Weather Review*, 93: 331–5

Pearce F 1986 A green unpleasant land. *New Scientist*, 111 (1518): 26–7

Peck A J 1978 Salinization of non-irrigated soils and associated streams: a review. *Australian Journal of Soil Research*, 16: 157–68

Penning-Rowsell E C, D J Parker and D M Harding 1986 *Floods and drainage*. London, Allen and Unwin: 199pp

Perry A H and J M Walker 1979 *The ocean–atmosphere system*. London, Longman: 160pp

Peschke G and M Kutile 1982 Infiltration model in simulated hydrographs. *Journal of Hydrology*, 56: 369–79

Peterson D J 1993 *Troubled lands: the legacy of Soviet environmental destruction*. Boulder, Westview Press/Rand Research: 276pp

Petts G E 1980a Morphological changes of river channels consequent upon headwater impoundment. *Journal of the Institution of Water Engineers and Scientists,* 34(4): 374–82

Petts G E 1980b Long-term consequences of upstream impoundment. *Environmental Conservation*, 7(4): 325–32

Petts G E 1984 *Impounded rivers: perspectives for ecological management*. Chichester, Wiley: 326pp

Petts G E 1990 Water, engineering and landscape: development, protection and restoration. In D Cosgrove and G E Petts (eds) *Water, engineering and landscape.* London, Belhaven Press

Pirt J and C M Simpson 1982 A study of low flows using data from the Severn and Trent catchments – part II: flow frequency procedures. *Journal of the Institution of Water Engineers and Scientists*, 36: 459–69

Postel S 1992 *The last oasis*. London, Earthscan: 239pp

Power J M 1985 Canada case study: water resources. In G J Young (ed.) *Techniques for prediction of runoff from glacierized areas*, International Association of Hydrological Sciences Pub. No 149: 59–71

Prince of Wales and C Clover 1993 *Highgrove: a portrait of an estate*. London, Chapman: 283pp

Pruppacher H H 1982 Cloud and precipitation physics and the water budget of the atmosphere. In H Plate (ed.) *Engineering meteorology*. Studies in Wind Engineering and Industrial Aerodynamics, Amsterdam, Elsevier, vol. 1: 71–124

Pugh D T 1990 Is there a sea-level problem? *Proceedings of the Institution of Civil Engineers*, part 1, 88, 347–66

Rang M C and C J J Schouten 1988 Major obstacles to water quality management. Part 2: Hydro-inertia. *Proceedings of International Association of Theoretical and Applied Limnology*, 23: 1482–7

Rang M C, C E Kleijn and C J Schouten 1986 Historic changes in the enrichment of fluvial deposits with heavy metals. In *Monitoring to detect changes in water quality series*. International Association of Hydrological Sciences Pub. No 157: 47–59

Ranzi R and R Rosso 1991 A physically-based approach to modeling distributed snowmelt in a small alpine catchment. In H Bergmann, H Lang, W Frey, D Issler and B Salm (eds) *Snow, hydrology and forests in high alpine areas*. International Association of Hydrological Sciences Pub. No 205: 141–50

Raymo M E and W F Ruddiman 1992 Tectonic forcing of late Cenozoic climate. *Nature*, 359: 117–22

Reisner M 1993 *Cadillac desert: the American west and its disappearing water*. New York, Penguin: 582pp

Reynolds B and S J Ormerod 1993 *A review of the impact of current and future acid deposition in Wales*. Natural Environment Research Council, Institute of Terrestrial Ecology: 179pp

Richards L A 1931 Capillary conduction of liquids through porous mediums. *Physics*, 1: 318–33

Rind D 1988 The doubled-CO_2 climate and the sensitivity of the modeled hydrologic cycle. *Journal of Geophysical Research*, 93: 5385–412

Rind D and M Chandler 1991 Increased ocean heat transports and warmer climate. *Journal of Geophysical Research*, 96(D4): 7437–61

Roberge J and A P Plamondon 1987 Snowmelt runoff pathways in a boreal forest hillslope: the role of pipe throughflow. *Journal of Hydrology*, 95: 39–54

Roberts G and T Marsh 1987 The effects of agricultural practices on the nitrate concentrations in the surface water domestic supply sources of Western Europe. In J C Rodda and N C Matalas (eds) *Water for the future: hydrology in perspective*. International Association of Hydrological Sciences Pub. No 164: 365–80

Roberts W O and H Lansford 1979 *The climate mandate*. San Francisco, W H Freeman: 197pp

Robinson M 1979 *Water for the West*. Chicago Public Works Historical Society

Robinson M 1985 The hydrological effects of moorland gripping: a re-appraisal of the Moor House research. *Journal of Environmental Management*, 21: 205–11

Robinson M 1986a Changes in catchment runoff following drainage and afforestation. *Journal of Hydrology*, 86: 71–84

Robinson M 1986b The extent of farm underdrainage in England and Wales, prior to 1939. *The Agricultural History Review*, 34(1): 79–85

Robinson M 1990 *Impact of improved land drainage on river flows*. Wallingford, Institute of Hydrology Report 113: 226pp

Rodda A 1994 *Women in the humid tropics*. Paris, UNESCO/International Hydrological Programme Humid Tropics Programme Series No 6: 48pp

Rodda J C 1967 The systematic error in rainfall measurement. *Journal of Institution of Water Engineers*. 21, 173–7

Rodda J C 1969 *Hydrological network design – needs, problems and approaches*. WMO/IHD Report No 12

Rodda J C 1970 Rainfall excesses in the United Kingdom. *Transactions of the Institute of British Geographers*, 49: 49–60

Rodda J C (ed.) 1985 *Facets of hydrology II*. Chichester, Wiley: 447pp

Rodda J C and N C Matalas (eds) 1987 *Water for the future: hydrology in perspective*. International Association of Hydrological Sciences Pub. No 164: 517pp

Rodda J C and S W Smith 1986 The significance of the systematic error in rainfall measurement for assessing atmospheric deposition. *Atmospheric Environment*, 20: 1059–64

Rodda J C, R A Downing and F M Law 1976 *Systematic hydrology*. London, Newnes-Butterworth: 399pp

Rodda J C, A V Sheckley and P Tan 1978 Water resources and climate change. *Journal of the Institution of Water Engineers and Scientists*, 32: 76–83

Rodríguez-Iturbe I and J B Valdes 1979 The geomorphic structure of hydrologic response. *Water Resources Research*, 15: 1409–20

Rodríguez-Iturbe I, M González-Sanabria and R F Bras 1982 A geomorphoclimatic theory of the instantaneous unit hydrograph. *Water Resources Research*, 18(4): 877–86

Rogers J C and H van Loon 1979 The seesaw in winter temperatures between Greenland and northern Europe. Part II: some oceanic and atmospheric effects in middle and high latitudes. *Monthly Weather Review*, 107: 509–19

Römkens M J M, S N Prasad and F D Whisler 1990 Surface sealing and infiltration. In M G Anderson and T P Burt (eds) *Process studies in hillslope hydrology*, Chichester, Wiley: 127–72

Rosseland B O and M Staurnes 1994 Physiological mechanisms for toxic effects and resistance to acidic water: an ecophysiological and ecotoxicological approach. In C E W Steinberg and R F Wright (eds) *Acidification of freshwater ecosystems – implications for the future*. Chichester, Wiley: 227–46

Rowley G 1993 Multinational and national competition for water in the Middle East: towards the deepening crisis. *Journal of Environmental Management*, 39(3): 187–97

Rowntree P R 1990 Estimates of future climatic changes over Britain. Part 2: Results. *Weather*, 45(3): 79–89

Royal Society for the Protection of Birds, D Ward, N Holmes and P Jose 1994 *Rivers and Wildlife Handbook*. Sandy, Beds., Royal Society for the Protection of Birds, NRA, Wildlife Trusts: 426pp

Rutter A J, K A Kershaw, P C Robins and A J Morton 1971 A predictive model of rainfall interception in forests, I. Derivation of the model from observations in a plantation of Corsican pine. *Agricultural Meteorology*, 9: 367–84

Saiko T and I Zonn 1994 Deserting a dying sea. *Geographical Magazine*, 66(7): 12–15

Schadler B 1989 Water balance investigations in Swiss Alpine basins – tool for the improved understanding of impacts of climatic change on water resources. In *Conference on Climate and Water*, vol. 1. Helsinki, Academy of Finland: 462–75

Schaefer V J 1969 The inadvertent modification of the atmosphere by air pollution. *Bulletin of American Meteorological Society*, 50: 199–206

Schachori A, D Rosenzweig and A Poljakoff-Mayer 1967 Effects of Mediterranean vegetation on the moisture regime. In W E Sopper and H W Lull (eds) *Forest hydrology*, Oxford, Pergamon: 291–311

Scheidegger 1965 The algebra of stream-order numbers. US Geological Survey Professional Paper, 525–B

Scheidegger A E 1974 *The physics of flow through porous media*, 3rd edition, Toronto, University of Toronto Press: 353pp

Scholefield D, K C Tyson, E A Garwood, A C Armstrong, J Hawkes and A C Stone 1993. Nitrate leaching from grazed grassland lysimeters: effects of fertilizer input, field drainage, age of sward and patterns of weather. *Journal of Soil Science*, 44(4): 601–14

Schouten C J J and M C Rang 1989 Ceramic processing of polluted dredged mud. In P G Sly and B T Hart (eds) *Sediment/water interaction*, special edition. *Hydrobiologia*, 176/7: 419–30

Schouten C J J, M C Rang and C E Kleijn 1988 Major obstacles to water quality management, part 1: hydro-schizophrenia. *Proceedings International Assoc. Theoretical and Applied Limnology*, 23: 1476–81

Schumm S A 1956 Evolution of drainage systems and slopes at Perth Amboy, N.Y. *Geological Society of America Bulletin*, 67: 597–646

Schumm S A 1977 *The fluvial system*. New York, Wiley: 338pp

Sharpe R and N Skakkebaek 1993 Are oestrogens involved in falling sperm counts and disorders of the male reproductive tract. *The Lancet*, 341: 1392–5

Shaw E M 1994 *Hydrology in practice*. 3rd edition, London, Chapman and Hall: 569pp

Shaw E M and P P Lynn 1972 Areal rainfall evaluation using two surface fitting techniques. *Bulletin of the International Association of Scientific Hydrology*, 19(3): 303–18

Sherman L K 1932 Streamflow from rainfall by the unit-graph method. *Engineering News Record*, 108: 501–5

Shiklomanov I A 1985 Large scale water transfers. In J C Rodda (ed.) *Facets of hydrology II*. Chichester, Wiley: 345–87

Shreve R L 1966 Statistical laws of stream orders. *Journal of Geology*, 74: 17–37

Shreve R L 1972 Movement of water in glaciers. *Journal of Glaciology*, 11(62): 205–14

Shumskii P A 1964 *Principles of structural glaciology*. New York, Dover: 497pp

Shuttleworth W J 1983 Evaporation models in the global water budget. In A Street-Perrott, M Beran and R Radcliffe (eds) *Variations in the global water budget*. Dordrecht, Reidel: 147–71

Simpson E.A. 1980 The harmonization of the monitoring of the quality of rivers in the United Kingdom. *Hydrological Sciences Bulletin*, 25(1): 13–23

Singer F S 1973 Is growth obsolete? Comment. In M Moss (ed.) *The measurement of economic and social performance*. National Bureau of Economic Research: 532–6

Smith R E and H A Schreiber 1974 Point processes of seasonal thunderstorm rainfall 2: rainfall depth probabilities. *Water Resources Research*, 10(3): 418–23

Snyder F F 1938 Synthetic unit graphs. *Transactions of American Geophysical Union*, 19: 447–54

Solomon S J, M Beran and W Hogg (eds) 1987 *The influence of climate change and climatic variability on the hydrologic regime and water resources*. Wallingford, International Association of Hydrological Sciences Pub. No 168: 640pp

Sommerfield R A and E Lachapelle 1970 The classification of snow metamorphism. *Journal of Glaciology*, 9: 3 17

Speidel D H and A F Agnew 1988 World water budget. In D H Speidel, L C Ruedisili and A F Agnew (eds) *Perspectives on water uses and abuses*. Oxford, Oxford University Press

Starkel L 1972 The modelling of monsoon areas of India in relation to catastrophic rainfall. *Geographia Polonica*, 21: 103–47

Starkel L, K J Gregory and J B Thornes (eds) 1990 *Temperate palaeohydrology*. Chichester, Wiley: 548pp

Steinberg C E W and R F Wright (eds) 1994 *Acidification of freshwater ecosystems – implications for the future*. Chichester, Wiley: 404pp

Steppuhn H 1981 Snow and Agriculture In D M Gray and D H Male (eds) *Handbook of snow: principles, processes, management and use*. Oxford, Pergamon: 60–125

Stetson T M 1980 Let's look at some legal implications of streamflow forecasting. In *Improved hydrologic forecasting: how and why*, Pacific Grove, California, American Society of Civil Engineers, American Meteorological Society: 312–19

Stewart J B 1988 Modelling surface conductance of pine forest. *Agricultural Forest Meteorology*, 43: 19–35

Stott T, R Ferguson, R C Johnson and M D Newson 1986 *Sediment budget in forests and unforested basins in upland Scotland*. International Association of Hydrological Sciences, Pub. No 159: 57–68

Strahler A N 1964 Quantitative geomorphology of drainage basins and channel networks. In V T Chow (ed.) *Handbook of applied hydrology*. New York, McGraw-Hill: section 4.11

Street-Perrott A and G Roberts 1983 Fluctuations in closed-basin lakes as an indicator of past atmospheric circulation patterns. In A Street-Perrott, M Beran and R Radcliffe (eds) *Variations in the global water budget*. Dordrecht, Reidel: 331–45

Street-Perrott A, M Beran and R Radcliffe (eds) 1983 *Variations in the global water budget*. Dordrecht, Reidel: 518pp

Sutcliffe J V 1966 The assessment of random errors in areal rainfall estimation. *Bulletin of the International Association of Scientific Hydrology*, 11: 35–42

Swanson R H, P Y Bernier and P D Woodward (eds) 1987 *Forest hydrology and watershed management*. Wallingford, International Association of Hydrological Sciences Pub. No 167: 625pp

Taylor A B and H E Schwartz 1952 Unit-hydrograph lag and peak flow related to basin characteristics. *Transactions of American Geophysical Union*, 32: 235–46

Taylor C H and D M Roth 1979 Effects of suburban construction on the contributing zones in a small southern Ontario drainage basin. *Hydrological Sciences Bulletin*, 24(3): 289–301

Taylor J A 1976 Upland climates. In T J Chandler and S Gregory (ed.) *The climate of the British Isles*, London, Longman: 264–87

Tejwani K G 1993 Water management issues: population, agriculture and forests: a focus on watershed management. In M Bonell, M M Hufschmidt and J S Gladwell (eds) *Hydrology and water management in the humid tropics*. Cambridge, Cambridge University Press: 496–525

Thomas C and D Howlett (eds) 1993 *Resource politics, freshwater and regional relations*. Milton Keynes, Open University: 210pp

Thomas H A and M B Fiering 1962 Mathematical synthesis of streamflow sequences for the analysis of river basins by simulation. In A Maass, M M Hufschmidt, R Dorfman, H A Thomas Jr, S A Marglin and G M Fair (eds) *Design of water resource systems*. Cambridge, Mass., Harvard University Press: 620pp

Thorne C R and A M Osman 1988 Riverbank stability analysis. II: applications. *Journal of Hydraulic Engineering*, 114(2): 151–72

Thornes J B and K J Gregory 1990 Unfinished business – a continuing agenda. In L Starkel, K J Gregory and J B Thornes (eds) *Temperate Palaeohydrology*. Chichester, Wiley: 521–36

Thornthwaite C W and J R Mather 1955 The water balance. Centerton, NJ, Laboratory for Climatology Publications in Climatology, 8: 1–86

Tickell O 1993 US rethinks flood control. *Geographical Magazine* 65(10): 7

Tickell Sir C 1977 *Climate change and world affairs*. Cambridge, Mass., Harvard Studies in International Affairs No 37: 76pp

Tickell Sir C 1991 The human species: a suicidal success? *Geographical Journal*, 159(2): 219–26

Tinsley B A 1988 The solar cycle and the QBO influences on the latitude of storm tracks in the North Atlantic. *Geophysical Research Letters*, 15: 409–12

Titus J 1992 The cost of climate change to the United States. In S K Majumbar, L S Kalkstein, B Yarnal, E W Miller and L M Rosenfeld (eds) *Global climate change: implications, challenges and mitigating measures*. Pennsylvania: Pennsylvania Academy of Science

Tiuri M and A Sihvola 1986 Snow fork for field determination of the density and wetness profiles of a snow pack. In A I Johnson (ed.) *Hydrologic applications of space technology*. International Association of Hydrological Sciences Pub. No 160: 225–32

Todd D K (ed.) 1970 *The water encyclopaedia*. New York, Water Information Center: 559pp

Trimble S W, F H Weirich and B L Hoag 1987 Reforestation and the reduction of water yield on the southern piedmont since circa 1940. *Water Resources Research*, 23: 425–37

Tsegaye F 1991 Technical and managerial aspects of environmental and health impact assessment of water resource development projects – the Ethiopian experience. In R. Wooldridge (ed.) 1991 *Techniques for environmentally sound water resources development*. London, Pentech: 290–305

UK Climate Change Impacts Review Group 1991 *The potential effects of climate change in the United Kingdom*. First Report, London, HMSO: 124pp

UN 1970 *Integrated river basin development: report of a panel of experts*. New York, UN Department of Economic and Social Affairs: 60pp

UN 1989 *World population prospects*. New York, UN Department of International Economic and Social Affairs

UN Conference on Environment and Development 1992 *The Earth Summit, Rio de Janiero 1992*. Regency Press Corp.

UNEP 1977 *World map of desertification*. UN Conference on Desertification, UNEP/FAO/UNESCO/WMO: 11pp

UNEP/UNESCO 1990 *The impact of large water projects on the environment*. Paris, UNESCO: 570pp

UNESCO 1978 *World water balance and water resources of the Earth*. Paris, UNESCO: 663pp

UNESCO 1992 *Water and health*. Paris, UNESCO: 48pp

UNFPA 1992 *Population, resources and environment: the critical challenge*. New York, UNFPA: 154pp

UN World Commission on Environment and Development 1987 *Our common future*. Oxford, Oxford University Press: 374pp

Unwin D J 1969 The areal extension of rainfall records: an alternative model. *Journal of Hydrology*, 7: 404–14

US Army Corps of Engineers 1956 *Snow hydrology*. Portland, Oregon, North Pacific Division, US Army Corps of Engineers: 437pp

US Army Corps of Engineers 1960 *Runoff from snowmelt*. Engineering Manual 1110–2–1406: 59pp

USDA Soil Conservation Service (SCS) 1968 *Engineering handbook*. Hydrology Supplement

USDA Soil Conservation Service 1970 Irrigation water requirements. US Department of Agriculture Technical Release, No 21

US Weather Bureau 1947 Thunderstorm rainfall. USWB Hydrometeorological Report 5

Van Bavel C H M 1966 Potential evaporation: the combination concept and its experimental verification. *Water Resources Research*, 2: 455–67

Van Loon H and K Labitzke 1988 Association between the 11-year solar cycle, the QBO, and the atmosphere. Part II: surface and 700 mb in the northern hemisphere in winter. *Journal of Climatology*, 1: 905–20

Venne D E and D G Dartt 1990 An examination of possible solar cycle–QBO effects in the northern hemisphere troposphere. *Journal of Climatology*, 3: 272–81

Viessman W, Jr, and G L Lewis 1996 *Introduction to hydrology*, 4th edition. London, HarperCollins: 784pp

Walker G T 1924 Correlations in seasonal variations of weather. Memoranda, Indian Meteorological Department, 24: 275–332

Walling D E 1974 Suspended sediment and solute yields from a small catchment prior to urbanization. In K J Gregory and D E Walling (eds) *Fluvial processes in instrumented watersheds*. Institute of British Geographers, Special Pub. No 6: 169–92

Walling D E 1977 Limitations of the rating curve technique for estimating suspended sediment loads, with particular reference to British rivers. International Association of Hydrological Sciences Pub. No 122: 34–48

Walling D E and B W Webb 1983 Patterns of sediment yield. In K J Gregory (ed.) *Background to palaeohydrology*. Chichester, Wiley: 69–100

Walling D E and B W Webb 1987 Suspended load in gravel-bed rivers: UK experience. In C.R. Thorne, J C Bathurst and R D Hey (eds) *Sediment Transport in Gravel-bed Rivers*. Chichester, Wiley: 691–723

Walsh J 1995 Man the dikes. *Time Magazine*, 145(6): 14–19

Wang H F and M P Anderson 1995 *Introduction to groundwater modeling – finite difference and finite element methods*. San Francisco, Freeman/New York, Academic Press: 256pp

Wankiewicz A 1979 A review of water movement in snow. In S C Colbeck and M Ray (eds) *Modeling of snow cover runoff*. Hanover, NH, US Army Corps of Engineers, Cold Region Research and Engineering Laboratory: 222–52

Ward R C 1971 Measuring evapotranspiration: a review. *Journal of Hydrology*, 13: 1–21

Ward R C 1975 *Principles of hydrology*, 2nd edition. London, McGraw-Hill: 367pp

Ward R C and M Robinson 1990 *Principles of hydrology*. 3rd edition, London, McGraw-Hill: 365pp

Waring E A and J A A Jones 1980 A snowmelt and water equivalent gauge for British conditions. *Hydrological Sciences Bulletin*, 25: 129–34

Warrick R A and G Farmer 1990 The greenhouse effect, climatic change and rising sea level: implications for development. *Transactions of the Institute of British Geographers*, 15(1): 5–20

Warrick R A, E M Barrow and T M L Wigley 1990 *The greenhouse effect and its implications for the European Community*. Brussels, Commission of the European Communities, EUR 12707 EN: 30pp

Wathern P (ed.) 1990 *Environmental impact assessment*. London, Unwin: 332pp

Weeks W F and W J Campbell 1973 Icebergs as a freshwater source: an appraisal. *Journal of Glaciology*, 12(65): 207–33

Wellburn A 1988 *Air pollution and acid rain: the biological impact*. London, Longman: 274pp

Wetherald R T and S Manabe 1988 Cloud feedback processes in a General Circulation Model. *Journal of Atmospheric Science*, 45(8): 1397–1415

Wheater H S, S Tuck, R G Ferrier, A Jenkins, F M Kleiser, T A B Walker and M B Beck 1993 Hydrological flowpaths at the Allt A'Mharcaidh catchment: an analysis of plot and catchment scale observations. *Hydrological Processes*, 7(4), 359–72

Whipkey R Z and M J Kirkby 1978 Flow within the soil. In M J Kirkby (ed.) *Hillslope hydrology*. Chichester, Wiley: 121–44

Whitehead P G and C Neal 1987 Modelling the effects of acid deposition in upland Scotland. *Transactions of the Royal Society Edinburgh*, 78: 385–92

WHO 1984 *Guidelines for drinking water quality*, vol. 1. Geneva, World Health Organization

Wicks J M and J C Bathurst 1996 A physically-based, distributed erosion and sediment yield component for the SHE hydrological modelling system. *Journal of Hydrology*, 175: 213–38

Wigley T M L and S C B Raper 1992 Implications for climate and sea level of revised IPCC emissions scenarios. *Nature*, 357: 293–300

Wilson E M 1990 *Engineering hydrology*, 4th edition. London, Macmillan: 348pp

Winand Staring Centre 1994 Fingered flow accelerates transport of water and solutes in soil. *Scan* (Wageningen), 5: 32–5

WMO 1986a *Manual for estimation of probable maximum precipitation*, 2nd edition. Operational Hydrology Report 1, Geneva, WMO No 332: 269pp

WMO 1986b *Intercomparison of models of snowmelt runoff.* Geneva, WMO Operational Hydrology Report 23

WMO 1994a *Guide to hydrological practices*, 5th edition. WMO Pub. No 168, Geneva, WMO: 735pp

WMO 1994b *Observing the world's environment: weather, climate and water.* Geneva, WMO: 42pp

WMO 1994c The American mid-west floods of July/August 1993. *WMO Bulletin*, 43(1): 70–2

WMO/UNESCO 1991 *Report on water resources assessment.* Geneva, WMO: 64pp

Woo M K 1996 Hydrology of northern North America under global warming. In J A A Jones, C M Liu, M K Woo and H T Kung (eds) *Regional hydrological response to climate change.* Dordrecht, Kluwer: 73–86

Wood J C, W H Blackburn, H A Pearson and T K Hunter 1989 Infiltration and runoff water quality response to silvicultural and grazing treatment in a long leaf pine forest. *Journal of Range Management*, 42(5): 378–81

Wooldridge R (ed.) 1991 *Techniques for environmentally sound water resources development*. London, Pentech: 332pp

Woolhiser D A, T O Keefer and K T Redmond 1993 Southern oscillation effects on daily precipitation in the southwestern United States. *Water Resources Research* 29(4): 1287–95

World Resources Institute 1992 *World Resources 1990–91*. Washington, DC, World Resources Institute: 383pp

World Resources Institute 1994 *World Resources 1994–95*. Oxford, Oxford University Press: 400pp

Worldwatch Institute 1984 *Water – the next resource crisis?* Washington, DC, Worldwatch Institute

Yair A and H Lavee 1985 Runoff generation in arid and semi-arid zones. In M G Anderson and T P Burt (eds) *Hydrological forecasting*. Chichester, Wiley: 183–220

Yang D Q, Y S Zhang, Z Z Zhang, K Elder and R Kattelman 1991 Physical properties of snow cover and estimation of snowmelt runoff in a small watershed in high alpine Tianshan. In H Bergmann, H Lang, W Frey, D Issler and B Salm (eds) *Snow, hydrology and forests in high alpine areas*, International Association of Hydrological Sciences Pub No 205: 169–77

Young G J (ed.) 1985 *Techniques for the prediction of runoff from glacierized areas*. International Association of Hydrological Sciences Pub. No 149: 149pp

Young G J, J C Dooge and J C Rodda 1994 *Global water resources issues*. Cambridge, Cambridge University Press: 194pp

Yperlaan G J 1977 Statistical evidence of the influence of urbanization on precipitation in the Rijmond area. In *Effects of urbanization and industrialization on the hydrological regime and on water quality*. International Association of Hydrological Sciences, Pub. No 123: 20–30

Zarjevski Y 1987 *A future preserved: international assistance to refugees*. Oxford, Pergamon: 280pp

Index